Early Investigations of Ceres and the Discovery of Pallas

Clifford Cunningham

Early Investigations of Ceres and the Discovery of Pallas

Historical Studies in Asteroid Research

Second Edition

Springer

Clifford Cunningham
Ft. Lauderdale, FL, USA

This book was published in 2004 under this title:

Jousting for Celestial Glory: The Discovery and Study of Ceres and Pallas.
Star Lab Press. ISBN 0-9708162-3-5

ISBN 978-3-319-80433-0 ISBN 978-3-319-28815-4 (eBook)

DOI 10.1007/978-3-319-28815-4

Softcover reprint of the hardcover 2nd edition 2016

Printed on acid-free paper

This Springer imprint is published by Springer Nature
The registered company is Springer International Publishing AG Switzerland

Foreword

It is my privilege and my pleasure to present this new book by Clifford Cunningham to intellectual readers. The book is devoted to one of the most dramatic moments in the history of astronomy—the discovery of the first two minor planets (asteroids) Ceres and Pallas. It is not the first book written by Cunningham—he is well known as a talented writer who came to science armed with knowledge of physics and having experience of astronomical observations. It is quite natural that his interests are concentrated mainly upon the history of astronomy and in particular upon investigations of minor planets. It is also not his first book on the discovery of Ceres—his earlier book has been well received.

In the present book the author pursues the theme traced in the preceding one. Therefore to look through *Discovery of the First Asteroid, Ceres,* can be recommended to any curious reader. Nevertheless, the present book possesses its own interest. Cunningham thoroughly reproduces the broad picture of events that agitated the educated world 200 years ago. He now focuses the readers' attention on such psychological problems as temptation by glory, envy and jealousy, grandeur and serenity of astronomers in answer to pinpricks by others. The differences between, and opposition of, the German and French schools of celestial mechanics are treated at some length. All these problems are considered against a background of events directly or indirectly connected with the dramatic story of Ceres' discovery, of its "near-loss" and rediscovery, and the unexpected finding of one more asteroid (Pallas) at approximately the same mean heliocentric distance that seemed to contradict the beautiful progression of the planetary distances from the Sun found by Titius and then by Bode long before this discovery.

A great deal of work has been done by the author in the search for new documents of the epoch; these constitute an important component of the book. Many of them are published here for the first time in English. The presentation of events is emotionally saturated. The author does not hide from readers his interpretation of what is going on. This makes a retrospective journey into the history of astronomy especially fascinating. The author's position reflects the displeasure and irritation of the astronomical community of that time caused by the sluggishness of Piazzi when informing other astronomers about his discovery.

Taken as a whole the book presents a serious and captivating historical investigation of the epoch and considered events, and it can be recommended for reading by professionals and amateurs alike.

St. Petersburg, Russia Victor Shor

Artist's Preface

What a fascinating book this is! Clifford Cunningham has done an extraordinary amount of research in preparation for detailing the discovery of two unknown planets in the early nineteenth century. His material is presented in a mosaic of illustrations and words that, for me, gave true feelings of being alive in the period.

Imagine how thrilling it must have been to live in a world that had just thrown off the shroud of a religion-dominated environment. A world where the Age of Enlightenment in the eighteenth century had made ordinary people like ourselves think "The world *is* understandable!" "Human beings can discover why we are here on this ball whirling through space!" "The leap of faith isn't necessary; we'll find out for ourselves what this is all about!" It must have been a blast.

Now we assume science will finally explain everything only to find out we still know little about our reasons for inhabiting this rapidly revolving orb. But for all the gentlemen in this interesting book, all decked out in their wigs, lace jabots, and frock coats, each new discovery that certain planets revolved around the Sun in a stately pattern, there for us to discover, must have been wonderfully exciting and full of meaning. Door after door on our own place in space was being thrown open!

Clifford Cunningham evokes this world and under their wigs and behind their jabots, these scientists become real. As real as that college professor you thought so strange and wonderful; your intellectual neighbor in your hometown; that smart guy you met at your cousin's wedding. Cunningham's subjects become just as real as people you have met yourself and their world with them. They are obviously so obsessed, so fascinated, so competitive. Science had real glamor then, and they were captivated by it.

I know little of science or its history, at least until this book, but I found the vitality of the world Cunningham recounts wonderful. It brought another time alive for me. Who needs science fiction when the very history of science can be so interesting, even for science-innocents like myself?

Miami Beach, FL — David Leddick
August 2015

Preface

The discovery of Ceres and Pallas represented a challenge. It was a challenge not only to mathematics and our vision of the cosmic order but also to the supremacy of France as the leading light of science. How those challenges were met, sometimes with failure and sometimes with success, is the story told in this book.

Tacitus wrote in *The Annals* (Book 15) "The spirit of a noble rivalry and the desire of glory" are "emotions which stir men in success." This book tells of the rivalry and desire for glory amongst those astronomers who sought to study and name the object Zach referred to as the "coquettish little Ceres."

The initial discovery of Ceres, covered in *Discovery of the First Asteroid, Ceres,* is extended here to examine in more depth the response at both the national and the personal level. Professional rivalries between individual researchers are central to this study, and all the scientific papers about Ceres are provided here.

The story of Pallas in correspondence is followed through the year 1802; the scientific papers on Pallas will be in another volume. The discovery of a second asteroid was seen by some, including Johann Bode, as a frustration of their scheme for the cosmos. This manifested itself as a refusal to admit the planetary nature of Pallas. In game theoretic terms, "scientific inquiry is a non-zerosum game of imperfect information in which the neutral universe is not antagonistic towards human exploration" (Swirski, 2000: 82). It was best put by Norbert Wiener (1954), the founder of cybernetics. "Nature plays fair, and if, after climbing one range of mountains, the physicist sees another on the horizon before him, it has not been deliberately put there to frustrate the effort he has already made." The challenge from physics posed by Ceres had no sooner been addressed than a new 'mountain' appeared. Pallas upset the orderly notion of a single 'missing planet' between Mars and Jupiter. How the difficulties posed by the existence of Pallas were explained by the astronomers of the day is explored here and in the next volume of this series.

This volume begins with a study of the philosophical underpinnings of mathematics and astronomy as the eighteenth century ended. This was to a large extent the work of the Philosophes, whose arcane knowledge helped unlock the secrets of the cosmos. Here is Carl Becker, Professor of History at Cornell University, writing in 1932.

> There must be some private passageway to the heavenly throne, some secret backstairs entry that all the *Philosophes* know of, some door, closed to us, that will open to them when they give it a certain understood succession of raps. We should like to enter this door.

Clifford J. Cunningham

About the Author

Clifford J. Cunningham earned his Ph.D. in the history of astronomy at the University of Southern Queensland in Australia, and he is a research associate with the National Astronomical Research Institute of Thailand. His undergraduate degrees in physics and classical studies were earned at the University of Waterloo in Canada. He has written or edited 14 books on the history of astronomy, and his papers have appeared in many major journals, including *Annals of Science, Journal for the History of Astronomy, Culture & Cosmos, Studia Etymologica Cracoviensia, The Asian Journal of Physics*, and *Renaissance and Reformation*. He is associate editor of the *Journal of Astronomical History and Heritage*, where several of his asteroid research papers have been published. He is also a contributor to *Encyclopedia Britannica*, and since 2001 he has been the history of astronomy columnist for *Mercury* magazine. Asteroid (4276) was named Clifford in his honor by the International Astronomical Union based on the recommendation of its bureau the Harvard-Smithsonian Center for Astrophysics.

Prologue: The Dual Challenges of Asteroids and Hieroglyphs

The phenomena are disemboweled
and embalmed with numbers and
signs, and on the scientific coffin
are painted bizarre figures.
—Goethe's comment on the mathematical approach, which reminded him of Egyptian tombs. From a letter of November 24, 1817.

Fig. 1 Johann Goethe in 1817

Deciphering the Symbols

Fig. 2 A sketch of Gauss by his pupil J. B. Listing

In 1801 the discovery of Ceres presented astronomers with a challenge from physical reality that could only be answered by mathematics: Carl Gauss faced the task of extracting, from a sequence of symbols in the form of numbers (positional data by Giuseppe Piazzi, shown below), the unknown orbit of the new planet. In this he succeeded, and by January of 1802 Ceres was no longer listed as "missing." The very next month, the Rosetta Stone arrived in England. Science was faced with another task of extracting hidden meaning from a sequence of symbols, in this case from a fragment of a stone that had lain hidden in Egypt for 2000 years. The decades-long struggle by Jean-Francois Champollion to decipher the ancient Egyptian language mirrors Gauss' titanic struggle in the years to come to determine the perturbations of Pallas, which was discovered just a few weeks after the Rosetta Stone reached London. These dual challenges can be understood in terms of consilience (Wilson, 1998).

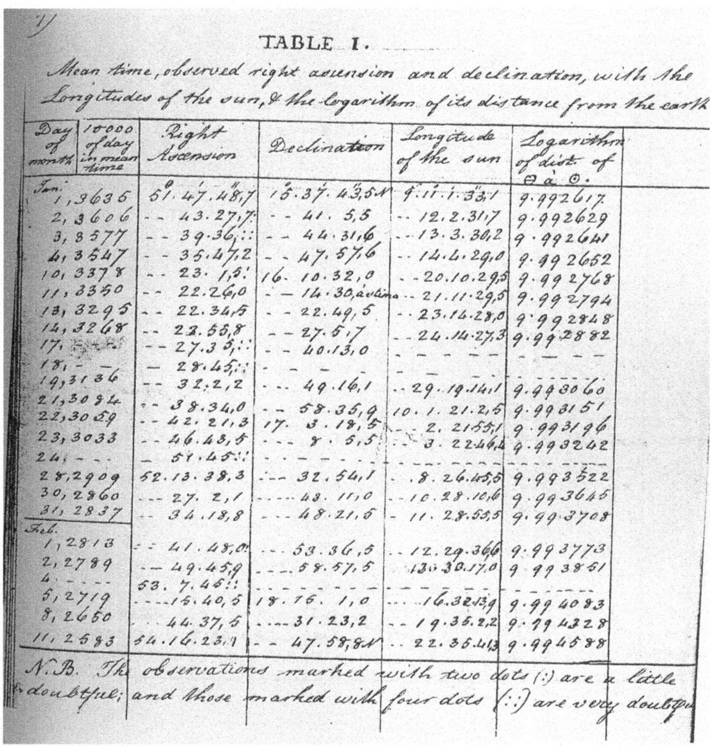

TABLE I.

Mean time, observed right ascension and declination, with the Longitudes of the sun, & the logarithm of its distance from the earth

Day of month	10000 of day in mean time	Right Ascension	Declination	Longitude of the sun	Logarithm of dist. of ⊖ à ⊙.
Jan. 1	3635	51°. 47′. 48″,7	15°. 37′. 43″,5 N	9. 11. 1. 33″,1	9.992617
2	3606	-- 43. 27,7:	-- 41. 5,5	-- 12. 2. 31,7	9.992629
3	3577	-- 39. 36,::	-- 44. 31,6	-- 13. 3. 30,2	9.992641
4	3547	-- 35. 47,2	-- 47. 57,6	-- 14. 4. 29,0	9.992652
10	3378	-- 23. 1,5:	16. 10. 32,0	-- 20. 10. 29,5	9.992768
11	3350	-- 22. 26,0	-- 14. 30, à estima	-- 21. 11. 29,5	9.992794
13	3295	-- 22. 34,5	-- 22. 49,5	-- 23. 14. 28,0	9.992848
14	3268	-- 22. 55,8	-- 27. 5,7	-- 24. 14. 27,3	9.992882
17	[illegible]	-- 27. 35,::	-- 40. 13,0	-- --	-- --
18	--	-- 28. 45,::	-- --	-- --	-- --
19	3136	-- 32. 2,2	-- 49. 16,1	-- 29. 19. 14,1	9.993060
21	3084	-- 38. 34,0	-- 58. 35,9	10. 1. 21. 2,5	9.993151
22	3059	-- 42. 21,3	17. 3. 18,5	-- 2. 21. 55,1	9.993196
23	3033	-- 46. 43,5	-- 8. 5,5	-- 3. 22. 46,4	9.993242
24	-- --	-- 51. 45::	-- --	-- --	-- --
28	2909	52. 13. 38,3	-- 32. 54,1	-- 8. 26. 45,5	9.993522
30	2860	-- 27. 2,1	-- 43. 11,0	-- 10. 28. 10,6	9.993645
31	2837	-- 34. 18,8	-- 48. 21,5	-- 11. 28. 55,5	9.993708
Feb. 1	2813	-- 41. 48,0:	-- 53. 36,5	-- 12. 29. 36,6	9.993773
2	2789	-- 49. 45,9	-- 58. 57,5	13. 30. 17,0	9.993851
4	-- --	53. 7. 45::	-- --	-- --	-- --
5	2719	-- 15. 40,5	18. 15. 1,0	-- 16. 32. 13,9	9.994083
8	2650	-- 44. 37,5	-- 31. 23,2	-- 19. 35. 2,2	9.994328
11	2583	54. 16. 23,1	-- 47. 58,8 N	-- 22. 35. 41,3	9.994588

N.B. The observations marked with two dots (:) are a little doubtful; and those marked with four dots (::) are very doubtful

Fig. 3 Piazzi's observational data on Ceres

Fig. 4 Jean-Francois Champollion

Consilience, or the unity of knowledge, was first mentioned by William Whewell in 1840. In this synthesis Whewell explained that, "The Consilience of Inductions takes place when an Induction, obtained from one class of facts, coincides with an induction, obtained from another different class. Thus Consilience is a test of the truth of the Theory in which it occurs." The scientific method has become universally accepted as the exclusive method for testing the status of any scientific hypothesis or theory. 'Inductions' which arise out of applications of the scientific method are, by definition, the only accepted indicators of consilience. From one class, Gauss deciphered the symbols of celestial mechanics to derive the orbits of Ceres and Pallas, while, from another class, Champollion worked to decipher the symbols of a forgotten language on the Rosetta Stone. These dual challenges of trying to gain knowledge from the study of symbols both arose in 1802 in a way that had never before happened in scientific inquiry.

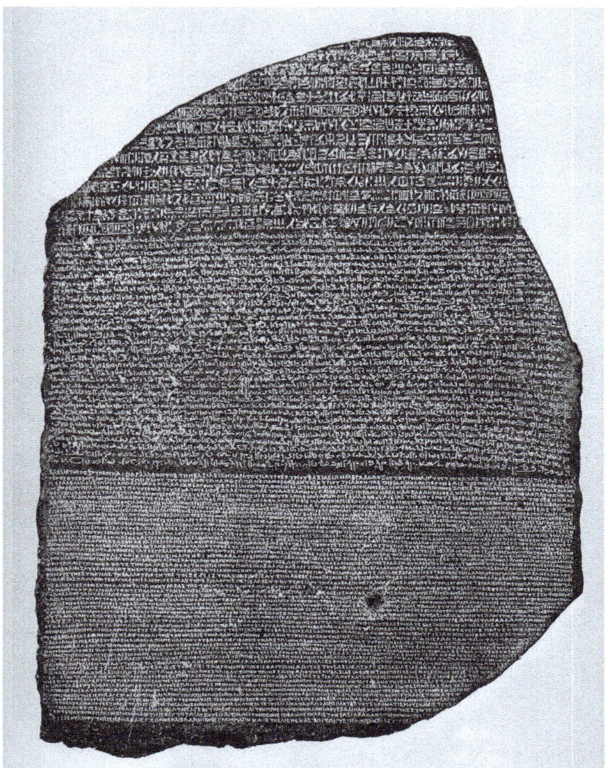

Fig. 5 The Rosetta stone

In addition to the temporal link, two other links exist between the asteroids and the Rosetta Stone. The Greek inscription on the Stone was first deciphered by Stephen Weston, who was asked in 1802 by Joseph Banks to coin a word to describe Ceres and Pallas. A literary connection exists between Champollion and Kepler, who first posited the existence of a planet between Mars and Jupiter. It comes from Edgar Allan Poe, who also wrote about Bode's Law. Poe (1848), in *Eureka*, writes about a "letter from the future" dated 2848 AD which concludes with Kepler's exclamation when he discovered his third law, and a comparison to the decipherment of the Stone:

> I care not whether my work be read now or by posterity. I can afford to wait a century for readers when God himself has waited 6000 years for an observer. I triumph. I have stolen the golden secret of the Egyptians.

Contents

Chapter 1
Mathematics and the Triumph of the Human Intellect

C. Cunningham, *Early Investigations of Ceres and the Discovery of Pallas*,
DOI 10.1007/978-3-319-28815-4_1

Fig. 1.1 A mathematician being inspired to solve a problem. Frontispiece of *Algebra* by Abel Burja (1786)

Mathematics: A Sterile Science?

The early twenty-first century is a comfortable vantage point to look back on the triumphs of mathematics, epitomized in the popular imagination by $E=mc^2$. It is now widely understood, even by those who have no deep understanding of higher mathematics, that all of our scientific accomplishments from the computer to space exploration rest on a solid foundation of mathematics (Fig. 1.1).

Thus it might come as a surprise to learn that in the eighteenth century, when all of the scientists associated with the discovery and study of Ceres and Pallas were born, mathematics was regarded skeptically, if not derisively, by some of the most famous *philosophes* (intellectuals) of the age.

Of these luminary figures of the Enlightenment, none wrote more persuasively than Denis Diderot (1713–1784), chief editor of the *Encyclopedie*. In his 1754 book *Thoughts on the Interpretation of Nature*, Diderot attacked the state of science. For him, there was an inordinate emphasis on the sterile science of mathematics, brought about by the influence of René Descartes (1596–1650). Diderot argued that mathematics is unable, in the final analysis, to say anything significant about the real world.

The reach of this thought has been surprisingly long-lasting. In the 1960s, physics Nobel laureate Murray Gell-Mann discovered the fundamental building block of nature, which he named quarks. In his initial discussion of the quark, he termed quarks that could not get out of a particle (e.g., a proton or an electron) as "mathematical," while a quark that was out he termed "real." He did so because he "was afraid of philosophers," and his terminology was used by many to suggest they did not believe particles were actually composed of quarks. The terminology now used to replace 'mathematical' is 'confined.'(Gell-Mann, 2002).

The relationship between math and nature was explored by the English political philosopher William Godwin (1756–1836; 1831: 392):

> *There are many circumstances, which are calculated to induce a circumspect enquirer to regard the affirmative positions of astronomy with considerable diffidence. They are founded next to the evidence of our senses, upon the deductions of mathematical knowledge. Pure mathematics are concerned only with abstract propositions and have nothing to do with the realities of nature. It is conceived therefore by the generality of observers, that mathematics is the science of certainty. But this is not strictly the case. Human mathematics, so to speak, like the length of life, are subject to the doctrine of chances. Mathematics may be the science of certainty to celestial natures, but not to man.*

The unreality attached to mathematics can be traced directly to Diderot, who, even if not directly responsible, was giving form to what many felt to be the case (Saur, 1705). Referring to two noted astronomers of the age, James Bradley (1692–1762) and Pierre-Charles Le Monnier (1715–1799), Diderot (1754: 36) posed a rhetorical question about the value of devising theories to determine the orbits of celestial objects: "What, they say, is the good of all these profound theories about the heavenly bodies, all these vast calculations dealing with rational astronomy, if they do not save Bradley or Le Monnier the trouble of observing the heavens?"

Here he uses the term 'rational astronomy' to mean astronomy expressible by numbers. Diderot (1754: 62) equated mathematicians with metaphysicians as people who "don't know anything." In his study of the *philosophes* such as Diderot, Becker (1932) wryly notes that "they scorned metaphysics, but were proud to be called philosophers." Fortunately mathematics was immune to such philosophical barbs. As the twentieth century mathematician Kurt Göedel writes, "Mathematics, by its nature as an *a priori* science, always has long withstood the spirit of the time that has ruled since the Renaissance." (Devlin, 2002; Göedel, 1961).

The Role of Analysis in Astronomy

Carl Friedrich Gauss (1777–1855) was 7 years old when Diderot died, and he, more than anyone, would show how profoundly mathematics could elucidate the nature of the real world. He was, as his friend Farkas Bolyai (1775–1856; 1851: 44) said, "the mathematical giant who from his lofty heights embraces in one view the stars and the abysses." For Gauss, Mathematics was the Queen of the Sciences. Algebra, inspiring the mathematician in the opening illustration of this chapter, was a key element in the solution of the orbit of Ceres. As explained by Edna Kramer, Professor Emerita of Mathematics at the Polytechnic Institute of Brooklyn, Gauss was the heir to a great mathematical tradition stretching back 4000 years:

> *It was possible to obtain only a very few exact observations of Ceres' positions in the course of its motion around the sun, and therefore the computation of its orbit from such limited empirical data presented a problem which aroused Gauss's interest. In formulating the issue, he had to solve (approximately) an algebraic equation of the eighth degree. Thus he provided a 'practical' application of the very equations which the Babylonians of Hammurabi's day had solved merely because they enjoyed algebra.* (Kramer, 1982: 475)

As a professor in Goettingen, Gauss first began giving a course in astronomy in 1808. In 1815 one of his students, Peter Merian, kept a careful notebook of Gauss's Inaugural Lecture (provided later). In it Gauss outlines the state of theoretical astronomy and its goal to determine orbital perturbations, which were key areas in early asteroid studies. Rejecting the philosophy of Diderot, Gauss explains that 'real knowledge' in astronomy derives from mathematical analysis. This was also considered central by Pierre-Simon Laplace (1749–1827; 1798):

> *Astronomy, considered in the most general manner, is a grand problem in mechanics, in which the elements of the celestial motions are arbitrary constants; its solution depends both on the accuracy of the observations and on the perfection of Analysis, and it is very important to banish all empiricism and to borrow nothing from observation except indispensable data.*

This emphasis on analysis derives from Newton (1704): "As in Mathematicks, so in Natural Philosophy, the Investigation of difficult Things by the Method of Analysis, ought ever to precede the Method of Composition. This Analysis consists in making Experiments and Observations, and in drawing general Conclusions from them by Induction."

As noted by Hahn (1998: 48), Jean-le-Rond d'Alembert (1717–1783) had set the stage for France's scientific agenda "by asserting that completing the Newtonian program was the most virtuous and urgent activity of fellow scientists." In this passage, d'Alembert (1766) specifically repudiates the notion that mathematics is sterile:

If astronomy is one of the chief sciences that ennobles the human mind, physical astronomy is one of those that honours modern philosophy the most. The search for causes of celestial phenomena, which has made such progress in our times, is no mere sterile speculation, but [an activity] whose value is gauged by the extent of its object and by the formidable effort required to comprehend it.

D'Alembert further states the requirement for fulfilling such lofty goals. "One could presume to have identified the authentic causes of the motion of planets once one could derive consequences from these causes that match what observation has disclosed." This French view of 'physical astronomy' contrasts sharply with the German view, as expressed by Gauss in his Inaugural Lecture. For him, seeking the *cause* of celestial phenomena is quite literally a waste of time. Gauss insists we must accept reality as we find it, and move on to analysis.

In this view he was joined by the Polish astronomer and philosopher Jan Sniadecki (1756–1830). Sniadecki, who made many observations of the asteroids, removed the search for first causes to the area of 'intellectual faith' (Bujarski, 1972: 7).

The Inaugural Lecture on Astronomy

Here is an excerpt from Gauss' Inaugural Lecture on astronomy from 1815:

All the heavenly bodies, insofar as we have scientific knowledge of them, are the subject of astronomy, or star-lore. The definition would be too foolhardy if we said unconditionally, all heavenly bodies are the subject of astronomy, for it is not only possible, but most probable that there are many celestial bodies of quite a peculiar nature, whose existence remains unknown to us. Therefore according to the present condition of the science as it is applied to the Sun, the ten major planets, to which even our earth comes as the eleventh insofar as it is quite similar to the other major planets, the Moon, the Moons or satellites of the other planets, the rings of Saturn, the comets, the fixed stars and nebulae. Of all these celestial bodies, however, only that which we really know, belongs genuinely in astronomy, that which is founded on reliable observation, inferred from it by maturely pondered reasoning and rigorous calculation and incontrovertibly verified by complete, never disturbed agreement. But not badly grounded suppositions, idle dreams and hypotheses snatched from the air. About the natural constitution of the celestial bodies, however, we know fundamentally only very little and astronomy as an exact science can therefore receive only few of the suppositions in this respect and only such as are formed with great prudence according to the rules of analogy. Opinion in astronomy stops right there and real knowledge begins in those topics that are capable of a mathematical treatment: and those are the magnitude and shape of the celestial bodies, their distances, their mutual situations, or in other words, the movements.

Physical astronomy shows how the laws of the movements, deduced a posteriori from the phenomena, are all only the consequences of one great natural force universally broadcast and everywhere active in the same manner. As it does not admit of desiring to explain further the cause of this power, but merely assumes its existence as incontrovertibly proved by the phenomena, it is the business of theoretical astronomy to evolve the effects of the same with the assistance of the finest analysis: first, to show the motion of celestial bodies in conic sections according to Kepler's laws found in theoretical astronomy as the necessary effect of the reciprocal attraction of two heavenly bodies, then the perturbations in those movements which are to be determined as a consequence of the reciprocal attraction of several heavenly bodies, and thus the perturbations of the major planets among themselves, the gradual advance of their orbits, the disparities in the motion of the Moon, finally also the shape of the celestial bodies whose parts have set themselves into a condition of equilibrium according to the laws of reciprocal attraction, just as disturbances in rotation are to be evolved by variation of the position of the axis of rotation. These investigations are the triumph of the human intellect, but they belong to the most difficult of astronomy, demand the application of the finest artifices of analysis, and are of such an extent that in our course time will scarcely remain to penetrate into the heart of them. [Translation from Dunnington (1937); the original text is in Gauss Werke XII.]

Gauss includes here the four asteroids (Ceres, Pallas, Juno and Vesta) as 'major planets,' a view diametrically opposed to the one advanced by William Herschel. Philosophically, we can see here elements of the work of Jakob Friedrich Fries (1773–1843); Fig. 1.2, one of the few philosophers at the time with a serious interest in astronomy. Gauss was "particularly fond of a history of philosophy which Fries wrote." (Buehler, 1981). Fries realised how many of the advances in mathematics since Newton's time found no place in German idealistic philosophy and the failure of all

Fig. 1.2 Jakob Friedrich Fries

systems of philosophy to convincingly hold together the expanding realms of mathematical and empirical knowledge had led to a crisis of scientific rationality (Pulte, 2006). This is reflected in Gauss's lecture where he delineates quite starkly what can and cannot be addressed by astronomy both observationally and empirically.

The topics Gauss lists as being subject to "mathematical treatment" (magnitude, i.e., size, shape and distance) are the very ones Fries considers. He believed that the construction of objects of sense perception in space and time comes by virtue of the power of the imagination. We are able to handle the ideas of figure and duration just as mathematics uses them. In applying these ideas to things outside us, such as planets or asteroids, our geometrical sketches seem to belong to sense and not to our imagination. But, said Fries, all our ideas of size, shape and distance are also used voluntarily (Gregory, 2006: 93).

Mathematics and the *Encyclopedie*

The *Encyclopedie* was an attempt by its co-editors, Diderot and D'Alembert, to collect and codify all knowledge. It ran to 17 volumes from 1751 to 1756. Johann Wolfgang von Goethe (1749–1832) summed up the feelings of many in the face of this overabundance of material: "When we heard somebody talking about the Encyclopedists, and opened one volume of their huge work, we felt as if walking among the innumerable moving bobbins and weaving looms of a vast factory." (Goethe, 1994: 361). In the classification of human knowledge that prefaced the *Encyclopedie*, mixed mathematics took pride of place. Pure mathematics, comprising geometry and arithmetic, was dwarfed by mixed mathematics, which embraced all of mechanics, optics, pneumatics and astronomy.

Mixed mathematics included not only applications but whole disciplines that would now be classified as integral parts of physics or astronomy, such as celestial mechanics. Mathematics defined by the *Encyclopedie* consisted of three parts: pure mathematics, mixed mathematics, and physico-mathematics, which studied quantity as it related to the causes of observed events. In the preface to Volume 1, d'Alembert (1751) argues that pure mathematics has nothing to do with reality and is merely concerned with manipulating symbols (Wells, 1978). D'Alembert's account of mixed mathematics explained how celestial mechanics was possible, which directly influenced those who would use it to plot the orbits of Ceres and Pallas (Aspray & Kitcher, 1988).

For eighteenth century mathematicians, mathematics did not exist without a real interpretation. Mixed mathematics did not simply translate the subject matter of physics or astronomy into the abstract, neutral language of pure mathematics: it was a genuine science in its own right, with subject matter as well as a powerful method (Daston, 1988: 54). In the context of this book, the subject matter was the orbit of the asteroids Ceres and Pallas, and the powerful methods were the competing schemes for determining those orbits based on Newtonian mechanics.

The Failure of Burckhardt

Johann Carl Burckhardt (1773–1825); Fig. 1.3, became known as Jean-Charles when he moved from Germany to work with Lalande in Paris (Débarbat & Dumont, 2014). Why did Burckhardt fail to compute the orbit of Ceres from the initial observations of Piazzi in 1801, while Carl Gauss succeeded? Even though the rediscovery of Ceres was a matter of great moment in the history of astronomy, it appears that no scholarly attention has been paid to this point.

The answer involves the differing mix of mathematics, philosophy and society in France and Germany. "The mathematical language has more to commend it than being the only language which we can speak; it shows that it is, in a very real sense, the correct language." (Wigner, 1960). If mathematics provided the symbols for creating language, then the French and German astronomers used them to create quite different languages. For the French, the application of probability theory was the syntax that enabled lengthy symbolic equations to explain the orbital motions of planets and asteroids. The Germans, while impressed by the mathematical virtuosity of their French counterparts, opted for a more prosaic language, shorn of the flourishes so characteristic of much of French society. Quickly arriving at a workable result was their goal.

It has been noted by Grattan-Guinness (1983) that "around 1800 German astronomers such as Olbers, Encke, Soldner, Bessel and Gauss began to develop compact and feasible methods for several areas of celestial and planetary mechanics. This approach contrasted sharply with the French love of long equations in celestial mechanics." Why did the French have this preference?

Fig. 1.3 Johann Carl Burckhardt

It derives from Leonhard Euler's "wonderful idea about the series, and its implementation by Lagrange and Laplace. Philosophically, it may be boosted by the I-can-know-everything view of Laplace, that if you take enough terms in the series you will get an 'exact' answer." (Grattan-Guinness, 2002). It was Euler (1744) who made the first real progress in understanding the orbit determination process from an analytical point of view. He introduced "infinite trigonometric series" to simplify the calculation of perturbations (Golland & Golland, 1993), but even so his method of orbit determination has been criticized as having "calculations that are long and tiresome" (Dubyago, 1961).

Laplace's support for lengthy series calculations also appears to stem from his belief that periodic forces produce periodic effects and therefore needed periodic functions in the mathematics. All of this was embodied in the *Mécanique céleste*. Burckhardt translated the first two volumes of Laplace's *Mécanique céleste* into German (published in Berlin in 1801, the year Ceres was discovered).

Burckhardt thus employed Laplace's method of orbit determination, even though it was unsuited to the case of Ceres (Marsden, 1995). With the experience of Ceres, it is thus even more surprising that he used Laplace's method on the newly discovered Pallas in 1802. As explained by Baron Franz von Zach in the September issue of the *Monthly Correspondence* (MC), this led his calculations astray again.

German astronomers adopted an alternative strategy. Since the phenomena are too complicated to handle 'exactly,' they opted to settle for making an approximation. "Two years before *Mécanique céleste* appeared Wilhelm Olbers (1797) had given a nice means of approximating the paths of comets that contrasts strongly with Laplace's lucubrations. Soon afterwards Gauss used a method of this kind to analyse the motion of the recently discovered minor planets." (Grattan-Guinness, 2002).

Laplace approached Napoleon about accepting a copy of *Mécanique céleste*. Someone had told Napoleon that the book contained no mention of the name of God; Napoleon, who was fond of putting out embarrassing questions, received it with the remark, "M. Laplace, they tell me you have written this large book on the system of the universe, and have never even mentioned its Creator." Laplace, who, though the most subtle of politicians, was as stiff as a martyr on every point of his philosophy, drew himself up and answered bluntly, "I do not need this hypothesis." Napoleon, greatly amused, exclaimed, "Ah! This is a good hypothesis, it explains a lot of things."" (Stenger, 2013: 436).

Written with the hindsight of more than 20 years, Zach's analysis of the methods of orbit determination confirms the value of Olbers' method, which was employed for both comets and asteroids:

> *All of the methods by La Place, LeGendre, Olbers, Gauss, Bessel etc. for calculating the elements of a cometary orbit, whether it be parabolic, elliptical or hyperbolic, are indirect or are methods of false positions. To tell you the truth, these methods are empirical. They are attempts, groupings, but an astronomer in making his calculations does not want to do amazing feats Secundum legem artis (following the law of art). Rather, he wants to reach his goal at very little cost, and he will willingly sacrifice the rigour of geometry for empiricism, which leads to the same goal by a shorter route. All calculations in astronomy are done in this way, and I think that an astronomer has never been required in this*

world who has calculated an eclipse by a direct method. They have all been calculated by inaccurate positions. No one has ever found the numerical solution to a 3rd, 4th or 5th degree equation using [Gerolamo] *Cardan,* [Rafael] *Bombelli or Scipion Ferri's rules. Is it not quicker to use two or three Suppositions to find the roots than with all the other rules of the art? The most efficient empirical method for calculating the orbits of comets is indisputably that of Mr. Olbers. I have always had the approximate elements within 4 or 5 hours. This would never be possible by any direct method. All of the work by astronomers and geometricians in this field is reduced to further simplifying these indirect methods, but nothing more can be gleaned in this field, and this is why we have no obligation to geometricians like Cardinali when they want to approach this problem with new methods. This has also been well recognised in England, and Doctor Young has translated Mr. Olbers' method, which I was the first to publish 23 years ago, into English. I think that he published it in the Journal of the Royal Institute* [it was published in *The Quarterly Journal of Science,* 1820] *since he only sent me the loose sheets.* [Letter from Zach to William Henry Fox Talbot; January 9, 1822]

The Ramist Interpretation

Although the failure of Burckhardt can be partially explained in terms of mathematical arguments, there is a deeper philosophical background to the methods used by him and his rival Gauss. Founded by the French philosopher Petrus Ramus (1515–1572); Fig. 1.4, the emphasis of Ramist logic on clarity, precision and testing encouraged the scientific spirit in Protestant countries—notably England and Germany. While it is true, as noted by Cajori (1897: 276), that "… his views respecting the basis of geometry controlled French textbooks down to the nineteenth century …" the full impact of Ramist philosophy was blunted in France, as he was actually barred from teaching philosophy in France in 1544 and his philosophical works were repressed.

"Nature," said Heraclitus (Patrick, 1889: 86), "loves to conceal herself." Nature's favorite hiding place, according to Parmenides (cited in Manley, 1980: 25), is in the differing beliefs, practices, and experiences of men. It is to this we must look to understand the way Burckhardt and Gauss tried to shine the light of science (*episteme*) on that hiding place where nature kept the secret of Ceres' orbit.

"There usually may be found many ways to climb the same hill and various paths to the same city." So wrote William Lewin in a prefatory letter to Gabriel Harvey's *Ciceronianus* (1577), the book that introduced the Ramist method to England. Harvey's interest in astronomy fostered his regard for Ramus (Johnson, 1937), who urged a return to observational astronomy of the Babylonians and Egyptians in an attempt to determine the nonhypothetical, directly observable regularity of the heavens. Ramus (1569: 50) wrote that he would relinquish his chair as professor of philosophy and rhetoric to anyone who could develop an "astronomy without hypotheses," by which he meant concentric spheres, epicycles, eccentrics, and equants that Proclus in the fifth century dismissed as nothing but "conceptual notions" that do not exist in nature. Kepler, in his 1609 book *Astronomia nova*, claims to have met Ramus's demands. He repeated the claim in his *Rudolphine Tables* (1627), where

Fig. 1.4 Petrus Ramus

he listed among the causes for the long delay in publication the "transfer of the whole of astronomy from fictitious circles to natural causes." His "main thesis is that only *true* hypotheses lead to true consequences." (Hooykaas, 1999: 232). Thus Kepler saw his response to Ramus' challenge as a central pillar of his new astronomy, which in turn was the bedrock upon which the astronomers of 1801 worked from in their study of Ceres. Leroy Loemker (a philosopher at Emory University) states that Kepler, who was the first to posit the existence of a missing planet between Mars and Jupiter, was influenced by Ramus in his tendency "to exclude the order of knowing, and the relations essential to it, from the harmonious order of nature to which it refers." (Loemker, 1972: 195). This has had a direct influence on mathematics in the twenty-first century, especially in the person of Benoit Mandelbrot (1924–2010). His hero and scientific model was Kepler, "who brought ancient data and ancient toys together and founded science." (Mandelbrot, 2012). The wish to emulate Kepler fueled the search for the mathematical structures he was to call fractals.

The greatest analyst of the influence of Ramus, Walter Ong (Professor of English at Saint Louis University), best explained his importance. "Ramism is a state of mind arising within a complex of established intellectual and cultural traditions and exhibiting them in new aspects." (Ong, 1958: 7). I contend that one of these new aspects arose in the study of celestial mechanics around 1800. My analysis does not suggest that Ramism was deliberately or consciously employed in this effort that found its greatest expression in the recovery of Ceres. For, as Ong (p 295) explains, "By the time method and logical analysis have established themselves firmly in the Western European psyche, the paramount role of Ramism in their establishment has been forgotten." This Ramist tradition was much stronger in Germany than in France. Ong traces 151 editions of Ramist dialectic and rhetoric in Germany from 1543 to 1700, but only 14 in France (and only one of those after 1580). Germany was "the real seedbed of Ramism," and it was there that its diagrammatic approach to knowledge fires the imagination of ..codifiers of all the sciences… Ramist method moves into the uppermost branches of the curriculum." (Ong: 298).

Ramus recognized the importance of mathematics in astronomy, as evidenced in his letter to 'the first Copernican' Georg Joachim Rheticus (1514–1574) of August 25, 1563. Ramus refers here to the 1551 book *Canon Doctrinae Triangulorum*, which extolled the usefulness of trigonometry and offered for the first time in history six-function trig tables (this book will be considered further in a subsequent volume in this series):

> *Although I wanted to compose a system of astronomy, I got adequate help from neither logic, nor books, nor people. All I could see was a discipline whose complexities and obscure hypotheses rendered it incomprehensible. But finally, finding your little book the* Canon,…*I read it through again and again…Wonderfully, it awoke in me a great hope. Particularly for any mathematician well versed in numbers, it promised an astronomy completely liberated not only from the problems discovered by the followers of Pythagoras and Jabir, but also from countless hundreds of useless tables.* [English translation by Dennis Danielson (2011: 165).]

German textbooks in arithmetic and geometry were directly influenced by Ramism, which was linked to the thought processes producing "the Newtonian revolution, with its stress on visually controlled observation and mathematics." (Ong: 318). The linkage between Ramus and Newton has also been noted by Danielson (2011: 166):

> *The practically-minded Ramus shared Proclus' longing for a system that did not force a separation of mathematics from actual bodies moving harmoniously in nature. Indeed, it is hard not to be a little Whiggish in this reading of science history and to say that Proclus, Copernicus, Rheticus, and Ramus were all somehow hoping for the practical fusion of mathematics, physics, and astronomy that we now usually see as being consummated in the work of Isaac Newton.*

Ramus's swipe at useless astronomical tables was echoed in poetry a century later by Claudius Quillet (1602–1661; 1655) in his disingenuous advice to a fair lady:

> *Start not, ye Fair, nor from my Verse retreat,*
> *Thinking the Study of the Science great,*
> *For all these mighty Volumes of the Sky,*
> *Explain'd in short, and easy Tables lie.*

Both Burckhardt and Gauss were trying to climb the same hill evoked by Lewin, but their approach differed greatly. The American Renaissance scholar Professor Hardin Craig (1875–1968; 1936) called Ramus "the greatest master of the short-cut the world has ever known." Gauss applied the rigid rationalism of the Ramist short cut. Burckhardt was steeped in the French tradition of Laplace's method—the gradual establishment of a path through use and repetition, which had been successful before but did not lead to the summit.

The framework in which Gauss thought was one that places the source of order and regularity in universal, natural principles. It articulated its principles through discursive reason. It was (in the words of Ong) this "set of mental habits" and the "stress on mathematics" based on Ramist logic that helped lead Gauss to success. Ong details how the spatialization and quantification of thought in dialectic and logic during the Middle Ages enabled 'a new state of mind' to emerge in print culture, a state of mind representing 'a real mathematical transformation of thinking' associated with the emergence of modern science (Ong, 1962: 72).

"In its long-term effect, Ramism, with the topical logic which it exploits, is favourable to the emergence of modern science, experiment included, because of the way it loosens up the field of knowledge in encouraging visualist approaches to this field." (Ong, 1958: 269). Nowhere was the application of spatialization more necessary than in the three-dimensional problem posed by the orbit of Ceres. Gauss was able to visualize the problem in terms of mathematics in a way no one else could. His rival in the race to calculate the orbit of Ceres, Burckhardt, was bound by established social and scientific usages and customs. His framework articulated its principles though tradition. In the face of the challenge from physical reality posed by Ceres, it failed (Manley, 1980).

The Sacred Ceres

Gauss always recognized the special role Ceres played in his life. He kept a notebook with his calculations of Ceres' orbit and comparisons with the observations of others. But for the historian the most singular element of the notebook comes under the heading "Cereri Ferdinandeae Sacrum." Here he wrote down five lines that were used by Zach in the *Monthly Correspondence* of April 1801. Zach told of receiving a letter from some far distant place in which an ignoramus made sport of the work of astronomers by advising that they should refrain from the building of air castles (particularly pertinent, as the search for the unseen planet could be considered just such an air castle). Zach wrote:

> *We cannot restrain ourselves from quoting here an excellent passage from a letter of our Dr. Gauss which indicated the noble qualities and attitude of this worthy scholar. 'It is scarcely comprehensible,' writes Gauss, 'how men of honour, priests of science, can reveal themselves in such a light. As for me, I look on such incidents only as tests of whether I work for my own sake, or for the sake of the subject concerned.' These are the onera of fame, and Gauss will experience more in the course of time. We admonished him, therefore, to persist and abide steadfastly in these noble maxims, which we also would do very well to remember, and to recall our always sprightly, happy and worthy old patriarch and teacher Lalande:*
>
> *There are a thousand million people living on this earth,*
> *Of these thousand million heads*
> *How many are wicked, foolish, bestial,*
> *But we cannot cure,*
> *We can only pity and serve them.*

Gauss was ridiculed by some for wasting his time on the computation of the orbits of Ceres and Pallas. The wits of the day said that while Ceres may be the goddess of the fields, no corn grown on the new planet would ever find its way into the Brunswick market of a Saturday afternoon. Gauss never replied publicly to these barbs (Bell, 1937: 242).

Chapter 2
France Versus Germany

C. Cunningham, *Early Investigations of Ceres and the Discovery of Pallas*,
DOI 10.1007/978-3-319-28815-4_2

Fig. 2.1 The Berlin (Prussian) Society of Science (1710) has a bust of King Frederick I of Prussia, a telescope and a roundel with a series of fractions

Mathematics and Astronomy in 1800

At the turn of the century there were only one or two mathematicians of note in Germany (Fig. 2.1). A mathematician was "essentially a pure drudge, whereas an astronomer was a scientific professional." (Fauvel, 1993: 9). This is in marked contrast to the situation in France (Fig. 2.2). "The ideology of the new French Republic from the 1790s onwards was one which promoted mathematics as tremendously significant. Mathematics was taken very seriously, both intrinsically and as a science in the service of the state. In all Europe, it was France where creative mathematics was primarily happening at the turn of the century." (Fauvel, p. 10).

During the 30 years after the discovery of Ceres, mathematics in Germany experienced a renaissance, but it was not an attempt to copy the French style. "Rather, it was the growth of a new institutional style, a new way of doing mathematics. The kind of mathematics promoted tended towards what we would call 'pure' mathematics." (Fauvel, p. 10).

While the mathematical vigor in France would seem to have given them the edge in tackling a new problem such as the orbit determination of Ceres, the exact opposite is what actually happened. The person who perhaps best put his finger on why this extraordinary situation occurred is Carl Gustav Jacobi (1804–1851). In his inaugural lecture as a professor at Koenigsberg in 1832, Jacobi criticized the French mathematicians for putting too much stress on applied mathematics, and for mixing up the true and the incidental causes of progress in science.

> *We are unhappy that most French geometers who originate from the school of the famous Laplace have presently fallen into this error. While they seek to obtain the only salvation for mathematics in physical problems, they desert that true and natural path of the discipline, which has brought the analytical art to the importance which it now enjoys. In this way it is not so much pure mathematics, but its application to physical problems, that suffers.* (Schubring, 1993: 29)

No failure to solve a "physical problem" was more prominent than the failure of Burckhardt and other geometers from the school of Laplace to determine the orbit of Ceres.

Science and Society

The exalted status of French science was never described more boldly than by Napoleon Bonaparte in a letter he wrote to Oriani on May 24, 1796: "All men of genius, all those who have achieved distinction in the Republic of Science, are French, no matter what their native land."

To explore the difference between astronomy in France and Germany—a difference that led to such a divergent response to the discovery of Ceres—we must also look at the societal difference that characterized the scientific establishment of the two countries. Their *genius loci* was quite different. When one speaks of French mathematicians and astronomers, one invariably means those who worked in Paris.

Fig. 2.2 The *Memoires* of the Academy (1699) has an oval depicting King Louis XIV of France

German scientists lived a much more fragmented existence, as clearly explained by Goethe in a conversation with Johann Eckermann on May 3, 1827.

> *To you it has not been so easy; and we others also, in Central Germany, have been forced to buy our little wisdom dearly enough. There we all lead a very isolated miserable sort of life! Our talents and men of brains are scattered over the whole of Germany. One is in Vienna, another in Berlin, another in Koenigsberg, another in Bonn or Duesseldorf – all about a hundred miles apart from each other, so that personal contact and personal exchange of thought may be considered rarities. But now conceive a city like Paris, where the highest talents of a great kingdom are all assembled on one spot, and, by daily intercourse, strife, and emulation, mutually instruct and advance each other. In addition to all this, conceive not the Paris of a dull spiritless time, but the Paris of the nineteenth century, in which, during three generations, such men as Moliere, Voltaire, Diderot, and the like, have kept up such a current of intellect as cannot be found twice on a single spot in the whole world.* (Moorhead, 1998: 200).

It is interesting that of all the great thinkers, Goethe mentioned Diderot in particular. Like Diderot, Goethe had a reputation for disparaging mathematics: "Mathematics has the completely false reputation of yielding infallible conclusions." (Quoted in Newman, 1956)

Civilization vs Culture

The development of French and German culture since the Renaissance, which was the web of incubation for the rise of science and mathematics in the early nineteenth century, is a factor to be borne in mind when examining this rivalry.

> *Civilization, the French asserted, was a progressive human accomplishment, evolving over centuries. It was a supreme achievement of Reason, culminating in the Enlightenment. But for German intellectuals, civilization was a danger to Kultur. For them, culture was everything civilization was not. It developed out of spiritual values rather than reason; it was particular rather than universal; it was reflective rather than progressive. Culture, not civilization, was the touchstone of the Romantic Counterenlightenment.* (Rothstein, 1999)

The German view of France was emphatically determined by the French Revolution, and the Napoleonic occupation of Germany: "By the time of Johann Gottlieb Fichte's (1762–1814) *Reden an die Deutsche Nation* (*Addresses to the German Nation*) in 1806, Germany was presented as a unique and original nation that, unlike the French, had not lost touch with its original genius." (Watson, 2010: 262). The rivalry between France and Germany, stoked by the race to calculate the orbit of Ceres, is explored further in another volume in this series, where the French 'up the stakes' with Pallas, and the Germans set out to capture the prize—a kilogram of pure gold.

Central to the point of this study is the role of mathematics. A foundational text in this regard was written in 1615 by Giuseppe Biancani (1566–1624), a Jesuit professor of mathematics at the University of Parma. He noted in his treatise that mathematical demonstrations are most powerful and the best means of attaining certitude in the physical sciences. "It absolutely follows," he wrote, "that mathematics is superior to all the other sciences, in the same way that truth is superior to opinions." This concept was adopted by the French, but not the Germans before Gauss.

The German View of France

> "Mathematics are a species of Frenchman; if you say something to them, they translate it into their own language and presto! It is something entirely different."
>
> – Goethe, 1826

For a contemporary German view of French mathematics, we need look no further than Gauss's own teacher, Abraham Gotthelf Kaestner (1719–1800); Fig. 2.3. In 1756 he went to Goettingen University as Professor of Mathematics, a post he held until his death. On the vexed question of the calculus of Newton versus Leibniz, Kaestner "… kept a middle position." (Bullynck, 2014). Even though Gauss himself learned little from Kaestner's popular lectures, the old man's attitudes and beliefs undoubtedly had some influence on the young Gauss (Dyck, 1980).

Fig. 2.3 Abraham Kaestner

In a letter of April 28, 1797, to J. F. Pfaff, Kaestner left no doubt about his opinion of the state of French mathematics:

> *France is famed for the learned institution where Lagrange teaches; it is said that the students arrive there with such knowledge that they start where students at German universities leave off. Now that is a stiff breeze of French republicanism, more disturbing than their royalism. French teaching in the rudiments of mathematics is much poorer than at any German university, and certainly produces nobody who could listen to Lagrange with profit. It is particularly ridiculous to listen to someone whom one ought to read – and that is what those Germans who have learnt the rudiments do by themselves, if they have the time and inclination.*

German researchers were very reluctant to apply highly-mathematical work in physics, which had been spearheaded by the French, because of its abstract nature (Jungnickel, 1986: 45). Lagrange in particular was criticized in this regard as the 'calculator' who had eliminated images. "What good is a mechanics without figures, such as Lagrange's?" asked Kaestner of Pfaff in 1798. "I have not found any applications of Lagrange's *Mechanics* yet." (Pfaff, 1853: 216).

The views of Kaestner were influential, and not confined to private letters. Through textbooks, articles and reviews, he was at the forefront of spreading mathematical thought in Germany, so his overtly critical opinion of the mathematics devised by the Frenchmen Laplace and Lagrange surely had an impact. His "... role as a populariser and propagator of the mathematical sciences cannot be underestimated." (Bullynck, 2006: 5)

Few people were better placed to evaluate the status of German science than the electrical engineer John Theodore Merz (1840–1922). Also a chemist and historian, he was born in England of a German father. He wrote this in his magisterial four-volume work on the history of European thought:

> *The general impression we receive from a perusal of the histories of science and learning in Germany at the close of the eighteenth century is that the university had not–with the exception perhaps of Goettingen–received into its pale the modern spirit of exact research, such as it had been developed by the great French Academicians...Although Gauss introduced the higher and abstract branches of exact science into the programme of a German university, and established a link between Paris and Germany in mathematics...fully a quarter of a century was to elapse before the spirit of exact research, and of the higher mathematics, really began to leaven the German universities... During these twenty-five years Gauss lived and soared in solitary height. But astronomy was not then within the pale of the universities.* (Merz, 1896: 178–184)

The early nineteenth century position of Germany in science was quite obvious to the medical theorist Johann Christian Reil (1759–1813); Fig. 2.4 This is from a speech delivered in Halle in 1810 (Reil, 1817: 318):

> *The period of my present teaching position coincided with the most noteworthy time in which the study of medicine, as well as that of the entire natural sciences, underwent an almost complete revolution. The effort at explanation has made place for living intuition; the idea has entered the arena of the mechanical principle; and observation has achieved a standpoint from which to view things in their natural relations. Indeed, the machine of the heavenly bodies has been animated. Only the German scholars have given birth to this renaissance of science.*

Fig. 2.4 Johann Christian Reil

He attributed this transformation entirely to German scholars, who "… have given birth to this renaissance of science." (Reil, 1817). His reference to the "machine of the heavenly bodies" undoubtedly refers to the contributions of Gauss to celestial mechanics, which were motivated by the discovery of Ceres.

Zach on the State of Mathematics in Germany

Baron von Zach made clear his opinion of the state of mathematics in Germany in this letter to Gauss on Feb. 21, 1802:

> *I need not to say how pleased I was to read that your noble Duke* [Charles William, Duke of Brunswick, 1735–1806] *had put you in such a position that allows you to devote yourself entirely and carefree to your sublime science.*
>
> *The Duke has not only done you but also science a good turn. Mathematics are not strongly supported in Germany these days. And you see the result. The famous Goettingen University has no mathematician since Kaestner's death. Let Kluegel die in Halle and*

Hindenburg in Leipzig and there will be no mathematicians except in Brunswick. Stahl in Jena is doing fine but will die of starvation, all my efforts have not been successful to give him that what your excellent Duke gave you voluntarily. Our good Burckhardt had to leave his country to make his fortune otherwise he would have rotted in his little attic room. Töpfer wastes away in Grimma, this excellent head has to teach snotty little brats. In what is France today mathematicians are Senators and Councillors of state. In England they head the payrolls. Only in Germany they are starving. It has to be said publicly how much the Duke of Brunswick protects and supports the mathematic sciences. He gave us Kluegel, is preserving Pfaff and he now gives us Gauss. If only he knew how grateful I am! He bestows his favour on these sciences because he is a great commander and understands how useful these sciences are for an engineer, artilleryman, sapper, and the general staff when you are at war. Friedrich the Great was not an expert on mathematics, but he anxiously made sure that his academy had enough mathematicians. He had the best heads of Europe. He had Euler, La Grange and Lambert. In the writings of this great king he sometimes made fun of mathematics! But no, he did not mock mathematics but the mathematicians. And not even this he only mocked Euler because in truth he was angry at him because he left and went to Russia. Hence all the sarcasm that Euler had gone to observe the Great Bear of the North. How well did he not treat Maupertius and La Grange. You only have to read his letters to D'Alambert and how hard he tried to recruit La Grange to know how much he wanted this great geometer for his academy. After this great king's death his excellence the corporal Count Hertzberg, 'who wanted to lead the academy like a regiment,' drove this excellent man to France, where Bonaparte, who protects the mathematic and astronomical science, too, made him Senator. La Grange and La Place are now des grands seigneurs, have equipages in Paris, and earn almost 8,000 Reichsthaler. The German mathematicians deem themselves lucky if they only had a tenth of that! But enough of this jeremiad! Long live the Duke of Brunswick! (Fig. 2.5)

Fig. 2.5 The Duke of Brunswick

Zach's Diatribe Concerning France

Zach's fiery temper was never more evident than in a letter he wrote to Gauss just a month later, on March 20, 1802. At the root of the French attitude towards Ceres, Zach believed, was the very fact that they did not discover it: "It is incredible how ridiculously these citizens behave: they consider it a disgrace that has been brought over the Grand Nation [France] that they neither observed nor calculated the planet first. But still they make a great fuss, stretch their backs, speak in a high voice: "'Let us do it, we arrange all this'."

In this remarkable letter, Zach accuses the French of distorting the truth, of implying that observational data from Germany is their own, and of such jealousy that Ceres was found by a foreigner that French astronomers are like one "who wants to cry his eyes out in front of Bonaparte like an old whore because he was not the first to rediscover the planet."

In the battle over the name of the newly discovered planet, Zach could hardly contain himself as Joseph-Jerome Lalande (1732–1807) gloated in a letter that "soon we [namely, the French] will have all satisfaction. And the name Juno is being used. The senator La Place uses it exclusively."

Zach was also dismissive of Pierre Méchain's attitude: "Méchain plays the diplomat and is still manoeuvring. He neither writes Juno nor Ceres, but only "the new planet"; it is ridiculous to see how anxiously and world-wisely he tries to avoid the *nomen proprium* [proper name]."

While the need for French astronomers to name the discoveries of others does strike one as rather pathetic, the psychology involved was not born with the discovery of Ceres. We read, in a November 1781 letter from Joseph Banks to William Herschel, that naming his great discovery had better be done quickly. If Herschel failed to move with dispatch, Banks warned, "our nimble neighbours, the French, will certainly save us the trouble of baptising it." In the end it was the German astronomer Johann Bode who dubbed it Uranus (Standage, 2000).

French jealousy did not abate with the discovery of Pallas, as Zach urges Gauss to calculate its orbit in an April 27, 1802, letter. "Let us show the high-spirited French, who like it so much to push the Germans into second place."

Germany, France and England

A distillation of what scholars in each of the three countries thought of one another is illuminating. The German Georg Hegel was among the first to see in the geographical triad of Germany, France and England an expression of three different existential attitudes: reflective thoroughness (German), revolutionary hastiness (French), and utilitarian pragmatism (English) (Žižek, 1997: 5). In 1854, the English philosopher John Stuart Mill declared aphoristically that "the characteristic of Germany is knowledge without thought; of France, thought without knowledge; of

England, neither knowledge nor thought." In 1842, the Frenchman Louis Raymond de Véricour (lecturer in the Royal Athenaeum of Paris and Professor of Modern Languages in Cork Queen's College) wrote "Germany stands pre-eminent for laborious research, France for activity and perspicuity in development, and England for practical adaptation. Their intercommunication and reciprocal influence form the basis and mainspring of universal civilisation."

There was a philosophical disagreement in both Germany and France with Newton. For Bernard de Fontenelle (1657–1757), secretary of the Paris Academy of Sciences, attraction at a distance was his bête noire. He believed "it smacked of occultism." (Paul, 1980: 31) Saint-Simon in France, as we have seen, appeared to be motivated by nationalism in his criticism of Newton. By contrast, Hegel kept his criticism on an intellectual plane. Even though his university teaching career began in 1801 with a dissertation on Newtonian astronomy, in his *Lectures on the History of Philosophy*, Hegel devoted only two pages to Newton (Hegel, 1805: 322–324), and "He certainly had no very high opinion of Newton's ability to deal with thoughts ... He was quite clearly of the opinion that Newton was not aware that when he was speaking about forces he was actually dealing with notions." (Wahsner, 1993: 81). Goethe also famously disagreed with Newton's classic theory of white light and colour, but again this does not seem to have been motivated by base nationalism (Sepper, 1993). Hegel (1817) also openly criticised Newton's theory of colour, and in a letter of July 1817 Goethe thanked Hegel for his remarks on the subject (Althaus, 2000: 139).

Finally, it is useful to ask what German astronomers thought of how valuable the contribution of English astronomers had been in the early nineteenth century. Fortunately we have just such an account from the pen of a Scotsman, James Finlay Weir Johnston (1796–1855), who attended the 1830 annual meeting of the GDNA (Society of German Natural Researchers and Doctors). That year it was held in Hamburg. Johnston, a co-founder of the British Association for the Advancement of Science, summarizes the presentation of the astronomer Friedrich Georg Wilhelm Struve (1793–1864); Fig. 2.6 of Dorpat Observatory, and makes his own displeasure clear to the English readers of his account:

After the reading of the laws, and the list of members already arrived, the rostrum was occupied by Professor Struve from Dorpat, who delivered a long oration on the history, the importance, and the present state of astronomy. After magnifying astronomy above every other science that either was, is, or ever will be cultivated, he adverted to its history during the last hundred years. From this review he concluded, that during that time the main advancement of astronomy was due to Germany;– that at the present day Germany cultivated it most assiduously, and made the best astronomical instruments – a circumstance we are supposed to acknowledge, by engaging Repsold [Johann Georg Repsold, 1770–1830] *of Hamburgh [sic], (whom they dignify with the name of immortal Repsold) to furnish a transit instrument for the Edinburgh Observatory; – that after Germany Russia came next as a patron of astronomical science, by the building and equipping of observatories; – then follow England and Italy, France being lowest of all, having only two observatories at Paris and Marseilles. This discourse was neither judicious, nor, I believe, in general well received. No one science needs now-a-days to be exalted at the expence of others. Every man naturally ranks highest that particular branch of science to which he has dedicated himself: but he cannot expect to take other men along with him when he depreciates the departments to which they have with equal ardour addicted themselves. Nor is it necessary to drag in every name to exalt the scientific character of one country above that of other countries. Granted, as Sir James South has done in the* Literary Gazette, *that Germany deserves better of astronomy than England does, – yet why claim for that country the honour of names and labours which other countries will not con-*

Fig. 2.6 Friedrich Georg Wilhelm Struve

cede? – "Why claim for Germany," said a Polish professor to me, "men who were countrymen of mine." And though the Herschels, we may add, be of German extraction, their labours at least are English. (Johnston, 1831: 219–220)

The account is particularly pertinent as it makes clear that the work and discoveries of William Herschel and his son John were considered—even by a Scotsman, not an Englishman—to be entirely English. Even though their ancestry was German, their scientific work could not be appropriated by Germany. This account also shows that at least some German astronomers regarded England's contribution to astronomy to be quite minimal, on a par with that of Italy.

Ceres in English Satire

Not every aspect of Ceres was regarded with utter seriousness. There was some comic relief thanks to English satire. In this scurrilous image by William Elmes (published by Thomas Tegg) from 1811, we see the infant king of Rome (son of Napoleon) farting his defiance at the Pope (Fig. 2.7). He is surrounded by the planets of the Solar System. At bottom left is Herschel's discovery of Georgium Sidus with an image of King George III. Just above that is Ceres denoted by a sheaf of wheat and a scythe, the emblems of the Roman goddess Ceres. Even though Pallas, Juno and Vesta were known in 1811, only Ceres is accorded full planetary rights in

Fig. 2.7 An 1811 satirical print that includes Ceres

this print. At lower right are figures in England and France peering through telescopes at the apparition in the sky.

The historical occasion for this print began when Napoleon annexed to his empire the papal states in 1809. Pope Pius VII excommunicated him for this act. At the birth of Napoleon's son on March 20, 1811, he at once created the baby king of Rome as a further slap against the papacy. In 1812 Napoleon then took an even more daring step: he sent his men one summer night to scale the walls of the papal palace, and carried the Pope as a captive to France.

Chapter 3
Professional Rivalries

C. Cunningham, *Early Investigations of Ceres and the Discovery of Pallas*,
DOI 10.1007/978-3-319-28815-4_3

Fig. 3.1 An 1808 painting of Piazzi and Urania, muse of astronomy by Francesco Farina. It measures 145 x 170 cm. ©Palermo Observatory

Piazzi and the Perils of Classical Allegory

To commemorate Piazzi's discovery of Ceres, a beautiful painting was commissioned by friends of Piazzi (Fig. 3.1). It was done by the portrait painter Francesco Farina (1778–1837), pupil of the famous Joseph Velasco (1750–1827). The 1808 painting shows the muse Urania looking directly into the eyes of Piazzi. He points to some sheets, including a topographic map of the valley of Palermo, while Urania points upwards to Ceres, who sits triumphant in a carriage or chariot. Between Piazzi and Urania rests a celestial globe in front of two large books representing his star catalog. Even though the catalog was published as a single volume, it appears artistic licence was taken to magnify its size. Alternatively these may represent the original logbooks from which the final printed catalog was published as a single book. Urania herself wears a blue gown named *ortostadia*, a Greek name denoting a straight tunic.

What was Piazzi trying to convey to his contemporaries and posterity with this stunning imagery? The only precedent for its creation was the discovery by William Herschel of the Georgian Planet, Uranus. Looking at the painting that commemorated that event 20 years previously, we see few similarities (see the beginning of Chap. 4 for this painting). The use of paper is evident. In the case of Piazzi a topographic map, two books and most importantly an orbital map of the Solar System with his finger on the orbit of Ceres (Fig. 3.2). Herschel also is depicted with an orbital diagram depicting the Georgian planet, and its two newly discovered satellites. What is most apparent is the absence of any classical allusions. While Piazzi is literally dominated by the standing figure of Urania, Herschel appears alone as

Fig. 3.2 A detail from the Piazzi painting, showing the orbit of Ceres

the embodiment of the Enlightenment that has no need of deities to give their stamp of approval. Herschel is directly engaged with the viewer, while Piazzi's gaze is averted. It is almost as if we are witnessing the apotheosis of Piazzi, a figure more at home with the gods than with mortals.

Diderot reserved his strongest criticism for paintings that mixed allegorical and historical elements. What we see in the painting by Farina is both a psychological reality (Piazzi's terrible intimacy with Urania) and a theatrical tableau with all the props appropriate to the sitter as an astronomer, one intent on projecting to future observers the importance of his discovery of Ceres. A discovery so important, indeed, that a mere chart of a planetary orbit (which sufficed for Herschel) was inadequate. But central to the portrait is the eye contact between Piazzi and the Muse. What are they sharing? Farina is here employing the notion of anecdote in the original meaning of 'things not given out' (from the Greek *anecdota*), in the sense of unpublished, secret, or private narratives. It invites us, the viewers, "to look for a deeper meaning, an original context for the anecdote thus published." (Grootenboer, 2012: 28). In hindsight we can see in Piazzi a vaguely confessional invitation. "But to respond to the invitation," writes literary critic Harry Berger (2000: 226), "is to encounter a diffidence that constitutes the observer as a voyeur peering into a new depth of privacy, a depth that comes to light in the specific fragment of life that the sitter devotes to being a sitter who publishes his secrets precisely *as* secrets." It was Piazzi's penchant for secrecy surrounding his discovery of Ceres, and its consequences, that is examined in this chapter.

The legacy that the astronomers who first studied the asteroids have left us is far richer than the heroic acts of observation that sought to discover the physical characteristics of tiny objects between Mars and Jupiter. That legacy shines a unique light on the way astronomy actually worked during the early nineteenth century in a way that no other event of the era did. Like the transit of Venus expeditions of the eighteenth century, it lets us see and appreciate how astronomers in different countries acted and reacted to each other and celestial events. In the case of the transit expeditions, they were dealing with an event that could be predicted years in advance. While still imbued with rivalry, it also showed how nations could cooperate for the advancement of astronomy (Wulf, 2012). The case of the asteroids was far different. The surprise discovery set off a train of competing investigations. Instead of nationally coordinated efforts, it was 'every man for himself' as astronomers scrambled to be the first to recover Ceres after it was lost by Piazzi in 1801.

The recovery of Ceres by Zach and Olbers was followed closely by the discovery of Pallas in 1802 and the naming of the new objects as 'asteroids' by William Herschel, a word given to him by Charles Burney Jr. as research for this book discovered. The study of the asteroids became embroiled in a series of controversies and mutual recriminations. Abuse was heaped on Giuseppe Piazzi, discoverer of Ceres, for not letting others know of his discovery at once. Another extraordinary volley of abuse was heaped on Herschel for the introduction of 'asteroid' and his determination that Ceres and Pallas were not planets. German astronomers, rife with jealousies amongst themselves over the totemic prizes of new planets, were not only annoyed with the English but upset with the French for naming the new discoveries as they saw fit.

With nearly every telescope or astronomical measuring instrument in the world focusing on Ceres and Pallas, the year 1802 in particular was a watershed. Even though the discovery of Uranus by Herschel in 1781 elicited a lot of observational and computational work, the directed and sustained effort granted to Ceres, Pallas, Juno and Vesta throughout the first decade of the nineteenth century had never been seen before. The discovery of the asteroids thus ranks as one of the greatest epochs in the history of astronomy.

In an analysis of the processes that maintain rivalries, Goertz (2005) found that (1) the most recent conflict in a rivalry does matter, and (2) the longer-term history of the rivalry is more important than previously recognized.

In the context of the discovery of Ceres and Pallas, this can be interpreted in stark terms. Recent conflicts as the race was underway to determine the orbit of Ceres were very fresh indeed. And as the history of embarrassments in astronomy shows, the rivalry French astronomers had with their colleagues both in England and in Germany was decades old.

The Planetary Chess Match

Theatricality was not really the forte of Giuseppe Piazzi, but the Italian monk and astronomer could hardly have set the stage better. Just as the curtain rose on the nineteenth century, he discovered the eighth planet of the Solar System from Palermo Observatory in Sicily. Here we will consider what happened in terms of game theory, illustrated in part by moves on a chess board (Cunningham, 2007).

The metaphor of a chess game is also appropriate for the time period, as evidenced by a satirical print showing the British General Lord Cornwallis (1738–1805) playing a match against Napoleon. Cornwallis was the chief British signatory to the Peace of Amiens (see Chap. 5). The satire relates to the inept attempts by Cornwallis to negotiate peace with the French leader (Fig. 3.3). Coincidentally (and appropriately), the image was published the very day Olbers recovered Ceres, January 1, 1802.

Here, Piazzi plays Black, while White is played by Europe's astronomers. Not surprisingly, Piazzi opens with a Sicilian Defense (Fig. 3.4). This lets him unbalance the position and play for a win, without having to take any unjustified risks. Unbeknownst to Piazzi, however, this was a very dangerous opening move that carried huge risks. He should not have been so combative.

Lalande (Fig. 3.5), in his annual paper *History of Astronomy*, leads off the list of accomplishments of the year 1801 with Piazzi's discovery. Remarkably, Lalande's brief survey of the discovery of Ceres gives equal weight to the calculations of Burckhardt and Gauss. Nowhere does he even allude to the fact that his protégé Burckhardt did not develop the orbital elements necessary to find Ceres!

> *The commencement of the 19th century was distinguished by an astronomical event, the discovery of a planet at Palermo in Sicily, by M. Piazzi, on the 1st of January. It was as small as a star of the 8th magnitude: he observed it during forty days. The observations he sent me*

Fig. 3.3 Napoleon plays chess with Cornwallis

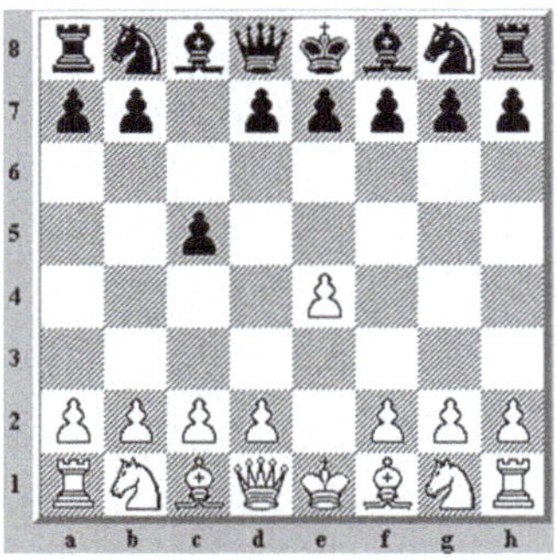

Fig. 3.4 The Sicilian Defense

arrived too late for us to be able to follow it, and we were obliged to calculate its orbit from his own observations alone. Burckhardt, Olbers, Bode, Piazzi and Gauss, have found that to represent these observations it is necessary to suppose that it revolves in four years. The following are the elements found by Burckhardt and Gauss (Fig. 3.6).

The difference of these elements appeared to me to throw some doubt on the reality of this orbit of four years; but in the beginning of the year 1802 it was perfectly confirmed and we now have an eighth planet. On the 25th of October we received a printed memoir of Piazzi,

Fig. 3.5 Joseph Jérôme Lefrançois de Lalande

Inclination - -	10° 47′	Inclination -	10° 36′ 57″
Node - -	2ˢ 20° 58	Node - -	2ˢ 21 0 44
Aphelion -	2 9 0	Epoch - -	2ˢ 16° 28′
Paſſage of the aphelion Jan. 1, 1801	8 hours	Mean anom. -	3 15 55
Eccentricity -	0·0364	Aphelion	10ˢ 26° 27′ 38″
Semi-axis - -	2·574	Eccentricity -	0·0825017
Revolution -	4·13 years	Equation - - -	9° 28′
		Diſtance - - -	2·7355

Fig. 3.6 Elements of Ceres derived by Burckhardt and Gauss

> *with his observations and calculations. As he hopes that this star will be acknowledged to be a planet, he has given it the name of Ceres Ferdinandea, in honour of the king of Naples; and Bode wishes it to be called Juno: as for my part, I shall call it Piazzi, as I gave the name Herschel to the planet discovered in 1781. The pagan deities are no longer interesting; and adulation pleases only the person who is the object of it.* (Lalande, 1802b)

An important issue raised by Lalande was the name that should be given to the new object. It was here that Piazzi made his next important move—the King's Gambit—in order to maintain the favoritism showed him by King Ferdinand (Fig. 3.7). Piazzi's plan at this stage of the game has become a choice: to consolidate his material advantage or give it back to neutralize White's pressure, pressure in this case being the demand from all astronomers for full disclosure of his data. He chose the former course—one that would have serious consequences.

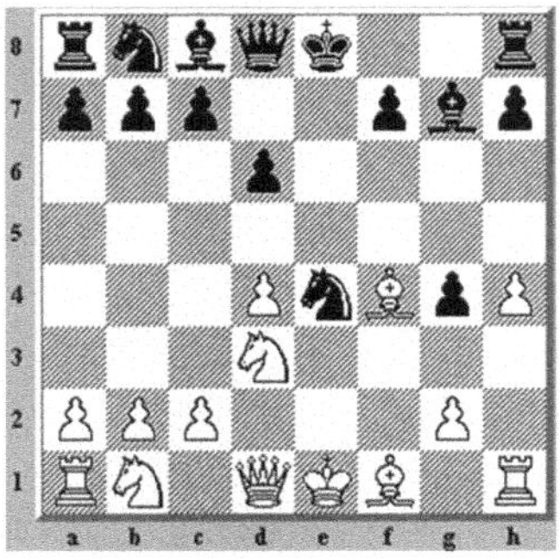

Fig. 3.7 The King's Gambit

In his first monograph on the discovery (presented in the author's book, *Discovery of the First Asteroid, Ceres*), Piazzi mentioned the monarchs who had previously been honored with a celestial object (Charles II of England, Sobiesky of Poland as John III, Frederick II of Prussia, and George III of England) and the great astronomers who bestowed those honors (Edmund Halley, Johannes Hevelius, Johann Bode and William Herschel). This museum of striking examples is made manifest by 'exemplum,' a rhetorical term from Aristotle onwards that means 'an interpolated anecdote serving as an example.' Subsequently, rhetoric added the 'exemplary figure,' which blends with historical narrative to form a 'pictura.' They are so joined "because pictura has the two epideictic functions of imitating an individual and creating a pattern that will arouse emulation or abhorrence." (Paul, 1980: 100) It is just such a pictura that Piazzi creates for the reader, and of course he intended it to arouse emulation in his push for acceptance of 'Ferdinandea.' But for Lalande, such 'adulation' aroused abhorrence. Even though his influence was not powerful enough to have the discovery termed Piazzi instead of Ceres (except briefly in France; see figures of the French orrery); Fig. 3.8 and Fig. 3.9, his voice was the loudest in the chorus of European astronomers that ensured 'Ferdinandea' would never achieve lasting acceptance.

The Covetous Monk

In chess, *zugzwang* is not merely a bad position but the state of being obliged to move when no move at all would be preferable. Here Piazzi, playing Black, has been maneuvered into an untenable position (Fig. 3.10). Piazzi obviously writes with a mixture of reluctance and shame to his mentor Lalande on April 10, 1801: "I had intended not to communicate my observations to anyone before having extracted the elements of the comet; however, since it is you who is asking for them, I have no objection anymore; you will find them herewith." Here we see Piazzi's own stark admission that he wanted the glory of deriving the elements for himself, which is why he deliberately withheld his data from all the astronomers of Europe. This is the untenable position he has maneuvered himself into. As in Zugzwang, he is

Fig. 3.8 A French orrery, c. 1803

Fig. 3.9 The roundel in the orrery denoting Ceres

obligated to make a move, and does so by sharing his precious data. His strategy has failed, and he must now move to face the endgame.

Piazzi's reluctance to promptly share his observational data quickly became a sore point with astronomers throughout Europe. As noted in *Discovery of the First Asteroid, Ceres,* even Piazzi's first letter to Oriani describing the discovery of Ceres contained an error in declination: the very first observation of January 1 was actually 30 arcminutes south of the position he wrote. Perhaps he should be given the

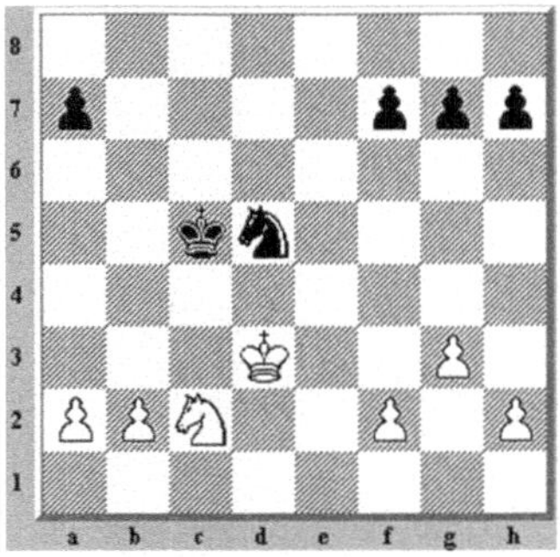

Fig. 3.10 The chess position *zugzwang*

benefit of the doubt, but could it be that throughout 1801 he practiced a deliberate series of delaying tactics and deception?

Piazzi's attitude towards his discovery was an extreme manifestation of covetousness. He admitted as much to Zach: "The surname of my planet, Ferdinandia, which so many astronomers deemed as unnecessary, brought me a splendid equatorial sector and a yearly salary increase." He also wrote Ceres was "like something I own." (my underlining). While wholly understandable in an age when 'priority of discovery' was already a major issue, it nearly resulted in the loss of the very object of his desire. All the goodwill Piazzi had built up during his visits to France and England dissolved before his eyes as astronomers in England, France and Germany grew increasingly upset as 1801 progressed. Even his closest friend Barnaba Oriani despaired.

Oriani's Balancing Act

Oriani must have felt like a member of the Italian *commedia dell 'arte*. Similar to the Pulcinelle on a tightrope, Oriani had to maintain a fine balancing act between his two famous correspondents and friends: Zach and Piazzi. Both astronomers confided in Oriani, but he was unable to let them know what they truly thought of one another in those difficult months of 1801 as Ceres remained 'lost.'

Consider just one line of Zach's diatribe against Piazzi, dated July 6, 1801: "It is truly unpardonable that this Sicilian has made a secret of his discovery for such a long time." As we now know, Piazzi did not receive any letters from Oriani between July 1801 and March 1802. As an excuse, Oriani gave his trip to Leon (he arrived back in Milan at the end of January 1802), but we know that Oriani was in communication with Zach during this period. He even sent Piazzi's monograph about Ceres to Zach in November.

Clearly, Oriani's balancing act was very uncomfortable for him. In none of his letters to Piazzi does Oriani say what Zach was writing and thinking about Piazzi. And even in March, when the correspondence with Oriani resumes, Piazzi takes the first opportunity to swipe at Zach: "Mr. Zach calculated from the RA I published an average time different than mine. The mistake is on his side." (March 2, 1802)

Reaction in Germany was also strong. Ferdinand von Ende wrote that Piazzi was acting "like a charlatan," and Bode said Piazzi had not undertaken any calculations "in order to falsify his statement it was a comet."

Piazzi's deception is clear. He had motive (keeping the discovery close to the vest, so that no one else could lay claim to it), method (delaying whatever error-prone data he did divulge), and opportunity (no one else had any competing information). The results are equally clear—every astronomer in Europe felt degrees of annoyance that verged on professional disgust unparalleled in modern astronomy (Cunningham, 2003).

Deliquium

In his handling of the discovery of Ceres, Piazzi sacrificed a great deal of his reputation. He found to his regret that the opinion and sentiments of other astronomers mattered a great deal more than winning the joust for celestial glory. In the end, it was checkmate for Piazzi (Fig. 3.11).

That Piazzi sought glory is unlikely, but its beneficent rays certainly shone on him as the discoverer of the 'eighth planet.' The cessation of this glory—the deliquium of this light from above that the painting by Farina was supposed to immortalize—came quickly as the delight of discovery turned to recrimination. Since the opinions of his contemporaries have now been laid bare, it behooves us to look further into how his actions fit into the moral philosophical outlook of the times. Jean-Jacques Rousseau (1750) offers a good starting point:

> *I would show how much this universal desire for reputation, honours, and preferment, which devours us all, exercises and compares talents and strengths; how much it excites and multiplies the passions; and how much, by making all men competitors, rivals, or rather enemies, it daily causes reverses, successes and catastrophes of all kinds, by making so many contenders enter into the same joust.*

Fig. 3.11 Checkmate

There were certainly several astronomers jousting for the prize—the discovery of the 'missing planet.' Whoever won the joust would be, like a knight of medieval times, wreathed in glory. But as Benichou (1948: 108) notes, "the worth attached instinctively to glory, far from saving the honour of man, is the most striking sign of his wretchedness." That Piazzi was wretched in 1801 and 1802 (he almost quit Palermo Observatory due to anxiety and disgust) can be traced entirely back to how he handled the discovery of Ceres. Piazzi should have heeded Alexander Pope in his *Essay on Man* (Pope, 1776: 9):

Trace Science then, with Modesty thy guide;
First strip off all her equipage of Pride;
Deduct but what is Vanity or Dress,
Or Learning's Luxury, or Idleness;
Or tricks to shew the stretch of human brain,
Mere curious pleasure, or ingenious pain.

According to Crocker (1959) "there was one congeries of ideas, centred around the notions of self-esteem, that acquired particular importance, in the eighteenth century interpretation of human nature and its moral components. Pride, the desire for approbation and self-approbation, and their more special forms, such as the search for reputation, glory and immortal fame, are present in most of these evaluations."

Was it really pride ("the most designing of all things" according to Philo in 40 CE) that got Piazzi into such trouble with his colleagues? Jacques Abbadie (1692); Fig. 3.12 divided pride into five branches: love of esteem, presumptuousness, vanity, ambition and arrogance.

Piazzi's letter to Lalande of August 25, 1801, concludes with "most respectful, most affectionate, and most grateful of your pupils." The need for esteem from Lalande is certainly evident, but perhaps the crucial element left unsaid here is that both men were Masons. That Lalande gave vent to his feelings over being ignored by Piazzi (not just a professional colleague but a fellow Mason) in the early months of 1801 is almost certainly an admonishment that reduced Piazzi to the level of a 'grateful pupil' in his response.

Presumptousness? Certainly: the word literally comes directly from Maskelyne, where he calls Piazzi imprudent. Both words imply a rashness—an inattentiveness to actions—that is all too evident.

Vanity originally meant 'lacking in sense.' The response of Baron von Zach, if asked whether or not Piazzi lacked sense, can scarcely be doubted. His charge against Piazzi of "puerility" makes the case clear.

If ambition is the inordinate desire for something, the case can easily be made that Piazzi was guilty on this count as well. Zach specifically accused Piazzi of "irrational jealousy," and it was this jealous desire to keep the details of the discovery to himself that so infuriated his colleagues.

Finally we come to arrogance. Again, Zach's analysis was right on the mark when he railed against Piazzi's conduct and secrecy as "very reprehensible." His main British detractor, Maskelyne, had also tarred him with the "covetous" epithet, another major element of arrogance. The philosopher David Hume on reputation:

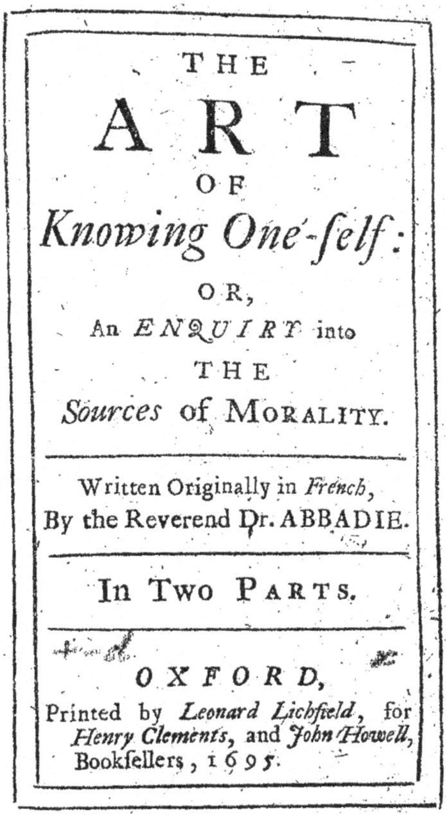

THE

ART

OF

Knowing One-self:

OR,

An *ENQUIRY* into

THE

Sources of MORALITY.

Written Originally in *French*,
By the Reverend Dr. ABBADIE.

In Two PARTS.

OXFORD,
Printed by *Leonard Lichfield*, for
Henry Clements, and *John Howell*,
Bookfellers, 1695.

Fig. 3.12 Title page of the 1695 edition of the Jacques Abbadie's book, Art of Knowing One-self

> *Our reputation, our character, our name are considerations of vast weight and importance; and even the other causes of pride: virtue, beauty and riches, have little influence, when not seconded by the opinion and sentiments of others.* (Hume, 1739)

Here is a relevant literary anecdote. Giuseppe Lampedusa's only novel, *The Leopard*, was published in Italian in 1958. It tells the story of a great Sicilian astronomer, Don Fabrizio. "The Sicilians," says Fabrizio, "never want to improve for the simple reason that they think themselves perfect; their vanity is stronger than their misery." (Lampedusa, 1986: 146). Vanity is the handmaiden of pride. Shades of Piazzi?

Bode's Gambit: Stealing Celestial Glory

What do impudent Sicilians, French whores, foolish Englishmen and German donkeys have in common? They are all astronomers who are now highly revered with the hindsight of two centuries. Few astronomers of the age escaped having their reputations sullied by the discovery of Ceres and Pallas. Piazzi was vilified by many for his reluctance to share data. Herschel was ridiculed for his introduction of the word 'asteroid.' As shown in Chap. 2, the leading French astronomers were lampooned for a multiplicity of sins, each one lovingly enumerated by Zach.

The leading astronomer of Germany did not escape either. As usual Zach was the judge, jury and executioner: Bode was guilty. His crime—attempting to "steal the glory from Piazzi" as being the first to recognize Ceres as a planet. Piazzi was not alone in playing a dangerous chess game. Bode launched his own gambit, too, one that was just as fraught with danger. Bode sent a letter about Piazzi's observations to Zach on April 14, 1801. It strained Zach's credulity, as Zach explained to Oriani on May 29, 1801:

> *Bode adds that these observations immediately struck him as very special and he made a small calculation and recognized on the spot that the comet was a planet and that it was the one he had long since suspected between Mars and Jupiter. You, my dear friend, who has calculated several orbits of planets, and are experienced, tell me, I am begging you, how can someone immediately tell from two positions they are planetary? I shook my head reading Bode's letter and said to myself this is fishy... He wanted to steal the glory from Piazzi being the first one to recognize the planet and to appropriate it.*

Zach baldly labels this a "deception." Bode's Byzantine machinations went one step further: Zach writes that Bode sent him a confidential letter in which he (Bode) stated that the name of the new planet should be Junon (Juno). Zach chose not to give Bode's gambit any publicity in the *Monthly Correspondence*. "I did not write anything of Bode's nice idea in my journal" writes Zach, but the proverbial cat was out of the bag when Zach shared this information with Lalande. He was immediately exposed by Zach in the June 1801 issue of the *MC*. After mentioning that both he and Oriani had concluded Piazzi's discovery was a planet, Zach lambasted Bode in the very public forum of his journal:

> *The honour, therefore, of not only first discovering this planet, but also first recognizing it as a planet, can accordingly not be disputed by Professor Bode. One ought almost think that he wanted to also reserve the honour for himself (which one should not hold against the first discoverer of this planet) of having first calculated the elements of its path, all the while remaining meagre if not incomplete in the announcement of his observations of this remarkable body. Professor Bode immediately reported his discovery and suspicion to the Royal Prussian Academy of Science, had the news of it printed in the* Berlin Newspaper *no. 57 of May 12, in the Intelligensia page of the* General Literary Newspaper of Jena *no. 90 of May 6, and in the* Hamburg Impartial Correspondence *no. 76 of May 13. Out of this, it came to the general public through several other political newspapers.*

As Zach relates Bode made certain that the announcement of the discovery, and his assertion that Ceres was the eighth planet, was widely published in German newspapers. To emphasize the importance of newspaper stories, Zach discusses the press coverage surrounding Ceres at great length in his journal for November 1801 (given in Chap. 10). Although Piazzi was given credit for the discovery, the carefully planted news reports asserted that Piazzi merely thought his discovery to be a comet. Even worse, Bode took it upon himself to preempt Piazzi's right to name the discovery. He did this even before Ceres was recovered in 1802. A sensational article appeared in the Berlin newspaper on December 8, 1801; it was reported as far away as Spain in the January 12, 1802, issue of the *Gaceta de Madrid:*

The astronomer Bode has received two letters from Piazzi of Palermo, who claims to be already persuaded, as was the same Bode, the star discovered on January 13 [sic] *of this year is a planet, though it was first doubted because of its faintness. Astronomers of Germany proposed to give the name of Juno by analogy with the other planets, but Piazzi wants to give the name Ceres Ferdinandea, referring to Sicily, the domain of Ceres, and the ruling monarch of the island.*

Another incendiary article appeared in the newspaper *Hamburger Zeitung* on January 12, 1802:

A new planet discovery that Mr. Piazzi had made 1 January 1801, and astronomers had sought unsuccessfully for four months, was observed again by Olbers in Bremen. Bode has proposed to give the name of Juno to this planet, which at the time of last observations was among the stars of Virgo. It seems as a star of the 9th magnitude, known to revolve around the Sun between Mars and Jupiter, and runs in 4 days an equal space to the moon's disc.

Bode made sure his name, and the name Juno for the discovery, was included in nearly every German newspaper report. It may reasonably be said he was the first astronomer who ever used the power of the popular press to advance a personal agenda. To ensure that the public regarded his presence to be at the root of the discovery of Ceres, a lengthy article in a Berlin newspaper on January 16 printed his full explanation of what we now call "Bode's Law." It was given wide publicity, appearing again on January 30 in the Munich newspaper *Kurpfalzbaierische Müncher Stattszeitung*. The following notice appeared on page 56 of January 26, 1802, issue of Stuttgart's political newspaper *Schwaebischer merkur (Fig. 3.13):*

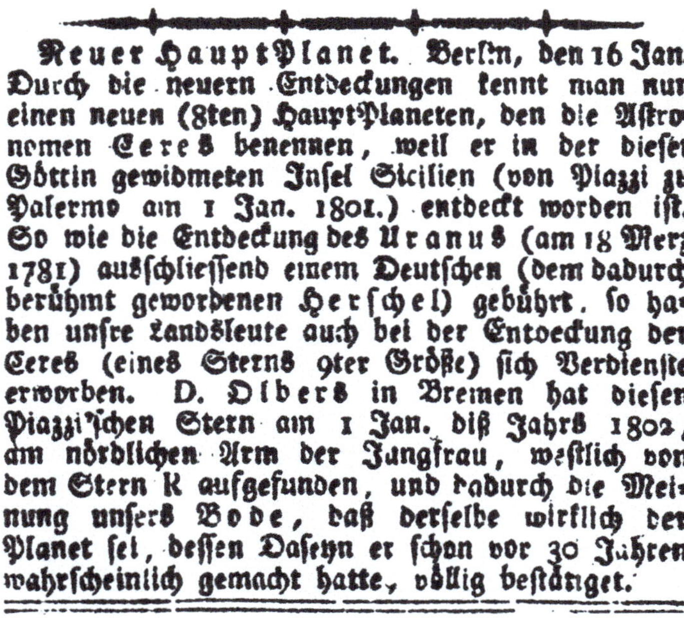

Neuer HauptPlanet. Berlin, den 16 Jan. Durch die neuern Entdeckungen kennt man nun einen neuen (8ten) HauptPlaneten, den die Astronomen Ceres benennen, weil er in der dieser Göttin gewidmeten Insel Sicilien (von Piazzi zu Palermo am 1 Jan. 1801.) entdeckt worden ist. So wie die Entdeckung des Uranus (am 13 Merz 1781) ausschliessend einem Deutschen (dem dadurch berühmt gewordenen Herschel) gebührt, so haben unsre Landsleute auch bei der Entdeckung der Ceres (eines Sterns 9ter Größe) sich Verdienste erworben. D. Olbers in Bremen hat diesen Piazzi'schen Stern am 1 Jan. diß Jahrs 1802, am nördlichen Arm der Jungfrau, westlich von dem Stern K aufgefunden, und dadurch die Meinung unsers Bode, daß derselbe wirklich der Planet sei, dessen Daseyn er schon vor 30 Jahren wahrscheinlich gemacht hatte, völlig bestätiget.

Fig. 3.13 Article in the January 26, 1802, issue of *Schwaebischer merkur*

New main planet. Berlin 16 Jan

The new discoveries have given us a new (8th) main planet, which astronomers call Ceres, because she was discovered in Sicily, the island dedicated to this goddess (by Piazzi in Palermo on 1 Jan 1801). The discovery of Uranus (on 18 March 1781) can be entirely ascribed to a German (Herschel who has become famous in the wake), also in this case our compatriots played a major role in the discovery of Ceres (a 9th magnitude star). D. Olbers of Bremen has rediscovered this star of Piazzi on 1 Jan 1802, in the northern arm of Virgo, west of the star R, and thus confirmed the opinion of our Bode that this same star is really the planet whose existence he already suspected 30 years ago.

After Olbers discovered Pallas, Bode again had articles published in the Hamburg newspaper, and these obviously incensed Olbers. Unlike his newspaper ploy with Piazzi and Ceres, Bode seemed oblivious to the fact that he was recreating the same negative scenario with Olbers and Pallas. On April 10, 1802, he actually told Olbers what he had done: "Today I announced your star as comet in the papers—what else can it be?" Bode's name appears right alongside that of Olbers in this newspaper article published in the *Berliner Hofzeitung* on April 10. After reporting Olbers had discovered a comet on March 28, it relates Bode spied it at the Berlin observatory on April 5 and 7. This was reprinted in other newspapers, including the *Allgemeine Zeitung* (issue 108) in Munich on April 18 (Fig. 3.14):

Bode did his best to repair relations with his friend Olbers in a letter of April 30, 1802, by lashing out at Zach:

Upon my return of an absence of seven days I found your letter of the 16th, whose content saddened me and caused the bitter feeling of loss. How is it possible that you can imagine that I wanted to misappropriate the discovery of your comet and even am responsible for the article in Hamburger Zeitung. *You have been set against me by several astronomers and Mr. v Zach and from this and some remarks in the MC I learn that I have a secret enemy in him through no fault of mine. He wants to make me smaller, looks out for my mistakes and censures them.*

Bode again aired his grievances with Olbers on May 4 and 15, 1802 (complete text of these letters are in Chap. 7):

I would like to repeat my fair complaints about Mr. von Zach's behaviour against me, maybe you can show me a way to remedy or you can contribute a friendly share. There is almost no issue of MC without clear expressions of his unfriendly attitude against me...

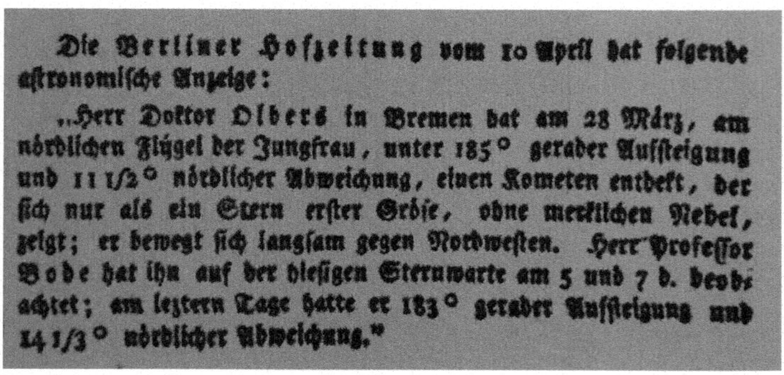
Die Berliner Hofzeitung vom 10 April hat folgende astronomische Anzeige:

„Herr Doktor Olbers in Bremen hat am 28 März, am nördlichen Flügel der Jungfrau, unter 185° gerader Aufsteigung und 11 1/2° nördlicher Abweichung, einen Kometen entdeckt, der sich nur als ein Stern erster Größe, ohne merklichen Nebel, zeigt; er bewegt sich langsam gegen Nordwesten. Herr Professor Bode hat ihn auf der hiesigen Sternwarte am 5 und 7 d. beobachtet; am leztern Tage hatte er 183° gerader Aufsteigung und 14 1/3° nördlicher Abweichung."

Fig. 3.14 Article in the April 18, 1802, issue of *Allgemeine Zeitung*

What am I to think of his not responding to my letters, maybe six or seven since November, and not reacting to my therein expressed excuses and justifications to the publicly made bitter reproaches? I dearly love peace, especially in a literary respect, and would prefer only to get involved with these accusations against me if necessary. Today I wrote him regarding this matter, do you consider my actions right and proper? Don't you, too, sometimes find the tone of the MC arrogant and presumptuous? What tribunal allowed Mr. von Zach's personal attacks against astronomers?

I am waiting impatiently for the next issue of MC. What will Mr. von Zach say about this peculiar star? And I might even be presented with a witty joke (i.e., another mean-spirited comment) for my doubts. I am asking you, dearest friend, not to publicise my last remark about Mr. von Zach if you want to contribute something to appease his unjust anger against me.

The feud between Zach and Bode became known to the entire astronomical community. In a letter of September 19, 1802, Olbers wrote to Gauss that Bode had recently received positional data on Pallas from Méchain in Paris. Bode relayed this data to Oriani, but asked him not to share it with anyone:

That ban to share his observations probably refers only to Zach. This is due partly because, as you know, Zach and Bode are now not friends, and partly because Bode wants to reserve these observations for his yearbook and does not want them made known beforehand in the M.C.

Thus we learn Bode regarded Zach's *Monthly Correspondence* as a rival publication. The Republic of Letters was, if not in tatters, at least being frayed at the edges.

The French, who (like Bode) were also keen to name the planet, used the incident to inflame Franco-German rivalry to astronomical heights. "Fleeced of the honour to be the parent of the new planet," Bode was dubbed by Lalande Baudet: the German donkey.

Zach Versus Maskelyne

As a prime nexus for the flow of scientific information, Nevil Maskelyne as Astronomer Royal assumed a pivotal role in garnering information about the asteroids from Continental researchers, especially Carl Gauss. Thus, his reputation cannot be overlooked. That he was held in great esteem by the French is clear from a letter written by Lalande to Maskelyne. In introducing him to the astronomer John Baptiste Joseph Delambre (1749–1822), Lalande (1787) writes that Maskelyne was for Delambre "… the god of astronomy." He was, however, not so highly regarded by the prime nexus of Continental astronomy, Zach. In this 27 April 1802 letter to Gauss, Zach first expresses astonishment that Maskelyne said nothing about Pallas in a recent letter, then accuses Maskelyne of keeping observations to himself, and finally says that Maskelyne is irritated by a difference between his positional measurements and those of Zach:

I do not intend to send you Maskelyne's observations, because he informed me he wanted to do it himself. "I have sent several observations of the new planet C.F. to Dr. Gauss and will send him more. He promises to make the best use of them in his power." In his last letter of April 15th he gave his last observation of Ceres on April 6th as follows: 10h 59′ 52″

RA Ceres 179° 28′ 19″ Decl. Ceres 18° 9′ 10″.2. I can read between the lines that he is not against a comparison with your elements and a publication in the M. C. So if you have a series please communicate it for this purpose. I have published two of your comparisons of Maskelyne's observations. The astonishing thing is that Dr Maskelyne did not say a word about Pallas in his letter of April 15th. And on the same day I got a letter from Sir Joseph Banks of April 16th that Gilpin, clerk of the Royal Society of Sciences had found Ceres on April 9th and that a very ingenious young astronomer M Lee had observed Olbers' heavenly body on April 13th at 11h 50′ 48″ m. t. RA Pallas 182° 24′ Decl. N 16° 27′. Apparently, the Astronomer Royal has not yet given up but keeps his observations to himself. It irritates him that my observations differ 10″ from his in declination. He believes my spider's threads, which he dislikes, are to blame. But why do my stellar declinations correspond so well to Piazzi's?

Here he mentions George Gilpin (1755–1810), who was Secretary to the Royal Society from 1785 to 1810; and Stephen Lee (fl. 1817–1834), assistant secretary of the Society. With the astronomical stakes so high, rivalries were rife across Europe. Oriani seems to have been the only major figure to escape criticism. Like an agony aunt (or advice columnist), he received all the complaints and apparently did a superb job at maintaining neutrality amid the swirls of controversy.

Chapter 4
Herschel's New Dynasty

C. Cunningham, *Early Investigations of Ceres and the Discovery of Pallas*,
DOI 10.1007/978-3-319-28815-4_4

Fig. 4.1 A pastel portrait of William Herschel by John Russell. (Courtesy of the Herschel Museum, Bath. Used with permission)

Herschel's Search for a Word, Part 1: William Watson

Herschel (Fig. 4.1) visited Paris in August 1802, where he met First Consul Bonaparte and the foremost astronomer in France, Pierre-Simon Laplace. It was Laplace who insisted on naming the new discoveries Piazzi and Olbers, in honor of their discoverers (Manara, 1997). Herschel did not concern himself—as the French did—with the naming of the new celestial objects individually. His concern was their collective appellation.

The search for a new name began on April 25, 1802, when Herschel turned to his friend Sir William Watson (1744–1824) for help. At the time of writing the relevant portion of the letter reproduced below, he was likely well aware that Isaac Newton (1726) had written an analysis of the motion of comets in the third book of the *Principia,* in which he shows that comets "are a sort of planet."

> *...I have now [to] request a favour of you which is to help me to a new name. In order to give you what will be necessary I must enter into a sort of history. You know already that we have two newly discovered celestial bodies. Now by what I shall tell you of them it appears to me much more poor in language to call them planets than if we were to call a rasor a knife, a cleaver a Hatchet, etc. They certainly move round the Sun. So do comets. It is true they move in ellipses; so we know do some comets also. But the difference is this: they are extremely small, beyond all comparison less than planets; move in oblique orbits so that, if we continue to call that the ecliptic in which we find them, we may perhaps, should one or two more of them be discovered still more oblique, have no ecliptic left the whole heavens being converted into ecliptic which would be absurd. I surmise (again) that possibly numbers of such small bodies that have not enough matter in them to hurt one another by attraction, or to disturb the planets, may possibly be running through the great vacancies, left perhaps for them, between the other planets especially Mars and Jupiter. But should there be only two surely we can find a name for them. The diameter of the largest of them (at present entre nous) is not 400 miles, perhaps much less as I shall know in a few hours but have not time to wait. Now as we already have Planets, Comets, Satellites, pray help me to another dignified name as soon as possible. If it could any way express the condition of a nimble, small, interloper going obliquely through the majestic orbits of the great bodies of the Solar System it would be just what is required. But pray, if you can, help me soon. I am writing a paper in which if possible I would propose a name, but as it should go to London by next Thursday I am hardly willing to press you so much for haste. However you will give it a thought, and if two or three names could be proposed it would give me some choice. Greek derivation such as planet from πλαναω would probably be best.* (Herschel, 1802a)

The word written in the letter in Greek, *planao,* is the verb "to wander." Trusting to the English postal service in 1802 as we can scarcely hope for today, Watson received Herschel's letter the next day and responded after a day of thought.

> *I received much gratification at the perusal of your letters – the discovery of a new species of heavenly bodies is truly surprising, and I agree with you that a new name ought to be given such bodies. The best name I can think of is Planetel as a diminutive of Planet, just as Pickerel or Cockerel (used by Shakespeare) is of a Pike and a Cock. The sportsmen too call a young stag stagerel. You may also use as the diminutive the word Planeret (sic), as baronet is of the word Baron – so we say islet tartlet tablet cygnet, the respective diminutives of island, tart, table, Cygne the French for Swan. But as these are made by the mere addition of et, except tartlet, the word should be Planetet, and that does not sound well. Diminutives are also formed by adding – kin as manikin, lambkin, so you may say*

Planetkin – or better Erratikin – being the diminutive of Erratic. I should like Planetine (pronounced Planeteen) best of all, but I find no example of that way of diminishing in English. The diminutives formed by adding –ling such as duckling will not have place here – we cannot say Planetling. So upon the whole I think the word Planetel the least objectionable. Perhaps you may be more happy in your research after a new name.

P.S. Since I wrote the above I recollected that after the Romans we make diminutives by adding –ule such as spherule, a little sphere. So Planetule may be a little Planet. (Watson, 1802)

One diminutive suggestion he did not make was to suggest the word planetella (as in novella, a small novel). Planetkin has entered the OED as a nonce word. It identifies the Scottish philosopher Thomas Carlyle (1795–1881) as the first person to use it in 1832 (Norton, 1887: 35).

As William Herschel stated in his April 25 letter, he intended to include the new name in his paper, which was due "by next Thursday." This date was May 6, which was in fact the date Herschel's paper was read before the Royal Society in London.

Herschel's Search for a Word, Part 2: Charles Burney

In a letter of "Monday night May 10th [1802]" from Dr. Charles Burney Sr. (1726–1814); Fig. 4.2 to his son Charles, Jr. (1757–1817), Burney (1802a; his underlining) writes (Fig. 4.3):

My dear Charles

Herschel came hither today, to ask me if I c^d. furnish him a Latin or Greek name for the small stars that have been lately found, & called by some planets, & by others Comets; but

Fig. 4.2 Charles Burney, Sr.

Fig. 4.3 Detail of the May 3, 1802, letter of Burney, Sr., to his son

he says they are neither one nor the other, but a new genus of erratic heavenly bodies within the ecliptic, that have orbits round the Sun: yet so small that they cannot be found by a Telescope. There are however 12 astronomers in Germany formed into a Society, who have divided the ecliptic into 12 parts, assigning one to each who is not to encroach on the other departments. The last new planet, as it is called, is not above 150 miles in diameter – Mercury or the Moon wd. make 1000 such - it has, however, a disk, and is in motion.– Now what can he call a star of this nondescript kind?

Does not Hadrian call his soul animula, vagula, blandula? and is there not a diminutive of the Greek word Αστηρ - ? Αστεριςκος - & *in Latin is not stellula the diminutive of stella?* Αστηρ *implies any kind of heavenly body, be it planet, satellite, or fixt star – asteriscos, or Stellula wd. be a pretty name for one of these little wanderers, that are taking a peep at us.*

The first line of Pope's imitations – "vital spark of heavenly flame" – suits this last little lady to a T– does it not? – if you say nay, send me a better for my friend, as soon as possible for it is to be given in to the secretary of the R.S. [Royal Society] tomorrow to be voted for reading on Thursday.

It must not be a big name for so small a star. C.B.

The first Greek word he uses is 'aster' Αστήρ (which he uses without the accent), and the second one 'asteriskos' Αστεριςκος. Burney has a ligature between sigma and tau, which was common in some types of Greek script. He was not using the ς form of sigma, which is only found at the ends of words, but a common ligature which somewhat resembles that form of sigma. The second vowel in the word is long, and so is written not with an epsilon but with an eta. Burney had to insert the letter ρ in the second word with a caret, and the medial sigma in Αστερισκος has the wrong shape.

The Latin Burney alludes to, with its use of the diminutive –ula, can be easily traced. According to the fourth century text *Historia Augusta* (Magie, 2014), the emperor Hadrian composed shortly before his death in A. D. 138 the following poem:

Animula, vagula, blandula
Hospes comesque corporis
Quae nunc abibis in loca
Pallidula, rigida, nudula,
Nec, ut soles, dabis iocos...
P. Aelius Hadrianus Imp.

Roving amiable little soul,
Body's companion and guest,
Now descending for parts
Colourless, unbending, and bare
Your usual distractions no more shall be there...

The reference to Alexander Pope (1688–1744) is to the first line of his 1712 poem *The Dying Christian to his Soul*. The words are based on the deathbed utterance attributed to the Roman emperor Hadrian: *Animula vagula, blandula, hospes comesque corporis*. Pope had been inspired by these words from an early age, as he relates to Richard Steele, co-founder of *The Spectator* magazine, what led him to pen his poem.

> *I was the other day in company with five or six men of some learning; where chancing to mention the famous verses which the Emperor Adrian spoke on his death-bed, they were all agreed that 'twas a piece of gaiety unworthy of that prince in those circumstances. I could not but differ from this opinion: methinks it was by no means gay, but a very serious soliloquy to his soul at the point of his departure; in which sense I naturally took the verses at my first reading them, when I was very young, and before I knew what interpretation the world generally put upon them.* (Pope, 1712)

From the cover of the letter from Burney, Sr., in Chelsea to Burney, Jr., in Greenwich (a distance of only 11 km), it can be read that Burney, Jr., was sent this letter by two penny post at 9 am on Tuesday morning. But was Burney, Sr., correct in dating this letter Monday the 10th of May? It would make sense if he dated it Monday the 3rd of May, because it was to be that Thursday (May 6) when the paper was read. We see here it was to be in the hands of the secretary the very next day, which would be the 4th of May. The author's own reading of letters from this period has revealed incorrect dates—sometimes the year is actually written incorrectly! It would certainly not be impossible for a person working by candlelight, late at night, to get the day of the week correct but the day of the month wrong by a week. It is also obvious by the way the letter is written that it has been done in haste—he twice had to use carets to insert a phrase or a Greek letter in its proper place, and the last line quoted above was written at the bottom of the letter after a paragraph of personal details—it was clearly an afterthought. He also wrote his son that the objects could not be found in a telescope, another indication he was tired and writing in haste, since clearly they were found using a telescope.

If we consider this letter as being in the hands of Charles Burney, Jr. (in Greenwich), on the afternoon of Tuesday, the 4th of May, we must conclude he supplied an answer to Herschel that very day. Since there were four country mail dispatches and deliveries daily in that era, he could have devised an answer that would have been in Herschel's hands that evening. It is certainly clear from this letter that Herschel had *not* chosen a word—he was also clearly in great haste to get an appropriate word, since he visited Charles, Sr., in person on May 3 instead of writing to him at leisure. Burney, Sr., uses the phrase "as soon as possible," which has entered modern parlance as ASAP, thus emphasizing how urgent it was.

Charles Burney, Jr. (Fig. 4.4) likely supplied his answer to his father in writing, and Burney, Sr., then gave the response to Herschel. This may have been done

Fig. 4.4 Charles Burney, Jr.

verbally, as no known letter exists. However, the 'smoking gun' letter was written later that year.

Dr. Burney's two-page letter (postmarked Dec. 7, 1802) to the political hostess Frances Crewe (1748–1818) is definitive. In this he tells her that his son furnished Herschel with the word 'asteroid.' He tells her about:

> *...a new vol. of the Philosophical Trans. in wch are two curious astronomical papers by Herschel. In one of wch he gives an acc^{t} of the 2 newly discovered celestial bodies, Ceres, & Pallas. The first in magnitude is only $\frac{3}{8}$ of the Moon; its Diameter no more, if I understand right, than 161 miles. The 2^{d}, Pallas, still less, about $\frac{3}{4}$ of Ceres– its Diameter about 147 miles – not an 8th p^{t} of Mercury – They are not allowed by Herschel to be either Planets or Comets, but asteroids, italick, a kind of star – a name wch my son, the Grecian, furnished.* (Burney, 1802b, his underlining)

Thus, after more than two centuries, it has been established beyond doubt that Charles Burney, Jr., invented the word asteroid! Every book, dictionary and reference that gives credit to Herschel for creating this appellation is incorrect, although he certainly deserves full credit for being the first to publish it and correctly recognize that Ceres and Pallas were in a separate category from planets or comets. Charles, Sr., deserves some of the credit for coining the word, as he chose the Greek word *aster,* and passed this idea along to his son, who added *–oid.*

It certainly appears that Charles, Jr., had no interest in publicly claiming his invention of 'asteroid.' Considering the great opprobrium heaped upon Herschel for choosing that word, it is perhaps not surprising. Why is there no letter in the Herschel archives of the Royal Astronomical Society about the extremely important creation of the word? There seem to be two possibilities.

First, there was *no* letter from Charles Burney, Sr., or Jr. to Herschel. It is probable that Burney, Jr., informed Burney, Sr., either by letter or in person, and that the word was given to Herschel by Burney, Sr., in person.

Second, there *was* a letter from Burney, Sr., to Herschel, but it was deliberately destroyed by Herschel. Again there seem to be two possible motives for this course of action. Either Herschel wanted to keep the credit for coining the word for himself, or, after he realized what great opposition the new word had created, he destroyed the letter to protect the Burney family from abuse. Since Herschel hardly needed any more fame than he already possessed it seems most likely that Burney, Sr. told Herschel about it in person, in which case there was no letter to be found (Cunningham, 2015).

According to the OED, the first use of the word 'asteroidal' is by the English astronomer Norman Lockyer (1836–1920) in 1868. But research for this book has established its first use by Herschel (1807b) in describing his observations of the asteroid Vesta: "The spurious nature of the asteroidal disk…"

Although there appears to have been no personal relationship between Herschel and Stephen Weston, the same cannot be said for Dr. Burney, Sr. An anecdote is related by his daughter Miss Burney in late 1786: "This morning my dear father carried me to Dr. Herschel. That great and very extraordinary man received us almost with open arms. He is very fond of my father, who is one of the council of the Royal Society this year, as well as himself." (Sime, 1900: 199). Dr. Burney has left vivid recollections of his visits to Herschel who, he wrote in 1798, "is one of the most pleasing and well-bred natural characters of the present age, as well as the greatest astronomer." (Sime, 1900: 201) They often met at meetings of the Royal Society, and Herschel frequently stayed at Burney's house, where he almost certainly met Charles Burney, Jr.

Burney, Jr., is referred to in the letter to Mrs. Crewe as "the Grecian." This was not just parental boast, as Burney, Jr. was one of England's preeminent Greek scholars in the late eighteenth and early nineteenth centuries. He was elected a Fellow of the Royal Society (1802), made Professor of Ancient Literature at the Royal Academy (1810), and elected to the Literary Club (1810).

Herschel's Search for a Word, Part 3: Sir Joseph Banks and Stephen Weston

Herschel still felt uncomfortable with his choice of 'asteroid.' Clearly unimpressed by Watson's ideas, Herschel turned for help to the president of the Royal Society, Sir Joseph Banks. One of the prime reasons for his choice of Banks was the fact that no one had a greater familiarity with the very problem Herschel was grappling with. To understand this requires a look into the parallels with botany.

In creating a separate classification for Ceres and Pallas, he was following in the footsteps of eighteenth century researchers such as Carl Nilsson Linnaeus (Linnaeus, 1735) in botany:

> *The first step in wisdom is to know the things themselves. This notion consists in having a true idea of the objects; objects are distinguished and known by classifying them methodologically and giving them appropriate names. Therefore, classification and name-giving will be the foundation of our science.*

It was actually Linnaeus (1767: 563) who first used the word 'asteroid' in a botanical sense, but this was in a book written in Latin. The place of Charles Burney, Jr. is secure, as a new creation of the word for introduction into the English language, derived from Greek; he had no knowledge of the earlier Latin usage. It has been noted that "The discovery of a new planet or comet has a similar significance in astronomy as does in botany the description of new types of grasses." (Savich, 1855: 3) A century earlier the parallel between studying plants and celestial objects had been mentioned by the famous French scientist Jean-Jacques Rousseau (1712–1778): "Plants seem to have been sown profusely on earth, like the stars in the sky, to invite man to the study of nature by the attraction of pleasure and curiosity." (Rousseau, 1782: 62).

The multiple names attached to the recent Solar System discoveries (Uranus = the Georgian Planet = Herschel; Ceres = Hera = Piazzi) also found its parallel in botany, where the same species might have several designations (Phillips, 1841). Naming a discovery after its discoverer was another commensurable link with botany (Lemmon, 1878). The English love for botany as an intellectual aspect of their lives was well known in Germany, as evident in this quote from Heinrich Gottlieb Reichenbach (1793–1879; 1822: ix):

> ... *the sensibilities of the inhabitants* [of England] *have always shown a general tendency to appreciate ... the ownership and knowledge of the plant world as an essential part of their* *Bildung* [self-cultivation].

In 1781 Erasmus Darwin began a translation into English of Linnaeus' *Systema Vegetabilium*. Darwin sent numerous letters to Banks for advice as he set out to create a new botanic language, "... creating vernacular compounds in English as Linnaeus had done in Latin." (Uglow, 2002: 380) When *System of Vegetables* was published in 1783 it was dedicated to Banks (Gascoigne, 1994). Banks had established his reputation at age 23 by publishing the first Linnaen descriptions of the plants and animals of Newfoundland and Labrador, which he collected and classified on an expedition in1766 (Lysaght, 1971). Nearly three decades later he called Linnaeus "... the God of my adoration." (Banks, 1792) With a lifetime of experience classifying and naming newly found objects in nature, he became the man both Darwin (in 1781) and Herschel (in 1802) turned to for sage advice. And as Banks knew better than anyone, "... the seemingly simple function of naming objects does not present a simple connection between a thing and a word." (Goldstein, 1948: 196) Despite his vast experience, the seemingly simple task of creating the word needed to describe Ceres and Pallas eluded Banks.

Fig. 4.5 Stephen Weston

Banks gave the task to Stephen Weston (1747–1830); Fig. 4.5, a Fellow of the RS since 1792 (Cotton, 1892: 6) and a great scholar of Classics, Persian, Arabic and Chinese:

> *I applied to Mr. S. Weston as I always do in these occasions to stand God Father to your new species of moving stars and [he] has sent me a card which I enclose. I really think Aorate a good name and much better than any that has been hitherto suggested and the more so as it is not probable that any of this new kind of wanderers are visible to the naked eye.* (Banks, 1802)

In this letter to Herschel, Banks favored 'Aorate' to describe Ceres and Pallas. The elements of the word are *a-* 'not', *(h/-)ora-* 'see', *-t-* passive participial suffix (i.e., making 'see' into 'seen'), *-e* = *-η* fem. termination unusual in a compound containing *a-* 'not'; ἀόρᾱτος is perfectly good classical Greek for 'invisible,' the very attribute of Ceres and Pallas that Banks highlighted in his letter.

The replacement of the termination -ος with the long vowel -η would automatically draw the accent on to the penultimate syllable. Weston may have finished his word with -η in imitation of several Greek names of goddesses that have the same ending. The most instructive is Persephone, likely a folk etymological modification of the original Persephatta, both forms being compounds with transparent etymologies and therefore not likely to form a fem. in -η.

Aorate and other words offered by Weston (on a card now apparently lost) were given by Banks to Herschel (1802e), who was clearly disappointed with the offerings:

> *The names you have done me the favour to send I have carefully examined, and beg leave to give you my remarks on them. The title of them, "Names for the new Planet," shews*

immediately that none of them can possibly be used for the new species of bodies which we have to christen: for they are not planets.

If Mr. Weston were to have a definition of the thing we want a name for, he might possibly find a better than that of asteroids, which is not exactly the thing we want, tho' still the most unexceptionable of any that have been offered by my learned friends. Will you do me the favour to consult him once more upon the subject, and mention to him that the bodies to be named are neither fixed stars, planets, nor comets, but have a great resemblance to all the three?

With this view before him he will probably succeed in an appropriate appellation.

In this extraordinarily frank letter, Herschel admits that the term 'asteroids' is not optimal—merely the best of an unremarkable suite of options. He is also being a bit pedantic in rejecting the suite of names because of the title. There is no evidence that Weston looked into the matter again. Perhaps Banks thought better of asking him a second time, or Weston simply did not offer any further ideas. Thus the word 'asteroid,' used in the May 6 paper, was the *de facto* choice to designate the newly discovered celestial bodies.

Herschel seems to be expressing exasperation that the title of the options given to him included the word "Planets." Even though on dynamical grounds asteroids do bear a great resemblance to planets, in the telescopes of the day they looked exactly like stars. It was only with careful study that Herschel was able to estimate their diameters, but he was using the most powerful telescope in the world, certainly on a par or superior to that used by Johann Schroeter (1745–1816) in Lilienthal. Any other astronomer at the time would have seen only a pinpoint of light. Thus star-like is an apt visual description as Ceres and Pallas bore an exact resemblance to stars, the only difference being they (like comets) moved against the starry background.

Holmes (2008: 509) erroneously claims that Rev. Steven (sic) Weston was actually the person who suggested the word 'asteroid' to Herschel, even though Herschel specifically says in the June 10, 1802, letter that none of the names suggested by Weston could be adopted! (Cunningham & Orchiston, 2011).

To understand the actual meaning of the word Herschel chose, we must look at its Greek etymology.

Greek has two words for "star": aster, which gives astero- in compound words, and astron, which gives astro- in compounds. The first means an individual star (usually a conspicuous one), whereas the second word is normally used in the plural to refer to "the stars" in general. This distinction is generally observed in compound words, whether by luck or design: thus asterisk means "a little star", and asteroids "like a star", whereas astrology, astrometry, astronomy and astrophysics all refer to study of "the stars" in general. (Fitch, 1987)

In ancient Greek we find πλανήτης (*planētēs*), a variant of πλάνης (*planēs*, "wanderer, planet"). The planets were called by the Greeks *asteres planetai* (wandering stars) or *planetai* (wanderers). The Latin term used in place of the Greek was *stellae errantes* (wandering stars); but late Latin borrowed the Greek term in the plural form, *planetae,* while the singular was *planeta.* The English word *planet* comes directly from the Latin *planeta.* In Greek, *aster* is α'στήρ. The word 'astyrred' is found in Old English as an adjective meaning starry (Borden, 1982).

In choosing asteroid over planet, Herschel was also undoubtedly aware of the recent French trend to use the word *planete* as a feminine noun, "contrary to analogy and to etymology, considering them as immediately derived from the Greek," in the words of English antiquarian Capel Lofft (1751–1824; 1798). Since the precedent had already been set to name the asteroids after female deities (a precedent that would be followed into the twentieth century), this left the planets firmly in the realm of male pagan deities, with the sole exception of Venus. This precedent was followed with the selection of the names Neptune and Pluto for future planetary discoveries.

Herschel's Chain Letter

Herschel wrote to Méchain on May 22, 1802. This is the extremely important "chain letter" that announced the word 'asteroid':

> *Regarding the two celestial bodies which were last discovered I am giving you a summary of my observations. In a memorandum, read to the Royal Society in London on the 6th and 13th of this month, I explained in detail my measurements of the diameters of these stars and I believe to have proven that that of Ceres, seen from the earth on April 22, was only 0″.216; and that of Pallas according to an equally good measurement 0″.17; but according to another even more accurate measurement only 0″.13.*
>
> *Calculating with these as much as our still imperfect knowledge of the orbits of these stars allows, I found that Ceres' diameter is about 162 English miles and that of Pallas only 70.*
>
> *I explained with the help of all my observations that these bodies cannot be called planets because of their small size and because they are beyond our zodiac. And, as I prove as well, they are not comets either and thus can only be regarded as a species between comets and planets which has been unknown to us and demands a name of its own. Since they resemble small stars and they are difficult to distinguish even with the best telescopes, I called them asteroids.*
>
> *Here follows the definition of this word: "Asteroids are small celestial bodies that revolve around the sun on ellipses more or less eccentric and whose plane can be inclined towards the ecliptic at any angle. Their motion can be direct or retrograde. They may or may not have considerable atmospheres, very small comas, disks or 9nuclei."*
>
> *You see, sir, that this definition leaves us a great deal of space and with tolerating these three species of heavenly bodies, planets, asteroids and comets, we make it much easier to classify future discoveries.*

On this date, Herschel also sent nearly identical letters to Lalande, Laplace, Bode, Zach, Olbers, Seyffer, Schroeter and Piazzi (Herschel: 1802d). Most of the recipients regarded the tone of the letter as the height of presumption. It was a fine example of optative etymology, but what was Herschel trying to do with this letter? By targeting all the leading astronomers, he was trying to build a consensus, but his reputation as something of a rebel, combined with his apparent proclamation of the term 'asteroid', foiled this approach. The concepts involved are neatly summarised by Smolin (2006: 295) in his discussion "What is Science?":

Science requires both the rebel and the conservative. This seems at first paradoxical. How has an enterprise flourished for centuries that requires the conservative and the rebel to coexist? The trick seems to be to bring the rebel and conservative into lifelong and uncomfortable proximity, within the community and, to some extent, within each individual as well. Science is a democracy, in that every scientist has a voice, but it is nothing like a majority rule. Still, whereas individual judgment is prized, consensus plays a crucial role.

Herschel's letter and his paper provoked an extraordinarily intense negative reaction amongst Continental astronomers. How much of this might have been due to a long-standing antipathy of Continental philosophers towards their English counterparts is unknown, as it was never explicitly stated, but the words of the Swiss mathematician Johann Bernoulli (1667–1748) need to be considered: "It is a characteristic of the English that they begrudge everything to other [nations] and attribute all things to themselves or their nation." (Bardi, 2006: 210).

The Reaction in France

Herschel's letter (specifically the one sent to Méchain) of 22 May 1802 was made public, and printed in French in the *Gazette Nationale ou Moniteur Universel* on 2 July 1802 (page 1164). The *Moniteur* was the official Government publication of the time. Laplace rejected the term 'asteroid' in a 17 June 1802 letter to Herschel:

As to the name you give to these stars [French: astres], *I still see no reason not to continue calling them planets. Ceres differs only by its inclination, which is a bit large, but it follows only that it is necessary to spread the width of the zodiac, and even include Pallas, if as seems likely its orbit is an ellipse, the eccentricity of which is only slightly greater than the orbit of Mercury.*

B. Voiron of Chambery, writing a decade later, was so dismissive of the word 'asteroid' that he did not even mention Herschel's name in connection with it:

Some scholars, who beheld them as appearancing like a small star, proposed to include them in a particular class under the name of asteroids, but the general opinion of astronomers put them among the planets which are distinguished from other stars, not by volume but by the nearly circular orbits around the Sun they describe. (Voiron, 1810: 82)

The Reaction in Germany

Herschel's term 'asteroid' was initially rejected by everyone on the Continent, with the single exception of Wilhelm Olbers:

I agree with you, honoured Sir, in your sagacious suggestion that Ceres and Pallas differ from the true planets in several respects, and the name asteroid seems to me to fit these bodies very well. Yet I would not lay too much stress on the difference in size, as the old planets differ from one another so much in this respect. Yet taking all the particulars together there seems to me to be a real difference between the asteroids and the true planets. (Olbers, 1802c)

Johann Bode, Director of Berlin Observatory, wrote to Herschel with no small temerity. In a letter of August, 1802, he questioned not only Herschel's diameter measurements but his choice of the term 'asteroid'. In mid-1802, he was still very uncertain about the true nature of Pallas, especially since it conflicted with the mathematical progression of planetary distances now best known as Bode's Law:

> *I believe or am still convinced that Ceres is the eighth main planet of our solar system and that Pallas is a neighbouring extraordinary planet (or rather comet) revolving around the sun. Thus there would be two planets between Mars and Jupiter where I have been expecting since 1772 only one and the known beautiful progressive order of the distances of the planets from the sun is only completely proven by this discovery since there was a gap at the distance 4+24=28.* (Bode, 1802c)

Baron von Zach adopted an extremely aggressive and fiery attitude when telling Oriani about Herschel's choice:

> *Mr. Herschel just wrote me that he had observed Ceres and Pallas with his large telescopes, neither has any satellites at all, he found the diameters much smaller than Schröter. For Ceres 162, for Pallas 70 English miles. Furthermore he does not want these corpuscles to be called planets, he invents a class of its own and calls it asteroids. But this is nonsense. For in this case Mercury is an asteroid as well in comparison to Jupiter. So this nomenclature means nothing, it is arbitrary and offensive.* (Zach, 1802h)

His tone was slightly more dispassionate when relaying the news to Gauss:

> *Dr. Herschel wrote and tried to suggest his term asteroids to me. He wants to introduce three distinct species. Planets, asteroids and comets. He wrote: "I hope this classification will meet with your approbation and that you will do me the honour to adopt it." But I have no inclination to do so because his definition of asteroids is not convincing. I rather stick to the name planet, together with you and Olbers. Only if there are several small planetulus between the older ones we can talk about a new classification, but today smallness, inclination and eccentricity do not decide on planetism or not planetism. Thus, Herschel's definition of asteroids is arbitrary.* (Zach, 1802g)

Here Zach has independently arrived at the term 'planetula', the same appellation suggested by Watson to Herschel.

Gauss rarely allowed strong opinion to intrude into his letters, but when it came to Herschel and the asteroids he let the mask of scientific imperturbability slip:

> *Mr. Herschel also gave me information on his 'Asteroids.' What surprises me is (1) that he doesn't announce it as being a modest proposal, but rather says simply "I call them", and (2) that his reason in Ceres' case consists in that it now "is out of the Zodiac." That shows a very biased and, it seems to me, unphilosophical outlook. It is likewise strange that he withholds his measured apparent diameter. Should it be correct, then, as it seems to me, a smaller mass is hardly important to be able to distinguish Pallas or Ceres from the remaining planets.* (Gauss, 1802c)

The word scientist had not yet been coined in 1802, so Gauss is characterising Herschel's outlook as unscientific when he uses the word 'unphilosophical'. He is also strongly objecting to what he perceives as Herschel's *de facto* pronouncement that the term asteroids must be adopted. Clearly, Herschel's attempt at consensus-building, noted earlier, had failed miserably. Gauss expanded on his views in a letter (dated 16 October) to Zach, who published it in his journal, the *Monthly Correspondence*:

> *Dr. Herschel still is not willing, as Prof. Huth told me and who visited him in England, to tolerate the new planets, although as far as I know not a single astronomer has approved of his suggestion* [of the term 'asteroids']. *Basically, and I agree with you, it depends only on our agreement whether we call Ceres and Pallas planets or not. And people are not saying whether they are planets or not, but whether it is proper and becoming to call these celestial bodies planets that partly resemble the known planets and partly not at all. That the latter is irrelevant you have shown sufficiently in your July issue and that astronomers believed in a circle-like orbit and a dependent perennial character seems to be proven by the circumstance that all astronomers accepted their planetism without hesitation as soon as they learned of the orbit. It even appears to me that, if future finds prove our excellent Olbers' hypothesis right—that Ceres and Pallas are only pieces of a destroyed planet, even then, we do not have to give up calling them planets. I believe it more important to study whether these celestial bodies are entitled to the name planet because of their fundamental characteristics than how they have become it.* (Gauss, 1802d: 503)

As Gauss mentioned, Johann Huth (1763–1818) had personally received Herschel's reaction to the general consensus that Ceres and Pallas should be thought of as planets. Huth (1802), a professor of mathematics and physics in Frankfurt an der Oder, did not meekly accept his refusal, but tried to reason the matter through with Herschel.

> *I think it unwise to introduce new names especially a general one, if we can avoid it. New categories entice us to see differences where there are none. The fact that Ceres and Pallas are smaller than the other planets cannot hinder us to expel them from the set of planets.*

Herschel was unmoved by this entreaty. He never replied to Huth, but this did not stop Huth (1804: 266) from coining his own term of 'coplanets' to denote Ceres and Pallas.

The Origin of the Word *Planetule*

Herschel did adopt Watson's final suggestion, although it only appears once in his notes. In the undated "Work to be done," he lists as the first task "To observe the 4 Planetules." (RAS, W.2/6, f. 25). It was written in 1816 during a conversation with his son about what studies he might continue to work on as William reached the end of his active career. The word *planetule* (meaning 'little planet') was current before 1845, as it is found in Bolles (1845: 567). Many sources attribute its first use in English to the English geologist William Daniel Conybeare FRS (1787–1857), who applied a superfluous adjective: 'little planetules' (Conybeare, 1836: 32). The printed version comes from a lecture he delivered in Bristol College in 1831. While he was the first to use it before an audience and in print, it was first coined by William Watson in 1802. It was subsequently used by the American astronomer Daniel Kirkwood (1888: 27) as a synonym for asteroid.

The most important historical point about planetules (or, in the case to be discussed, 'planetulas') is the confusion that was caused in Germany about what word Herschel chose to describe Ceres and Pallas. Nowhere in Herschel's seminal paper (Herschel, 1802c) are Ceres and Pallas termed 'planetulas,' but as we can see in the correspondence between Gauss and Olbers in 1802, they thought he had. This is an extract from a letter Olbers sent to Gauss, in which he is quoting from a letter by George Best (1756–1823) in England sent to Schroeter on 7 May 1802. Schroeter then forwarded it to Olbers, who on May 23 relayed it to Gauss: "Herschel's observations of Ceres and Pallas were read in the Society yesterday (May 6). They go to the 2nd or 4th of May… He denies they have any cometary and planetary characteristics and wants to name them *planetulas,* without thereby detracting from the discovery in the least." (Olbers, 1802a; underlining probably by Best)

Gauss (1802) replied to Olbers, and correctly made the point that planetula is the diminutive of planeta: "To want to distinguish between 'planeta' and 'planetula' seems to me to be almost pedantic. Mercury, Venus, Earth and Mars are also 'planetulae' compared with Jupiter, and perhaps our Sun compared with other fixed stars would just be a tiny 'solculus.'"

By May 24, Olbers was aware that Herschel was using the term asteroid to denote Ceres and Pallas, as he was the first Continental astronomer to use the new word in private correspondence, as evidenced by his letter to the French astronomer Joseph Jérôme Lalande (Olbers, 1802b). Based on what he read from Best via Schroeter, Olbers may have believed Herschel used 'planetulas' in his RS paper of May 6, so it was Best's account of the reading of Herschel's paper that was the source of the confusion.

The Origin of the Term *Planeto-Comet*

The true nature of Pallas was a matter of great debate. In this handwritten French note by Baron von Zach, probably dating to 1802, he speculates about the newly discovered planet: "Pallas does neither move on a circular orbit nor a parabolic orbit. It moves in a very eccentric ellipse which means it is a new species of celestial body—a planeto-comet (Fig. 4.6)."

Even though Zach rejected the term 'asteroid,' he was ready to admit to Banks on May 1, 1802, that the discoveries of Ceres and Pallas presented astronomers with a situation that demanded some sort of new categorization:

> *This latter heavenly body is a very remarkable one, and certainly a Middle-Thing between a Planet and a Comet…. The Pallas cannot be deemed a comet, if we understand by a comet a hairy blazing star, moving in a parabolical orbit. For she has not all the appearance of a nebula, or a dark gloomy star, not the least trace of a tail, bush, or pencil of spurious light. She looks rather clearer than Ceres; is about of the same size. She moves not in a parabolical curve. The Pallas cannot be deemed a planet, if we understand by planets, heavenly bodies revolving in little eccentric ellipses round the Sun, and pursuant to the law of distances, completed now by the discovery of Ceres, and extending to the Georgian planet, and perhaps beyond. The Pallas has no assigned place as a planet according to this law in our solar system. She moves in a too eccentric ellipsis, and has a too great inclination of the orbit, as that she might be ranked amongst our primary planets. This body gives us therefore the indication of a new species, that we might call Planeto-Comet, so we'll have, fixed Stars, Primary Planets, Secondary-Planets, and Planeto-Comets.* (Zach, 1802d)

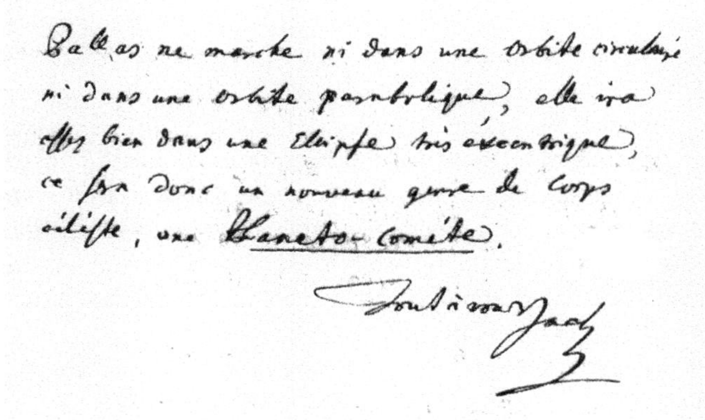
Pallas ne marche ni dans une orbite circulaire
ni dans une orbite parabolique, elle ira
assez bien dans une Ellipse très excentrique,
ce sera donc un nouveau genre de corps
céleste, une Planeto-Comète.
Tout à vous Zach

Fig. 4.6 Zach's note about the term *planeto-comet*

A few days earlier, April 27, Zach also proposed the term to Gauss in a letter about Pallas: "I am very eager to know what you will find, now that neither ellipse nor circle obey, most likely a very eccentric ellipse, ergo a planeto-comet a new kind of heavenly body." Maskelyne was also treated to this new term by Zach on May 4, 1802: "No doubt that several other such Planeto-Comets might exist in the heavens, especially in the great and immense spaces between Jupiter and Saturn, Saturn and the Georgian Planet."

Even though Zach promoted the term, he did not invent it. In the *Monthly Correspondence* (May 1802, p. 494) Zach says Wurm thought of the term: "One of Prof. Wurm's first thoughts regarding Ceres was the following question: 'What is to become of this celestial body? A planeto-comet or a cross between both? [i.e., a 'middle-thing']'" Zach first quotes this in the *MC* (October 1801, p. 370) so it shows here that the term originates with Wurm. The term has been used recently: "…just as the undoubtedly planetary origin of some long-period comets arriving from the *joint planeto-cometary cloud* beyond Neptune…" (Drobyshevski, 2008; italics in original)

The Origin of the Words *Planetoid* and *Cometoid*

Piazzi, the discoverer of Ceres, rejected the asteroid terminology. His overweening pride would not allow his discovery to be anything other than a primary planet. The tone he took in expressing his opinion to his friend Barnaba Oriani (1752–1832), director of Brera Observatory, was quite contemptuous. "Soon we will see dukes, counts and marchesi in the sky as well" (Oriani, 1802). The quote by Piazzi is contained in a letter to Zach, in which Oriani added his own thoughts: "You [Zach] have already successfully proven the planetism of Ceres and Pallas; consequently, it is useless to ponder Herschel's new dynasty."

With great glee, Zach repeated this information to all and sundry. On September 15 he told Gauss what Piazzi and Oriani thought, and 2 days later he told his friend Jan Sniadecki (1756–1830), director of Vilnius Observatory in Cracow, that "The Italians make fun of the asteroids." (Zach, 1802a). The same day, Zach jovially responded to Oriani. Echoing the etymological construct of "Herschel's dynasty," he directed Oriani to read the new issue of his journal *Monatliche Correspondenz*. It was the only one in the world at the time devoted entirely to astronomy. "Piazzi's remark about the celestial deities made me laugh, this bon mot is brilliant. You said I had successfully proven that the two stars were planets but you will be even more content to hear what I said about this matter in my September issue. Herschel's dynasty is not popular in Germany either." (Zach, 1802b).

Piazzi (1802a) wrote to Oriani, asking his opinion of Herschel's proposed word 'asteroid.' His first point is the relevant one to this study. "What do you think? It looks to me 1st Whatever the name given to this new star doesn't really matter. Are they moving stars? You can call them planetoids or cometoids, but not asteroids."

While his candid opinion about dukes in the sky was given to his friend Oriani, Piazzi (1802b) was much more courteous to his supposed friend Herschel, as he sugar-coats a bitter pill. "…could we not establish as a distinctive mark between the planets and comets the intersection of their orbits reduced to the ecliptic? And for the naming, could one not call the little planets Planetoids? Because I confess the name asteroids seems to me more appropriate for the small stars." While most continental astronomers were airily dismissive of Herschel's choice, this letter shows Piazzi trying to reason with Herschel. In these important letters, Piazzi coins a word that has become widely used ever since to denote small planets such as asteroids, namely 'planetoid.'

The first printed example of the words planetoid and cometoid comes from the pen of the critic Henry Brougham (1778–1868; 1803), who in later life became Lord Chancellor of Great Britain. He could not possibly have seen the private letter from Piazzi to Herschel, so it is he who must be given credit for the introduction of these words into the English language.

The *OED* regards cometoid as an obsolete word, and gives its origin as W. Taylor, 1805. It was used by Capel Lofft (1805) in the *Monthly Magazine*, but the *OED* (using the same publication and page number) erroneously gives the name Taylor as the originator. Thus, the *OED* entry is wrong both in citing its first use in 1805 instead of 1803, and attributing the 1805 use to Taylor instead of Lofft. Despite its obsolete status, there are many instances of the use of 'cometoid' in the modern literature as an object that exhibits the properties of both an asteroid and comet (e.g., Chaikin, 2003). After excoriating Herschel for bringing the word asteroid into use, Brougham wrote:

> *To us, that name presents the idea of some body resembling fixed stars; whereas the two new planets have no one circumstance in common with those distant bodies. If a new name must be found, why not call them by some appellation which shall, in some degree, be descriptive of, or at least consistent with, their properties? Why not, for instance, call them Concentric Comets, or Planetary Comets, or Cometary Planets? Or, if a single term must be found, why may we not coin such a phrase as Planetoid or Cometoid?*

The derivation of planetoid is also from Greek and Latin, and one wonders how Herschel would have responded to this suggestion had it come from his friend Watson. Compared with the outrage that greeted the word asteroid, it seems highly likely that the word planetoid would have raised far fewer objections. The suffix is used in mathematics (rhomboid, trapezoid), biology (arthropoid, humanoid), and chemistry (alkaloid), so its extension into astronomy would have raised few hackles. The *–oid* suffix was used once again by the IAU, when it named all spherical objects beyond the orbit of Neptune "plutoids." (IAU, 2008). However, the term plutoids is hardly ever used , the most-used terms being Kuiper-belt objects (KBOs), transneptunian objects, Centaurs, Damoclids and plutinos.

The suffix *–oid* is derived from the Latin suffix *–oides*, which in turn came from the Greek. It possesses the meaning "having the likeness of." In some words *–oid* has a slightly extended meaning—"having characteristics of, but not the same as," and it would be in this sense that Piazzi suggested the word because he uses the word 'little.' Thus he is signifying that the smallness of Ceres and Pallas is a distinguishing criterion for applying a different appellation to them. It might also be noted that the prefix *aster–* is used in science as well. Just drop the letter 'o' from asteroid and we have the word asterid, which denotes such flowering plants as daisies, sunflowers and potatoes.

Herschel actually used the designation 'planetoid' in 1803 in a paper published by the Royal Society, but in attributing the creation of the term to 'an eminent astronomer' he fell short of mentioning Piazzi's name. This appeared after the *Edinburgh Journal* article by Brougham in the same year, so Brougham made the word public before Herschel.

> *As the solar system presents us with all the particulars that may be known, respecting the arrangement of the various subordinate celestial bodies that are under the influence of stars which I have called insulated, such as planets and satellites, asteroids and comets, I shall here say but little on that subject. It will, however, not be amiss to remark, that the late addition of two new celestial bodies* [Ceres and Pallas], *has undoubtedly enlarged our knowledge of the construction of the system of insulated stars. It is not in the least material whether we call them asteroids, as I have proposed; or planetoids, as an eminent astronomer, in a letter to me, suggested; or whether we admit them at once into the class of our old seven large planets.* (Herschel, 1803; 339–340)

At the IAU meeting in 2008, the draught of Resolution 5A called median bodies such as Ceres and Pluto 'planetoids,' but the plenary session voted unanimously to change the name to 'dwarf planet'.

Words Proposed in 1801/2 to Categorize Ceres and Pallas

Name of Proposer	Word Proposed	Date of Proposal/Notes
Johann Wurm	Planeto-comet	Oct. 1801; in the *MC*
William Watson	Planeret	April 27, 1802
	Planetel	
	Planetet	
	Planetkin	Used by Carlyle in 1832
	Erratikin	
	Planetine	

	Planetule	Used orally by Conybeare in 1831; in print 1836
Charles Burney, Sr.	Stellula	May 3, 1802
Charles Burney, Jr.	Asteroid	May 5, 1802
Stephen Weston	Aorate	June 8, 1802
Giuseppe Piazzi	Planetoid	July 2, 1802
	Cometoid	
Henry Brougham	Planetoid	Late 1802, published 1803
	Cometoid	Next used by Lofft in 1805

Herschel's visit to Paris during a short interval of peace between Britain and France (Grainger, 2004) happened in August 1802. Just two months later a new publication in Scotland would appear that proved to be a platform for a major attack on Herschel and his choice of "asteroids" to denote Ceres and Pallas. It was the first of several attacks of his home turf.

Henry Brougham

First to attack was an anonymous reviewer in the pages of the *Edinburgh Review*, whose argument was conceived as a refutation. "Dr. Herschel's passion for coining words and idioms has often struck us as a weakness wholly unworthy of him. The invention of a name is but a poor achievement in him who has discovered whole worlds." He added the insinuation that Herschel had devised the word 'asteroid' for the express purpose of keeping Piazzi's and Olbers' discoveries on a lower level than his own of Uranus. Apparently Herschel's only response was to characterize it as "… the illiberal criticism of the *Edinburgh Review*." (Clerke, 1901: 90)

How could such an attack have been launched, and by whom? The first issue of the *Edinburgh Review* was published October 10, 1802; the attack on Herschel appeared in the second issue in early 1803. The *Review* rapidly assumed the foremost place among English critical journals (Sydney, 1898: 228). The original staff were all very young, with Henry Peter Brougham at the age of 23 the youngest of all (Fig. 4.7).

All these were young men full of talent and ambition, to which the *Edinburgh Review*, at its commencement, was a vent for feelings and theories that had been accumulating for years. Above all, it enabled them to give full utterance to those political principles that were so obnoxious to the rulers of the day, and so doubly proscribed in Scotland. Each individual no longer stood alone, but was part of a collected and well-disciplined phalanx; and instead of being obliged to express his opinions in bated breath, and amidst an overwhelming uproar of contradiction, he could now announce them in full and fearless confidence, through a journal which was sure of being heard and feared, at least, if not loved and respected (Chambers, 1856).

Fig. 4.7 Henry Brougham, 1st Baron Brougham and Vaux, by James Lonsdale. ©National Portrait Gallery, London. Used with permission

The very publication of this new journal, the *Edinburgh Review*, was castigated by its intellectual opposite number, *The Anti-Jacobin Review and Magazine*: "As the object of the *Edinburgh Review* is the depreciation of whatever tends to elevate, or to support our country, a natural and obvious branch of their plan is to vilify every writer who supports constitutional loyalty, patriotism and order." (J. B., 1803) Herschel, as the king's astronomer, is certainly covered under this broad umbrella, so the attack against him in the pages of the *Edinburgh Review* can be seen not as an isolated one but as part of a wider assault on British writers.

In the same vein, we read in Volume 14 of the *Cambridge History of English and American Literature* (Ward & Waller, 1907–1916: 37) that in the early issues of the *Review*, Brougham "…developed a policy of hostile criticism, of which English educational institutions were the object." Volume 12 of the *Cambridge History* (Ward and Waller, 1907–1921: 16) gives us further valuable insight into Brougham's character:

> *Henry Brougham, the youngest of the three, was to become, in a few years and for a time, by dint of extraordinary energy and ability, one of the most powerful political leaders in England. His services to the* Review, *in its early days, had been quite invaluable. Hardly any public man of the nineteenth century approached more nearly to the possession of genius. But his great gifts were weighted with very serious faults of character and temper; and, as the years went on, he earned for himself universal distrust among his fellow-workers – editors of, and contributors to,* The Edinburgh, *or statesmen engaged in the wider field of British politics. It was long a tradition among* Edinburgh *reviewers that, on one occasion, a complete number of the* Review, *with its dozen or more of articles, was, from cover to cover, written by the pen of Brougham, and the story, whether true or not, is illustrative of the universality of capacity generally attributed to him.*

Who was this precocious young man who dared castigate the great Herschel? Henry Brougham was born in 1778 into an intellectually charged Edinburgh that had become known as the 'modern Athens' because of the many noted philosophers and scientists who lived there. By the tender age of 18, the Royal Society had already published his first scientific paper (Brougham, 1796) on the properties of light. One of the founders of the *Edinburgh Review*, he was described at this stage of his life as an "... uncommon genius ...," although by outward appearances tall, thin and quite ugly. This did not hinder his career, as he became Lord High Chancellor of England and one of the greatest statesmen of the nineteenth century. Lord Brougham lived to be 90 (for a general biography see Hawes, 1957; for his early life see New, 1961).

The fact that Brougham became the most brilliant legal mind of the age is a crucial factor in understanding his mindset as he approached the Herschel-asteroid matter. Many of the reasonings of lawyers, wrote the Scottish philosopher David Hume (1711–1776); Fig. 4.8, are of an analogical nature. This was of the very nature of jurisprudence, which was

> *... in this respect, different from all the sciences ... [in that] in many of its nicer questions, there cannot properly be said to be truth or falsehood on either side. If one pleader bring the case under any former law or precedent, by a refined analogy or comparison; the opposite pleader is not at a loss to find an opposite analogy or comparison: and the preference given by the judge is often founded more on taste and imagination than on any solid argument.* (Hume, 1748: 221)

Fig. 4.8 David Hume

In his *Edinburgh Review* article, Brougham was arguing—in part—from analogy, which Hume specifically identified as inappropriate in the field of science. A quotation from the German philosopher Friedrich Schiller (1759–1805) seems apt: "The basis for your complaint seems to me to lie in the oppression of your imagination by your intellect." (Schiller, 1788). One might say Brougham was 'too clever by half.' Herschel first learned of the review article in a letter from his friend Watson:

> *The same day I received your letter I received one also from our excellent friend Sir Joseph Banks. Among other things he tells me that there has come out an* Edinburgh *review of which two numbers have appeared, which appears to him to be written with a very caustic spirit, decrying our literary men, in order to raise the merits of his own countrymen. He tells me you have not escaped his lash, and indeed, if depreciation of merit is his aim, you must be the first to aim at. You, I am persuaded, will regard his darts with due indifference, and trust as you have hitherto done, in a calm and dignified reliance that nothing can affect and overthrow truths and discoveries founded on experience and observation.* (Watson, 1803)

Herschel's correspondent David Brewster wrote a letter from Edinburgh, in which he scathingly told Herschel about the identity of his anonymous assailant: "I do not know if you have seen the review of some of your papers, in the first numbers of the *Edinburgh Review*. It may perhaps be interesting to you to be informed that the gentleman who reviewed them, and also Dr. Young's papers, was Mr. Brougham, a member of your Society who seems to take pleasure in holding up to ridicule those characters only who are esteemed and revered." (Brewster, 1805) Brewster alludes to the famous controversy mentioned in *Discovery of the First Asteroid, Ceres* (Chap. 5) with Thomas Young (1773–1829), where Brougham savaged his optical memoirs (Cantor, 1971).

Brougham did one great service to astronomy: he persuaded Mary Somerville (1780–1872) to translate Laplace's book *Mécanique Céleste* into English. The resulting book, *Mechanism of the Heavens* (1831), became a standard text for the rest of the century.

Thomas Thomson

Brougham's opinion had a major influence on his fellow Scotsman, Thomas Thomson (1773–1852); Fig. 4.9. Thomson had been elected a Fellow of the RS in 1811, and the following year published his *History of the Royal Society*. In this tome he chose to launch a bitter personal attack on William Herschel, who had been elected a Fellow in 1781:

> *The distance between Mars and Jupiter is so great when compared with the other planetary spaces, and the distance from the sun, that astronomers had been looking for some primary planet in that position. Of late years these expectations have been more than accomplished by the discovery of no fewer than four planetary bodies almost all in the same place; but so small that Dr. Herschel refuses to honour them with the name of planets, and chuses [sic] to call them asteroids, though for what reason it is not easy to determine, unless it be to deprive the discoverers of these bodies of any pretence for rating themselves as high in the*

Fig. 4.9 Thomas Thomson

list of astronomical discoverers as himself. These four bodies have received the names of Ceres, Pallas, Juno, and Vesta. They were discovered by Mr. Piazzi, Dr. Olbers, and Mr. Harding. (Thomson, 1812: 358)

Thomson's attack was much more influential than Regnér's swipe at Herschel for using 'unjust influence,' as his treatise was written in Latin and read by very few. Sir Richard Phillips (1836: 162) reinforced the scurrilous opinion of Herschel's motives offered by Thomson, stating in one of his later books that "Ceres ought to be taken in the measures of Schroeter, for Herschel seemed disposed to underrate the new discoveries of other astronomers." The celebrated astronomer François Arago wrote:

I should require nothing further to annihilate such an imputation than to put it by the side of the following passage, extracted from a memoir by this celebrated astronomer (Herschel), published in the Philosophical Transactions *for the year 1805: "The specific difference existing between planets and asteroids appears now, by the addition of a third individual of the latter species [Juno], to be more completely established, and that circumstance, in my opinion, has added more to the ornament of our system than the discovery of a new planet could have done.* (Arago, 1871: 217)"

Without mentioning Thomson by name, Herschel's noted biographer James Sime (1843–1895; 1900: 191) poured vitriol on the attack:

Strange to say, the friend of Piazzi and Olbers, who discovered these small bodies, was charged with intending, by the suggestion of this diminutive [asteroid], to cast a slight on the achievement of his friends, in comparison with his own glory as the discoverer of the great planet, Uranus. A more stupid slander of a most generous heart could scarcely be imagined. That one scientific man should attack, or rather slander, another for giving to these small bodies a scientifically appropriate name, on the ground that he thereby intended to derogate from the credit of his own friends, whom he publicly extolled as 'celebrated discoverers,' seems incredible. Yet it was done.

The attack was also dismissed by James Smith (1815: 555). "This insinuation appears too gratuitous to be considered either generous or just." Like Brougham, Thomson achieved great distinction. At the age of only 23 he became editor of the *Encyclopedia Britannica*. In 1798 he introduced the use of symbols into chemical science, and he became legendary in his post as Regius Professor of Chemistry in the University of Glasgow (Crum, 1855).

The antipathy towards William Herschel did not prevent either Thomson or Brougham from corresponding with his son John. The Royal Society archives contain letters from both of them. In any case, criticism of William Herschel was nothing new: "William was always a controversial figure, despite his personal charm. The controversies began in his first contacts with the Royal Society, where his claims as to the magnifications of his eyepieces were met with incredulity and Fellows said he was fit for Bedlam (that is, Bethlehem, the hospital for the insane). In the first decade of the nineteenth century, his ill-judged insistence on publishing papers on colored rings also brought a lot of criticism." (Hoskin, 2003: 50). Indeed, in 1802—the very year he employed the word 'asteroid' for the first time—Herschel was accused by the English author John Corry (1802: 81) of "philosophical quackery."

The German Connection

The motivation for Thomson's personal attack is obscure, but it is interesting to note that both of the people who criticized Herschel were Scottish, born just 5 years apart. Even though we regard Herschel as an English astronomer, he was German-born, an ethnic element that may have played a factor in their disdain, according to J. Arthur Thomson, Professor of Natural History, University of Aberdeen: "Britain is wont to be proud of William Herschel, who extended Newtonian methods to the study of the stars and recognised the occurrence of vast developmental changes in the heavens. But William Herschel was a Hanoverian." (Thomson, 1915: 143).

Even though Herschel, as a church organist, entirely conformed to the Church of England and thus fitted into the English scene without overt controversy, it is possible his Scottish detractors held this perfect melding into British society in light regard. How foreign Protestants should be integrated into English society had been a topic of controversy just 10 years before Herschel emigrated. For example, this is mentioned in two letters written by the Swiss mathematician Leonhard Euler (1707–1783) to Johann Wettstein (1693–1754). Wettstein was a foreigner (from Basel) who became closely associated with the British royal family as its chaplain: "The newspapers are saying a great deal about Parliament's plans to naturalise the foreign Protestants." (Euler, 1748a), and "I am amply aware that even after one has allowed foreigners to establish themselves in England, it will still be a long time before granting them subsidies (pensions)." (Euler, 1748b). Herschel became a naturalized British citizen in 1793 (by Act of Parliament, Private and Personal Acts c.38). As a result when the Hanoverian Guelphic Order of Knighthood was established and Herschel appointed, he was one of the British appointees announced in London, not one of the Hanoverian appointees announced in Hanover. In many ways Herschel was more English than the English.

Upon appointment as Astronomer to His Majesty, Herschel received a royal pension of 200 pounds per year (agreed to in July 1782) from King George III (this compares to the salary received by Maskelyne, the Astronomer Royal, of 300 pounds). It was Sir Joseph Banks who induced the king to confer this appointment on Herschel (Ball, 1985: 246). The king (Fig. 4.10) invented the informal post near his residence at Windsor Castle so that he himself could further his own education in astronomy and also because he could think of no other way to fulfill his obligation to Herschel as his patron. An ancillary benefit was an opportunity to entertain his guests after dinner by showing them Herschel's telescopes.

George III had another beneficent role in the saga of the asteroids: he granted Johann Schroeter funds that were used in part to hire an assistant, Karl Harding, who discovered the third asteroid. Following along the pattern set by Herschel in commemorating the discovery of a planet with the name Georgium Sidus to honor King

Fig. 4.10 King George III

George III, the third asteroid was named Juno Georgia by Schroeter to honor his patron—George III. (*The Eclectic Review*, 1807: 183). The name Juno was actually proposed first by Olbers and accepted by Harding (1804a), but Schroeter added Georgia to curry favor. This may have been a contributing factor in the appointment of Schroeter, in July 1816, to the very same Hanoverian Guelphic Order of Knighthood that had been bestowed upon Herschel on March 22 of the same year. The order was of recent mintage, having been founded just the year before as a result of Hanover's elevation to the status of a kingdom by the Congress of Vienna (Vick, 2014).

Herschel was also elected to the Royal Society. Numerous Germans had been elected to the Royal Society during the eighteenth century, but many never visited England and "... therefore had little personal contact with English thought and culture." (Davis, 1969: 51). It is possible that Herschel's elevation, and physical presence at the very seat of power, caused some (such as Brougham and Thomas Thomson) to harbor some ill feelings towards him, even though by the time of their attack in 1802 Herschel has been thoroughly integrated into English society. Thus the assertion by J. Arthur Thomson that Herschel "was a Hanoverian" does not reflect the legal standing of Herschel as a British citizen, or his status as a paragon of British astronomy, but rather an underlying realization that he was not born in England. While this was of no relevance to Herschel's personal friends or multitude of admirers among the wider populace, it may have been one element that influenced his detractors.

The German connection here was quite strong (Beuermann, 2005). Herschel had been born in Hanover in 1738. It was none other than George III who was Elector of Hanover, and even though he never visited his Royal domain on the Continent he personally managed the diplomatic affairs of Hanover independently of England (Blanning, 1977). Was Herschel a part of the dual track method of ruling exercised by the king? According to Sime (1900: 94), he was: "In fact George III and his advisers dealt with Herschel, not as an Englishman but as a German." The accuracy of this assumption is open to question, as Herschel did everything he could to distance himself from his Germanic roots. He dropped his Hanoverian name of "Friedrich Wilhelm" the moment he set foot in England, adopting the anglicized "William." He almost never spoke or wrote German again, and in his correspondence he wrote to Germans in English. Thus modern scholarship about Herschel (e. g., Hanham & Hoskin, 2013) casts doubt on Sime's assertion.

Herschel's relationship with Maskelyne is also of significance, as they often shared observational data on asteroids. As evidenced by a letter that Herschel (1782b) sent to his sister Caroline (1750–1848), he believed his instruments were superior to those used at the Royal Observatory: "These two last nights I have been star gazing at Greenwich with Dr. Maskelyne and Mr. Aubert. We have compared our telescopes together, and mine was found very much superior to any of the Royal Observatory." Herschel was certainly correct here, but the duties of Maskelyne at Greenwich were related to positional astronomy, and for this purpose Herschel's telescopes were useless. Maskelyne was therefore not in possession of inferior instruments with regard to the work his post required.

That Herschel had superior optical telescopes was no idle boast (Bennett, 1976). Its reality was known far and wide: "I am still of the opinion that I have seen the two most wonderful things that have ever been seen in this Planet: the French Revolution and Dr. Herschell's telescope." This was in a letter dated October 7, 1791, from the Scottish natural philosopher John Anderson (1726–1796) to Herschel's friend James Lind M. D. (1736–1812). Herschel actually made much of his money from the sale of his telescopes, often made-to-order for wealthy clients (Maurer, 1998). Maskelyne himself asked Herschel to make him a 7-foot telescope, which he did, although he never quite managed to replicate the perfection of the mirror used to discover Uranus in 1781.

Since Maskelyne had invited Herschel to "star gaze" at Greenwich, Herschel extended a reciprocal invitation to Maskelyne to observe with him at Windsor. The close and warm friendship between Herschel and Maskelyne is best illustrated by Maskelyne's letter of August 8, 1782, two months after Herschel's visit to Greenwich (quoted above) and the first after Herschel's appointment as astronomer to His Majesty.

> *I hope you have had some good nights at Windsor for seeing your newly discovered double stars with your admirable telescope. I thank you for having shewn me many of them, which my own telescopes, tho' reputed excellent, would not discover…Astronomy and Mechanics are equally indebted to you for what you have done; the first for your shewing to artists to what a degree of perfection telescopes may be wrought; & the latter for your discovering to Astronomers a number of hitherto hidden wonders in the heavens, which could not be explored before for want of telescopes equal to yours; and they are both likely to receive equal improvement from it in the construction of better telescopes, and in the application that may be made of them to the heavens for repeating and extending your observations. I hope you will do the astronomical world the favour to give a name to your new planet, which is entirely your own, & which we are so much obliged to you for the discovery of.* (Maskelyne, 1782)

Over the years the two men were in constant contact (as evidenced by 66 letters in the RAS archives), and Maskelyne was especially close to Caroline (Higgitt, 2014: 128). He invited her to visit him, took her to meet other astronomers and gave her astronomical presents. He even persuaded the Royal Society to have Caroline's book published at the society's expense.

The Links Between Brougham and Thomson

There were certainly links between the Scottish philosopher and mathematician Dugald Stewart (1753–1828); Fig. 4.11, Brougham and Thomson, the attackers of William Herschel. The noted historian and philosopher James Mill (1773–1836), while studying at the University of Edinburgh, received his greatest intellectual impulse from Stewart, the Professor of Moral Philosophy, and Thomson was also a student of Stewart (Maas, 2003: 338–360)! When Mill began *The Literary Journal* in 1803, Thomson took charge of the science section. In 1808, Mill began writing for *The Edinburgh Review*, where Brougham wrote on mathematical subjects and

Fig. 4.11 Dugald Stewart

Thomson was employed on antiquarian matters. The philosophical and social mindset that gave birth to *The Edinburgh Review* has been explored by Flynn (2002). Also of import is that Charles Babbage, who wrote about the link between asteroids, meteorites and comets (see *Discovery of the First Asteroid, Ceres*) was deeply affected by the philosophy of Dugald Stewart (Rice, 2001: 159). The influence Stewart had on his pupils was examined by Winch (1983).

At the turn of the century, the young Scottish Whigs included Henry Brougham, Thomas Brown (1778–1820) and Francis Jeffrey (1773–1850)—another key figure in this mix. Jeffrey was another student of Stewart's (1753–1828), but in the 1804 issue of *The Edinburgh Review*, he savaged Stewart's philosophy so severely that subsequently "… all of Stewart's writings were coloured by the fact that he believed moral philosophy was under siege." (Tannoch-Bland, 1997: 308) It was yet another example of the attack-dog attitude shared by the trio—Herschel was just one of several prominent targets.

These critics were the product of the Scottish enlightenment that made Edinburgh one of the chief centers of learning in eighteenth century Europe. When Sydney Smith (1771–1845) met them at the Academy of Physics in Edinburgh, the three polymaths were already deep into "… the investigation of nature, the laws by which her phenomena are regulated, and the history of opinions concerning these laws." (Welsh, 1825: 77). Smith suggested they collaborate to publish a critical review. Through the winter of 1801–1802, plans for *The Edinburgh Review* were developed at the very same time William Herschel was studying Ceres and Pallas. For Brougham, Ceres and Pallas were a ready-made subject for his exposition, giving full vent to his thoughts on the laws of nature and opinions concerning them.

Raised in this atmosphere of criticism—at both *The Edinburgh Review* with Brougham as an exemplar—and later within a London society noted for its personal attacks on prominent men, it is no wonder that Thomson had no compunction about firing a broadside against Herschel in 1812. Indeed, he was merely following the precedent set by Adam Smith (1723–1790), Professor of Moral Philosophy at Glasgow University, who wrote for the short-lived *Edinburgh Review* in 1755. Smith actually accused Dr. Samuel Johnson (1709–1784), in his *Dictionary*, of being insufficiently grammatical (Buchan, 2003: 133)! To use a word Dr. Johnson employed in his *Dictionary*, one might say Brougham exhibited 'hebetude'—a bluntness that exhibited a particular insensitivity.

Temperamentally, Thomson was also well suited to this form of attack. He wrote a stinging piece of satire in 1799 (Larder, 1970: 296), and his own personality was scathingly parodied by the well-respected scientist David Brewster (1806): "With a little judicious comprehension, the credendum of this philosopher [Thomas Thomson] might be easily imprisoned in the cavity of a nut shell. It is however of the gaseous kind, and would probably require a ground stopper to prevent evaporation."

Herschel received plaudits as well as barbs. In one of the most prestigious publications of the age, *The Monthly Magazine* (1796), we read this account of him, in equal parts exalted pride in his accomplishments and snobbish sympathy for everyone else: "Astronomers, in different parts of the world, may be discouraged from continuing their observations, when it should seem, that their discoveries must be anticipated by our observer (Herschel); but, though he has so much the advantage, much is left to their labour and industry."

Sir Joseph Banks

Although Herschel was scarcely acquainted with Brougham or Thomson, he was a close associate of Sir Joseph Banks, president of the Royal Society. One would have thought that Herschel would get a favorable hearing from Banks, but such was not the case. Banks wrote to Zach on June 7, 1802, and an extract was printed in the M*onthly Correspondence* (1802: 90):

> *Dr. Herschel still persists in his opinion in view of the small size of these two new planets, and continues to maintain that these must be strictly differentiated and classified especially from the planets and comets, except when these questionable comets are found in a quiescent state. I believe that he wishes to name them asteroids, because they are not visible to the naked eye. We see no difficulty in the requirement that the light from such small bodies should reach us.*

That Zach took great delight in publishing this 'barbed arrow' in his journal can hardly be doubted. A review of papers in the *Monthly Correspondence* regarding Ceres and Pallas shows that this is in fact the *only* extract from a letter Zach received from Banks to be printed in the journal. What Herschel thought of this is unknown, but he may not have read it since he was not a subscriber to the journal. While still avoiding use of the word asteroid, Banks finally wrote Herschel that "It gives me much pleasure … that the Germans should so readily and properly have adopted the distinction which you have made between them and planets." (Banks, 1807).

The Attack in *The Critical Review*

One of the most influential publications of the time was *The Critical Review, or, Annals of Literature.* The May issue (1803a, pg 15–19; his emphasis) includes a detailed and damning critique by an anonymous reviewer of Herschel's 1802 paper *Observations on the two lately discovered celestial Bodies.* After printing the substance of the paper where Herschel gives his seven criteria for distinguishing Ceres and Pallas from the other planets, the writer launches his first attack:

> *This reasoning is, however, too rigorous. By a similar argument, it might be contended that there should be no more than seven planets, seven colours, etc.: to which we may add, that the vacant space may be as aptly filled by two smaller bodies as by one larger. Had we found a large planet, three times the united diameter of the two now under our eyes, we should not have contested its title; and we see not, as we shall presently show, that we ought, from any considerations, to combat the claim of either Ceres or Pallas. The other objection is still weaker. If we admit bodies, it is said, of such great geocentric latitudes, we must resign the zodiac. But what power fixed its limits? – the motions of planets, which did not wander beyond it; and now some more eccentric are found, its limits must be, for the same reason, extended. If, however, these bodies be not planets, we may ask, What are they? We know only of three kinds of celestial bodies; planets revolving about the sun, deriving their light from it, with a determined annual parallax, and a diameter subtending a sensible angle; fixed stars shining with a light peculiarly their own, without any parallax, and subtending no sensible angle; and comets, deriving their light from the sun, which they seem to convey in a peculiar form, that of a coma, and a tail projected in a direction opposite to the sun, with a very considerable geocentric latitude – in other words, moving in a plane greatly inclined to that of the earth's orbit. Ceres and Pallas are certainly observed with comae: are they not, therefore, comets?* (p. 17).

The reviewer then quotes Herschel's five distinguishing characteristics of Ceres and Pallas, which give him a platform to launch his second attack:

> *In fact, the smallest coma of a comet exceeds that of Ceres or Pallas above a hundred times; and neither movies in orbs even approaching the eccentricity of a parabola, or is distinguished by a tail. It is also highly probable that the nuclei of comets are very small: they never disturb the planetary motions, though often disturbed by them.*
>
> *Why then are not these bodies planets? We see no reason for any distinction: they revolve round the sun, and are not comets. We must discover another system, before we are allowed to change the appellation. Mr. Herschel would call them asteroids; but he labours for a distinction, which, in the end, will fail him.* (p. 18)

He finally quotes Herschel's definition of the term asteroid, and concludes his critique of the *Philosophical Transaction* paper:

> *We shall not extend our article by enlarging on our own original idea, that these bodies may have been comets constrained to revolve within less eccentric orbits; because, in reality, we know little of the nuclei of comets, and have no criterion by which we can measure their density, nor indeed, very correctly, their diameters. The suspicion may remain on record, to*

> *be tried by future observations, with little solicitude, in the author, respecting its truth or fallacy.* (p. 19)

The critique has one critical weakness. In the first passage quoted, he believes Ceres and Pallas have a coma, and are therefore comets. Unfortunately for his thesis, Ceres and Pallas do not exhibit comas, as they have no atmosphere at all. His belief that Ceres and Pallas are former comets is not true, although we now know that some asteroids were originally comets, so he glimpsed part of the real situation. If he had not held to the false belief that Ceres and Pallas exhibit comas, he may have been more reticent in criticizing Herschel's idea to discriminate between comets, planets and asteroids.

Neologism: The Philosophical and Social Context

The criticism leveled at Herschel must also be placed within the context of late eighteenth century/early nineteenth century British society. Although we may view these remarks of Brougham and Thomson simply as misplaced criticism, they actually hold a place within the satirical framework that shaped so much of literary society at that time. Dr. Johnson, in his famous *Dictionary* of the 1780s, defined lampoon as a "personal satire" that aimed "not to reform but to vex." According to the English satirist and scholar of Italian, Thomas James Mathias (1754–1835; 1801: 7): "…all publick men, however distinguished, must in their turns submit to satire … [and] satire can never have effect, without a personal application… [since] it must come home to the bosoms, and often to the offences of particular men." Why did Brougham write anonymously? According to Mathias (1801): "It [satire] never has its full force, if the author of it is known or stands forth; for the unworthiness of any man lessens the strength of his objections."

As a leading light of the scientific establishment, it is thus not surprising to find Herschel one of its victims. Herschel was one of the astronomers targeted in a vicious satire. It was written by the famous John Wolcot (1738–1819; 1816: 455), who wrote under the name Peter Pindar:

> *The fame of Herschel is a dying blast:*
> *When on the moon he first began to peep,*
> *The wond'ring world pronounc'd the gazer deep:*
> *But wiser now th' un-wondering world, alas!*
> *Gives all poor Herschel's glory to his glass;*
> *Convinced his boasted astronomic strength,*
> *Lies in his tube's, not head's prodigious length.*

A footnote to this portion of the satiric verse is the real plunge of the knife into Herschel's abilities, comparing him to the amateur creator of mirrors, John Mudge (1721–1793):

> *We would not detract from Mr. Herschel's real merit. – By a true German cart-horse labour, he made a little improvement on Dr. Mudge's method of constructing mirrors: such are this*

gentleman's pretensions to a niche in the temple of Fame – As for his mathematical abilities, they can scarcely be called the shadows of science.

This final 'swipe' goes to heart of the fact that no astronomer in England performed any serious mathematical calculations on the orbits of the asteroids. Wolcot actually began a long series of satires on scientists—and Banks in particular—in 1788 (Jones, 1966: 195). Wolcot's best-known poem is "The Apple Dumplings and a King," which makes a mockery of King George III. He alleges that George, when visiting Herschel, was not interested in the majesty of the heavens, but only in the everyday objects that he imagined he saw through Herschel's telescope while looking at the Moon (Robertson, 2009).

Another important context that must be kept in mind is that to the early modern mind, words were a system of exchange to which arbitrary value was attached, a value we agree to honor in the interests of establishing meaning (Maguire and Smith, 2012: 193). The phrase 'coining a new word,' which we still use in the twenty-first century, is directly related to this concept as coins were (and are) a means of exchange, but their value was (and is) arbitrary. The word asteroid was 'coined' by Herschel to arbitrarily establish the value of Ceres and Pallas as a denomination separate from the planets. The Welsh scientist William Grove (1811–1896; 1843: 524) used the analogy between language and money to decry the proliferation of new words in science: "... generally coinage of words has been the undoubted prerogative of the kings of science. But now there is no 'pudor' in the matter. Every man has his own mint."

One must also ask why it mattered so much to Brougham? Why did the identification of a new kind of object offend his sensibilities so much? The answer can be found in the very use of the word sensibility to frame the question. In the latter eighteenth century arose "... the doctrine of sensibility, which judged images, objects, texts, and experience in terms of their emotional effects." (Bermingham, 2005: 12). It derived from the 'science of sensibility' that was founded on the optics of Isaac Newton and the psychology of John Locke (1632–1704); Fig. 4.12, who said everything we know derives from what the senses tell us, i.e., from our sensibility (Vermeir & Deckard, 2012). Locke (1690: 249) had this to say on what he termed "the abuse of language." He enumerated three "ends of language" as:

First, To make known one Man's Thoughts or Ideas to another. Secondly, To do it with as much ease and quickness, as is possible; and Thirdly, Thereby to convey the Knowledge of Things. Language is either abused, or deficient, when it fails in any of these Three.

For Brougham, Herschel's use of 'asteroid' certainly failed the third test. Had Brougham been aware of this? Quite likely. His relationship with Dugald Stewart has already been noted. When Stewart was at the University of Edinburgh from 1765 to 1769, he came under the influence of John Stevenson (1694–1775). As Professor of Logic, Stevenson was one of the first Scots to introduce his students to Locke (Howell, 1971: 410).

Brougham's own use of the word 'invidious' (a word meaning 'unfair,' and 'likely to arouse resentment or anger in others') immediately flags his tirade not as a dispassionate critique but as an emotional response. His *sensibilities* were offended

Fig. 4.12 John Locke

by the new word 'asteroid,' not entirely on the subjective grounds he chose to couch the essay in but through the very doctrine of sensibility that was so much a part of the culture he was raised in. Like Plato, he believed words should perfectly correspond to the essence of the (signified) thing. For him, 'asteroid' failed that litmus test. To paraphrase University of California Professor of Literature Richard Terdiman (1985: 243), the determinants of his censure were much more structural and ideological than chronological. Whether Ceres and Pallas had been discovered 20 years earlier or 20 years later would make no difference to his implacable opposition of a new astronomical term for a celestial body. "Words *do* express and *can* refer to the moment of their difficulty, their inadequacy, or their disability as they reach the edge of the garden, the margin between language and materiality…*Words do badly at bodies*." (Terdiman, 2005: 173; his emphasis). The reviewer in the *Critical Review* evoked the word 'bodies' several times, and certainly the year 1802 was a supreme moment of difficulty for a word–any word–that was created to classify Ceres and Pallas.

Philosophically, the whole controversy can best be understood as an early example of the nineteenth century attitude towards neologism. Words we now regard as embedded within language, such as altruism, society, police and virtue, were *avant-garde* at that time, and changed the way people thought. Resistance to linguistic innovation in general was undoubtedly an element, not just to neologism for its own sake but to broader intellectual and social change. Viewed in this way, the specific case of the new word 'asteroid' being introduced into the scientific lexicon may be put in context. Brougham was not merely being satirical or pedantic—he was voicing an underlying concern that new words may not just be engines of change but may actively remold the concepts to which they refer (Dixon, 2008: 33). This was certainly the case with 'asteroids.' If they could not be called planets or comets, then people would be faced with a new metric to deal with, one that seemed to most both unnecessary and dangerous. In the aftermath of the political upheavals of the French Revolution, any challenge to the *status quo* was regarded as suspect—in the words of Brougham (1803: 428), "… perplexing our ideas." As perceptively stated by the English author G. K. Chesterton (1874–1936), the importance of the French Revolution in England can scarcely be overstated (see also Whatmore, 2008): "It is not idle Hibernianism to say that towards the end of the eighteenth century the most important event in English history happened in France. It would seem still more perverse, yet it would be still more precise, to say that the most important event in English history was the event that never happened at all—the English Revolution on the lines of the French Revolution." (Chesterton, 1913: 17).

Stability is what people sought, and the clockwork regularity of the Solar System was a bedrock in the minds of most people, one of the few things they could still rely on in age rocked by revolution. Another planet they could accept, a new *kind* of object was a step too far. Or so Brougham thought when he was writing in 1803, but after a generation passed, and two more similar objects (Juno and Vesta) had been found, the term 'asteroid' entered common parlance. Whatever professional astronomers, and opinionated critics such as Brougham might have thought, 'asteroid' became imbued with a certain cachet that proved irresistible. Its very existence remoulded concepts of the Solar System (no, Ceres was not the eighth planet, it was an asteroid!) and raised questions about its very formation, just the sort of intellectual revolution Brougham was trying to forestall.

According to Simon Schaffer, Professor of History and Philosophy of Science at Cambridge University, Herschel called upon a conceptual framework to name these new and novel objects, even as it repelled the cognoscenti:

> *Brougham's critique represented a Regency commonplace. Language's stability represented a commonsense, ordered world. Terms should refer to unproblematic objects. So Herschel's neologisms and ambiguities, his desperate coinage of a literary technology designed to stabilize irreproducible and unreachable cosmological classes, scarcely carried conviction among the polite society of the literati.*
>
> *Calling these bodies "asteroids" fitted them into a large-scale investigative scheme centered in Germany in which Herschel then took part, and fitted them, too, into the pattern of the solar system. Herschel's work demonstrated that naming novel objects required appeal to networks of familiar concepts, and extended that network in one direction rather than another. Double stars were fitted into the program of parallax measures; asteroids into the regime of celestial mechanics.* (Schaffer, 1998: 194)

"Any functioning practical, ideological, or conceptual system has a repertoire of elements that appear licit and admissible within it. Elements extrinsic to these then appear, initially, as incomprehensible or meaningless." (Terdiman, 2003: 29). The duelling notions of a 'network of familiar concepts' (such as planets) and 'extrinsic elements' (such as asteroids) also finds expression in the following passage about a system of established ideas:

> *What we come to call the truth or validity of some statement – historical report, scientific explanation, cosmological theory and so forth – is best seen not as its objective correspon-*

dence to an autonomously determinate external state of affairs but, rather, as our experience of its consonance with a system composed of already accepted ideas, already interpreted and classified observations, and, no less significantly, the embodied perceptual and behavioural dispositions that are thereby engendered and constrained. (Smith, 2010: no page number)

The 'accepted idea' of what a planet is, based on observations spanning thousands of years, was in state of flux in the early eighteenth century, just as it is in the early twenty-first century with the discovery of Kuiper Belt objects and the controversy over the status of Pluto. One element of the controversy arises from sloppy linguistics. In an analysis of William Van Orman Quine's critique of modal logic, John P. Burgess, Professor of Philosophy at Princeton (2008: 207) uses Ceres, asteroid and planetoid as an example to examine "… a priori truth and linguistic truth." He states:

Quine often complained that others were sloppy about distinguishing use and mention. If one is sloppy, quibbles and confusions can result if, as was commonly done, one uses 'linguistic' interchangeably with 'analytic' or 'a priori' and 'empirical' interchangeably with 'synthetic' or 'a posteriori' respectively. For consider:

(1) Planetoids are asteroids.

(2) Ceres is the largest asteroid.

(1′) In modern English, 'planetoids' and 'asteroids' refer to the same things.

(2′) In modern English, 'Ceres' and 'the largest asteroid' refer to the same thing.

As to (1), discovery that planetoids are asteroids requires (for a fully competent speaker of modern English) mere reflection, not scientific investigation. As to (2), discovery that Ceres is the largest asteroid requires natural-scientific investigation of the kind engaged in by astronomers. Discovery that (1′) is the case requires social-scientific investigation of the kind engaged in by linguists. Discovery that (2′) is the case requires both kinds of scientific investigation. Since linguistics is an empirical science, using 'Linguistic' and 'empircal' for 'analytic' and 'aposteriori' can be confusing when dealing with meta-level formulations like (1′) and (2′) rather than object-level formulations like (1) and (2), but such usage was common.

In a broader philosophical context, the issues raised by Brougham relate to the very basic issues of how we perceive change, either as an evolutionary or a replacement process:

Whether we think of change as, at one end of the spectrum, replacement or, at the other, an unfolding of an essence or core forever present, our conception of change is intrinsically tied to our conception of entity or identity. Evolution-change may or may not assume an unfolding kernel or essence. Replacement-change permits newness and difference but tends to make its appearance, no matter how well prepared for, arbitrary and ultimately inexplicable. Unless there is some connection, or nexus, between what was and what comes after, we tend to think we have not a change but merely two things. (Bynum, 2001: 20)

In this framework of thought, Herschel's letter to the leading astronomers was his way of preparing the intellectual groundwork for replacement-change: the newness and difference of 'asteroid.' But, as Professor Caroline Bynum of Columbia University notes, this rarely prevents change from being regarded as arbitrary, which is precisely the way his choice was received by his contemporaries. Largely blind to the connection between asteroids and planets, they merely perceived two things, and decided to maintain the terminological *status quo*.

The Ancient Greek Precedent and Biblical Roots

Brougham's fulminations were promulgated in the nineteenth century, and were rooted in the eighteenth century. But one thing Brougham could not do was appeal to ancient precedent, which clearly refuted his point of view:

> *While large areas of Greek science manifest a certain conceptual vagueness, there are important exceptions to this, cases where technical terms are coined and given clear working definitions. Anatomy, zoology, harmonics and astronomy all provide examples. Astronomy, especially, developed a wealth of technical terminology, clearly defined words for zenith, meridian, apogee, perigee, parallax, colure, station, retrogradation and many others, let alone geometrical terms such as homocentric, epicycle, eccentric.* (Lloyd, 1987: 206)

The issue of distinguishing comets from planets was not born with the discovery of Pallas. Indeed, it has ancient roots. Aristotle mentions, but with evident contempt for the notion, that the Pythagoreans taught that a comet is a planet, which appears after a long interval of time and which, at the apex of the hyperbola which it describes, approaches as near to the Sun as Mercury. This notion was apparently one which the Pythagoreans obtained from the Chaldeans (*The Classical Journal*, 1821: 282). Aristotle's study of psychology led him to conclude that the mind generally investigates a new thing by looking at what it resembles, by noting differences, and by attempting to classify it. It was just this thought process Herschel engaged in to classify Ceres and Pallas.

Harkening back to the quote about nature by Heraclitus (given in Chap. 1), it had been understood in England for a long time that the issue of naming something had serious theological implications:

> *...when things have a true being, we have a care to give names and titles, agreeable to the nature and quality of them, that the act and nature of the thing, may be made manifest in the name of it, as written in the forehead: for as a man draweth good Liquor out of the Cask, so out of the meaning and signification of the Word, and denominations given by God, we may draw out the hidden nature and knowledge of a thing, for* nomen est symbolum rei....*it doth not avail us, that things be distinct in nature, if there be a confusion of names.* (Andrewes, 1657: 33–35)

The Role of Classification in Science

Closely tied to nomenclature is classification. The urge to classify celestial objects has distant roots in English thought:

> *Divine differences have been gathered by learned men, to shew the severall distinctions, and most honorable places, held to eternity amongst those heavenly Bodies: It is questionless, that the number of them are infinite (according to our apprehension) Nay, infinite their distinctions and places; yet, for our better understanding, and the strengthening of our sicke Capacities, they are drawne and contracted into these certain numbers of Hierarchies & Orders.* (Markham, 1625)

It was not until the Enlightenment provided the scientific framework that this necessity to place things in hierarchies and orders came to be applied with precision and with universal coverage to the natural world:

> *To master nature by fixed patterns is not the task of philosophy, and whenever philosophers have tried it, the result was a utopian scheme and a failure. The task of the sciences, however, is just this kind of mastership, and in order to become masters they have to limit their search for truth, have to resign themselves to a symbolically reduced pattern. This reduction will, as every ritual, be partly arbitrary and contingent.* (Foss, 1949: 104)

It was just this arbitrariness that Herschel was vilified for, both in his choice of the term 'asteroid' and in his criteria for placing the planets, comets and asteroids into his symbolically reduced pattern. But in doing so he was far ahead of his time, as it is just this sort of methodology that has been used in science in modern times—the starkest example being the vast array of subatomic particles whose relationships to one another can only be understood through a symbolically reduced pattern. In astronomy it is widely used in such things as classifying types of galaxies, stars and nebula. Each one of the schemes has its own degree of arbitrariness and contingent factors, but each is essential for humans to make sense of the universe as revealed to us by our instruments and our senses (Dick, 2013).

In Herschel's own time, classifying plants based on affinities was championed by the Portuguese naturalist Correai da Serra (1751–1823) (Diogo, Carneiro, & Simoes, 2001). This concept in biology of using a resemblance in structure that suggests a common origin was mirrored in astronomy by the asteroids, which most researchers believed had a common origin. This concept of similitude was explored by Capel Lofft, who also invoked the name of Linnaeus, as detailed in the next section.

What Is a Planet?: A View from the Eighteenth Century

The noted writer Capel Lofft (Fig. 4.13) had a great deal to say about the planets. Certainly no one in England embodied the Romantic spirit more than Lofft. His "Eudosia: a poem on the Universe" encompassed much of what was known about astronomy when he laid down his pen on Sept. 29, 1780:

> *The PERIODS of the PLANETS to discern,*
> *Nice is the toil; but not inglorious.*
> *First, winged Mercury in haste revolves;*
> *Three months, with little space beyond, include*
> *His circuit: Venus, in an ampler orb,*
> *Measures eight months: Mars, free, expatiates*
> *Two lunar years: imperial Jupiter*
> *Nearly twelve years, such as our annual course*
> *Numbers, within his mighty orb includes:*
> *While Saturn, to the dreary marge exil'd,*
> *Requires thrice ten to fill his slow career.*
> *Nor Poetry these numbers will disdain,*
> *Since Harmony, her sister, these approves;*
> *In perfect scale most musically true:*

Fig. 4.13 Capel Lofft

So sweet a concert regulates the spheres.
Justly, O KEPLER! Are the ides of May
Rever'd, which taught to thee this wondrous truth.

Lofft here alludes to the fact that Kepler discovered the Third Law of Planetary Motion on May 13, 1618. Kepler "assumed that the discovery of scientific laws was a penetration into the harmony of nature." (Loemker, 1972: 194) The concept of the 'harmony of the spheres' was kept alive in the long centuries of the Dark Ages, reappearing in medieval times. In his first work *Se artibus liberalibus,* Robert Grosseteste, the Bishop of Lincoln (c. 1175–1253), gave a treatment of music. For him, the laws of harmony applied not only to the human voice and musical instruments but to celestial bodies, bodies made up of the four terrestrial elements, and the harmonic relation between body and soul in man (Freely, 2013; Grosseteste, 1514). It is in this context that we must view the work of Kepler as translated into poetry by Lofft (1798), who had more to say about the use of the word planet.

May I be allowed to remark on a use, which appears to be stealing into the French language of making Planete, Comete, and such words, feminine nouns, contrary to analogy and to etymology, considering them as immediately derived from the Greek; beside, though we are used to it in ships *of war, there is no great elegance in making the male deities of the Pagan mythology migrate into a female appellation. This ill suits, Mars, Jupiter, and Saturn. And with respect to the only planet in the system (except our moon) where it is proper that the feminine personification should be retained, it is easy to avoid the word Planete.*

When Boyer wrote, Planette was the orthography; and this almost compelled the word to be construed with a feminine adjective; still, as he very justly observed, astronomers employed it as a masculine substantive. And indeed, if they had not, there would have been a strange confusion, beside the other objections, in passing from astronomical papers in the French language to those of Halley and Newton in the Latin. At present, when the right spelling and pronunciation is restored, there is no more necessity to consider these nouns as feminine, and scarcely more propriety, than in making Athlete so. I remain yours sincerely, C. Lofft.

Lofft refers here to Jean-Baptiste de Boyer, Marquis d'Argens (1704–1771). He was a French philosopher and friend of both Voltaire and Euler. Lofft's first foray into the subject of asteroids was published in October 1802, just months after the term 'asteroid' had been proposed by Herschel:

I wish to discuss with the respect due to a name to which ASTRONOMY is much indebted, the propriety of introducing a new order of bodies into the vocabulary of that science, under the appellation of Asteroids, and the principle on which the two celestial Bodies discovered by Professor Piazzi and Dr. Olbers are proposed to be thus characterised.

If the distinction be not plainly necessary and founded on facts, I apprehend it most clearly ought not to be admitted.

Now, to try whether it be necessary or not, we have to enquire whether these two bodies have no proper place assignable to them in astronomical description under the forms already in use.

A Planet, I think, is understood to be a body revolving round a Sun as its centre in an ellipsis not very greatly deviating from a circle; and accordingly capable of being seen, when its orbit has been once ascertained, either by the eye or with a telescope, and usually in the whole of it.

A Comet, I apprehend, is a body revolving in an ellipsis, very greatly deviating from a circle, and if returning (which both analogy and observation appear to indicate) so liable to perturbation, and so long for the most part in its period, and having generally so short a part of its orbit within reach of observation, and passing that part of it with such velocity, and so obliquely, for the most part, to the paths of the ordinary planets, that it may easily revolve without being observed at all.

The other circumstance of its being accompanied with a diffused light or coma, though it originally gave the name, is not universal of comets. May it not be inferred therefore, that a body revolving round the sun, if with great eccentricity and obliquity, so as to cut the orbits of the other planets or some of them, may be accounted a comet; and if revolving with moderate eccentricity and obliquity, so as not to cut the orbit of any planet, a planet – A difference derived from its lying out of the limits of the Zodiac seems not sufficient: the limits having been assigned by the early Astronomers, with mere reference to the observation of eclipses of the Sun and Moon; whence the Zodiac is also termed the Ecliptic.

A difference of magnitude can hardly exclude a celestial body from the order of planets: much less can it place it as an Asteroid, if on account of its smallness it is disputed whether to call it a planet. The difference of the magnitude of the planets is great and various: but the greatest of them is comparatively as nothing to a fixed Star. Indeed a large planet between Mars and Jupiter was greatly improbable, as it would much have disturbed Mars. Names of similitude, Linnaeus has observed, are too vague to be well suited to science. Even if in this instance a new name were required, this would be no slight objection to the choice of the name, and beside, how slight the similitude? It consists merely in resemblance to a telescopic star. But some comets have had the same resemblance: some of the satellites of Jupiter and Saturn have this resemblance: and the Herschelian Planet itself. The points of dissimilitude between a small body shining by reflected light and revolving round a sun, and a fixed star or sun, are incomparably greater than the single point of resemblance; faint and imperfect as it is in that solitary particular itself.

The Piazzi seems to have every claim to the title of a planet.

The small revolving body discovered by Dr. Olbers seems answer better to the idea of a comet.

1. By the very great eccentricity and obliquity of its orbit.

2. By its intersecting the orbit of the Piazzi planet.

3. By its very small distance, if an ordinary planet of the system, from the Piazzi: a circumstance incompatible with the beautiful harmony of distances, suggested by Bode, and with which the Piazzi so well agrees. The maxim that names in science are not lightly to be multiplied, Nomina non sunt temere multiplicanda, *seems most forcibly to apply here.*

We should see very clearly and determinate differences, before we admit other Bodies in Astronomy than FIXT STARS, primary and secondary planets, and comets.

I can hardly dwell on the observation, that the Piazzi would not fill its place between Mars and Jupiter with sufficient dignity as a planet.

Much better our immortal MILTON: – that great Infers not excellence – P. L. B. Viii. And otherwise ill-fares it with the Herschelian planet, only about the tenth of the magnitude of Jupiter. (Lofft, 1802a: 199)

Lofft evokes both the names of Linnaeus and the English poet John Milton (1608–1674). Lofft refers to lines 85–91 in Book 8 of *Paradise Lost*:

Alreadie by the reasoning this I guess,
Who art to lead thy ofspring, and supposes
That bodies bright and greater should not serve
The less not bright, nor Heav'n such journies run,
Earth sitting still, when she alone receaves
The benefit: consider first, that Great
Or Bright inferrs not Excellence.

This is one of the many passages in *Paradise Lost* that refer to astronomy, but the true meaning of Milton's astronomical allusions has only recently been discovered (Cunningham, 2016). For Milton, see Fig. 7.2.

The next issue of *The Monthly Magazine* included a correction by Lofft: "In my paper, which you have obligingly inserted, on Asteroids, as a term lately introduced into Astronomy without, I apprehend, sufficient reason, there is an inadvertence of mine at the end of it. The Herschelian planet is, I believe, about one third of the diameter of Jupiter; and so I should have expressed it." (Lofft, 1802b: 375).

His second article, quite brief, was published in *The Monthly Magazine* in January 1803. Here, in an article titled "On Asteroids, as a Class of Astronomical Bodies," he clearly lays out what he believes to be an adequate definition to distinguish planets from comets. At the conclusion he employs a rhetorical flourish by arguing from contraries in an anaphoric doublet:

In addition to what I said before, I would remark that the Piazzi Planet is much larger compared with Mercury than Mercury is to Jupiter; indeed more than three times that proportion: and consequently the Star discovered by Dr. Olbers, whether Planet or Comet, is not below the scale of relative difference already ascertained in the Planetary System.

If in any celestial body, *shining by* reflected *light, and* revolving *round the Sun as a centre, the* least *axis is* shorter, *and the* greatest longer, *than that of a given planet, is not such body a Comet? And may not this be accepted as an astronomical* definition? (Lofft, 1803a: 479, his italics,).

Lofft's third article, titled "Whether it would be preferable, according to their Phenomena, to call Ceres, Pallas, and Juno, simply Planets, or Asteroids, or

Cometoids," was published in *Monthly* magazine*; or British Register* in July 1805 (Lofft, 1805; his italics):

> *On the planets Ceres, Pallas, and Juno, I formerly took the liberty to remark, that when the term* asteroids *was applied to the first of them by a great astronomer, it was liable to objection, as its* analogies *to the planets of our system, notwithstanding its smallness and eccentricity, were still such as to correspond far better with that denomination than with the denomination of* stars. *But the relative situation of them, now* three *have been discovered, to each other and to the sun, does make a difference. If the intersection, therefore, of the Ceres and Pallas, and the non-ascertainment of a solid* nucleus *to any of the three, and the thin* nebulose *[sic] light which has been observed to surround them, together with the circumstance of their being all of them small, and pretty nearly equidistant from the sun, should be thought sufficient to take them out of the denomination of* ordinary planets*; and if their eccentricity, so much less, I believe, than that of any known comet, though so large compared with the ordinary planets, together with their being visible during nearly the whole of their revolution, so far as can yet be judged, should be thought a reason against strictly classing them with comets, would not the term* cometoids *correspond best with the* phenomena, *as they resemble comets in many more particulars than they do any other celestial body, and differ from them in fewer and less material.*

Lofft had no academic credentials to critique Herschel, and never made any serious astronomical observations. His reports to the English periodicals were primarily meteorological observations and poetry (Cunningham & Oestmann, 2013). Motivating his writing on the subject of Ceres and Pallas was his friendship and correspondence with Johann Schroeter, who publicly contradicted both the observations of Ceres and Pallas by Herschel, and his choice of the word 'asteroid' to set them apart from comets and planets. There is no direct evidence Schroeter encouraged Lofft to write articles in the English press to support his case at the expense of Herschel, but one may reasonably believe there was an animating influence at work here.

> *When we meditate on the comparative diameters of Uranus, Saturn, Jupiter, and the Sun, we are astonished; but our curiosity is much more excited by the diminutive proportions of the Asteroids. They best suit the limited compass of our understanding. Man most admires the great; but he most loves the little.*

This observation by Charles Bucke (1823: 231) encapsulates the quintessence of how the asteroids were viewed in England in the early nineteenth century. As this book has shown, their study by just a few astronomers in England was the subject of tremendous curiosity and widespread interest by the reading public. Virtually everything written about them in a professional context was quickly disseminated via magazines, newspapers, popular books, textbooks, almanacs and encyclopedias so that everyone had access to what was known or merely conjectured.

There is no question that the studies done in England in the early nineteenth century were an essential element in understanding the four new 'planetary' objects. At the forefront of these contributions were the first truly scientific studies of Ceres and Pallas by William Herschel in 1802, and his decision to distinguish them from the seven planets already known. Whichever side of the argument one was on regarding the wisdom of denominating them asteroids, it effected a paradigm shift in our understanding of the Solar System as the whole idea of what a planet or comet is came under detailed scrutiny for the first time. With his discovery of Uranus, Herschel enlarged the physical bounds of the Solar System, and with his decision to regard Ceres and Pallas as a new class of object, he set astronomers on a new path about the nature of the Solar System. For all the theoretical and observational work done by astronomers in Continental Europe, none had the leap of intellectual insight into the workings of the cosmos to effect this change. Only Herschel, with his reputation and determined purpose of mind, possessed the gravitas to make this happen. Herschel's papers on Ceres and Pallas, and additional analysis of them, are included in the next book in this series.

Herschel's Vindication

We do not know with certainty what Herschel thought of the firestorm he had raised by his new terminology, but his likely reaction was resignation. He had become all too familiar with his views being either derided (lunar volcanoes), or misrepresented (a relationship between sunspots and the varying price of corn; Love, 2013). Herschel's attitude to his critics is best summed up by David Brewster (1823: 220), writing about the reaction to Herschel's discovery of infrared rays:

> *An individual whose speculations* [about] *the discovery of Invisible Solar Heat had cast into the shade, attacked Dr. Herschel with an asperity far beyond the limits even of severe criticism; but though that venerable man often spoke, with suppressed feelings, at the attempt which was thus made to discredit and depreciate his labours; yet he never condescended to repel the charge.*

Whatever his views may have been about the criticism, Herschel never backed down. Indeed, he felt more vindicated than ever with the discovery of Juno in 1804:

> *It will appear, that when I used the name asteroid to denote the condition of Ceres and Pallas, the definition I then gave of this term will equally express the nature of Juno, which, by its similar situation between Mars and Jupiter, as well as by the smallness of its disk, added to the considerable inclination and excentricity of its orbit, departs from the general condition of planets. The propriety therefore of using the same appellation for the lately discovered celestial body cannot be doubted. The specific difference existing between planets and asteroids appears now, by the addition of a third individual of the latter species, to be more completely established, and that circumstance, in my opinion, has added more to the ornament of our system than the discovery of a new planet could have done.* (Herschel, 1805: 64)

This was reinforced even further with the discovery of Vesta in 1807:

> *As cloudy weather has prevented an immediate continuation of my observations of Dr. Olbers' new star, and its increasing distance from us will soon put it out of the reach of telescopes that are directed to it for no other purpose than an examination of its physical condition, I have sent you the inclosed paper, which indeed appears to me quite sufficient to determine that the new star is a fourth asteroid.* (Herschel, 1807a)

The word 'asteroid' was in common use in England by 1830. One early adopter was William Phillips (1775–1828; 1817: 68), previously the author of two books on mineralogy. His 1817 book was intended as "… an introduction to the science for the use of young persons, and others not conversant with the mathematics."

"It is a remarkable fact," he wrote, "that some irregularities observed in the motions of the old planets, induced some astronomers to suppose that a planet existed between Jupiter and Mars; a supposition that arose previous to the discovery of the four new planets, or Asteroids." His suggestion that astronomers looked for a planet between Mars and Jupiter due to "irregularities" in their motions is curious, as the stated motivation by the Celestial Police and others such as Capel Lofft was the large gap between the planetary orbits.

The word 'asteroid' also was used in a popular magazine by a certain Thomas Cooke of Draycoth, near Derby (Cooke, 1830). The use was by no means universal, even in England. Miss Christian Cann (1828: 79) is quite clear that "Under the denomination of planets, are now comprised, Mercury, Venus, the Earth, Mars, Jupiter, Saturn, the Georgium Sidus, or Herschell, Ceres, Pallas, Juno, and Vesta; the four last named planets are recent discoveries."

There was still reluctance in the United States to regard the four objects as worthy of a special category. An anonymous member of the United States Naval Lyceum, in an essay on astronomy, included the four objects in a list of the *primary* planets (Anon, 1837: 119). But Herschel finally achieved professional vindication in the United States when Benjamin Apthorp Gould (1824–1896; 1848: 28), founder of the *Astronomical Journal*, gave his stamp of approval to the term, offering at the same time a straightforward definition:

> *By the common consent of astronomers, they have received the name of "asteroids," a name proposed by the elder Herschel, in consequence of a theory of his own. The word asteroid, in its present signification, may be defined as "a small planetary body, which revolves around the sun between the orbits of Mars and of Jupiter."*

By the end of the century, the distinction between asteroids and planets was so accepted in English society that it could even be used to measure literary merit:

> *A group of dramatists and lyrical writers, among whom Beddoes is by far the greatest, link the generation of Keats and Shelley with that of Tennyson and the Brownings; but most of them are nebulous, and the most eminent mere asteroids in comparison with the planets which preceded and followed them.* (Gosse, 1897: 332)

The Gentleman Astronomer: Alexander Aubert

By the late eighteenth century the divide between amateur and professional astronomers was becoming more clearly defined, although gray areas still existed. If one defines a professional astronomer as one employed at a state-run observatory, the list would include the following: Johann Bode (Berlin), Franz von Zach (Gotha), Nevil Maskelyne (Greenwich), Giuseppe Piazzi (Palermo), Barnaba Oriani (Milan), Joseph-Jerome Lalande (Paris), Marcin Poczobut (Vilnius) and Jan Sniadecki (Cracow). Notable amateurs, namely those who operated their own observatories without state support, included Johann Schroeter, Friedrich von Hahn, Gottlieb Schrader, Heinrich Olbers and Ferdinand von Ende (all in Germany), and the Duke of Marlborough, Hans von Bruhl, Sir George Shuckburgh, William Larkins, Charles Greville and Alexander Aubert (in England). William Herschel occupied a unique position in that his funds came partially from the state and partially from his private sale of telescope mirrors. He can be ranked as a professional, even though his

Fig. 4.14 Alexander Aubert

observatory was mixed pro/private enterprise (Cunningham & Orchiston, 2015). Hutton (1815: 129) lists 20 private observatories in England. Of the ones just listed as active in the late 1700s, only the observatory of Marlborough survived to be included in the 1815 list.

Aubert (1730–1805); Fig. 4.14 was one of the few men in England who saw Ceres, and, as a measure of his prominence, he was also savaged in the same satiric verse by Peter Pindar (1816: 105) that featured Sir Joseph Banks as the man trying to defend the honor of Herschel and others. Pindar lists Aubert as one of those flatterers "who say soft things" to Banks, president of the Royal Society. Pindar sets the scene with Banks at breakfast at his home in Soho Square, surrounded by his "royal sycophants." Banks was given some very bad news at breakfast, upon which he roared "There goes, then, my hypothesis to hell!"

Dead-struck sat Aubert, Blagdon, Planta, Woide,
Whose jaw-bones in the mumbling trade employ'd,
Half open'd, gap'd, in sudden stupor lost;
Whilst from the mouth of ev'ry gaping man,
In mazy rill the cream-clad coffee ran,
Supporting dainty bits of butter'd toast.

At the other end of the spectrum, a laudatory verse was penned in 1794 by a certain Dr. Kelly, who had an observatory at Finsbuy Square, London. *Lines on Practical Astronomy, addressed to Alexander Aubert, in reference to his splendid Observatory at Highbury:*

Thy *dome*, Aubert, with reverence I view,
And hail its noble use and learned store;
Such as Egyptian temples never knew,
Nor Greece nor Rome, with all their boasted lore.

Here truths sublime, and sacred science charm,–
Creative arts new faculties supply,–
Mechanic powers give more than giant's arm,
And piercing optics more than eagle's eye.
Eyes that explore creation's wondrous laws,
And teach us to adore the great designing Cause.
Borne on these wings, we mount ethereal space,–
The wide expanse of heaven minutely scan;
God's wisdom, power, and handiwork we trace,–
The noblest study of aspiring man.
New systems open to us as we climb;
Each glittering star gives law to circling spheres,
Which run eternal rounds in faithful time,
Nor err one moment in ten thousand years.
Perpetual motion heaven's high works maintain,
So often sought on earth, but ever sought in vain.

It was at the observatory lauded by Kelly that Aubert made positional measurements of Ceres in early 1802, which he communicated to Banks (Fig. 4.15). This verse is also quoted here to emphasize the point made earlier that "the clockwork

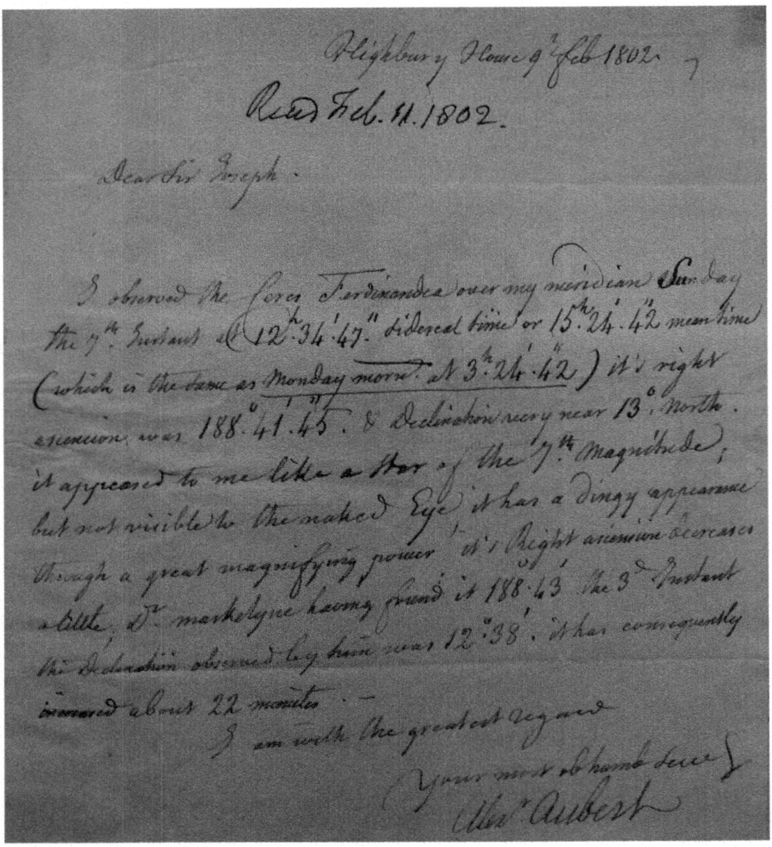

Highbury House 9th Feb 1802
Recd Feb. 11. 1802.

Dear Sir Joseph.

I observed the Ceres Ferdinandea over my meridian Sunday the 7th Instant at 12h 34' 47" sidereal time or 15h 24' 42" mean time (which is the same as Monday morng. at 3h 24' 42") it's right ascension was 188° 41' 45". & Declination very near 13° North. it appeared to me like a Star of the 7th magnitude; but not visible to the naked Eye, it has a dingy appearance through a great magnifying power its Right ascension decreases a little, Dr. Maskelyne having found it 188° 43' the 3d Instant the Declination observed by him was 12° 38'. it has consequently increased about 22 minutes. –

I am with the greatest regard
Your most obt humble Servt
Alexr Aubert

Fig. 4.15 Letter of Aubert to Banks, dealing with Ceres Ferdinandea, February 9, 1802

Fig. 4.16 Aubert's Highbury Observatory

regularity of the Solar System was a bedrock in the minds of most people." The perpetual motion Kelly mentions, where the clockwork of the heavens does not err in the slightest even after the passage of 10,000 years, is what people of the era regarded as a given. The sudden introduction of new objects and the new designation 'asteroid' was seen by some as a threat to this perpetual order.

As a man of means (he was governor of the London Assurance Company), Aubert actually built two observatories. The first was designed by John Smeaton (1724–1792) and erected at Loampit-hill near Deptford, and furnished with instruments by the foremost craftsmen in London: James Short, John Bird, Peter Dollond and Jesse Ramsden. At his home beside the observatory Aubert (1786) entertained in grand style. At one dinner party he hosted Smeaton, Herschel, the chemist Henry Cavendish (1731–1810), the natural philosopher John Michell (1724–1793), secretary of the Royal Society Charles Blagden (1748–1820) and Lord Palmerston (1739–1802). His Highgate observatory near London, built in 1788, was the finest private observatory in Great Britain (Fig. 4.16):

> *He erected near to the house a lofty and spacious observatory, which he furnished with a very complete collection of astronomical instruments, particularly a very fine reflecting telescope by Short* [24 inches focus and 6 inches aperture], *being the largest ever made by that artist, and which was purchased out of Topham Beauclerk's collection* [Beauclerk, 1739–1780, was a celebrated wit and friend of Dr. Johnson]. *The largest and most important instruments, in order to prevent the effect of vibration, were insulated from the floors by being placed upon a pier of solid stone carried up from the earth and rising through the center of the building.* (Lewis, 1842: 185)

Aubert published little (Lynn, 1900) and he "seemed to have had no personal ambition in astronomy but only a passion for it and a standard of excellence." (Jungnickel & McCormach, 1996: 300). His reputation was such that even Herschel appealed to Aubert to confirm his own observations so that they would be taken seriously (Herschel, 1782c).

Chapter 5
The Discovery of Pallas

C. Cunningham, *Early Investigations of Ceres and the Discovery of Pallas*,
DOI 10.1007/978-3-319-28815-4_5

Fig. 5.1 The 1788 painting of Wilhelm Olbers, discoverer of Pallas, by Jacob Fehrmann © Focke-Museum, Bremer Landesmuseum fuer Kunst und Kulturgeschichte

Mysteries Still Lying in the Dark

Spring of 1802 saw one of the most important astronomical discoveries ever made. Only a week had gone by since the official end of winter, but to Wilhelm Olbers (Fig. 5.1) in his northern German city of Bremen the promise of spring was still just that—a promise. Another cold evening at the telescope beckoned as he saw the Sun descending. Like so many nights before it, Olbers was intent on one thing on March 28, 1802.

He wanted to find something only two other astronomers had ever found—another planet. He longed to join the ranks of the illustrious William Herschel, discoverer of Uranus, and Giuseppe Piazzi, who just a year before had found Ceres. On this night all his hopes and dreams would be fulfilled. On this night he found Pallas. Two days later he wrote excitedly to Johann Bode: "I hasten to communicate an important astronomical discovery to you. Since March 28, I've been observing "another moving star" in the northern wing of Virgo besides Ceres, which is similar to Ceres in all qualities, cannot be differentiated from a magnitude 7 fixed star with my Dollond, is retrograde like it, but only more strongly increases in northern declination." Olbers and Bode thought it might just be a comet.

Once it became obvious that Pallas had a highly inclined orbit, Schroeter (1802: 139) described the importance of the discovery:

> *If this very peculiar planet would have been created by procreation like the Sicilian Ceres, I would consider it a bastard from a mesalliance, whose father was a planet and whose mother was a comet. Certain it is, I think, that this discovery was made at exactly the right time, because we have already learned a lot about the atmospheres of celestial bodies and that it will probably help us solve other mysteries still lying in the dark.*

Unlike the polemics that surrounded the naming of Ceres, little controversy ensued over the naming of the second asteroid. Olbers chose to call it Pallas. As usual, Laplace rejected classical names and insisted it be called Olbers. His colleague Lalande (1804: 217) wrote acerbically "I see with regret, that the Germans have no more respect for Dr. Olbers than they have for Dr. Herschel. The name of Pallas has no foundation: jealousy, perhaps, is the cause of this injustice." Immediately after its discovery, Schroeter called it "the Olbersian star" so in Germany it was called Pallas Olbersiana. As a snub to Piazzi, the surname of his discovery was deleted in the same sentence that we find Pallas Olbersiana (Stilling, 1805: 54). According to the Director of Berlin Observatory Johann Franz Encke (1791–1865; 1846), "Pallas was at first called Olbersiana, until Olbers himself pronounced strongly against the adjective." Both names were first proposed for the satellites of Jupiter— Jean Dominique Cassini proposed Pallas, Juno, Themis, and Ceres be given to the four satellites discovered by Galileo.

Comets at the Turn of the Nineteenth Century

Schroeter, Zach and Bode devoted much thought to the relationship between comets and the asteroids Ceres and Pallas. David Brewster (1802b) had this to say: "...since the star Pallas resembles a comet in its *motion*, in its *smallness*, in its *orbit*, and in the *inclination* of that orbit, we are authorized to rank it among the number of these heavenly bodies." A prominent educator and promoter of science in England at the turn of the century was Margaret Bryan (1760–1851), a personal acquaintance of William Herschel (Saridakis, 2014: 251). This excerpt from a book she wrote gives a good idea of what the nature of comets was believed to be at this time:

> *These extraordinary bodies are found, by their reflective power, to be opake* [sic]. *The matter of heat and light darts from them like fiery tails; – as when an insulated jar is receiving a full charge from the electrical machine, it throws off its redundancy, so do the comets emit a stream of fire from their bodies on the side opposed to the Sun, from which they receive their superabundant fire; –therefore, if I may be allowed to reason from analogy, I should suppose that by this mode do the comets throw off the redundant heat which they must receive from the Sun. I venture to intimate the possibility of this being the cause of the effect we perceive of these motions called fiery tails; – that the additional heat thus issuing from these bodies prevents an accumulation unfavourable to animal existence, supposing these bodies to be inhabited, as we have great reason to do.* (Bryan, 1799: 126)

The Cosmology of Kant and Laplace

In the author's previous book, *Discovery of the First Asteroid, Ceres,* we have encountered Christian Wolff in connection with Bode's Law. But the most important aspect of Wolff's work is that his ideas were passed on to his brilliant young disciple, Immanuel Kant (1724–1804); Fig. 5.2. Although Kant began in the Wolffian school, he later broke decisively with it. Kant (1755: 163) emphasized the great distance between Mars and Jupiter:

> *The distance between the orbit of Jupiter and that of Mars is so great that the area contained in it exceeds that of all the lower planets combined; but this is worthy of the largest of all the planets, the one that has the greater mass than all the others combined. We cannot attribute this distance of Jupiter and Mars to the intention that their attractions should hinder each other as little as possible.*

In part three of his book, Kant (1755: 178–179) mentions the gap between Mars and Jupiter in a discussion of what is now termed the extraterrestrial life debate:

> *If the composition of a celestial body establishes natural barriers against its becoming inhabited, then it will not have inhabitants, even though in and of itself the planet would be more beautiful if it had its own creatures. The excellence of creation loses nothing in such a case, for among all large quantities the infinite is the one which is not diminished by the subtraction of a finite part. It would be as if one wished to complain that the space between Jupiter and Mars was unnecessarily empty and that there are comets which are not populated.*

Fig. 5.2 Immanuel Kant

This passage is apparently the source of the idea that Kant had suggested there might be another planet between them, but nowhere does he explicitly say this. It did not prevent an early biographer from making a grand claim however:

> *At the time of the French Revolution, he threw out many conjectures, and what then passed for paradoxical anticipations, especially in regard to military operations, which were as punctually fulfilled as his own memorable conjecture in regard to the hiatus in the planetary system between Mars and Jupiter, the entire confirmation of which he lived to witness on the discovery of Ceres by Piazzi, and of Pallas by Dr Olbers. These two discoveries, by the way, impressed him much; and they furnished a topic on which he always talked with pleasure; though, according to his usual modesty, he never said a word of his own sagacity in having upon a priori grounds shown the probability of such discoveries many years before.* (Wasiansky, 1804; quoted in DeQuincey, 1873: 112)

Kant himself paid homage to his Greek predecessors for inspiration. "I will therefore not deny that the theory of Lucretius, or his predecessors Epicurus, Leucippus and Democritus, has much resemblance with mine. I assume, like these philosophers, that the first state of nature consisted in a universal diffusion of the primitive matter of all the bodies in space." (Kant, 1755: preface). For Lucretius, it was not just the *origins*, but the *orbits* of the Sun and Moon that he attempted to explain. Kant, in his preface, further aligned himself with those who sought the 'music of the spheres,' when he wrote that "It is usual to signalise and emphasize in nature the harmonies, the beauty, the end of things, and the perfect relation of means adapted to them."

Intrigued by the immense difference in size between Mars and Jupiter, Kant (1755: 55) wrote that "It is probably the neighbourhood of the very large planet Jupiter which, by attraction on its side, has robbed Mars of the particles needed for its formation." Thus, while he noted the gap between Mars and Jupiter, he did not posit the existence of a 'missing' planet there.

Laplace, as explained in this quote from Simon Newcomb (1887: 508), was led to the Nebular hypothesis by considerations very similar to those presented by Kant a few years earlier:

> *How would these rings of vapour behave? As they cooled off their denser materials would condense first, and thus the ring would be composed of a mixed mass, partly solid and partly vapours, the quantity of solid matter constantly increasing, and that of vapour diminishing. If the ring were perfectly uniform, this condensing process would take place equally all around it, and the ring would thus be broken up into a group of small planets, like that which we see between Mars and Jupiter.*

The Kant-Laplace hypothesis forms the basis of all modern work on planet formation (Schaffer, 1989), but, as A. J. von Oettingen cautions, the differences between them is sufficiently great to keep the views distinct and separate: "Kant makes suns and planets arise out of certain regions of space, through gravitation; Laplace makes masses and rings detach themselves from the central body, through centrifugal force." (Quoted in Hastie, 1900: lxxx)

A. Meydenbauer, a strong advocate of Kant, points out several contradictions of Laplace's theory with the facts. One of these states that "with the exception of the asteroids, and in their case very conditionally, there is no example known in which a plurality of satellites move at nearly the same distance from the central body." (Quoted in Hastie, 1900: lxxxiv)

Laplace, in the third edition of his book *Exposition* (1808: 389), added the motions of the first four asteroids to those of the planets. He estimated the odds at 4 billion to one that such a planetary arrangement "is not the effect of chance". But others were not convinced. In a letter to Friedrich Bessel of 10 July 1812, Olbers points out Laplace's reluctance to consider in their true weight the large deviations of Ceres and Vesta from the ecliptic. Stanley Jaki (1924–2009; 1978: 129) concluded that "Laplace, the cosmogonist, was handling the facts by making the most of the apparently favourable ones and glossing over the unfavourable findings." In the event, he made no effort to derive Bode's Law from his contracting solar atmosphere.

Even though it was regarded with suspicion even in the early nineteenth century, fascination with Bode's law has scarcely abated. Indeed, theoreticians are still battling with the siren song of the orbital period relation it embodies. Patterson (1987) suggests that the present nearly resonant orbital periods of the planets can be explained in terms of past two-body resonance capture of planetesimals in the solar nebula. "It might now be possible to reconstruct the evolutionary history of the planets from their nearly commensurable orbital periods and, hence, provide an explanation for the Titius-Bode law. However, one must view the success of the Titius-Bode law in 'predicting' the positions of Uranus and Ceres as fortuitous." Hayes and Tremaine (1998) used it to study hypothetical planetary systems, but the discovery in the early twenty-first century of real exoplanetary systems has opened yet another chapter in the Titius-Bode law (Bovaird, 2015).

Olbers' Hypothesis

The visceral excitement engendered by the discovery of Ceres, and shortly thereafter Pallas, is evident in the writings of Olbers, Gauss and Zach. It seems they could almost taste a revolution of thought on the entire question of the origin and development of the Solar System.

This is how Zach, in obvious elation, described the discovery of Pallas, and the apparent vindication of Olbers' hypothesis (Zach, 1802e: 598; my emphasis):

> *...which grounds for speculation as to the origins and history of our planetary system will this remarkable planet not give? What a completely unexpectedly great inclination for a planet? And how striking the position of its path compared to that of Ceres? This brought Dr. Olbers to the thought that both planets might be but rubble of a single one destroyed by the collision of a comet. Who would have expected something like this in our planetary system, and to which new, important and great information shall and will this planet not lead us, which numbers* ***among the most important astronomical discoveries ever made*** *in more than one respect.*

Olbers, in a letter to Gauss dated May 15, 1802, writes "The determination of the mutual perturbations of Pallas and Ceres will give rise to completely new and intriguing investigations. Indeed, I still can't wholly abandon the idea that Ceres and Pallas are maybe just fragments of a former planet." Gauss waxed philosophical in his response of May 18, which is printed in full in Chap. 7:

> *Imagine the shock, the spiritual struggle, the incredulity, the defence of and opposition to providence we would see develop if the possibility that a planet can be shattered be verified as fact! What will those, who base their framework of knowledge so readily on the unshakeable stability of the planetary system, say if they see that they have built on sand, and that everything is entrusted to the blind and fortuitous play of the forces of nature!*

Olbers' theory was also at odds with one propounded by Huth. He believed that "I think it very probable that these little planets are as old as the others and that the planetary mass in the space between Mars and Jupiter has coagulated in many little spheres, almost all of the same dimensions, at the same time in which happened the separation of the celestial fluid and the coagulation of the other planets." (Bode, 1804) A century later, opinion was still divided.

> *It is generally considered by astronomers that the numerous minor planets between the orbits of Mars and Jupiter are the fragments of a large planet which had formerly revolved*

> *in an orbit about the same distance from the Sun as Ceres, and had subsequently been shattered by some internal convulsion.* (Wilde, 1910: 7)
>
> *Olbers' explosion theory, long accepted by astronomers, has been proven open to fatal objections. The minor planets are now believed to represent a ring of cosmical matter, cast off from the solar nebula like the rings that went to form the major planets, but prevented from becoming aggregated into a single body by the perturbing mass of Jupiter.* (Williams, 1904: 41)

Olbers' idea of a fifth planet that exploded long ago was imaginatively employed in Raymond Z. Gallun's 1950 science fiction story *A Step Further Out*. He envisioned that the original planet was home to an advanced civilization, whose artifacts could still be found by prospectors.

Our Pallas Is a True Planet

In 1754 Horace Walpole coined the word 'serendipity' in a January 28 letter to Horace Mann. It came from the title of the folktale *The Three Princes of Serendip*, in which the heroes "were always making discoveries, by accidents and sagacity, of things they were not in quest of." Serendipity was aptly used to describe the discovery of Ceres, but not so with the discovery of Pallas. Its capture was the result of careful and deliberate detection. The relationship between serendipity and the scientific method is neatly explored by Glashow (2002).

The discovery sparked a fierce debate: What was Pallas? Just 2 days after its discovery, Olbers (1802d) wrote to Johann Bode, editor of the *Berlin Astronomical Yearbook:* "What shall I think of this new star? Is it a strange comet or a new planet? I do not dare to judge it yet. It is certain that it does not resemble a comet in the telescope; no trace of nebulosity or atmosphere around it can be seen."

In nearby Lilienthal, Johann Schroeter began observing Pallas just 2 days after Olbers had found it. He concluded, by the end of April, that "undoubtedly Pallas does not describe a circular but parabolic orbit. In the strict sense it cannot be called a planet." But Gauss disagreed, and told its discoverer he "would certainly have no scruples to call Pallas a planet." This verdict was greeted with unalloyed delight by Olbers: "Your letter of April 20 (1802) brought me much enjoyment. You can well imagine how much I'd like to reserve the dignified appellation 'planet' for my discovery and your letter gives me the highest hope to do so." But Bode was having none of it. At stake was nothing less than his cherished arithmetical progression of planetary distances—the famed relation known as Bode's law.

The discovery of Ceres in 1801 was hailed as the long-sought 'missing' planet between Mars and Jupiter, where a large gap led many to believe a planet should be. Its discovery fit nicely into the progression Bode had been writing about for many years. Was there room for two missing planets? Ludwig Wilhelm Gilbert (1769–1824; 1802); Fig. 5.3, editor of the journal *Annalen der Physik*, certainly thought so. Gilbert, who attended the world's first astronomical congress hosted by Zach (see *Discovery of the First Asteroid, Ceres*) plugged the new objects Ceres and Pallas

Fig. 5.3 Ludwig Wilhelm Gilbert

into a table to show that the logical progression of the relationship indicated the existence a planet named Ophion beyond Uranus. But for Bode (1802b), there was no room for two missing planets; he thus dismissed the discovery of Pallas: "I consider it a very distant comet, maybe to be found beyond Ceres' orbit."

Laplace weighed in on April 30, 1802, in a letter to Barnaba Oriani in Milan: "One always follows here the star discovered by Olbers. I believe it to be a comet, but the continuation of the observations will teach us what we must think."

On May 7, Olbers told Bode in no uncertain terms that "Pallas has just as great a right to the honour of being a planet as Ceres." Bode (1802a) remained resolute, as he told William Herschel on August 5: "I hold myself still convinced that Ceres is the eighth primary planet of our solar system and that Pallas is a special exceptional planet—or comet—in her neighbourhood."

Olbers was totally exasperated with Bode. He graphically expressed his feelings to Gauss on May 23: "If Bode is still in doubt over the planet-like orbit of Pallas, if he still imagines every curve will fit the thus far traced arc, then you and I will have to attribute this solely to his lack of expertise with this type of calculation." Gauss wrote to Olbers in despair 2 days later: "He (Bode) absolutely doesn't want to hear anything about a planet. He believes it to be contrary to all celestial laws, and he imagines a comet at perihelion far beyond Ceres, and that I should just attempt to assume a greater distance and in this way fit the observations to a parabola."

In July, a scathing indictment of Bode's Law was published in Paris by Lalande (1802e):

> *A conjecture as easy to make as useless to the progress of astronomy, had raised a presumption that a planet existed in the wide space between Mars and Jupiter; but the law which had been imagined to prevail being founded upon the relative distances of the planets hitherto known, no sooner seemed to be verified by the discovery of Mr. Piazzi's planet, than it was contradicted in the most formal manner, by the discovery of a second new planet, very near to the first. This is an instance of the overturning of opinions, grounded merely upon deceitful analogies, and on the false ideas we entertain of what ought to be denominated regularity and order in the designs of nature.*

Bode eventually bowed to the inescapable conclusion that more than one object filled the space between Mars and Jupiter. He would not, however, relinquish his hold on the law that bears his name:

> *If this celestial body will show itself after its return from the Sun next year at exactly that position where it can be expected according to its calculated elliptical motion – we have to recognise it as an extremely extraordinary planet which revolves around the Sun within and close to Ceres' orbit with an exceedingly great inclination on a considerable eccentric orbit – unheard of among the eight major planets until now – and on an ellipse extremely different from that of Ceres but within the same time, if one is not inclined to declare it a planet-like comet, however curious these circumstances might appear. It would then be settled that there, where I expected only one planet between Mars and Jupiter, two of them race around the Sun on equally sized orbits needing the same amount of time, and the beautiful progression of the distances of the planetary orbits would remain intact.* (Bode, 1805)

In the end, the authority of Gauss prevailed. "This numerical relationship holds true, contrary to the nature of all truths deserving the name law, only approximately."

How the Public Learned About the Discovery of Pallas

The Edinburgh Journal; or Literary Miscellany published articles about Pallas, usually written by David Brewster:

> *The readers of the* Edinburgh Magazine *will be surprised to hear that another new planet has been discovered at Bremen by Dr Olbers, the same astronomer who rediscovered the planet Piazzi, after it had been lost sight of by the Italian philosopher. From the little information which we have been able to obtain concerning this interesting subject, it appears that the right ascension of this new planet upon the 28th of March, was 184° ..56′ ..49″, with 11° ..33′.. 30″ of north declination; and on March 29th, 184°..46′..36[″], with 11°..53′..0″ of declination. It is situated, therefore, within a few degrees of the planet Piazzi, and may be seen, in all probability, with an instrument whose magnifying power is sufficient for rendering Piazzi visible.*
>
> *From a similarity between its variation in right ascension, and that of the planet Piazzi, in a given time, Dr Olbers conjectured that the distances of these bodies from the sun were nearly the same. This opinion, however, is evidently untenable, for several reasons, which will occur upon the slightest reflection.*
>
> *Some eminent astronomers are firmly persuaded that this body is a planet, while others maintain, with perhaps equal plausibility, that it is a comet, on account of the similarity of distance between it and Piazz[i]. We forbear, however, to indulge in conjecture, upon a question which depends solely on future observation, and which, from the small number which have already been made, it is impossible to decide.* (Brewster, 1802a: 287)

Only 3 days after Brewster penned the article just quoted, William Walker in London wrote an article about the discovery for *The Gentleman's Magazine*:

> *I do myself the pleasure of communicating to you the discovery of another new planet, by Dr. Olbers, at Bremen, on the 28th of March last. It is situated extremely near to the place which the Ceres is noted to have been in, on the little configuration of stars printed in your Magazine for that month. It is invisible to my naked eye, but evident through a night-glass; and, with a magnifying power of 100 times, on a good telescope, appears of a sensible magnitude, but of a feeble, pale, red light. I think it less bright than the Ceres, although the last admits no disc, with any magnifying power I can use. It is at present about as far again from the Sun as we are. The Ceres is nearly three times as far from the sun as ourselves, and Mars about one and a half. The Ceres is very near the star Beta, in the Lion's Tail.* (Walker, 1802c)

On the same day (April 26), Walker penned another short article that was sent to *The Scots Magazine*. It contained spurious details about Pallas that did not appear in his *Gentleman's Magazine* piece:

> *The planet discovered by Mr. Olbers, at Bremen, on the 28th Mar. is now, in a very small degree, higher than the place of Ceres, on the 25th of March – and will be found near this place for some evenings to come. It is not visible to the naked eye, and through a telescope appears more faint than Ceres, and of a pale colour. It seems probable, that it is about as far again from the sun as the earth – whilst Ceres is near three times as far off – but I acknowledge that I have much hesitation in believing it a planet. The Ceres has advanced near to Beta Leonis, and each of these objects, by a night glass, may easily be discovered* (Walker, 1802b)

This same letter by Walker was published in the *Monthly Magazine* (Walker, 1802c). Where he got the notion Pallas was much closer to the Sun than Ceres is unknown.

In Germany, the public read about the discovery in the Berlin newspaper *Vossische Zeitung* (April 10, 1802): "Dr. Olbers in Bremen discovered on March 28, in the northern wing of Virgo, right ascension 185° 11.5° declination north, a comet, appearing as an 9th magnitude star without any perceptible nebulosity."

In France, Burckhardt (1802b) wrote about it in the April issue of *The Moniteur* :

> *The star, discovered on March 28 by Olbers resembles a planet to such an extent that it appears only natural to assign it an eccentric orbit. Consequently, I placed it between Mars and Earth, between Mars and Piazzi's star, between the latter and Jupiter. Thus I am far away from wanting to exclude elliptical orbits and I continue to work on those as we have a great number of observations; all the while I wanted to satisfy the curiosity which the possibility of a new planet seems to cause in the general public.*

Back in England, the May issue of *The Monthly Magazine* ran a one-paragraph article about the discovery by a correspondent signed 'Astrophilus'. Based on archival research, this can be attributed to Maskelyne. A report in *The Edinburgh Magazine* (Brewster, 1802b) shows not only the level of detail being imparted to the public but the possible association of Pallas with comets that was being considered at the time. In this important article (printed in Chap. 10 of this volume), David Brewster sets himself up as the prime opponent of Herschel in England. Brewster here uses the same criteria Herschel uses to distinguish planets from comets, but Brewster comes to the opposite conclusion—that Pallas is a comet. He appeals here to the wholly spurious concept of 'uniformity' to bolster his viewpoint, and he never uses the word 'asteroid,' even though he read Herschel's paper in which it was first introduced.

The importance of his writing, however, should not be underestimated. As a respected man of science, his numerous and very public expositions were seriously considered. As Herschel himself never wrote for the popular journals or newspapers, he never rebutted these disagreements, and many people must have been left with the impression that Herschel was wrong. *The Monthly Magazine* in August kept its readers informed about the study of Pallas in great detail:

> *Baron de Zach, in a letter to Sir Joseph Banks, observes, 'that Pallas is a planetary heavenly body, that moves between the orbits of Mars and Jupiter, with a very great eccentricity and inclination, and whose orbit comes very near to the orbit of the planet Ceres, perhaps touches it, perhaps even cuts it like two links in a chain, this way* *, which can not yet be asserted with certainty, the observed area run over by this planet being too small.' Another very remarkable circumstance is that the mean motions of Pallas and Ceres are very nearly, perhaps absolutely, the same: in this case, small as the masses of Ceres and Pallas may be, they will, nevertheless, exert a very sensible action upon the other, and thereby give occasion to very curious and interesting investigations in the mechanics of the heavens.* (*The Monthly Magazine*, 1802b)

The following month, *The Monthly Magazine* (1802c) devoted nearly six pages to the new discovery. The article began with a mathematical equation, which indicated the level of sophistication expected from its readers. It then went on to describe the elements derived by Gauss based on observations at Gotha's Seeberg Observatory. This section, like the following where Olbers' hypothesis of a planetary explosion to explain the origin of the asteroids is first broached, was in fact a virtual copy of Zach's paper in the June issue of his journal, *The Monthly Correspondence. The Monthly Magazine* nowhere states that it 'lifted' the article nearly verbatim from this source. *The Monthly Magazine* then switches to an account of Karl Seyffer's investigations. Seyffer (1762–1822) was a Professor of Astronomy at Göttingen University, where Carl Gauss also was studying Ceres and Pallas. This account reads as if it may have been sent to the magazine—it certainly was not in *The Monthly Correspondence*. By calling Seyffer 'intelligent' and giving him credit for his role in the recovery of Ceres and Pallas, it appears the editor of *The Monthly Magazine* was embroidering the account as a thank-you to Seyffer for communicating the information:

> *The intelligent Professor Seyffer, of Goettingen, who is one of the most zealous observers of the new planet, and may justly claim the merit of being one of the first rediscoverers of Ceres as well as of Pallas, proceeds to communicate his observations made on Pallas at the Royal Observatory of Goettingen, by which the ellipsis of Dr. Gauss is perfectly confirmed, and neither a parabola, nor a new larger ellipsis, to be admitted, as has of late been pretended to be found by French astronomers. Dr. Gauss, who has received part of his education at the celebrated University of Goettingen, and whom Professor Seyffer remembers, with the greatest satisfaction, as one of his friends and pupils, has made a third attempt of finding out the most possibly accurate ellipsis for the orbit of Pallas, the result of which he has communicated in a letter to Professor Seyffer. On comparing his observations with these new elements of Dr. Gauss, M. Seyffer found them perfectly agreeing, except in some trifling differences, and they even correspond with the newest observations made on the 19th, 20th and 21st of June, so that there seems to be no occasion for making, for the present, any further corrections in them, as Professor Seyffer thinks they will be quite sufficient for re-discovering Pallas in the year 1803, provided the planet has light enough as to be seen, as it*

might be possible that Pallas, on account of its great distance from the earth, is not visible during the years 1803 and 1804, or, at least, is only to be seen by means of the most exquisite instruments, and that it is likely to appear again in the year 1805.

The Monthly Magazine article then continues with the substance of a letter written by Herschel to Seyffer of May 22, 1802, in which Herschel recounts his paper about Ceres and Pallas read before the Royal Society "... on the 7th and 13th of May." It is here that readers of the magazine first see the word 'asteroid.' The magazine then goes into a paean about Herschel: "Observations on the nature of the new planet from the masterly pen of the great Herschel, are entitled to the most distinguished attention of astronomers; and it is to this accurate and great observer, that we shall most probably be indebted for new and interesting discoveries relative to the nature of those stars." The main body of the article concludes with notice of the diameter differences of Herschel and Schroeter. But far from being finished, the magazine then publishes four pages of orbital elements, an ephemeris from May 24 to June 29, and observations by Olbers, Seyffer, Zach and Oriani. This article, the most extensive of any about Pallas in the publications of the day, promises more in the future: "Thus far the observations of the new celestial body, Pallas, are published, but we *shall* not omit communicating, in future numbers of this Journal, many new observations and discoveries relative to the nature of so remarkable a body as this appears to be 'among the radiant orbs, that more than deck, that animate the sky, the life-infusing suns of other worlds.'" The article ends with a poetic quote from "The Seasons," a work from 1730 by James Thomson (1700–1748).

The Critical Review, although it did not publish any detailed information about Ceres and Pallas, did give notice about the publication of Johann Bode's annual publication *Astronomisches Jahrbuch*. We read that the *Yearbook* contains 34 memoirs on astronomical subjects, many of which relate to the "... two new planets." *The Critical Review* (1803b) considers this so important that it goes on to describe some of these, while completely ignoring the others aside from listing the names of some authors: "The first memoir, by the editor [Bode], is designed to show that the movable star (Ceres) is really a planet, long supposed to exist between Mars and Jupiter. The second relates to the re-appearance of Ceres, observed by Olbers; the third, the account of the discovery of another movable star (Pallas)."

Tilloch's Journal made a point of publishing fresh material from France. In his annual report on astronomy, Lalande (1802b) began with the discovery of Pallas.

Continuing its report of asteroid observations from France, *The Philosophical Magazine* (1803d) informed readers of a unique occurrence, the appulse of an asteroid with a star. This account, which will be in the next book in this series, did not appear in any other English publication. Also in France, the publication *Societé Philomath* reported on the discovery of both Ceres and Pallas.

The Philosophical Magazine also kept readers apprised of information from Germany. It published a full report about Pallas by Zach, and in a series of issues (April, May, and June) *The Philosophical Magazine* (1803a, 1803b, 1803c) printed the positional data taken from Zach's journal for Ceres and Pallas.

A detailed description of Herschel's 1802 paper about Ceres and Pallas (Herschel 1802c) was widely covered in the English press: *The British Critic* (1804) devoted three pages to it; *The Gentleman's Magazine* (1803) devoted just one paragraph to it; *The Monthly Magazine* (1803a) devoted nearly a page to a summary of the paper; but Nicholson's *A Journal of Natural Philosophy* (1803a and 1803b) trumped them all by publishing Herschel's entire paper.

The English magazines even informed the public of communication between astronomers on the Continent:

> *Piazzi wrote to M. Seyffer on the 2d of February, that he had sought for the planet Ceres in vain during the month of December; through the greatest part of January, the weather had been unfavourable, and he had not found it again down to the instant of his writing; he was then proposing to seek for it with the elements of M. Gauss. M. Piazzi announces afterwards, that with those elements he found Ceres again, but it was only on the 23d of February, on account of the bad weather; and, he adds, that he is principally indebted for it to the ellipsis of M. Gauss.* (*The Monthly Magazine*, 1803a)

In 1803 Seyffer published a monograph that reprinted Piazzi's memoir on the discovery of Ceres together with his own reflections about Ceres. This publication (given in *Discovery of the First Asteroid, Ceres*) was summarized in *The Monthly Magazine* (1804a). The public was kept informed about Pallas long after it was discovered. This letter to *The Times* newspaper was reprinted in *The Mechanics' Magazine* (1830):

> *We extract the following interesting piece of astronomical information from a letter in The Times: –*
>
> *By reference to the Berlin or Encke's Ephemeris, the planet Pallas was detected here last night. She has the appearance of a star of the 8½ magnitude, is of a bluish tint, and bears but a very feeble illumination. Her right ascencion when on the meridian of my observatory (54′ 21″ west of Berlin) was 15 hours 22′ 49″ and 52/100; and her northern declination was 17° 43′ and 45″. Hence her place, as given in the Berlin Ephemeris, affords data sufficient for finding her; but as that excellent work cannot, I believe, now be purchased in this country, the following extract will be useful to those who take interest in such matters.*

The correspondent, F.R.S., who dated his letter April 7, 1830, then printed the ephemeris from April 6 to May 24.

Knowledge of the asteroids was taken to India by Mirza Abu Talib Mirza (1752–1805/6), who was in Europe when Pallas was discovered. In 1802 he composed a manuscript, two copies of which exist. One is in New College Library, Edinburgh; the other in Lucknow. His tract is a prose commentary on his 65-couplet poem about astronomy. In this text the planet Uranus is called Jaj. He writes "Within 2–3 years two other planets by the side of Jaj have been discovered; their speed is still unknown. One of them is called Siras [Ceres] and the other Palas [Pallas]." (Ansari, 2002: 136).

Ghulam Hussain Jaunpuri (1790–1862) was the court astronomer of the Raja of Tikari. His Encyclopedia of Sciences, in Persian, was completed in 1833 and published lithographically in 1835 in Calcutta (Fig. 5.4). On page 472 he writes about planetary discoveries, saying that "contrary to all ancient scholars, who believed that planets could not be more than seven, Christian scholars had found the following:

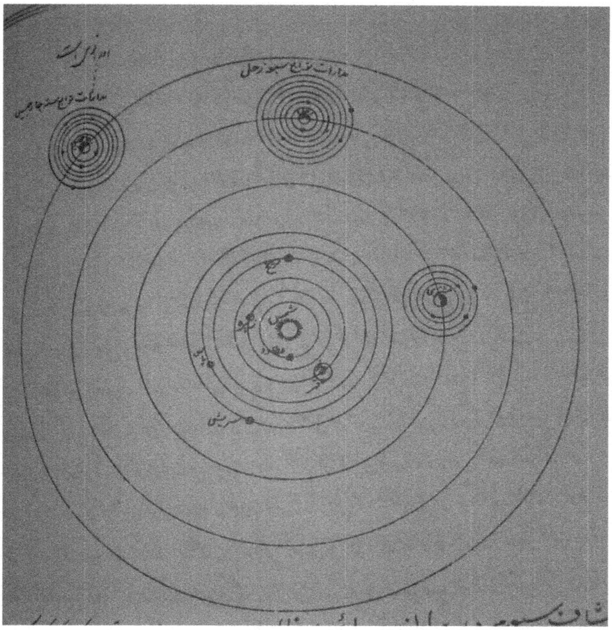

Fig. 5.4 A map of the Solar System from Ghulam's 1833 book, showing the orbits of Ceres and Pallas along with the orbits of the major planets

(a) On March 13, 1781 Piyârlas [Piazzi]—in accord with Dr. Hunter Sahib in the city of Fâlar Monî—found a planet moving above [the orbit of] Saturn, which was moving with more speed than fixed stars. It was named Jargis, after the name of the contemporary king.
(b) On January 1, 1806 a team [of astronomers] observed a planet, moving below [the orbit of] Jupiter but above [that of] Mars, with the speed 12.8533 min/day. It was called Sarish [Ceres].
(c) A scholar named Bâktar Albars [Wilhelm Olbers] discovered on March 28, 1802 another planet between Mars and Ceres with the speed 12.6835 min/day and which was named Pallas."

The English translation of item (b) (by Ansari, 2002: 138) apparently refers to an observation of Ceres in 1806, although the chronological order of the three items would indicate it was referring to the discovery of Ceres in 1801. Ghulam Hussain also confused the discovery of Uranus, saying it was found in 1781 by Piazzi instead of Herschel. Jargis of course is how the name George was written in Persian. As an astronomer, Ghulam Hussain made at least one telescopic observation of Pallas. This took place in Mirzapur (northern India) in 1826, under the guidance of a certain Captain Daranis.

The Discovery of Pallas by Sniadecki and Messier

Unbeknownst to Zach, Olbers and Gauss, Pallas had been independently discovered by Jan Sniadecki (Fig. 5.5) in Cracow, Poland. The secret discovery was not revealed until 1804, and has never before been published in English (this account comes from Balinski, 1865: 283):

> *In the summer of 1802, a new discovery in the sky gave Sniadecki the opportunity to send to the Society of the Friends of Nature yet other observations which enriched the Annals. We are speaking here of the discovery of the planet Pallas, by Doctor Olbers in Bremen, and we cannot at the same time ignore the curious event in this case that has to do with Jan Sniadecki, who also had a certain involvement in so important an event. Continually observing for some time the planet Ceres, until it got close to the Sun, and immersed in this tedious work, he noticed at one time in April 1802 a completely new little star, which he did not know until then, but not yet trusting himself, he started repeating his observations with the utmost diligence, not daring to mention anything about it in letters to Zach and Triesnecker* [Franz Triesnecker, Director of Vienna Observatory]. *But finally after a few* [the exact word used here, "kilkanascie," means a number between 11 and 19, inclusive, which presumably could be used to constrain the date of the first of these observations given the date of his letter to Zach] *observations having convinced himself conclusively that it was a new and unknown planet, he immediately wrote a report about this event to Zach, appending to it his observations; and when he was about to seal the letter, he was handed a list from Zach dated the 9th of April, which informed him joyfully that Olbers discovered a new, and specifically that same planet, and called it Pallas. Sniadecki, having compared the calculations that were sent to him with those of his own, found that his discovery was one and the same as that of Olbers. And thus fate, by a month, gave to Olbers ahead of the Polish astronomer the fame of the discovery that both of them made. Many a time Sniadecki told me of this event, adding that he did not tell of it to anyone with the exception of only Zach, when he met him personally in the year 1804. Observations on this planet Sniadecki sent on 3 June to the* Society of the Friends of the Sciences in Warsaw, *which did not fail to announce it in its* Annals.

The first sighting of Pallas was actually made in Paris by Charles Messier (1730–1817); Fig. 5.6, not Wilhelm Olbers. It happened during his study of a comet that had been discovered by Bode on January 6, 1779 (modern designation C/1779 A1, Bode). Messier (1779) observed the comet from January 19 to April 6, and on the evening of April 5, 1779, he noted a 'star' near the comet which was in fact Pallas. Unfortunately he only noted its position that one night, thus missing the fact it was a moving object (Bourtembourg, 2012).

Pallas: The French Perspective

Lalande's 5 July 1802 address to the French National Institute, on the discovery of Pallas, illuminates how the French astronomers responded to yet another 'foreign' discovery of a Solar System object. Unlike the text regarding Ceres, which was considered earlier, Lalande does not airbrush Gauss out of the story. He does not,

Fig. 5.5 Jan Sniadecki

Fig. 5.6 Charles Messier

however, quote the elements found by Gauss but those found by Burckhardt. These can be compared to the elements Gauss derived, printed at the end of this chapter. Lalande's brief discourse on irradiation is also valuable, considering the observations by Schroeter which were obviously skewed by this effect. He uses the now-obsolete length measure league, which was typically equal to 4 km. His estimate of the size of Pallas, 400 km, is remarkably close to the true value of 544 km.

When we announced, in the last public sitting, the discovery of a planet by M. Piazzi of Palermo, we were far from thinking that, in three months, we should have to make known a discovery of the same kind. It was also by a fortunate accident that this tenth planet was discovered; but accident could favour none but an intelligent and indefatigable astronomer.

On the 28th of March, at nine in the evening, Dr. Olbers of Bremen was observing Piazzi's planet, with which astronomers have been engaged for a year. He was examining with his telescope all the small stars in the Virgin's wing, to ascertain their positions, that he might be better able to establish the place of the planet, and had come to the 20th star of the Virgin, near which he had observed the planet in the month of January. He was surprised to see near this star, which is of the 6th magnitude, another smaller of the 7th magnitude. He was very certain that it had not been there at the time of his first observations: he therefore hastened to determine its position; and, having continued to view it for two hours, he perceived that it had changed its place in the course of that interval. The two following nights afforded him the means of being certain of its motion, which was 10 minutes per day. On the 28th of March, at 9h 25′ mean time, at Bremen, it had 184° 56′ right ascension, and 11° 33′ north declination.

Astronomers have been accustomed to consider as comets, all stars that have motion. This was the case with the planets of Herschel and Piazzi at the time when they were discovered. That of Dr. Olbers had no more resemblance to a comet than the rest. With an achromatic telescope, the magnifying power of which was 180, it could not be distinguished from stars of the 7th magnitude. It was better defined than the planet of Piazzi; and, with a telescope of 13 feet, which magnified 288 times, it seemed to have a diameter of 4 seconds: but this was an effect of irradiation, or of the dispersion of the rays of light, which always makes the diameters appear too large; for the satellites of Jupiter appear much larger than the new planets, and yet we know that their apparent diameter is not a second.

Dr. Maskelyne, by means of diaphragms placed before the object-glass of his telescope, ascertained that the light of Piazzi's planet is stronger by one half than that of the new planet.

Dr. Olbers, having observed the new star for four days, he sent notice to different astronomers; and on the 10th of April, C. Burckhardt, when he received his letter, went immediately to the military school to search for it, and next day sent his observation to the Institute.

He began to calculate its orbit, trying first a circle, and then a parabola known to be that of comets; but, at the end of three days, his elements were found to err 30 seconds. He tried also ellipses of different dimensions.

On the 15th of May we were informed, by a letter from Baron von Zach, the celebrated astronomer of Gotha, that Dr. Gauss, an astronomer of Brunswick, had found an ellipsis which corresponded to the first observations. On the 22nd we received the details. He found the revolution to be four years seven months, and the inclination 3°. This great inclination seemed to remove it from the order of planets, and some astronomers called it a comet; but its proximity, and continual appearance, will not allow of its being placed among the number of those stars of which we often lose sight for so long a time, and which go to enormous distances.

C. Burckhardt, on his part, made similar researches; he made several trials with ellipses very much elongated, which gave him a result very near that of Dr. Gauss.

On finding that this planet, like that of Piazzi, was between Mars and Jupiter, and that its motion must be affected by the attraction of Jupiter, C. Burckhardt undertook to calculate these perturbations. The calculation is long and difficult, but it is indispensably necessary to obtain the orbit with more exactness.

At last, on the 4th of June, he finished these laborious calculations, and found the following elements:

Distance 2.791, or 95,890,000.

Revolution, 4 years, 8 months, and 3 days.

Eccentricity, 0.2463; equation of the orbit, 28° 25′.

Epoch of 1802, 4Z 23° 50′; aphelion 10Z 2° 3′;

Node 5Z 22° 28′; inclination 34° 50′ 40″.

These elements corresponded to five observations of the 4th, 16th, and 27th of April, and the 7th and 20th of March; the last two made by C. Burckhardt, and Lalande's nephew, who, as well as C. Méchain, Messier and Delambre, continued to observe it as long as it could be seen in the meridian, because such observations are the surest. After the 21st of May, other instruments and other stars were necessary; but it still passed through some included among the 15,000 stars which we have published. On the 15th of June these elements corresponded, within a few seconds, with the observations of Méchain and Messier; which confirms the exactness of the elements found by C. Burckhardt, and assures us, that the motion of the new planet is already known. Baron von Zach has published a great many observations respecting it in his Journal.

C. Cabrol de Murol [see FN] *has calculated for us an ephemeris, which gives the situation of this planet to the 21st of October, on which day it will have 227° 7′ of right ascension, and 6° 8′ of declination. It will then set at 7h 51′: there is therefore reason to think that it may be still observed. It will be above Libra near the Serpent, after passing the legs of the Cow-herd (Aquila). He finds that, in 1806, it will have 33½° of south declination, and that it will then be difficult to see it at Paris; but C. Vidal, who has already observed it this year, will then be better able than we to follow it.*

Its greatest northern declination will not exceed 26½ degrees, a term at which it will be a year hence. It will be easier to be seen, but its distance will be double, and its light four times less than the present year. In the month of March 1804, it will be at three times the distance; its light will be nine times less, and, in all probability, it will be difficult to observe it.

As the orbit of this new planet intersects that of Piazzi, I was curious to know whether the two planets might not meet; but I found that, when they are in the same plane, there will be an interval of about 19 millions of leagues between them.

The planet of Dr. Olbers is very small. If we suppose its apparent diameter to be half a second, I find that its real diameter cannot be more than 100 leagues. Dr. Herschel, in a paper which he read before the Royal Society on the 7th of May, makes it be four times less. He says, that on the 22nd of April Piazzi's planet was only 22 hundredths of a second, and that of Olbers 13 hundredths; but it appears to me, that we have no means of determining, with certainty, quantities so small.

Dr. Olbers calls his new planet Pallas; but, as I see no sufficient motive for this fabulous denomination, I prefer giving it the name of the person to whom we are indebted for this valuable discovery.

Dr. Olbers distinguished himself in 1797 by an excellent treatise on comets, and was worthy of the good fortune with which his labours have been crowned. (Lalande, 1802c)

[Note: Michel Chabrol de Murol was born in Riom, France, on November 18, 1777, and worked as an astronomer at the Paris Observatory. He published a method to calculate eclipses. It is not known where or when he died (Zach, 1818: 525)].

The following year, Lalande (1803) offered additional information about the French study of Pallas:

The astronomers Delalande junior and Burckhardt observe with great assiduity the planet discovered by Dr. Olbers on the 28th of March 1802. Its longitude on the 1st of July at 11h

45′ was 9 signs 7° 14′ 25″, and its latitude 46° 23′ 18″. Burckhardt has thence deduced its revolution to be 1682 days, or four year seven months and twelve days; which is a day less than what was found some months ago, as may be seen in my Bibliographie Astronomique *just published; but at present there is scarcely an uncertainty of a few hours. He is employed in calculating the derangements it must experience from the attraction of Jupiter, and which are very complex; but he has presented to the* Institute *a learned memoir which leads to this research.*

Verses to Commemorate the Discovery of Pallas

The discovery of Ceres, the first object found between Mars and Jupiter, naturally elicited a substantial outpouring of commemoration in verse. Pallas received very little notice by comparison. In addition to his verse about Ceres, Marcin Poczobut (1728–1810) director of the observatory in Vilnius, wrote about Pallas: (MC July 1802: 74)

Falx Cereris signum esto; tu ut taere laboris
Sideribus sacros, aegida Pallas habe.

Oh sickle, be the sign of Ceres; and have with you
Pallas who has the Aegida, so that you can protect
the sacred works in the skies.

Each of the major planets had long been assigned a symbol, often used as shorthand in planetary tables. When Piazzi's discovery Ceres was named, it was suggested by Zach that a sickle be used as its sign, in accordance with the role of Ceres as goddess of agriculture. The Aegida is the protective shield made of goat skin and the Medusa head, used by Pallas Athena.

Zach rhapsodized about the discovery of both Ceres and Pallas, inserting what appears to be a Latin verse of his own design. "It is easy to immortalise such an epoch-making occurrence in the history of astronomy. The heavens will proclaim these works to all people and for all time to come." (MC, June 1802: 589)

Videbo coelos tuos, opera digitorum tuorum,
Cerem et Palladem, quae tu fundasti.

I will see thy heavens, the works of thy fingers,
Ceres and Pallas, which thou hast founded.

Even though Pallas had been discovered in Bremen, the following verse by an anonymous author is lacking in any nationalistic sentiment. This could possibly be due to the fact that Germany was not yet a unified country.

Der neue Planet Pallas.
Endlich erschienest du Pallas, und mit dir der Oelzweig des Friedens; Göttin der Weisheit, warum nahtest du leider so spät?
Mögest du heller noch leuchten, daß heller auf Erden es werde! –
Siehe wie funkelt so hell Venus vor allen hervor!
Zünd' o Weisheit dein Licht am Schwesternaltare der Liebe!
Weisheit mit Liebe vereint - so nur beglückt sie die Welt.

The new planet Pallas.
At last you appeared, Pallas, and with you the olive branch of peace;
Goddess of wisdom, why did you come so late?
May your light shine even brighter so that it might be lighter on Earth! –
See how bright Venus sparkles!
Light, o wisdom, your fire at the sisterly altar of love!
Wisdom combined with love – it only gladdens the world this way.

A footnote told the reader that Dr. Olbers discovered Pallas on March 28, one day after the Amiens peace treaty (which marked the end of the French Revolution), which explains the peace reference in the second line. The writer then implores Pallas to join with Venus, the goddess of love, so that their combined radiance (the wisdom of Pallas and the love of Venus) may salve the wounds of a lengthy war. The verse was published in the Berlin newspaper *Vossische Zeitung* on September 16, 1802.

A very early Italian poem that mentions both Piazzi and Olbers (and Ceres and Pallas) appears in a book of celestial poetry by Giuseppe Saverio Poli (1805: 25). Poli (1746–1825) was a physicist, biologist and natural historian, and he dedicated the book to none other than King Ferdinand, the patron of Piazzi:

Or per opra di Olbers, di Piazzi illustre
E pur de'Numi divenuto il Regno...
Pallade saggia, che I bei di rimena,
Con Cerere diviso ha gia l'impero.

Now thanks to Olbers and to famed Piazzi
It has become the reign of the gods...*
Wise Pallas, who brings fair weather back,
Has already shared her power with Ceres.

* *i.e., the 'empty/barren space' between Mars and Jupiter, according to what the poem says.*

The discoveries of Ceres in Palermo and Pallas in Germany were alluded to in a lengthy astronomical poem by Charles-Julien Lioult de Chênedollé (1807):

Déjà même, a sa voix, les prêtres d'Uranie
S'eveillent dans Palerme et dans la Germanie

Even now, in his voice, the priests of Urania
Awaken in Palermo and in Germany

The First Solar System Maps Depicting Ceres and Pallas

The first map of the Solar System that showed Ceres as an object orbiting the Sun appeared in 1638. That is not a typo. John Wilkins (1614–1672) was an Anglican clergyman and one of the founding members of the Royal Society. In 1638 he published his first book, *The Discovery of a World in the Moon*, in which he presented a world of truths ready to be uncovered by the diligent explorer. It was this text in

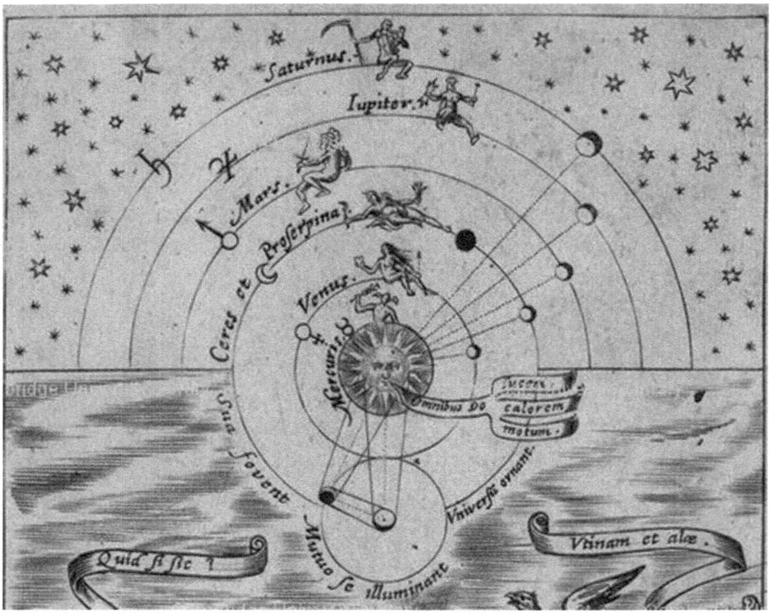

Fig. 5.7 Frontispiece of the 1640 book by John Wilkins

an expanded form that comprised the first book of his *A Discourse Concerning a New World and another Planet in Two Books*, published in 1640; the second book provided a point-by-point defense of the Copernican hypothesis.

Occupying the upper portion of the frontispiece is a cosmological diagram imported from the title page of his earlier work. Utilizing the familiar geographical line drawing typical of cosmological diagrams of the period, this figure represents a heliocentric Solar System surrounded by an unbounded region of fixed stars (Fig. 5.7). Each orb is accompanied by a mythological figure representing the relevant planet. In the case of Earth, the orb is labeled *'Ceres et Proserpina,'* representing Earth and the Moon respectively, recalling the mythological quest of Ceres to find her daughter, Proserpina, who had been abducted by Pluto. Wilkins interprets this myth as "the longing desire of men, who live upon Ceres, Earth, to attain a place in Proserpina, the Moon, or heaven." (Kaoukji & Jardine, 2010)

As for Ceres, the first book to depict it in a Solar System map was the one by Jean Henri Hassenfratz (1755–1827), Professor of Physics at the École Polytechnique in Paris (see The Discovery of the First Asteroid, Ceres). A hand-colored map of the same period shows an orbit inserted between Mars and Jupiter with the words 'Piazzi and Pallas.' This is particularly interesting, as both objects are denoted by a single orbit, the only map that shows the two new objects in this way (Fig. 5.8).

Another early map is one hand-drawn by Gauss in his Handbook no. 4 (Fig. 5.9). At the top of the map we see on the right the text that Ceres was discovered on Jan. 1, 1801, and another dot on the orbit just to its left shows the small movement it made by the time Piazzi lost sight of it in February. Figure 5.10 shows a map prepared by Lalande.

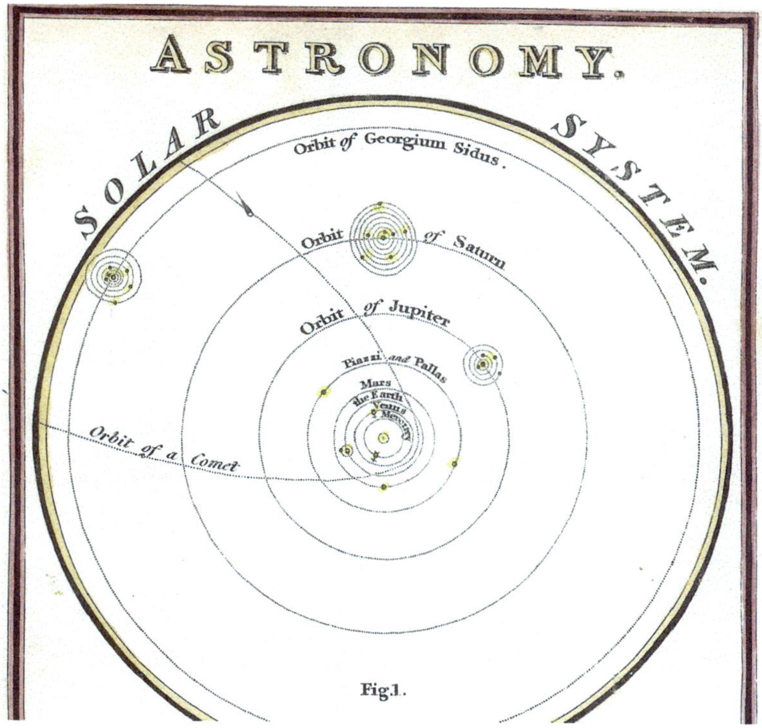

Fig. 5.8 A map of the Solar System, circa 1803, showing a single orbit of both Pallas and Ceres (Piazzi)

The First Sightings of Pallas

Messier	Paris	April 5, 1779
Olbers	Bremen	March 28, 1802
Schroeter	Lilienthal	March 30
Zach	Seeberg	April 4
Bode	Berlin	April 5
Gilpin	London	April 9
Sniadecki	Vilnius	April 11 (independent discovery)
Lee	London	April 12
Burckhardt	Paris	April 20
Herschel	London	April 22
Maskelyne	Greenwich	April 23
Oriani	Milan	April 25

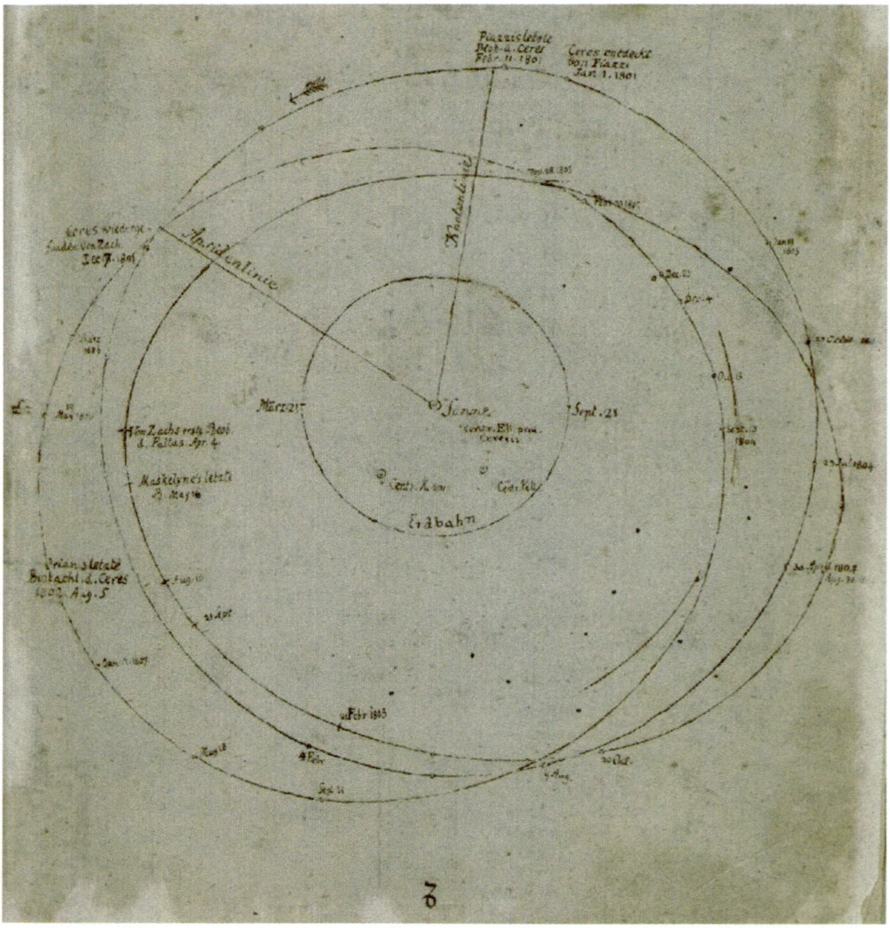

Fig. 5.9 A map showing the orbits of Ceres and Pallas, drawn by Gauss. (Courtesy of SUB Göttingen, Cod. Ms. Gauss Handbuch 4, folio 1r)

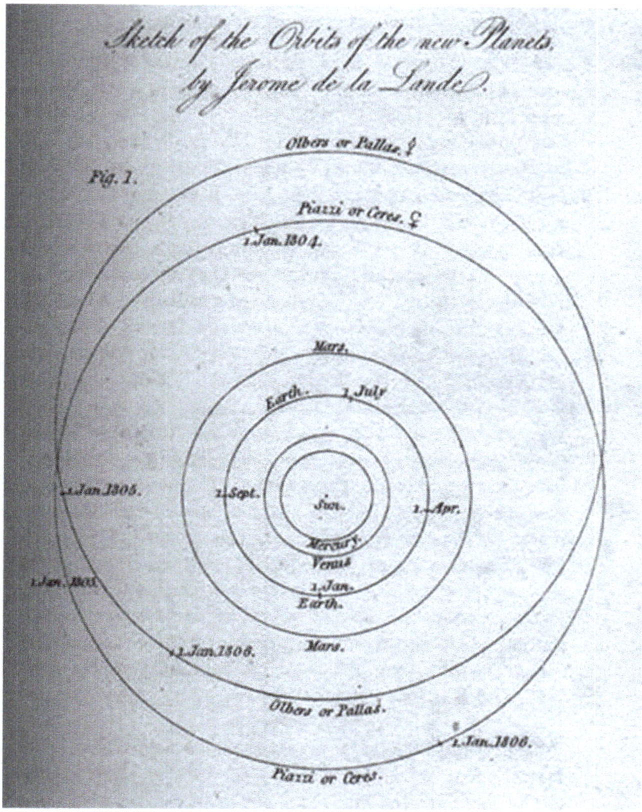

Fig. 5.10 A map prepared by Lalande in which he calls the objects both by their discoverer names, and the names that were accepted by everyone else— Ceres and Pallas

The Elements of Pallas

I and II come from June/1802 MC
III comes from July/1802 MC
IV comes from Oct/1802 MC
V comes from Dec/1802 MC

The Elements Derived by Gauss

	I	II	III	IV	V
Aphelion	304° 36′ 30″	300 5 4	300 58 47.7	301 38 42	301 28 24.0
Node	172 9 58	172 34 35	172 28 17.9	172 26 31	172 27 3.0
Incl	33 39 16.6	35 0 42	34 39 10.7	34 36 59	34 37 40
Eccentricity	0.215708	0.2591096	0.2476402	0.243888	0.244976
Daily motion	800″.770	757″.166	769″.547	769″.7263	769″.583
Log	0.4310494	0.4472636	0.4425664	0.4424992	0.4425529

In addition to these "official" elements, Gauss sent Olbers interim elements in letters dated:

April 20, 1802
May 19, 1802
June 8, 1802

On June 4, Burckhardt published his own elements of Pallas:

Aphelion	302° 3′
Node	172 28 57
Incl.	34 50 40
Eccentricity	0.2463
Daily motion	760″.98
Log	0.4457560

Chapter 6
The Original Logbooks: England, France, Germany and Italy

C. Cunningham, *Early Investigations of Ceres and the Discovery of Pallas*,
DOI 10.1007/978-3-319-28815-4_6

Fig. 6.1 Illustration from *The Village Astronomer*, translated from German by Matilda Wrench (1852). (Wertheim and Macintosh, London). An astronomer with his elbow on a logbook consults a star chart

In the original logbooks of the astronomers in England, France, Germany and Italy, we can see the actual observations and calculations being made on the extraordinary new objects, Ceres and Pallas (Fig. 6.1).

Lodged in the archives of the Paris Observatory are the handwritten observations made at the l'Ecole Militaire. In the late eighteenth century and the early nineteenth, the main places devoted to official astronomy in Paris were the Observatoire de Paris, the Observatoire de la Marine and the Observatoire de l'Ecole militaire (Débarbat, 1998). Lalande was the director at the l'Ecole Militaire until his death in 1807, when he was succeeded by Burckhardt (Bigourdan, 1887).

The extract dated January 25, 1802, shows the first observation of Ceres from the observatory (Fig. 6.2). In this entry, it is referred to (in the right-hand column) as Piazzi's planet (Fig. 6.3). This was just 3 weeks after Olbers had announced his recovery of Ceres.

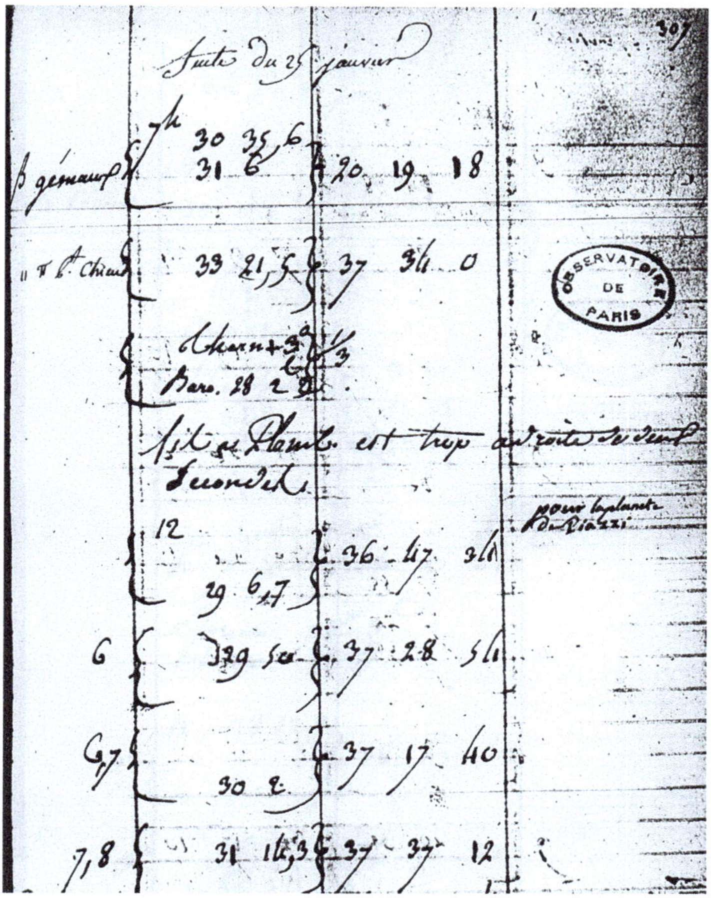

Fig. 6.2 Observations in Paris of 'Piazzi's planet' on January 25, 1802

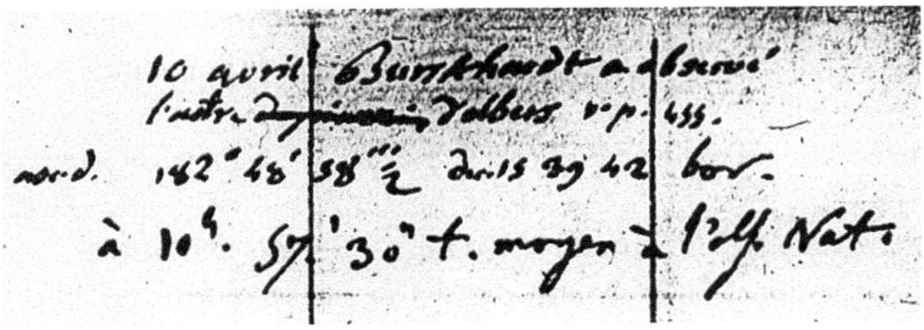

Fig. 6.3 Burckhardt's observations of 'Olbers' star' on April 10, 1802

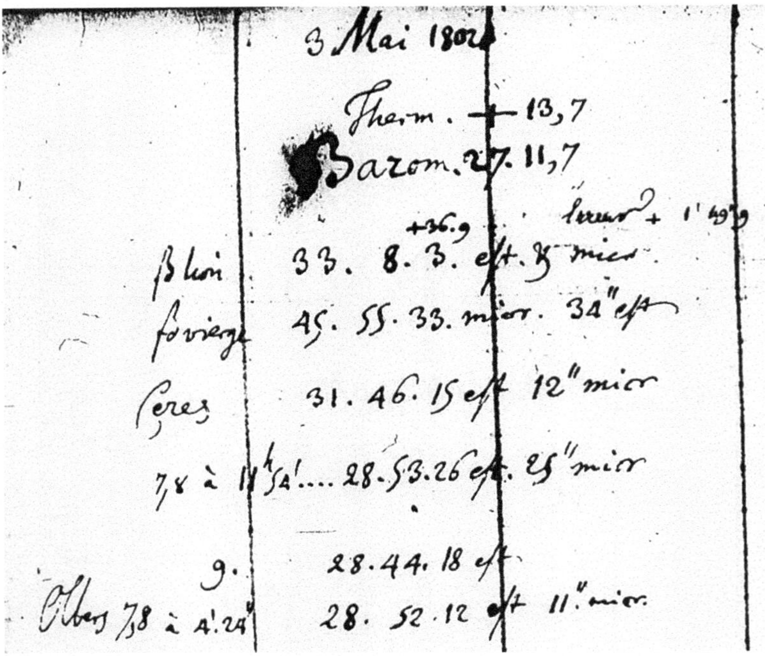
3 Mai 1802
Therm. + 13,7
Barom. 27. 11,7
β lion 33. 8. 3.
Ceres 31. 46. 15
9. 28.44. 18
Olbers 7,8 à 4'.24 28. 52.12

Fig. 6.4 Observations in Paris of Ceres and 'Olbers' on May 3, 1802

Shown below is the first note in the logbook about Pallas. Here, on April 10, 1802, it is referred to as Olbers' star.

On May 3, 1802, we see the first asteroid identified as Ceres, while Pallas is just referred to as Olbers (Fig. 6.4).

The very next night, however, reveals the full gamut of confusion. Ceres reverts to its former designation as Piazzi's planet, Pallas is again called Olbers' star, and Uranus is referred to as Herschel's planet (Fig. 6.5).

This nomenclature comes direct from Lalande, who insisted on attaching the names of the discoverers to the planets.

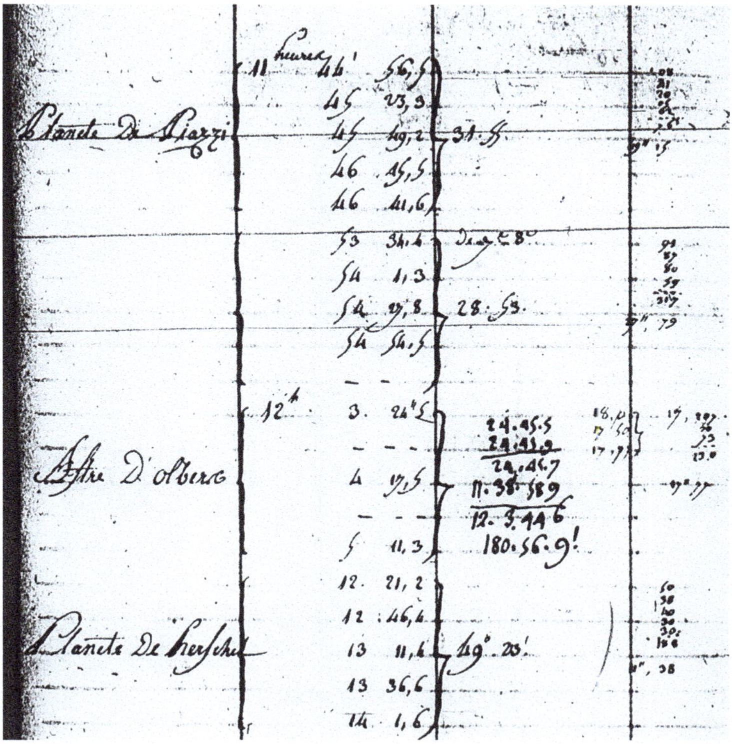

Fig. 6.5 Observations in Paris of 'Piazzi's planet' and 'Olbers' star' on May 4, 1802

Finally, Pallas was recognized as a planet (Olbers' planet). This did not happen until June 25, 1803. Note also that Burckhardt himself signed this page along the top, firmly identifying the observations as his (Fig. 6.6).

Across the English Channel, in Greenwich, the Astronomer Royal Nevil Maskelyne was also busy observing Ceres and Pallas.

In this detail from his observational logs, we read an entry from April 22, 1802:

> *From the time of Olber's planet passing the meridian till near 11 pm m.t. I looked alternately at this planet and Ceres, and they seemed about equal in brightness. Whichever was observed last was thought to be the brightest. At the transit instrument tonight, Olber's planet seemed equal in size and brightness to* [the way] *Ceres appeared when I observed it at the meridian last night. Mr. Gilpin found that on April 22 and 23 the two new planets Ceres and Pallas appeared to him exactly of equal brightness. The French astronomers on 10th and 12th found the same* (Fig. 6.7).

The relative brightness of Ceres and Pallas was the subject of many measurements by Maskelyne and his assistant, Thomas Firminger (abbreviated as T. F. in the notes): "Observations of the apparent magnitudes or brightness of the two new planets (Fig. 6.8)."

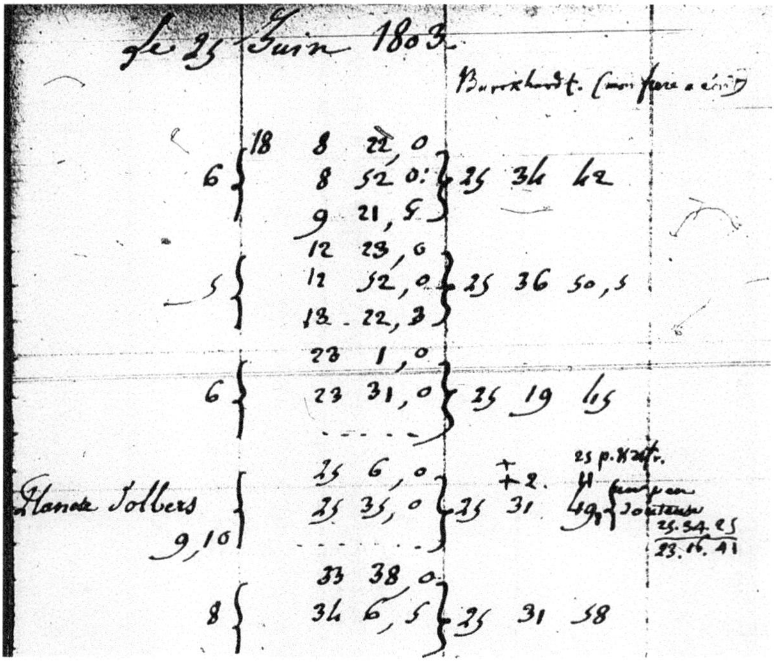

Fig. 6.6 Observations in Paris of 'Olbers planet' on June 25, 1803

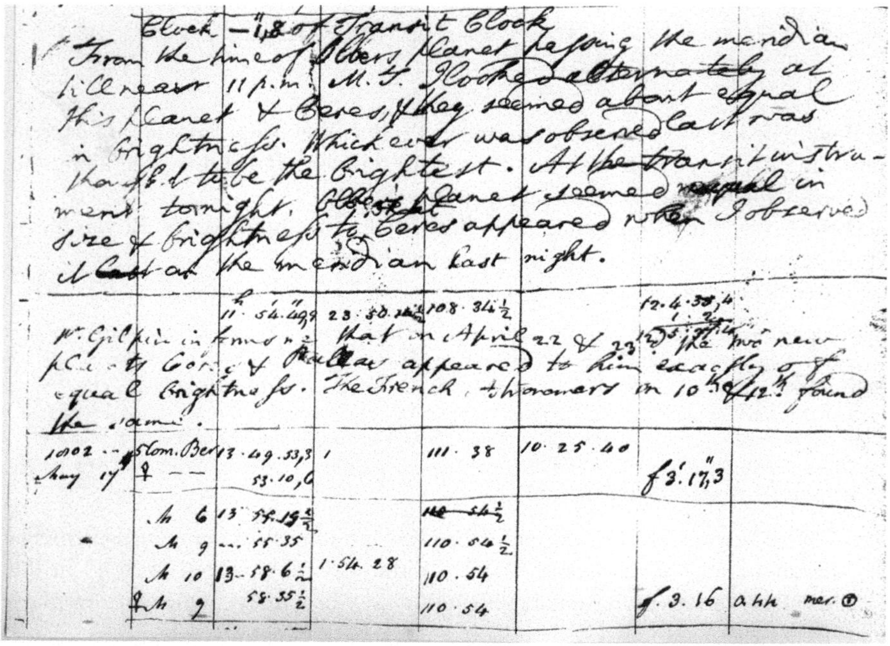

Clock −1,8 of Transit Clock

From the time of Olbers planet passing the meridian till near 11 p.m. M.Y. looked alternately at this planet & Ceres, & they seemed about equal in brightness. Whichever was observed last was thought to be the brightest. At the transit instrument tonight, Olbers planet seemed equal in size & brightness to Ceres appeared when I observed it at the meridian last night.

Mr Gilpin informs me that on April 22 & 23 the two new planets Ceres & Pallas appeared to him exactly of equal brightness. The French Astronomers in 10th & 12th found the same.

Fig. 6.7 Notes and observations of Pallas by Maskelyne on April 22, 1802

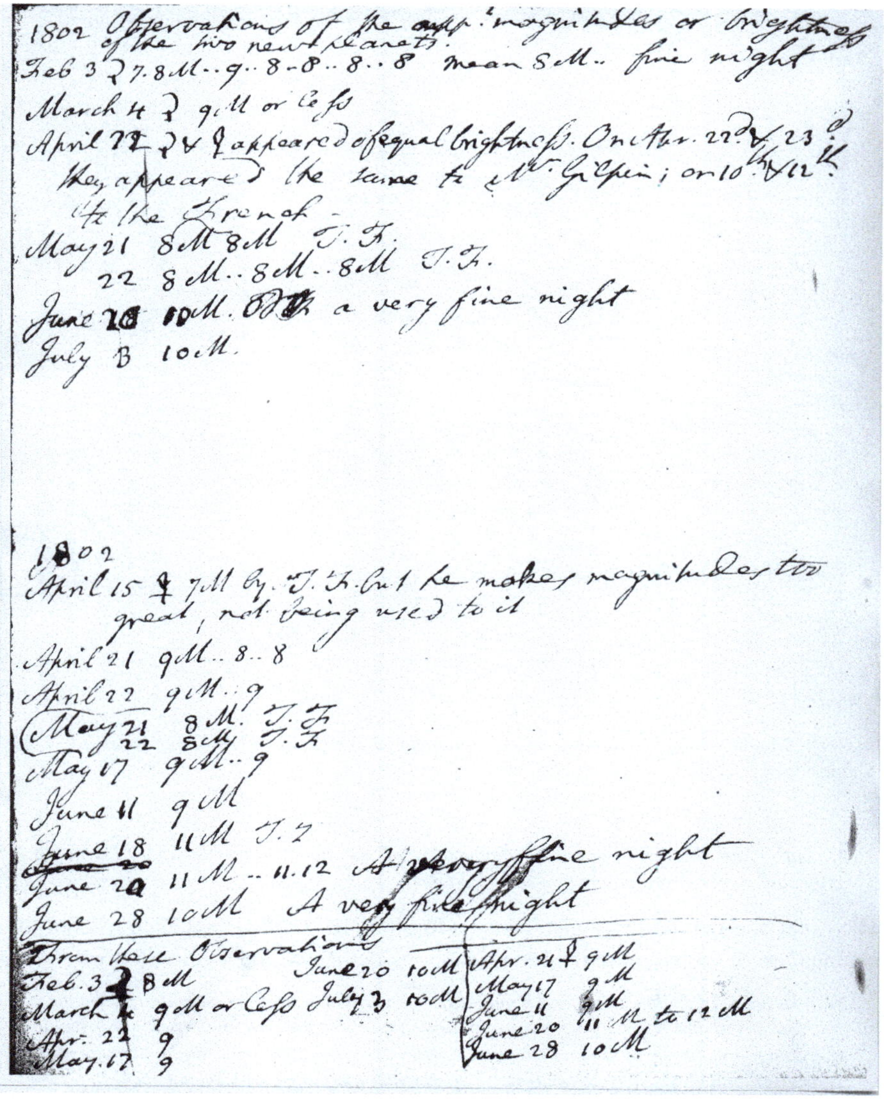
1802 Observations of the app. magnitudes or brightness of the two new planets.
Feb 3 7.8M .. 9 .. 8 - 8 .. 8 .. 8 mean 8M .. fine night
March 4 9M or less
April 22 appeared of equal brightness. On Apr. 22d & 23d they appeared the same to Mr Gilpin; or 10th & 12th of the French.
May 21 8M 8M T.F.
22 8M .. 8M .. 8M T.F.
June 20 10M. a very fine night
July 3 10M.

1802
April 15 7M by T.F. but he makes magnitudes too great, not being used to it
April 21 9M .. 8 .. 8
April 22 9M .. 9
May 21 8M T.F.
22 8M T.F.
May 17 9M .. 9
June 11 9M
June 18 11M T.F.
June 20 11M .. 11.12 A very fine night
June 28 10M A very fine night
From these Observations
Feb. 3 8M
March 9M or less
Apr. 22 9
May 17 9
June 20 10M
July 3 10M
Apr. 21 9M
May 17 9M
June 11 9M
June 20 11M to 12M
June 28 10M

Fig. 6.8 Observations of the apparent magnitudes of Ceres and Pallas by Maskelyne. (Cambridge University archives)

Some of the laborious calculations Maskelyne performed to derive the diameter of Pallas to be 116.56 English miles (Fig. 6.9).

This is the calculation of the distance of Pallas from the earth April 22, 1802 at mean midnight in the meridian of Greenwich. I have however taken the planet's longitude for m.t. at midnight in the meridian of Seeberg as great accuracy in the result was not required. The calculation was made from Dr. Gauss' elements (II). By my obs. the long. should be 5h 23°

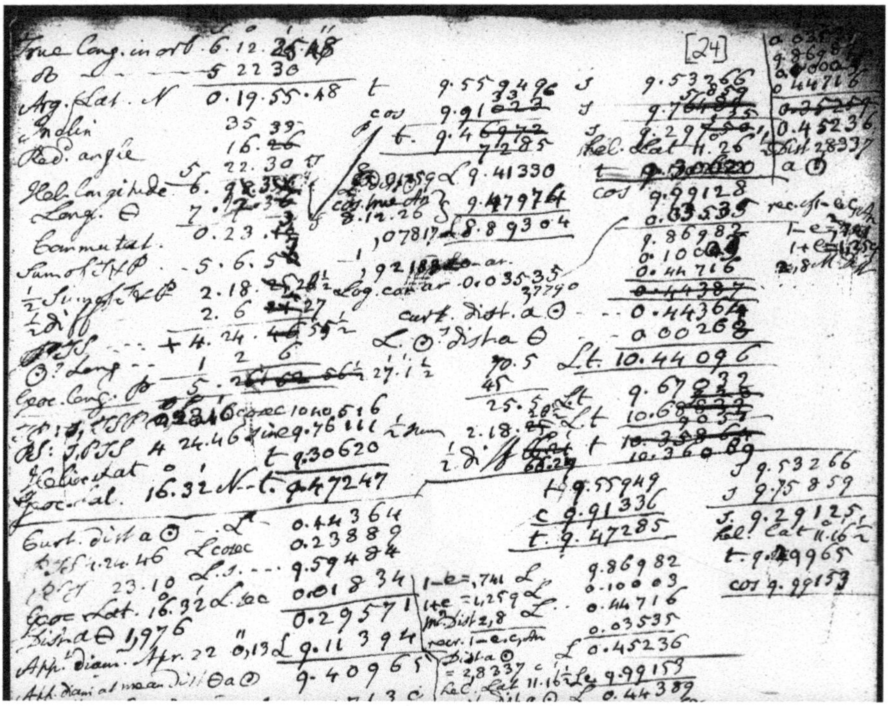

Fig. 6.9 Mathematical calculation of the diameter of Pallas by Maskelyne, June 23, 1802. (Cambridge University archives)

43′ Lat 17° 38′ 4″ N. The longitude computed is +2° 44′ and the latitude -1° 6′. The cause of this difference I cannot imagine. June 23, 1802.

The great Scottish psychologist Thomas Reid (1710–1796; 1785: 265) offered this explanation of 'apparent magnitude' in a book dedicated to Dugald Stewart:

No sooner is the visible figure and magnitude of an object seen, than immediately we have the conception and belief of the corresponding tangible figure and magnitude. We give no attention to the visible figure and magnitude. They are immediately forgotten, as if they had never been perceived; they have no name in common language; and indeed, until Berkeley pointed them out as a subject of speculation, and gave them a name, they had none among philosophers, excepting in one instance, relating to the heavenly bodies, which are beyond the reach of touch. With regard to them, what Berkeley calls visible magnitude was by astronomers called apparent magnitude.

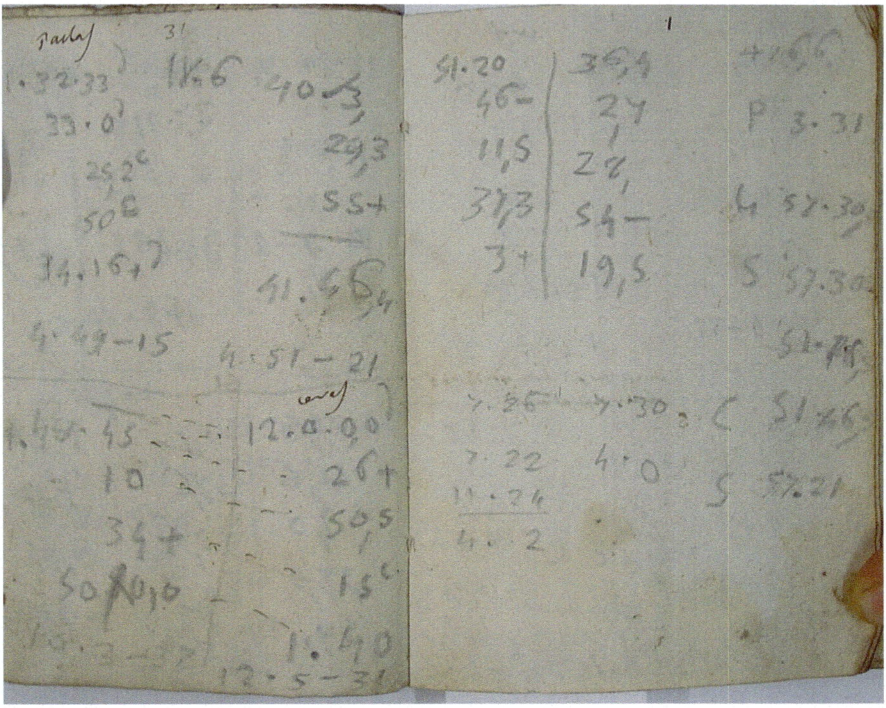

Fig. 6.10 A page from Barnaba Oriani's logbook, with observations of Ceres and Pallas. (Brera Observatory archives)

In Italy, Oriani was making concurrent observations of Ceres and Pallas, as evidenced by this page of his logbook from 1802. On the top left is written "Pallas," with "Ceres" written in black ink below (Fig. 6.10).

In Germany, Gauss was working on the mathematical calculations to determine the orbit of what he writes here as "Ceres Ferdinandea (Fig. 6.11)."

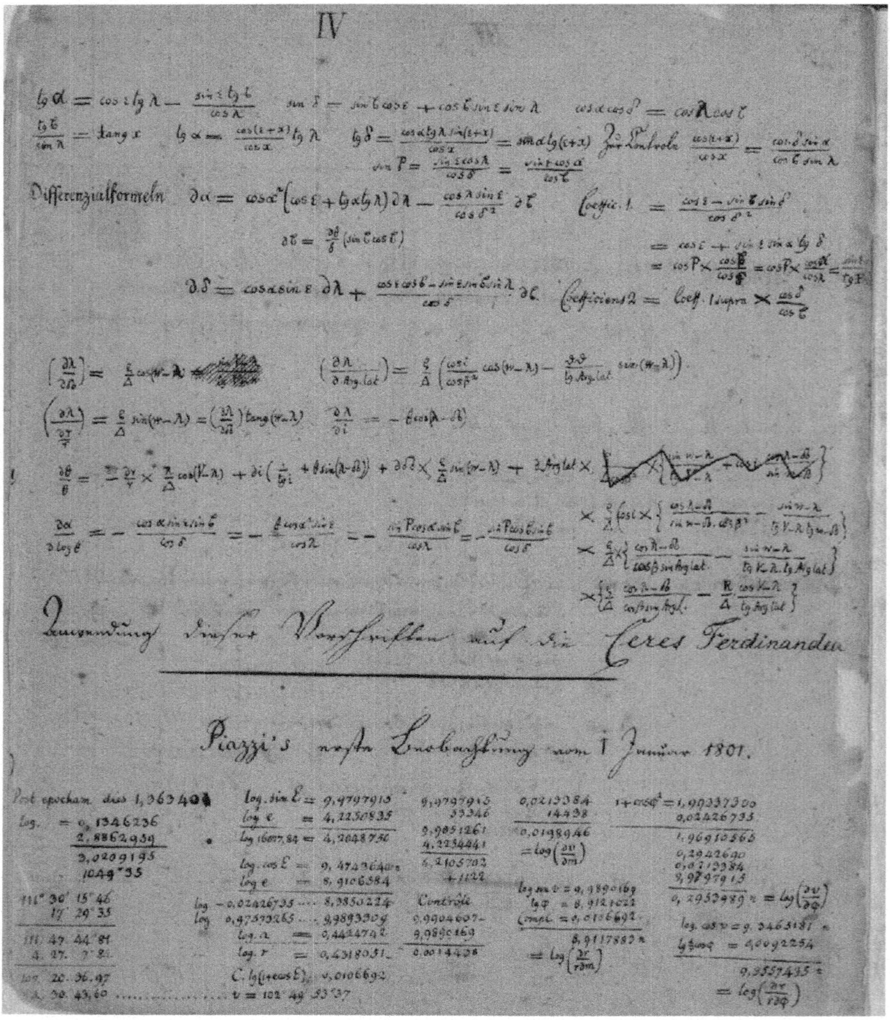

Fig. 6.11 A page from the notebook of Carl Gauss, showing his study of Ceres. (Goettingen University archives)

Chapter 7
The Ceres and Pallas Letters of 1802

C. Cunningham, *Early Investigations of Ceres and the Discovery of Pallas*,
DOI 10.1007/978-3-319-28815-4_7

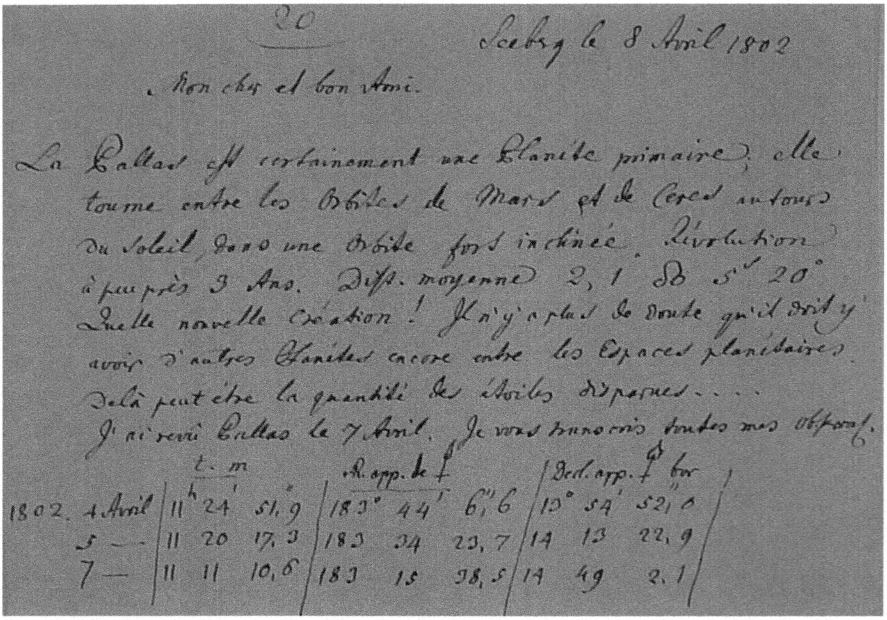

20

Seeberg le 8 Avril 1802

Mon cher et bon Ami.

La Pallas est certainement une Planète primaire; elle tourne entre les Orbites de Mars et de Ceres autour du Soleil, dans une Orbite fort inclinée. Révolution à peu près 3 Ans. Diff. moyenne 2, 1 8 5° 20° Quelle nouvelle Création! Il n'y a plus de doute qu'il doit y avoir d'autres Planètes encore entre les Espaces planétaires. Delà peut être la quantité des étoiles disparues....

J'ai revu Pallas le 7 Avril. Je vous donnerai toutes mes observations.

	t. m	AR. app. de ⚴	Decl. app. ⚴ bor
1802. 4 Avril	11h 24' 51,9"	183° 44' 6"6	13° 54' 52"0
5 —	11 20 17,3	183 34 23,7	14 13 22,9
7 —	11 11 10,6	183 15 38,5	14 49 2,1

Fig. 7.1 First portion of an April 8, 1802, letter from Zach to Oriani, in which he triumphantly announces that "Pallas is certainly a primary planet; it revolves around the Sun between the orbits of Mars and Ceres… What new creation!" (Brera Observatory archives)

The idea that life gets into letters more intimately, more unwarily, than into other writings makes letter-reading seem the closest we might come to the writer's thoughts.

Angela Leighton, 2003

Introduction

The correspondence of the asteroid pioneers revealed more about their thoughts and aspirations than their scientific papers. Many of their letters were published in the author's book *Discovery of the First Asteroid, Ceres,* and many more are published here for the first time in English.

Among the largest correspondence about the asteroids was that between Olbers and Gauss. Their first letters to one another dealt with Ceres, and those about Pallas from 1802 are published here. In total they span 16 years, 1802–1818, but this is short compared to the span of Pallas-related letters between Gauss and Bessel, 1804–1843. For more on the discovery of the letters between Gauss and Maskelyne, see Cunningham (2004). Only letters from 1802 are in this book; letters from 1801 were included in *Discovery of the First Asteroid, Ceres.*

The majority of the letters are from Baron Franz von Zach (Fig. 7.1), Carl Gauss, Johann Bode, Wilhelm Olbers, and Jan Sniadecki. Many letters have been lost. For example the letters of Zach to Oriani exist, but Oriani's replies do not.

The Olbers-Gauss letters were originally published in German by Schilling (1900), and his numbering system is retained here. Their correspondence dealing with Ceres was published in *Discovery of the First Asteroid*, Ceres. Letters between Oriani and Piazzi were published in Italian by Cacciatore and Schiaparelli (1874), and their numbering system is also retained. In these letters F.N. stands for footnote.

Newspaper articles are also mentioned in this chronological survey, including the Berlin publication *Koeniglich privilegierte Berlinische Zeitung von Staats- und gelehrten Sachen,* more popularly known as *Vossische Zeitung* (see Chap. 5).

Maskelyne's memoranda about Ceres and Pallas are also included here.

Before looking at these erudite letters from 1802, it will not be amiss to consider the state of society at the time. Even though the asteroid pioneers looked to the heavens for inspiration, they all lived in a world two centuries removed from our own, and it was not always a pretty sight. This description of England is taken from Hawes (1957: 34):

We must pause for a moment to visualize just how things stood in the year 1802. First and foremost there was a stifling blanket of fear over the whole intellectual and political life of the country, fear of the French, fear of the masses, above all fear of change. The criminal law was still in its old state of savagery: stealing five shillings was a capital offence, the poacher was transported [to Australia], the prisons were dungeons of corruption and disease. The mountain of stupidity and delay and extortion which was the Court of Chancery lay immovably in the path of Civil Law. Catholics, Jews and Dissenters were debarred from holding any place under government and from entrance to the Universities. Popular education was practically non-existent, and the mere suggestion of such a thing was considered highly dangerous. The press was gagged and public meetings prohibited. People were afraid to talk, almost afraid to think. The periodical press, what there was of it, was moribund, without vigour or talent.

A Poem by Wordsworth: London, 1802

MILTON! thou should'st be living at this hour:
England hath need of thee: she is a fen
Of stagnant waters: altar, sword, and pen,
Fireside, the heroic wealth of hall and bower,
Have forfeited their ancient English dower
Of inward happiness. We are selfish men;
Oh! raise us up, return to us again;
And give us manners, virtue, freedom, power
Thy soul was like a Star, and dwelt apart:
Thou hadst a voice whose sound was like the sea:
Pure as the naked heavens, majestic, free,
So didst thou travel on life's common way,
In cheerful godliness; and yet thy heart
The lowliest duties on herself did lay (Fig. 7.2).

Notable French Publications

Histoire Céleste Française

Mention of this publication recurs many times in the following correspondence. It has a most interesting connection with Baron von Zach. In 1778 the English instrument maker John Bird (1709–1776) made an 8-foot quadrant (with an aperture of 7 cm). An instrument of this size gave readings accurate to about half a second of arc (Ventura, 1990). It was sent to Bergeret, Receiver-General of the Finances of France. He loaned it to the observatory of the Ecole Militaire. When Bergeret died in 1785, the observatory faced demolition, and Zach saw his chance of procuring the

Fig. 7.2 John Milton

quadrant for his new Seeberg Observatory. Unfortunately for him the instrument was purchased by the council of the Ecole Militaire who housed it in a new observatory. It was then used by J. J. Lalande, between 1789 and 1799, to determine the positions of 47,390 stars. This work was published as the *Histoire Céleste Française* in 1801 (Couteau, 1981).

Connaissance de Temps

The *Connaissance de temps (Knowledge of the times)* first appeared in 1679. In 1795 its production became the responsibility of the original members of the Bureau des longitudes in Paris: Jean Delambre, J.-J. Lalande, P.-S. Laplace and J.-L. Lagrange. The journal served as an astronomical yearbook, and it included new observations and scientific works as well. It did not begin publishing reports about Ceres until 1803.

Gazette Nationale ou le Moniteur Universel

This was the official government publication of the time, now called the *Journal Officiel*. The *Moniteur* was first issued on May 5, 1789, the very day the Estates General held a meeting at Versailles. Lalande (1802a) and Burckhardt (1802a) issued reports on Ceres in the *Gazette.* Olbers and Gauss often quoted from this publication. The first pages were reserved for political matters, but more than in any other newspaper, scientific articles were published. The results of the meetings of the Institut National and the Institut de France were covered in detail. It was paginated in numbered volumes every 6 months.

Notable English Publications

Nicholson's Journal

William Nicholson, a writer and teacher of chemistry, brought out his *Journal of Natural Philosophy, Chemistry and the Arts* in 1797. *Nicholson's Journal* (as it was popularly known) presented an accessible mix of original research reports and news of lectures and meetings of scientific societies. It established the model for a scientific periodical accessible to a broader readership than the somewhat intimidating *Philosophical Transactions of the Royal Society.*

Tilloch's Journal

The *Philosophical Magazine*, named for its editor, Alexander Tilloch, was popularly known as *Tilloch's Journal.* Tilloch founded the magazine in 1797 and remained its sole proprietor until 1822.

The Monthly Magazine

The Monthly Magazine, or *British Register,* was founded by Richard Phillips in 1796 and was located at No. 6, New Bridge-Street in London. This periodical was a prime conduit of information to the reading public in England about the discovery of Ceres and Pallas.

Edinburgh Review

The *Edinburgh Review*, from its first appearance in 1802, reflected the outlook of the Scottish universities, in which the sciences were treated as part of general letters and discussed in connection with political and moral issues. Geology, botany, chemistry and exploration were all treated in the *Edinburgh Review*, which subjected scientific writings—as much as literary ones—to informed criticism.

Gentleman's Magazine

Founded in 1731, this publication lasted until 1907. It offered the public "more in quantity and greater variety than any book of the kind and price." (Fader & Bornstein, 1972)

A later work by this author will provide a table listing all the asteroid articles in every British periodical during the early nineteenth century, as well as a list of all the asteroid-related correspondence with their archival locations.

Important Letters and Articles

Schroeter to Best* Lilienthal March 31, 1802

On 1802 March 28, Dr. Olbers after having observed the planet Ceres, accidentally casting a look upon the star No. 20 in the northern wing of the Virgin (near which he had rediscovered the Ceres the 1st of January last), discovered, to his great surprise, a star of the 7th magnitude, forming nearly an equilateral triangle with No. 19 and 20 of the Virgin, and which he was persuaded had not been visible there at that time. He compared it several times with the 20th of the Virgin, but always found the right ascension less and less and the declination greater. The new star appeared to him as perfectly resembling Ceres in his Dollond telescope, without either atmosphere or nebula and could not be distinguished from a fixed star. What a singular circumstance!

March 30: With magnifying power of 288 applied to the 13 feet reflector, I saw it much more striking and planet-like than Ceres, with something of a planetary disc, and tho' not altogether sharply defined, but rather hazy, yet with its limb more distinct than that of Ceres, so that I was able to measure its diameter. With the disc-micrometer with which I had measured the apparent diameter of Ceres on the 28th of March to be 4″.391, I found the diameter of the new planet to be 4″.635, consequently much larger than that of Ceres and Georgium Sidus, the diameter of which last I found on the 20th of March to be 3″.973.

The light of the Olbersian star was in comparison with that of Ceres pale and white but rather more intense, because its disc remained visible in the midst of the illumined disc of projection, and less of the hazy boundaries became invisible. Notwithstanding all this, I was not able to see the last trace either of this or of Ceres with the naked eye, although I could always see the Georgium Sidus as hitherto I had always done, even in a low situation.

The planet was followed westward by a very small darkish star, only visible in the 13 feet reflector. On the 1st of April it was again accompanied by a small dark star, from which it lay S.W. exactly as on the 30th of March, and seemed to be the same, which I suspect to be a satellite.

It is remarkable that the moveable and principal star, perhaps on account of the heavy dew, had no longer the planet-like appearance of the 30th of March; it appeared with the magnifying power of 288 in the 13 feet reflector, less in a stronger light. and could not well be distinguished from a fixed star. When it appeared with a small disc its apparent diameter was only 3″.244 instead of 4″.635 as on the 30th of March.

*[George Best (1755–1823). He was Chamber Secretary for the Electorate of Hanover, and later Privy Councillor to the Royal Court of England and a Fellow of the Royal Society. Best was a longtime friend of Schroeter (they both studied law at Goettingen), and the only one in England who knew that all the instruments of Lilienthal Observatory had been sold in 1799 to the University of Goettingen.]

Article by Seyffer in the GGA April, 1802

The details of this discovery (namely, Pallas) show at first glance the tireless master acquainted with the heavens (namely, Olbers) and if it is possible to exceed his contributions to astronomy, and lately to Ceres, and to increase his fame – this discovery is his crown. Due to bad weather Piazzi was not able to see Ceres until February 23. Piazzi wrote that "I am foremost indebted to Mr. Gauss' ellipse" in enabling him to find Ceres again.

Zach to Oriani Seeberg April 3, 1802

I have just this instant received your letter of the 17th of March, but it finds me as sick as a dog with a bad head cold. Thus, I cannot answer your letter, but since I had already prepared a sheet of observations to send you later I see that I am even less able to do it today. I see that you are busy with the calculation of the perturbations for Ceres.

That is why I am hurrying to send you the first copy of a letter from Burckhardt which showed the same calculation. I believed that this would please you to see what he had done. Another thing I am doing for you is to send the observations to Mr. Piazzi, to which I cannot answer today because I can hardly raise my head. I received two letters from the 2nd and 17th of February. He has not yet observed the planet at this time; he had only learned from the letters that it was found in Germany.

Zach to Ernst II Seeberg April 5, 1802

It is my honour and pleasure to inform you officially about the discovery of the most surprising that has ever been made in the realm of nature – the existence of a new and 9th planet. It is my friend Dr. Olbers who made this fortunate discovery and only now at this very moment I am able to announce this discovery with certainty. Here is the history of this singular appearance. On March 28 Dr. Olbers was observing as usual the planet Ceres; and while browsing through this celestial region with his telescope he noticed in the same area where he had observed Ceres on January 1 a small star of 7th magnitude which he could not remember having seen when he observed Ceres. This star was in no catalogue and at first he considered it one of those small stars that change their light and sometimes disappear and believing it was now in its visible period, he tried to determine it by comparing it to the nearest star no. 20 of Virgo. Being occupied doing so from 8 to 11pm, he saw to his utmost astonishment that this small star was changing its place; he examined it at a higher magnification, but he was still saw nothing but a well determined small star without any nebulosity, line or tail. While waiting he marked down the position of this star. The following day, on March 29 in clear skies he searched for his star as soon as dusk permitted. And he found it to his great surprise to have moved considerably, having a retrograde motion in right ascension in its declination moving north like Ceres. He tried to determine this star's position and impatiently awaited the third day in order to verify the true motion of this celestial body. He sky was favourable on March 30 again and he found, as on the previous days, that the star had moved further retrograde by 10′ and advanced 20 min. in declination to the north. He no longer had doubts regarding this small star but since Mr. Olbers does not possess fixed instruments to precisely observe true motions of stars he asked me to search, observe and certify the path of this strange body. The letter arrived on Sunday, April 4 at 11 am. So five days have passed since his last observation and it needed a little calculation to find the region where this new star could be met. I did this calculation on the hypothesis of a circular orbit and on the basis of this calculation I found on Sunday, April 4 at 11h 9′ 24″.1 mean time the RA of this star = 6z 3° 44′ 6″.6 and decl. 13° 54′ 52″.0. The appearance of this star was as Olbers had described it of the seventh order of magnitude without any trace of nebulosity. But one single position was not sufficient to state the identity of this vagabond in the Heavens; it could be a fixed star, but since no catalogue indicated one in that region it was necessary, in order to be certain, to make a subsequent observation that would prove that the star had changed place and that is what I am certifying right now. For the star that I observed yesterday appeared at 11h 4′ 42″ m. t. at 6z 3° 34′ 2″.4 in RA and at 14° 13′ 26″ declination and the position where I observed the star yesterday was perfectly empty since the body had moved already retrograde to the orient 10′ 4″.2 and 18′ 34″ to the north. The existence and motion of this new celestial body is thus certified. And it is my pleasure to advise Your Highness with certainty and with conviction. And it is the ducal observatory Seeberg that provided the first exact positions of the two most remarkable and extraordinary celestial bodies of this century – of Ceres and Pallas, for this is what Dr. Olbers wants to call his new star if it proves to be a planet. But is this a planet? We still do not know enough to call it that; but it seems to be the case. I tend to believe it is Lexell's comet of 1770 [see footnote] *whose period of revolution is 5 years but there are also objections to this assumption. It would be too boring and long for Your Highness to read the details here and I reserve the pleasure of doing so verbally; while waiting it needs to observe this star attentively, calculate its orbit and wait and see. Today I am only hastening to advise Your Highness of the first certain fact and I am laying Pallas and myself with utmost respect and submission at Your Highness's feet.*

F.N.: The Finnish-Swedish astronomer Anders Lexell (1740–1784). The comet of 1770 was discovered by Charles Messier, but after Lexell computed its orbit, it became known as Lexell's Comet. It features several times in the correspondence in this book.

Zach to Banks Seeberg April 5, 1802

In haste I have the honour to acquaint you with a most extraordinary discovery, which will as much astonish you, as it will surprise all English astronomers. My friend Dr. Olbers in Bremen discovered, March 28th in the northern wing of Virgo besides the well-known Ceres Ferdinandea, another little moveable celestial body, resembling in magnitude and light to Ceres, with a retrograde motion like Ceres except that its northern Declination is increasing quicker. This new heavenly body looks like a star of the 7th magnitude, without the least appearance of a nebulosity. Here is the history of this very extraordinary discovery. As Dr. Olbers is not provided with fixed instruments, but now observes Ceres to his amusement with a loose refractor, and a circular micrometer, to look out for Ceres with a nightglass, and to be in quest of her amongst the little stars, the more thorough. Doing this the 28th March, he found, that he could compare our new planet's brilliance with the star No. 3 Coma Berenices. Having done with this observation he made with his nightglass a review of this part of the heavens, for the purpose, to get better acquainted with all the little stars scattered thereabout, to the end that he might with greater ease find Ceres in his future observations. Sweeping by chance the spot where stands No. 20 Virginis, he with some surprise remarked a star of the 7th magnitude which with No. 191 of Mr. Bode's Catalogue formed very nearly an equilateral triangle. As this was the very same spot where Dr. Olbers saw first, January 1 the Ceres Ferdinandea, this part of the heaven was thoroughly known to him, and he recollected very well, that in January and February no such star had been visible there. His first surmise, was therefore, that this star might be, in the number of these, called stellae mirabilis, like Mira Ceti, or other of the kind discovered in the latter time, in greater multitude, by Mr. Goodricke [John Goodricke, 1764–1786], *and Mr. Pigott* [Edward Pigott, 1753–1825]. *In the meanwhile Dr. Olbers compared this new guest with No. 20 Virginis and continued so from 8:45 till 11 pm, when the weather began to be overcast. But this little interval of time was sufficient to show, that the subsequent observations of the Right Ascension grew shorter, and the Declination greater. These alterations were too regular, as to impute them to the errors of the observations. Thus Dr. Olbers was convinced the same evening, that his new guest, was a vagrant, and certainly had a proper motion. Very fortunately the following day March 29th the weather cleared up, and Dr. Olbers with as much pleasure as surprise perceived at the first glance of the eye that this heavenly body had considerably changed its place; he compared it with No. 20 Virgo, and found that he had moved 10 minutes in RA; and 20 minutes in Declination to the North.*

March the 30th the weather favoured again. The new rambler was already too remote from No. 20 Virgo to be compared with this star; Dr. Olbers took therefore his refuge to two little stars of Mr. Lalande's great Catalogue, in Connassaince des temps *Anneé X, page 254, the two last stars of the page. By this means Dr. Olbers found these three rough positions:* [Zach prints the data from March 28-30.]

The 31st of March Dr. Olbers was so kind, to send me this intelligence, begging that I might endeavour to catch this new star, and observe it with accuracy with my fixed meridian instruments, which he could not do. I received his letter Sunday morning, April 4th, and the very same evening, being very fortunately favoured by a very fair and clear sky, I immediately found after the first dusk with my nightglass, our strange host. I then, without delay prepared for a careful, and nice meridian observation, and was indeed so lucky to get the first very good, I dare say, very excellent, observation of the Right Ascension, and Declination of this new vagabond. I had the moral certainty, that it was Dr. Olbers' new star, but I wished to get the physical conviction of its being the very same body, in comparing its diurnal proper motion, and so I expected (as you will easily guess), with the greatest eagerness the following night, which very happily turned again very clear and serene. Looking to the place where I saw the day before, the star I had observed, the bird was flown, and I observed with delight, that it had moved just according Dr. Olber's indication; I again had a very nice meridian observation, and having thus two very accurate positions of this new rambler, and today being holiday for England [Easter], *I do not longer put off, to give you, Dear Sir, this most extraordinary intelligence taking the liberty to send to you here, my two very good observations, in hopes, you will take this attention to you kindly of me.* [Zach prints his data of April 4 and 5.]

I hope this will be sufficient to point out to the English astronomers this new guest, and that they will find him out soon, this letter being by this season only 12 days upon the road.

To my opinion it appears to me that this new body looks just as Ceres in magnitude, but a little bit brighter than Ceres. Dr. Olbers on the contrary estimates Ceres brighter than the new star. Dr. Schroeter in Lilienthal, who gazed at it with his 13 feet reflector and a magnifying power of 288, finds the star greater, brighter and better terminated than Ceres. He found the new rambler's diameter the 30 March = 4″.635 and this of Ceres the 28 March = 4″.831.

But now, pray Sir, what do you think, of this strange heavenly body? Is it a comet? The appearance and the constant regular motion is against this opinion. Is it a planet? What an immense and paradoxical inclination must this orbit not have! Where to range this planet? Is it perhaps Lexell's comet of 1770, that has a period of 5 years? This opinion is contradicted by the little inclination of this comet's orbit, which is only 1 ¾°. I don't know, what to say, to this strange appearance. Time will tell us, what it is; let us observe now with great accuracy and exactness this curious body, which Dr. Olbers wishes to call (to abridge the denomination) in the meanwhile, <u>Pallas</u>. I shall exert all my power and industry in observing this new candidate for planetisme; and I shall have the honor to give you from time to time my report of the success.

Ceres is not neglected by this new occurrence; and here I take the liberty to send the following observations of this planet. [Zach prints his data from April 1 to 5.]

Ende to Olbers Celle April 6, 1802

But now you, my dearest friend, discovered a planet that is impudent enough to roam a little bit differently than we predicted while picking up the solar system. The possibility of several planets on orbits that we did not even dream of is thus decided and this must become a new drive to study and search those diligently. – And consequently your discovery appears to me as the <u>only</u> means to revive the extinct fervour and to get our society going, which is utterly important for astronomy in many aspects. And also in a different aspect the strange chain of circumstances is utterly noteworthy. – Ceres must have been discovered by a typo a few days before her standstill. — One week later and we would not have heard anything of her. – If it was not discovered then, it would not have been found again in a region where also your new planet was roaming and this one also escaped us most likely. – Finally if Gauss had not calculated so accurately and we in Germany believed his calculations as little as France did, or if some more, like Oriani, whose departement it was, had given up the search completely. – Since your Pallas or Olbers, as de la Lande will certainly call it, *will finish its revolution faster than Ceres it was possible to find it for someone searching for Ceres, and mistaken for it. What contradictions must not arise between the new and old elements! What confusion did not originate from this, what conclusions of a totally perturbed orbit were not drawn – to cut a long story short: only experience or the accidental discovery of the true Ceres made it possible to escape out of this labyrinth. Thus, your discovery is extremely important at this point, because both planets will hide together in the rays of the Sun, in order not to mistake one with the other at their reappearance. – Furthermore now we can easily explain some of Mars' irregularities, whose orbit they must considerably perturb etc. And it is also remarkable that there where we expected one planet, we find two very small ones if we can trust the hitherto perceptions. – One almost wishes to say: once those two small planets had formed a bigger one; at least a comet shock is not more unlikely than throwing a comet against the Sun and the planets splinter off. Maybe, if de la Place's hypothesis of a contraction of the Sun's atmosphere is correct, a comet could be seized at exactly that moment which consequently would, as your Pallas does, complete its revolution on a very inclined orbit. – But these are only dreams that belong into the bedroom and not good society.*

With the greatest of pleasures I will try to observe Pallas and would like to assign to her the symbol [⚴].

I was looking for Pallas yesterday night and believe to have found her in the neighbourhood of the stars no. 85 & 86 of the catalogue. Today will tell whether it was her. Yesterday evening I weakened my eyes observing the occultation by the Moon of the Pleiades and was unable to resume the observation on Pallas. And I was just as unsuccessful with Ceres since I must illuminate the fine spider silks so much and Ceres does not tolerate this illumination. And consequently, I am doubting that I might observe your Pallas at the mural quadrant.

[This letter is the first written record of the idea that the asteroids may have resulted from a cosmic catastrophe. Olbers never gave credit to Ende for the idea.]

Bode to Olbers Berlin April 6, 1802

At 8 o'clock I found a 7th magnitude star at the position 'a' where according to the calculations your star was supposed to be (Fig. 7.3). Did I see the right one? I also observed it with the 3 f Dollond and found it with bright light and of 7th magnitude without nebulosity. Then I wanted to determine its position with the circular micrometer, but yesterday we had a busy evening here at the observatory, which is unfortunately not rarely the case. It had become known that the moon covered the Pleiades and thus spectators had come who actually always annoy me although this is not their intention. But I had to interrupt the studies on your star and its observation at the circular micrometer was impossible so I postponed a more accurate determination of its position until its culmination. I tried to determine it preliminarily by means of the chart as exact as possible and believed it sufficient enough since the field of the wall quadrant covers ¾°. But when I had positioned the wall quadrant to that altitude I was annoyed to see that it did not appear in the field. And meanwhile the air had become hazy. The stars appeared far paler and even Ceres appeared considerably less bright than usually at its culmination. I had let θ 447 and δ Leonis pass beforehand and hoped to determine the position of your star and Ceres' quite well, previous observations were mostly very successful. But your star did not appear despite all carefulness I applied, e.g. the illumination of the hairs was too great, I did see Ceres very distinctly and I do not grasp the reason and am very sorry that I cannot give you any better determination of its position. Today will show whether I hit the mark; I will observe the culmination at the circular micrometer, determine its position accordingly and then assume it for the wall quadrant. I believe it cannot escape me thus, with the following mail I will announce my success. Where are we to search, hardly one new discovery had been announced and caused trouble, the second takes place which also causes work – if only time would increase proportionally to all new tasks! I honestly do not know what to think and say of the new wonderful body, it

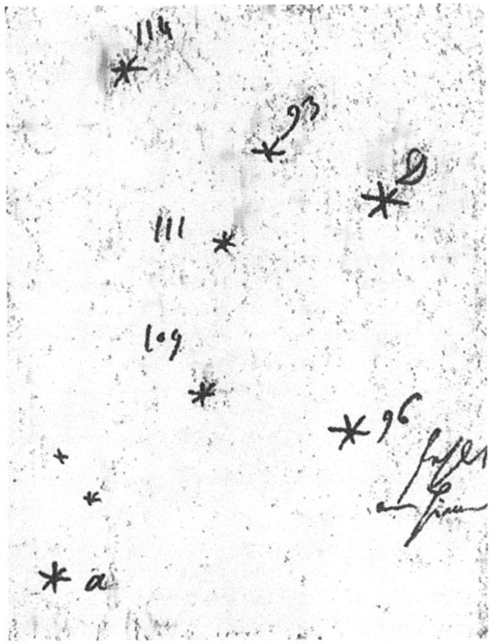

Fig. 7.3 Bode's finder chart of April 6, 1802

moves too much northwards and thus is rather a planet. Maybe and most likely a very distant comet without a perceptible nebulosity because of the distance with a shining core. Where should there be space in our solar system for a further planet, all positions are taken. Is it a shining sphere at an immense distance? Also Prof. Huth discovered in December a small star west of θ and δ Leonis and noticed a retrograde motion and which disappeared soon after, he sent me this sketch of its motion. The search for this star refrained me from looking for Ceres for a while.

Why did our friend von Zach not announce in December his rediscovery of Ceres? Last year he told me everything he received and learned of that planet. But I want to give you the following observation on Ceres at the wall quadrant which I consider good [The last of these observations, April 3, does not appear in Bode's French-language paper.]:

	mean time	apparent RA	apparent Dec	app Long	app Lat
27 March	11h 47' 21"	181° 29 15	17 50 47	5 24 2 14	16 54 54N
1 April	11 23 35	180 26 57	18 2 48	5 23 0 42	16 41 12
3 April	11 14 7	180 3 13	18 6 14	5 22 37 41	16 34 55

PS: Also Mr. von Hahn wrote me yesterday that he discovered above Vindemiatrix a body showing a small silver-white disk, with distinct boundaries without nebulosity. At a magnification of 600 times it became greater and brighter. At first he believed it was Ceres which should have been farther west. I did not yet have the time to check. I will announce your discovery of the moving star to the Academy and scientific society with pleasure. [Bode is referring here to the Prussian (Berlin) Academy of Sciences.]

Sniadecki to Zach Cracow April 6, 1802

It was my honour to communicate in my last letter of March 22 my observations of the new planet from March 1–20 inclusively. The clouds that have covered the sky for five days only allowed me on the 25th to resume my observations that I am hastening to transcribe for you. (Here follow the observations of March 25, 27, 28, 31, April 2, 3, 4, 5 as they are inserted in the great Protocol page 61.)

The declination of the 27th is dubious since I think I mistook a small planet for the planet to which it was close but its RA differed 47″ in time from it. The right ascension of 66 Virgo is erroneous in Bode's catalogue, it needs to be decreased by 3′ 5″.7 of the arc as I could convince myself by a large number of observations I made of this star comparing it to η Leo; this work was essential for the planet was close to this star for a couple of days. Yesterday I saw it close to 57 Virgo which it passed and I will compare it to that star, weather permitting but it has already started to get overcast. In order not to confuse the planet with any small star it passes closely I map every day that region of the sky where the planet is in comparing it to 66 and 57 Virgo. I observe these stars before and after the time of the passage with the greatest attention and I notice every day in an almost palpable manner the motion of the planet, changing place. I received your journal for March a couple of days ago and with pleasure I saw that my observations do not differ much from the planet's position you gave for the months of March and April. I believe to have noticed the considerable change of light Mr. Schroeter mentioned and that it is more to be seen with the help of excellent instruments, which he possesses. Piazzi, to whom I was very close in London in 1787, must feel like any other astronomer what is due to your zeal and journal, since without these the planet would not have been found again so early by continental astronomers, since some of them did not only doubt its existence but denied it, a fact of which I learned from my correspondence. If you can also show us the identity of this body with any body observed previously, we would owe you our fast progress in making this planet's tables as we owe you already the observation of its motion.

Zach to Méchain Seeberg April 6, 1802

The Dean [Laplace] *has probably announced the discovery of a new planet by Dr. Olbers which he calls Pallas. Here are its positions:*

	m. t.	RA	Decl. N.
March 28	9h 25′	184° 57′	11° 23′
29	8 49	184 46	11 53
30	8 3	184 36	12 13

I am sending you today my very exact observations of this star.

	m. t. Seeberg	RA of Pallas.	Decl. N.
April 4	11h 24′ 51″.9	183° 44′ 6″.6	13° 54′ 52″.0
5	11 20 17.3	183 34 23.7	14 13 22.9

Pallas appears to me fainter than Ceres.

Zach to Ernst II Seeberg April 7, 10am, 1802

The importance of the object is so great that I cannot refrain from informing Your Most Serene Highness of everything concerning Pallas. I have just now received the attached letter from Mr. Olbers confirming the conjecture which I had the honour of explaining to Y.M.S.H. yesterday: i.e., that there are probably many more planets in the gaps between those that we have known for centuries. Hence the number of small stars that have disappeared and whose path we have not followed. The road has opened up and there is no doubt anymore that, very soon, we will have masses of new planets, which must necessarily exist between Venus and Mercury, Venus and Earth, Earth and Mars etc.

His Highness is already holding one proof in his hands: I observed Pallas on Sunday, April 4; and Olbers has sent me his most recent observation of April 3:

	RA 183 54' 32"	*Decl. 13° 34' 16"*
I observed on April 4	*RA 183 44 7*	*– 13 54 52*
Diff. ...	*10' 25"*	*19' 36"*

The above difference in positions is in perfect accordance with those of the previous days where Olbers observed the planet at RA 10′ 35″ Decl. 19′. 48″.

The motion is slightly decreasing – as I had the honour to say yesterday. I hope to see Pallas this night, the wind is easterly and the barometers have prodigiously risen. I would like it to rain die tota, redeunt spectacula noctu [It can rain all day as long as the skies are clear at night so that I can watch the stars.]

With submission at Your Highness' feet
your very humble servant Zach

Zach to Gauss Seeberg April 7, 1802

I will not spend a long time on telling you the story of Pallas' discovery because I know that our mutual friend Dr Olbers did this. Thus, I content myself with giving you my observations of this peculiar heavenly body so you can practice your art, diligence and astuteness.

On March 30th Dr Olbers told me for the first time of his extraordinary discovery. I received his letter in the morning of April 4th and that very same evening I was observing Pallas myself. I observed this strange celestial body on April 4th, 5th and 7th; on the 6th it was overcast. I do not send you Olbers' observations since you most likely have them already but with every mail I will send you my continued observations. Sometimes Pallas appears brighter and then fainter than Ceres; they are probably similar. I am praying that we somehow will be able to get a sufficient number of observations to determine the path roughly in order to find it again. I will not neglect Ceres, I am still observing her, but soon Ceres and Pallas will be in collision and of course the least known of those two has priority. Here are the observations of Pallas. [These three observations were included in a table published in the May 1802 issue of the *MC*.]

Continued observations of Ceres

[This table covering the period March 27 to April 7 was published in the April and May issues of the *MC*.]

Gilpin to Maskelyne London April 7, 1802

By a happy accident the present letter has been detained one post day. I say happy, for this delay gives me an opportunity to join to it an advertisement which appears to be of the highest importance, viz. nothing less than another New Planet discovered by Dr. Olbers at Bremen. As he has charged me expressly to inform you if it and in the same time assure you of his most perfect regard, I transcribe here the import of his letter of April 2, which I received in this moment.

'Since March 28 I observe in the northerly wing of the Virgin, besides Ceres, another moving small star, perfectly similar to Ceres with respect to its light and exterior appearance. In my extremely good refractor of Dollond, magnifying 180 times, it may not be distinguished from a fixed star of the 7th magnitude, without any nebula, retrograde as Ceres, the northerly declination motion only increasing much faster.'

Since March 28 Dr. Olbers has made the 4 observations following [these are Mar 28 through April 1, 1802]: *Mr. Schroeter at Lilienthal observed this new star (which Dr. Olbers proposes to denominate 'Pallas' if it should prove to be really a new planet) since March 28; he distinguishes its disc still better than that of Ceres; also he estimates the first somewhat greater than the second.*

Dr. Olbers wishes ardently, that astronomers provided with fixed instruments (which he is not) may commence as soon as possible to observe this new star, as it has already past its opposition. It will hardly be observable longer than June. However, I hope we shall get a number of observations sufficient to compute its orbit, nearly enough to find it again in 1803 if it shall prove to be a course rentrante. *I do not doubt, but this important discovery will obtain all your attention, and you will oblige me infinitely, if to your observations of Ceres you will join your observations of Pallas.* [Appended are three observations of Pallas by Zach made on April 2, 3 and 7, 1802.] [Olbers also wrote a similar letter to Burckhardt on April 2]

Zach to Banks Seeberg April 8, 1802

The great importance of the object, will make my apology for teasing you thus, with my correspondence. But your ardent affection in promoting sciences, which are so much indebted to you, is a sure warrant of your indulgence.

The very extraordinary heavenly body, whose existence, and discovery, I had the honour a few days ago to signify to you, is certainly a primary planet, that moves according to the general laws of gravitation in a very inclined orbit round the Sun, between the orbits of Mars and Ceres. Its revolution is about 3 years, the mean distance 2.1. The ascending node 5° 20′. The 6th of April the weather was overcast. The 7th of April I had again a very exact observation of

Pallas Mean Time	*Apparent RA*	*Apparent N. Decl.*
April 7 11h 11′ 10″.6	*183° 15′ 38″.5*	*14° 49′ 2″.1*

The same night observation of Ceres

April 7	*10 55 21.3*	*179 17 39.7*	*18 9 47.1*

Dr. Olbers' observations of Pallas are as follows: [these data from March 28 to April 3 were published in the MC.]

[Zach's letter to Herschel, dated April 8, is the same as this letter to Banks, except the first paragraph was deleted. Zach also wrote a letter to Oriani on this date. The first portion of that letter is the initial image in this chapter.]

Zach to Schedius* Seeberg April 9, 1802

Now, dearest friend, something extraordinary! Read the printed announcement. Nothing less than a new planet, again! Oh, why isn't our Bogdanich around any more? [Bogdanich had recently died]. *Did Taucher* [Franz Taucher; he worked in Buda in 1802] *observe Ceres, then? Will he find Pallas? In any case, I enclose my observations for you, with which he can look up the newest planets.* [a table of observations from Apr. 4-8]

Oh, if no one in our fatherland discovers and observes Ceres and Pallas, I will feel ashamed to be a Hungarian baron, because they're being observed in Poland! From Sniadecki I received the observation of Ceres from Cracow and now I hope to receive some about Pallas from there. But won't any arrive from Ofen, Tirnau, Erlau, Carlsstadt? [see footnote] *Will Ceres and Pallas remain invisible constellations for my fellow wise and worthy countrymen, just as for the Lappen and Samojeden? Oh! If I could only get to Ofen on a magic wish, just for one minute, so that I could set up and focus the telescope on these two curious visitors of heaven! You see, dearest friend, I've still got the blood of an original Hungarian running through my veins. To each his own! Cicero said pro domo sua [in my own house], and I would like that my dear fatherland could still play a role in astronomy. It can do it, it has the means; it has the brains. Things just need to be shaken up a bit.* [Lajos Schedius, 1764–1847, a Hungarian scientist]

[F.N.: Ofen was the German name for Buda (modern Budapest). At this time (from 1777–1815) the observatory was located in the Castle of Buda. Since Zach was writing in German, it was natural for him to refer to the city by the name Ofen. The Hungarian language by educated Hungarians was not at all encouraged in those days. The other places mentioned are also in Hungary.]

Banks to Maskelyne *London* *April 9, 1802*

Dr. Olbers discovered on the 28th of March another Ceres apparently larger (in brightness) than the Georgium [Uranus]. *Mr. Schroeter and Harding saw it on the 30th and on April the 2nd Harding saw a star near it which he thinks may prove a satellite.* [Here he prints the data from March 28 to March 30.] The two first observations by Dr. Olbers, the 3rd by Mr. Harding.

[Banks wrote an identical letter to Herschel on this date.]

Zach to Sniadecki Seeberg April 9, 1802

I have received your letter of March 22 in which you are so kind as to send me your observations of Ceres Ferdinandea. Your trust is a great honour and I am very grateful and begging you to continue to send all what you have done and worked regarding this planet because I collect every observation and calculation for my journal and I do not want there to be a gap. But before talking about your observations, I must say a word of the negligence of my correspondence. It is absolutely and physically impossible to satisfy my needs, my arms are breaking; my correspondence has increased to such a point that I am often obliged to send printed pages to my correspondents. And that is what I am doing – sending you the printed note and the Ephemerides of the rediscovered Ceres. A few words would not add anything to the knowledge. And I can also see from your letter that my goal has been reached and that you found and observed Ceres. Today it is a notice of an even greater importance that I am sending you in a hurry which will astonish you incredibly, but the matter is true: It is the discovery of a new ninth planet in our Solar System. Since there is no time to lose if you want to find and observe it, I am sending you here the printed notice of this amazing discovery together with Dr. Olbers' observations and mine. Before closing this letter I hope to be able to enclose a position of Pallas, for I am observing it. My positions will help you find it for its course is rather regular. More so if you use the interpolation for the differences in seconds. Now a word on your observed positions of Ceres. We do not always agree in our observations as you can learn from my April issue or the pages enclosed. There is a great difference on March 2 and 19. But this is not your fault. You praise Bode's catalogue in your letter, but beware, my dear colleague, of that terrible catalogue, which is awful and causes mistakes, it has mislead all astronomers and you will see this in my April issue. Look at this example, the stars you have used (Fig. 7.4).

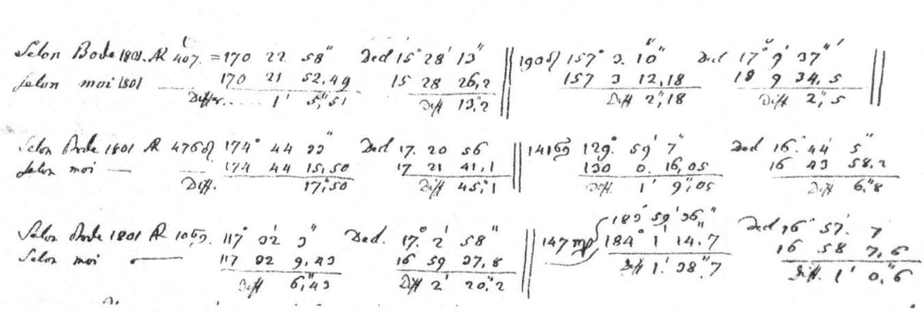

Fig. 7.4 Star positions according to Bode and Zach

In each set, Fig. 7.4 reads:
According to Bode
According to me 1801

*You can see after this sample what trust you should show Mr. Bode's catalogue, if you employed my positions you will find that your observations are in perfect accordance with mine. Send me another time all the details of your observations, I will reduce them then according to my positions for it would be too much to send you all my positions of small stars that you can use. You do certainly also feel my dear Sir and Colleague that the observations of Ceres should be of the utmost precision if they should be useful to correct its orbit. Consequently, I took the liberty to improve your observations according to the errors of the positions of the ** (stars) and we agree perfectly. I will publish them in my journal hoping that you won't contradict, it would take too long to wait for your improvements and I am certain that your corrections agree with mine. Here are Pallas' positions according to Mr. Olbers' observations (Fig. 7.5; The text of the figure reads:)*

Here are mine made at the meridian

Seeberg mean time RA of Pallas Decl. of Pallas

On April 9 the skies did not permit any observation, neither on April 10. I waited for Pallas at 182° 49' 3/4 RA and at 15° 38' 1/2 of Declin.

Bode to Olbers Berlin April 10, 1802

I correctly recognised on the 5th your star (comet). On the 6th we had a very dark evening, rain and a distant tempest. But on the 7th it was fair and I noticed with pleasure the star's motion closer to 109. The air was changeable. At 10 o'clock I let 51 and delta Virginis pass at the wall quadrant which matched your star, but when it culminated, it was not visible due to hazy air, so again my effort was in vain. On the 8th it was completely overcast again. Yesterday it cleared up in the evening, hazy air, moonlight, clouds and mists. At around 10

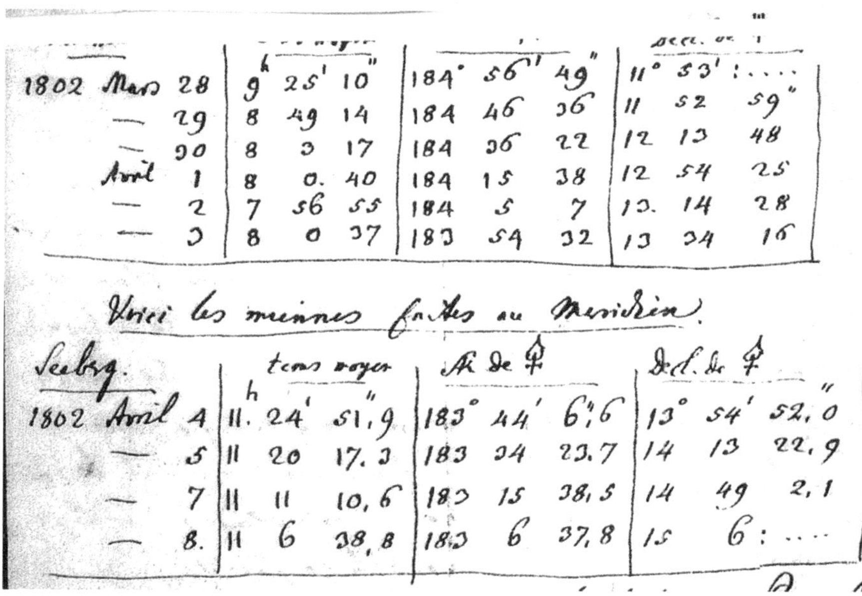

1802 Mars 28	9h 25' 10"	184° 56' 49"	11° 53' : ....
— 29	8 49 14	184 46 36	11 52 59"
— 30	8 3 17	184 36 22	12 13 48
Avril 1	8 0. 40	184 15 38	12 54 25
— 2	7 56 55	184 5 7	13. 14 28
— 3	8 0 37	183 54 32	13 34 16

Voici les miennes faites au Meridien.

Seeberg.	tems moyen	AR de ⚴	Decl. de ⚴
1802 Avril 4	11h 24' 51",9	183° 44' 6",6	13° 54' 52",0
— 5	11 20 17. 3	183 34 23.7	14 13 22, 9
— 7	11 11 10, 6	183 15 38, 5	14 49 2, 1
— 8.	11 6 38, 8	183 6 37, 8	15 6 : ....

Fig. 7.5 Positions of Pallas according to Bode (top) and Zach (bottom)

o'clock I let 109, 111 and the comet pass at the circular micrometer (109 had unfortunately a wrong ascension in La Lande, too far west, approximately 12′) despite clouds and with a great effort I obtained two observations at the circular micrometer (I left out 109 and only kept 111), this one gave me two quite corresponding results, at 11 h 23′ 40″ m.t. an average apparent right ascension of the comet 183° 23′ 3″ decl. 15° 26′ 11″. Today I announced your star as comet in the papers – what else can it be? It must have a very bright core and its fine nebulosity is not discernible at the moment, it is rather paler than a star of similar size, most likely it is moving away from us or we from it. We do not want to hope that the fixed stars become mobile? In order to answer all the disputed matters regarding Ceres I am planning on publishing a generally understandable history of its discovery and print all studies and observations on it until now. The finer details might be saved for the yearbook. Gauss' latest elements are probably sufficient to calculate tables of Ceres?

[The society referred to in the following letter by Seyffer is the Goettingen Society of Sciences.]

Seyffer to Olbers Goettingen April 11, 1802

After having received your kind letter of April 4 in the afternoon of the sixth, I immediately communicated its content to the Society and showed it to Mr. Mayer and Prof. [Johann Christian Daniel] Wildt. I am writing only today, because I wanted to enclose the proof sheet of the announcement I made to the paper. I observed the Olbers Planet twice (I wish you would keep this name) and I would like to hear your judgement whether it was the correct one. The declinations cannot be very accurate since the planet did not tolerate any sufficient illumination at the mural quadrant, which is, as you know, not very powerful. Since the 7th I have not had clear skies but today it seems to be clearing up and I am worried to find the planet again. I would appreciate it if you could send me your latest observations. Tomorrow I will write Piazzi about your discovery. From his letter from of the 2nd I learn that he discovered Ceres only on January 23 and he also wrote: 'I principally owe it to Mr. Gauss' ellipse.'

Blagden to Banks Paris Apr 12/13, 1802

On Saturday last a letter arrived here from Dr. Olbers of Bremen, mentioning that as he was looking at some stars near which Ceres had been last year, he discovered a new star of nearly the same magnitude, which appeared to have a proper motion, and being without nebulosity or tail, or any of the common appendages of a comet, but rather exhibiting a defined disc, though very small, he was inclined to consider it as a real planet. The night between Saturday and Sunday being tolerably clear here, most of the astronomers employed themselves in searching for this new heavenly body, and yesterday evening three of them read to the Institut *a notice of their having found it, nearly in the place where they expected it according to the account of its motion given by Dr. Olbers; but somewhat less distant from the place where he observed it, because it is now retrograde, and approaching to its stationary point. M. de Lambre's notice, read to the* Institut, *is, that when it passed the meridian in the night between the 10th and 11th April, its right ascension was 12h 11′ 15″.9, and its declination 15° 38′ to 40′ North. About 4 o'clock in the morning of April 11th its RA was found to be about 17″ less, and its declination had increased 4′. These gentlemen could perceive no disc: it looked to them like a star of between the 7th and 8th magnitude, less luminous than they expected: it was near the 6th star of Berenice's hair, which is of the 5th magnitude. If I learn anything more of this before the post goes away to-morrow, you shall have it.*[section about Buerg's Tables of the Moon]. *Laplace thinks them* [the Tables] *much more exact than any former ones. He wishes this to be communicated to Dr. Maskelyne, as also the observations of Olbers's new star (which De Lambre proposes to call Juno) and likewise to Dr. Herschel. Both these astronomers, no doubt, have Dr. Olbers's own observations already. I believe M. Mechain will send the French ones, by this post, to Sir H. Englefield and perhaps to others.*

Mr. Bouvard has just sent me from the Observatory the following observations of Olbers's new star. It passed the Meridian of the Observatory at 12h 10′ 8″½ sidereal time: its distance from the Zenith by the great mural quadrant was 32° 38′ 34″5. From these observations and the latitude and longitude of Paris, the RA and Declination are easily calculated; but I have no book here to do it. Mechain and Bouvard say it appears of the same magnitude as Piazzi planet, but they do not here venture to assert, as yet, that it is a planet.

Zach to Gauss Seeberg April 13, 1802

This letter from Prof. Wurm [Johann F. Wurm (1760–1833), pastor of Gruibingen in Wurttemberg, was the first to put Bode's law of planetary distances in the form of an equation: see The Discovery of the First Asteroid, Ceres] *just arrived; there is only time for a few additional words. It seems to me that Mr Wurm has started too early and with inaccurate elements the calculation of Ceres' perturbations. By the way, Kluegel's method, which he used is not accurate enough and his equations for Mars differ very much from Oriani's, Schubert's and Burckhardt's; but all four were wrong and the true equations of Mars that La Place has sent are completely different. But I am not to publicise them since La Place reserves the right to publish those in the* Méc. céleste. *I thought we leave them to Burckhardt, who makes them right in front of his eyes and La Place's supervision. But I am willing to publish Wurm's work in the M.C., although it is useless but just to show that Wurm is diligent and willing to do his share. There are so many astronomers who sit back and do nothing regarding Ceres. For this reason you have to encourage those who work to the best of their ability. Burckhardt has not sent me his perturbations formulae and tables yet, he is still polishing. He sent me his opposition of March 17, 1802 at 3h 46′ 8″ mean Parisian time Obs. National. Longitude 176° 21′ 26.5″ and 17° 7′ 57″.5 geoc. latitude. The mean error of his tables in longitude is –5″.4 and in latitude 21″.8. He subtracted from Delambre's solar tables 11″ of the longitude. Here are his observations:* [This table was published in the May 1802 issue of the *MC.*]

Burckhardt has changed the elements slightly.

ascend. node 81° 2′ 20″

Inclination 10° 37′ 17″

He wants to decrease the inclination by 12″ and increase the ascending node by +10″. We have a new winter here. Detestable weather. But on the 12th I saw Ceres and Pallas. Soon more of this.

Bode to Olbers Berlin April 13, 1802

Let me thank you for your continued observations of your moving star, which can most likely only be a very distant comet with a shining core. I believe it is already decreasing in brightness, is noticeably paler, and the coming full moon will render it completely unrecognisable. I could imagine that an observation of the 9th made in very hazy air, so that in a 3 f Dollond the small stars and the comet remain hardly visible, would not be exact. Yesterday I found that not no. 109 but 111 Virginis according to l. L. was incorrect. The latter was thus 15′ 18″, actually 1′ in time greater in asc. How many mistakes are there in l. L.'s catalogue? I noticed this already while reducing his stars to my great catalogue and had to face and conquer difficulties and doubts, what I could I tried to correct. Now my calculation shows that on the 9th the right ascension of the comet taken from 109 and 111 and two other observations as corresponding 182° 57′ 4″, but the decl. differs by 1° 57′ if I take both 109 and 111, consequently there must still be an error in one or the other. On the 11th I saw the comet at the wall quadrant and compared it to o and β Leonis. I found for 10 pm 53′ 9″ m.t. apparent right ascension of the comet from β Leonis 182° 40 20 from o Leonis 182° 40 29, the decl. from β Leonis 15.55.28 from o Leonis 15.55.20.

Yesterday evening was unsettled weather, every other moment thick clouds alternating with moonlight, I wanted to compare the comet with the stars of Virgo and o Leonis. The stellar observation was successful but then I had the comet only accurate at the 1st, 2nd and 3rd hour hand but when I wanted to bring it to the main hair, a cloud advanced and spoiled the observation of the declination. But I obtained the ascension of the comet 182° 33′ 25″ and o Leonis 182° 33′ 22″ at 10 h 44′ 33″ m.t. It is peculiar that the comet precisely rambles through the apparent path of Ceres, yesterday I had it together with Ceres in the field of the searcher. Ceres, the comet and Uranus culminate now so close together that one does not know how to arrange the observations regarding the stars culminating in between which should match one or the other in order not to get confused or overhasty.

Zach to Oriani Seeberg April 13, 1802

I expect you to say: What a terrible astronomical bombardment! Every post day a new bomb from Seeberg! But this only proves that I am a very humble and obedient servant, always ready to serve you, while you let me languish for your news. Are you one of those mysterious astronomers who like secretiveness? Knowing from your last letter that you are working on calculating Ceres' perturbations I thought you would appreciate it to have everything that has been done regarding this matter. Since the French are keeping their formulae secret, I am unable to tell you anything about their work. But I am sending you here what Mr. Wurm found out according to Klügel's method. I have sent you Ceres' orbital elements according to the perturbation equations. Since then he told me he had been forced to improve a little bit the node and the inclination using two more perturbations which he had initially neglected, knowing the separated terms that depend on the arguments $nt+\varepsilon$ *and* $nt+\varepsilon'$*. With these two additional equations he found the* ☊ *81° 2′ 20″ and the inclination 10° 37′ 17. He also wanted to increase the* ☊ *by 10″ and decrease the inclination by 12″. The opposition was observed in Paris on March 17 3h 46′ 8″ m.t. at 176 21 26.″5 real long. and 17° 7 57.″5 lat. geoc. In my calculation there is a slip of the pen of 10″ for the opposition in the lat. geocentr. Instead of 17° 8′ 9.″0 there must be 17° 7 59.″0. We agree in the longitude of the opposition of Pallas. We have terrible weather at the moment. I did not see Pallas again. But I saw Ceres yesterday.*

12 April 10h 32′ 18.″0 m.t. RA 178° 26′ 34.″0 [Decl. 18° 8′ 54″]

I am finishing my daubing in a hurry and assure you of everything you already know. I do not send you the observations of Triesnecker in Vienna, Sniadecki in Cracow or Bode in Berlin. You will find them in my journal. [Here follows a section on Ceres perturbations.]

Blagden* to Maskelyne Paris April 14, 1802

At the desire of Mr. Laplace, and in conformity to my own wishes, I send you the enclosed observations of the new star discovered by Dr. Olbers, made at the Observatory of Paris by Mr. Méchain and Bouvard; the former taking the zenith distances by the mural quadrant (of Bird) and the latter taking the passage over the meridian by the transit instrument.

	Mean time	apparent RA	appar. Decl.	Appar long.	Appar. Lat N
Apr. 10	13^h 58′ 40″	182° 45′ 10″	15° 41′ 30″	176° 6′ 58″	15° 19′ 16″
12	10 48 34	182 32 31.5	16 10 58	175 42 47	15 49 11

These observations are not corrected for parallax or aberration.

Mr. Bouvard has given me the last of these observations and that of last night, April 13, in the following form

April 12 RA 12^h 10′ 10″.2) sidereal time(32° 39′ 10″.5) zenith
13 12 9 39.4) (32 24 2.5) distance

You will see how far they agree. If this star be a planet, they have rudely calculated its distance from the sun as 1.9, somewhat more distant than Mars; but the great obliquity of its orbit to the plane of the ecliptic raises doubts. It appears here as nearly as possible of the same brightness as Piazzi's planet.

*Charles Blagden was Secretary of the Royal Society (1784–97), which awarded him its Copley medal in 1788.

Stephen Lee to Maskelyne Grove Hackney April 14, 1802

Sir Joseph Banks having obligingly written to inform me, that Dr. Olbers had discovered another erratic star on the 28th March last, I looked for it on Sunday night, but without success. On Monday, I looked for it again, and in RA 182° 36′ Decl. 16° 17′N by estimation discovered a small star, which from its appearance and position being certain I had not seen one in that place before, led me to suspect it must be the one in question. I was prevented from repeating by observations, as I intended to have done, two or three hours afterwards, by the cloudiness of the weather, but last night saw it again at 11h 50′ 48″ mean time in RA 182° 24′ Decl. 16° 27′ North. It is of the 7th magnitude, of a dull red colour, but appears brighter sometimes than at others, being from its faint light more easily observed by vapours floating in the air than the fixed star.

Banks to Maskelyne London April 16, 1802

The news of the new planet reached Paris on Saturday the 10th inst. On Sunday the 11th 3 memoirs were read to the Institut which happened to meet on the day by persons who had seen it; by desire of M. Laplace I communicate to you the following observations from Paris.

The planet which M. Delambre proposes to call Juno was when it passed the meridian on the night between April 10th and 11th in RA 12. 11. 15.9 North Dec. 15. 38 to 40. About 4 in the morning of the 11th its RA was found to be about 7″ less and its declination had increased 4″. On the night between the 12th and 13th of April it passed the meridian at the Paris Observatory at 12° 10′ 8″.2 sidereal time observed by the planet its distance from the zenith was 32° 38′ 34″.5 by mural quadrant.

[The account of Burckhardt receiving news of Pallas on April 10 is related in *La Decade Philosophique* (1802), no. 30, 129–132. Delambre (1802) also kept the *Société Philomathique de Paris* (created in 1788) informed of asteroid developments, including a pre-recovery report on Ceres in November 1801.]

Banks to Herschel London April 16, 1802

Mr. Gilpin saw the new planet on the 9th and has kept it ever since. Mr. Lee saw it on the 12th. No one else has seen it here as far as I know. [The next section repeats the last two paragraphs of the letter he wrote to Maskelyne on this date.]

[Earlier Olbers-Gauss letters were printed in *Discovery of the First Asteroid, Ceres*. The numbering system established by Schilling (1900) is retained for all the Olbers-Gauss letters:]

No. 13 Olbers to Gauss Bremen April 18, 1802

In the hope that you're occupying yourself with the orbit of my small and odd planet Pallas, I'm sending you the observations of two additional days apart from the reduction of the observation on the 13th: [These observations from April 13-17 were published in the May issue of the *MC*.] *Mr. Bode has sent me two meridian observations:* [These observations from April 11-12 were published in the May issue of the *MC*.] *Mr. Zach writes me that he will inform you of his observations directly. Pallas, in spite of a full Moon, is still easily visible as an 8th magnitude star and I hope the observations will continue for a considerable time.*

Zach to Sniadecki Seeberg April 19, 1802

I am hastening to answer your letter of April 6. Thank you very much for your continued observations of Ceres. Of all my correspondents you are the most indefatigable observer of the new planet and you can see from my journal that it is you who gave the best and longest sequence of observations of this planet. I have told you that I took the liberty and corrected the first sequence of your observations, which I publish in the May issue of my journal and I am glad you saw for yourself how erroneous Bode's catalogue is and I believe that you won't dislike to see in my journal of May on p. 386 how much we agree on the estimate of the error of the RA of no. 66 Virgo. You said that the RA had to be diminished by 3′ 5″.7. As you will see I took away 3′ 10″.2 and thus our difference is 0″.3 in time over which we will not quarrel since my error in this observation was +0″.15 and yours –0″.15, which means extremely in accordance and I am surprised how you could come this close with your instruments. In your last sequence of observations, which you kindly sent me, you again used badly determined stars, e. g. no. 415 Ω Bode gives for

RA = 171° 4′ 25″ Decl. 1801 17° 53′ 49″.21

Bode 17° 53′ 43″

6″.21

I found........................... 171 4 58.56

Differ. 33″.56

Consequently it is impossible that your observations agree with mine but here you will see them again using the true positions of the stars. The same thing happened to Triesnecker and David as you can see in the April issue. Their observations were impossible and they were obliged to correct them and this must happen to all those who compare the planet to small stars in the parallel, for our old catalogues are really a shame. I learn from a letter from Mr. Triesnecker that you sent him your observations of Ceres, for he sent me an identical copy of what you kindly sent me and I am asking you to send him our improved observations so that we agree on what he can publish in his ephemerides and I put into my journal. I really appreciate that it is you who gives him the corrections and I told him only that I treat you as I treat him, meaning that I gave you the new positions of the star. I gave you my reason why I took it into my head to make the corrections myself for without it your observations would have been delayed by one or maybe two months before I would have been able to publish them in my journal and the astronomer-calculators would have been alone far too long with these observations. I also corrected your second sequence of observations and you will find it at the end of this letter but because my May issue is complete I can only put it into the June issue and until then it is three weeks, so please keep sending me your observations if you agree with my corrections and if so I will print them the way I am sending them here and you will send them to Triesnecker for his ephemerides. I will do the same if your letter will arrive too late especially at present where I am obliged to arrange my June issue in advance since I will be travelling and want to accompany and see my friend Buerg home to the Bohemian frontier; he leaves for Vienna after having spent ten months at my place. We will make this journey in small daily stages since we want to work on geography on the way and we do the entire trip with my horses. It might therefore be very likely that going there and returning will take me three weeks and in every case I leave the entire manuscript with the printer and your last observations will be printed after my corrections on which you can count with all certainty and I will not compromise your ability and dexterity in observing, known to all astronomers in Europe. In order that this shall not prevent you from writing I have arranged some places to where all my mail will be forwarded and it might be that your letters reach me and I can give my orders in writing to the printer. His majesty the Duke had the kindness to send me a parcel by a hussar, my communication with Gotha is constantly established and as fast as the post.

Do not forget to notice that in your observation of March 1 and 2 there is a mistake due to copying and that in the declination of these two days you gave 16° instead of 15°. That jumps to the eye but often one does not notice it for it is not degrees but seconds one has to be afraid of. If you reply to me, please tell me if I can and shall send you the first page proof of my journal with the regular post – this would save me a lot of writing and you would receive all observations and astronomical news at least four weeks earlier; it goes without saying that I would send you only what naturally must and could interest you the most. So you know Piazzi personally. I had the bad luck to miss him as you did by one year. You were in London in 1787 and I left after having passed four weeks in England in 1786. But this does not prevent our close friendship. You described him quite well in your letter since before congratulating me on the first discovery of his little planet he adds: "I can never express what feelings of gratitude and attachment all of your efforts have aroused in me. I find in you such a brother and friend I could not possibly have hoped to find any better." I am very flattered by this confession for Piazzi is a man of merit, who does and works a lot for the promotion of our science and such scientists are always hard to find. Piazzi is of an astonishing assiduousness as you can see from his letters of which I will give extracts in my journal. You can be sure to get the Log. tables of Taylor, whom I knew personally; he is dead now but this should not prevent you having your copy. I looked at mine and I found you on the list of subscribers in every letter. Mr. John Sniadecki Professor of Astronomy in the University of Cracow. It was a good idea to send me your receipt of subscription and I forwarded it to Sir Joseph Banks who honours me with his correspondence. Dr. Maskelyne is director now; but you know as well that the doctor of Greenwich is a Niezwycz [see footnote] *(I write badly, but dear Sir you understand!). The president of the Royal Society (Banks) will know how you can get your copy and the surplus will be paid to my orders and I will reimburse you.*

Did you find Pallas? This enigmatic celestial body lets us work furiously. Is it a planet or a comet? I do not know – the more we observe the less we know. I had it the previous night and I send you all my observations as well as Olbers'. The star appears to me always of the same size and well determined, its motion has become a bit more irregular just as it is supposed to be moving to its halt. But the first attempts of a circular orbit were not successful and it is even proven that a circle would be wrong; it remains to know if it is an ellipse we need and if so it would be a very eccentric one. Patience! We still must gather a greater number of observations to decide. It is a very singular star! But what luck! Judge for yourself. If we had found by chance Pallas first before Ceres wouldn't we have taken her in the course for Ceres? What confusion. We would have combined Piazzi's observations with those of our Pallas – what Galimathias! [senseless twaddle, or confused gossip] *The worst thing would have been that the search for the real Ceres would have been neglected and before we could have learned of our mistake the real Ceres would have been lost in the rays of the sun and maybe for ever for us. And maybe we would have never discovered, nor even suspected this mistake, and made hypothesis after hypothesis until lost out of sight.*

Goodbye to beautiful Ceres – poor Piazzi! What a bad situation for him. They would have cloaked his observations in doubt, as they permitted themselves already to do. He would have been accused of fraud and lying, he would have been questioned disagreeably, and their mocking at him would have been directed at us tutti quanti, who had believed in the existence of this planet. How nicely we have escaped, thanks to the diva Ceres that has protected us and saved us from that abyss that we have approached so closely, for if one had not known the diversity of a Pallas and a Ceres that gather like a drop of water and that have appeared in almost the same celestial region where we waited for Piazzi's star, these strange observations would have upset current and future astronomers, and it may have taken centuries to disentangle this confusion. The famous Abbé Makry, at present cardinal, made his sermon short because he did not have enough saliva and I make my mine short because I do not have enough paper.

[F.N.: In Polish, *niezwyczajny* means 'unusual,' 'atypical,' so 'Niezwycz' (which is likely an abbreviation of Niezwyczajny) means someone who is weird or strange or difficult to deal with. The fact that the word is capitalized suggests that this may be a description or a name that the correspondents have adopted to refer to this particular person (a nickname, if you will). In any case, it's not a flattering name or description, in keeping with the fact that Zach does not seem to like Maskelyne.]

No. 14 Gauss to Olbers Brunswick April 20, 1802, 6 p.m.

My heartfelt thanks for both of your letters of the 13th and 18th and your observations of your Pallas which is becoming stranger day by day. Please do continue to communicate these observations to me.

Shortly after having received your first observations I attempted to fit an arc through the positions of March 29 and April 1 and, like you, always found the motion to be too quick. The same happened to me on the 14th when I received Zach's observations of April 4, 5 and 7 and tried to combine them with yours of March 29. Using my own method I tried, independently of any hypothesis, to fit an arc to the observations of March 29 and April 4 and 7, and immediately found this to be totally impossible due to the proximity of these observations. I therefore thought it best to leave the matter temporarily and wait for further observations. I might have re-attempted this after having received your penultimate letter (April 17) if some previous engagements hadn't prevented me. However, when I received your latest observations last night I couldn't resist any longer.

I chose your observations of March 29, Zach's of April 7, and the mean of your last two of April 17 (the irregular motion over a few hours isn't at all understandable even considering the possible observational errors). At the first attempt I immediately obtained the following orbital elements. I'm sending these just for you and request that you examine them as a token of my admiration and extra-ordinary interest in your remarkable discovery. I certainly hope I haven't rushed my calculation; but the effect of the smallest change in the observations is still so large that the true orbital elements could well still differ considerably from the following ones. Nevertheless, every day brings us closer to the truth, and I hope soon to be able to send you improved results. [These elements precede the official Elements I as printed in the *MC*.]

Elements of Pallas
The orbit of a non-retrograde ellipse

Epoch March 31, 1802, noon	
Seeberg meridian	*153° 24′ 14″*
daily (sidereal) motion	*673.33″*
log. of semi-major axis	*0.4811840″*
eccentricity	*0.334920*
aphelion	*295° 40′ 33″*
inclination	*37 24 52*
ascending node	*173 15 43*

These elements almost exactly reproduce your two outermost observations, but parallax and aberration haven't yet been taken into account. I've just now computed Zach's observations using these elements and find:

	calculated longitude	*error*	*calculated latitude*	*error*
Apr. 4	*177° 45′ 49.9*	*−12″.4*	*14° 13′ 29″.5*	*−7″.5*
5	*177 29 8.9*	*−10.9*	*26 30.9*	*−7.3*
7	*176 56 51.6*	*−13.2*	*51 38.6*	*−2.3*

Tomorrow I hope to receive further observations from Zach. Because of the large uncertainty still associated with these elements, one should, to be fair, refrain from draw-

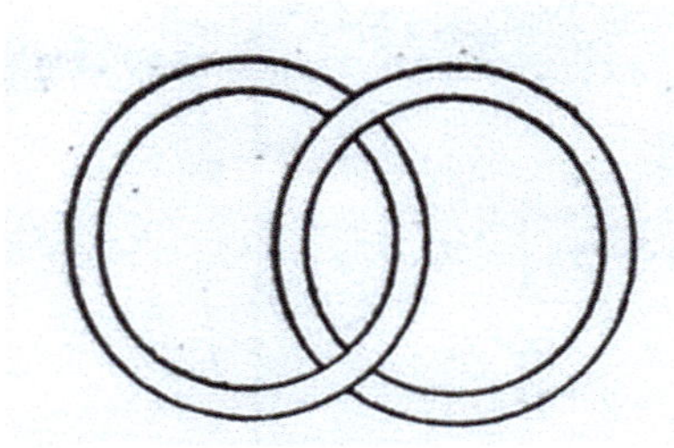

Fig. 7.6 Gauss' sketch of the interlocking orbits of Ceres and Pallas

ing any conclusions. Should, however, the calculated orbit not be totally dissimilar to the true one, then its position compared to Ceres' orbit, is one of the most remarkable celestial phenomena. These two paths interlink with each other, which I can most easily clarify using this figure (Fig. 7.6).

Both paths would come frightfully close together at a place not far from the area where the two stars are. Our descendants could perhaps some day be spectators of the most terrible phenomenon: the collision of the two celestial bodies! This time they'll miss each other. Nevertheless, these are now only dreams that perhaps will be shattered within 14 days as a result of better orbital information.

Should the previous elements, which are nevertheless surely possible, resemble the true ones, then one would certainly have no scruples to call Pallas a planet. Indeed the greatest distance from the Sun is double that of the smallest, but in the case of Mercury the ratio is 3:2. The path of Pallas would differ little from a circle except that the Sun wouldn't lie at the centre. But the familiar law, so splendidly introduced by Bode and so nicely confirmed by Ceres, would be instantly nullified. – On that point I wouldn't at all be surprised. I have, speaking confidentially, never thought highly of it and must make an observation here which I already have 'in petto' since 12 years and am surprised that someone hasn't yet made it. It is briefly this: The series 4, 4+3, 4+6, 4+12, 4+24, 4+48, 4+96, 4+192 is not a continuous series. One has only to be aware in order to see that 4+1.5, and not 4, should precede 4+3; that Mercury hence doesn't fit into the series, or that between Mercury and Venus there should still be innumerable planets. These one would certainly not expect. I'd like your opinion on this matter.

P.S. Seyffer [Carl Felix von Seyffer (1762–1822)] *was enormously pleased by your letter and sends you his highest regards. He also started to observe Pallas.*

Aubert to Herschel London April 20, 1802

I received yesterday, my very dear friend, your letter of the 18th and send you now my observations of last night (or this morning) both of the Ceres and of Dr. Olbers' planet

Ceres at 14h 26′ 37″.6 sidereal time
RA 11h 49′ 49″.6 Dec. N. 17° 58′ 14″.5

Dr. Olbers' at 14h 52′ 48″.6
RA 12h 5′ 14″.1 Dec. N. 18° 1′ 57″.5

By the observations communicated to me by Dr. Maskelyne and Mr. Lee I judge the motion of Dr. Olbers' planet to be diminishing in RA daily about 32 seconds of time and increasing in Dec. north near 17 min. of a degree daily. The above observation of Dr. Olbers' planet is the first I have been able to get, it having proved cloudy for several nights before; the moon does not hinder the observation of it. It is larger than the Ceres, of a reddish hue and rather more distinct in its appearance.

Memorandum about Pallas, by Maskelyne April 21, 1802

On the 20th of March 1802 Dr. Olbers of Bremen accidentally looking at the star No. 20 in the northern wing of the Virgin, near which he had rediscovered Ceres on the 1st day of January last, to his great surprise saw a star of 7th magnitude, which had not been there before, by which means and attending to its motion he found it was another new planet. It appeared to him, with his Dollond's telescope, perfectly resembling Ceres, without either atmosphere or nebula, and not to be distinguished from a fixed star. Mr. Schroeter of Lilienthal on the 30th looked at it with his 13 feet reflector, and thought it more striking than Ceres, with something of a planetary disc, and measured its apparent diameter 4″.635, whereas on the 28th of March he had found that of Ceres only 4″.021, and on the 20th of March that of Georgium Sidus 3″.973. The light of the Olbersien planet was paler and whiter than that of Ceres, but rather more intense; yet he could not see either of them with the naked eye, tho' he had never failed to see Georgium Sidus so, even in a low situation. Both on 30th of March and 1st of April, the Olbersien star was followed by a very small darkish star, which he suspects to be a Satellite, from which it lay S.W. Here at Greenwich it has appeared of a dullish light, less bright than Ceres, but not so bright as Georgium Sidus. At the Paris Observatory, from 10th to 13th April, it has appeared, as near as possible, of the same brightness as Ceres. Considering this star as a planet, they have roughly calculated its distance from the sun as 1.9 that of ours, consequently more distant than Mars; but the great obliquity of its orbit to the ecliptic (for its latitude on the 12th was 15°.49N) raises doubts.

Maskelyne to Herschel Greenwich April 23, 1802

Yesterday I received the ephemeris of the new planet Ceres Ferdinandea from Dr. Gauss who calculated its elements, which perhaps you may not have received, therefore I present you with a copy of them. They will probably find its place within a minute. Mr. Méchain in his letter of 27 January expressed his wish that you might make some discoveries about it with your great telescopes; and mentioned that the same of theirs was not yet finished. In his letter of 16 March, received only yesterday, he asks me whether you have measured the diameter of the planet, or seen anything remarkable about it, its disc or atmosphere, or any satellite of which there is some suspicion? The new planet, discovered by Dr. Olbers at Bremen, appears like to and of equal brightness with Ceres. I have sent you its places below, observed here, in the hope you will make some discoveries concerning it also. There is a suspicion of a satellite about it. [Positions from April 15 to 22 are printed here.] *Considering this star as a planet Mr. Méchain writes that they have calculated its distance from the sun as 1.9; but its great inclination raises doubts. They mention the observation on April 10 was 15° 19′ 16″ and 12h 15° 49′ 11″ both North.* [An ephemeris of Ceres from April 21 to June 29 ends this letter.]

Banks to Herschel London April 23, 1802

Mr. Best has received another letter from Mr. Schroeter in which he says Dr. Olbers wishes his new star to be called Pallas. From further observations he believes that this new heavenly body moves between Jupiter and Mars as Ceres does, is bigger than Ceres, about the size of Ceres including its nebula, and like Ceres has an orbit more oblique than the planets. On the 2nd of April he measured the diameter 3 days and made it as follows

April	*2*	*4″.735*
	3	*4″.671*
	4	*4″.680*

It seems remarkable to me that two objects similar to each other moving nearby on the same path should differ from all other Primary Planets so much, the nebula with which both of them are surrounded is quite different from what has been observed in those Primary

Planets. The inclination of their orbits also resemble those of comets. Their size also is so trifling that I cannot help thinking that on mature consideration and new identified observed stars, astronomers will not consider these strangers as Primary Planets but as another sort of revolving body such as have in fact before been discovered and of which many more here after be found. [According to Lynn (1904), Herschel suggested the asteroid family may number 30,000.]

No. 15 Olbers to Gauss Bremen April 23, 1802

Your letter of April 20 brought me much enjoyment. You can well imagine how much I'd like to reserve the dignified appellation 'planet' for my discovery and your letter gives me the highest hope to do so. I admire you, dear friend, and your method, your promised communication of which I am impatiently awaiting. I haven't at all tried to calculate the ellipse because the observations still seemed to me to be spaced much too closely. Shortly after the observation of April 10, I sought to obtain a parabola using the observations of March 28 and April 4 and 10. The calculation, based on my method and without any corrections, gave the following:

ascending node	*5z 23° 42.5′*
inclination of orbit	*38 28.5′*
log. of perihelion distance	*0.156206*
time of perihelion	*3z 18° 8.5′*
temp. perihelion	*Nov. 7 18^h 40.1801^m*

Yet, as I again computed the mean observation from these elements, the errors in longitude and latitude turned out to be +11′ 27″ and –2′ 42″ respectively. I couldn't rely much on the elements, because 'M' or the ratio of the distances from Earth could be found only with much uncertainty. This was so partly because of the small geocentric motion and partly because of the position of the geocentric apparent path that coincidentally cuts the ecliptic almost at the point that lies opposite the Sun in the middle observation (April 4). However, it already seemed to me that both the longitude and latitude in the middle observation could not simultaneously completely satisfy a parabola and that hence the true path could perhaps like wise differ just as much from a parabola as I previously found it to deviate from a circle. Your more rigorous calculations will confirm these rough speculations in a splendid and, for me, very unexpected way. There could well be more changes to the orbit you discovered. The arc is too small, and my observations are unfortunately not of such quality as can be obtained by a Piazzi. Suffice it to say that the orbit of Pallas is an ellipse which is much more eccentric than any planetary orbit, but still, as you rightly note, qualifies Pallas as a planet and will pass by in 5 or 6 years. – Please continue with your keen investigations of Pallas, which interests me more than ever. I'll daily send you my further observations, which I'll try to make as accurate as I possibly can. I almost feared that the mean of the two observations of the 17th might also be incorrect because both were taken singly between clouds. I believe the following ones to be better: [This table of observations from April 18-21 was in the May issue of the *MC*.]

Pallas is still very recognizable, is only slightly less bright than Ceres, and still surpasses that of an 8th magnitude star. I thus still hope to observe it for a fairly long time. However, it seems to me that the brightness, as was noticed with Ceres, is often subject to peculiar fluctuations, which appear to arise from its shape or atmosphere. Yesterday, the 22nd, was cloudy; but maybe today I'll still obtain a good observation which I'll then add at the end.

Fig. 7.7 Equation

$$=\frac{a\sqrt{\frac{3}{2}}}{2R}=\frac{a}{R}\sqrt{\frac{3}{8}}$$

There is something unusual about the position of Pallas' orbit. In particular, the very close approach of the paths of Ceres and Pallas at some point gives rise to much thought. How? If Ceres and Pallas were fragments of a former larger planet that had been destroyed by colliding with a comet? It's still too early to indulge in such dreams; but a collision with a comet isn't totally impossible. A number of years ago I found the following solution to what was initially a seemingly difficult problem using probability theory. If I knew nothing more about a comet other than it approaches the Sun closer than a planet, and the mean distance of the planet from the Sun is 'R', then the probability that it approaches the Sun before or after its passage through perihelion is as 'a'

This equation, for small values of 'a' and hence for large approximations, is strictly exact: that is to say, exact as long as a/R = the sine of an arc of such shortness that one may interchange the arc itself with its sine or tangent. – This probability will of course be very small for each single case; yet since a number of comets annually reach perihelion within Ceres' orbit, this probability can become very large over several million years. – Might we then still discover more pieces of this previously existing planet at its appropriate position? The point on Ceres' path, to which Pallas' path, as you note, comes so terribly close and hence almost cuts it, will be very odd; for just at this point all remaining pieces of the destroyed planet must cross Ceres' path, neglecting the gradual and small displacements caused by the perturbation of forces of the remaining celestial bodies, especially Jupiter. Thus one could search for these fragments in the region between Leo's tail, Virgo's northern wing, and Berenice's hair during the first 4 or 5 months of the year. I suspect you are probably smiling at my dreams!

Your comment about the mean planetary distances of Bode's law is quite correct. But can one really expect that this so-called law, for which one cannot give any physical foundation, is mathematically rigorous? There must certainly be an underlying and completely different still-to-be-discovered law, maybe simpler, maybe more complicated, for which that empirical progression of numbers, or series, only somewhat holds true. According to La Place, as is well known, the distances of the 3 inner satellites of Jupiter offer a good analogy, and I still hope something similar will be found for the distances of the main planets.

I've tested your patience long enough today. Do continue to attend to our Pallas, dear Gauss and inform me of the results of your thoughtful research as soon as possible. You can imagine my impatience waiting for the confirmation and improvement of your orbit.

P.S. I was able to observe Pallas only twice tonight since it quickly became overcast; there is good agreement between both observations.

Apr. 23 9h 41m 2s 181° 23′ 59″ 18° 32′ 11″

Here are the latest observations of Mr. Zach, should they not yet be known to you. [These observations from Apr. 15-19 were published in the *MC*.] *The right ascensions, writes Baron Zach, are all good. There is some confusion with the declination; Buerg probably mistook a fixed star for Pallas on the 18th.*

[Bode's law is now seen as one of many power laws that occur throughout nature. Zipf's law, for example, concerns word frequencies in any language. It was promulgated by Harvard scholar George Zipf (1902–1950). Although some of these laws may have a physical basis, others do not. The debate has raged for centuries.]

Sniadecki to Zach Cracow April 24, 1802

In reporting on my first observations of the new planet, I made an error in declination of March 1 and 2: I wrote 16° instead of 15°, this error happened while copying hastily my draft which is too gross not to be noticed by yourself, of which I forgot to tell you in my second letter of April 6 where I had the honour to send you my observations until April 5 inclusively. Here are all together with my calculation of the opposition of the planet.

	Mean time	appar. RA Ceres	app. Decl.n.	Compared star
March 1802			of Bode	
1	13h 50′ 24″	186° 40′ 17″	15° 30′ 0″	
2	13 45 50	186 31 30	15 36 43.8	165Leo 304 Leo 407
15	12 44 53	184 2 56.7	16 57 34	107Leo 147Virgo

(So the observations of March 16, 17, 20, 25, 27, 28, 31 and April 2, 3, 4, 5, 6, 7, 10, 14, 22, 23 were copied among 21.) For the following results I used your Solar Tables and without the planet's declination on March 16, I interpolated it from the preceding and following days.

	M. t. Cracow	App. long. Geo. of Ceres	Geo. latitude n. Ceres	Position of Sun
March 1802				
15	12h 44′ 53″	5ᶻ 26° 44′ 57″	17° 7′ 32″	11ᶻ 24° 41′ 32″.25
16	12 40 8	5 26 30 45	17 7 42	11 25 40 52.03
17	12 35 22	5 26 17 0	17 8 0	11 26 40 21.84
19	12 25 47	5 25 49 47	17 6 52	11 28 38 56.09
20	12 21 1	5 25 35 38	17 6 4	11 29 38 15.19
25	11 57 4	5 24 28 19	16 58 24	0 4 34 18.13

On March 17 the hourly motion of the Sun was 2′ 29″.03, hourly motion of the planet in longitude 34″.37 so the comp. hourly motion 3′ 3″.4. It follows that the opposition of Ceres to the Sun took place on March 17 at 4h 56′ 45″.8 m.t. Cracow. In the moment of opposition the Sun's position was 11z 26° 21′ 22″.78. The geo. long. of Ceres 5z 26° 21′ 22″.78, geo. lat. n. 17° 7′ 54″. All my observations were made during the passage of the planet with a meridian telescope and a quadrant placed on the plane of the meridian, comparing the planet with stars that were almost on the same parallel during the observation. There are mistakes in the positions of some, like 66 Virgo 415 Leo, 493 Leo that I tried to correct preliminarily. We will have to see whether such a precious work like Mr. Bode's needs corrections and refinement by the efforts of astronomers. I hope to see by your journal how much my work differs or agrees with that of others. The planet's brightness starts to decrease noticeably; yesterday with slightly illuminated hairs of the telescope I saw it only with difficulty.

Gilpin to Herschel London April 24, 1802

Agreeably to your desire I send you all that is yet known of the elements of Olbers new planet which I have procured from Sir Joseph Banks since I received your letter, and which Sir Joseph had just received from Mr. Zach. Dr. Maskelyne whom I have also seen this morning does not know more about it – the following is a copy of Zach's letter. My observation of Olbers: RA 181° 23′ Decl. 18° 33′N last night 10h 0′ m. time.

Blagden to Banks Paris April 24, 1802

Burckhardt has been engaged in calculating the orbit of Olbers's new star; but, not having found any circle or nearly circular ellipsis which answered to the observations, he had recourse to a parabola, and found the following elements. Inclination of the orbit 54° 58′ 30″; Node 5z 26° 45′ 34″; Perihelion 3z 23° 52′ 3″; the 29th Sept. 1801 16h 48′; perihelion distance 1.8432; motion direct. Burckhardt is, however, still trying to find an ellipsis which will agree with the observations; and the astronomers here do not yet absolutely

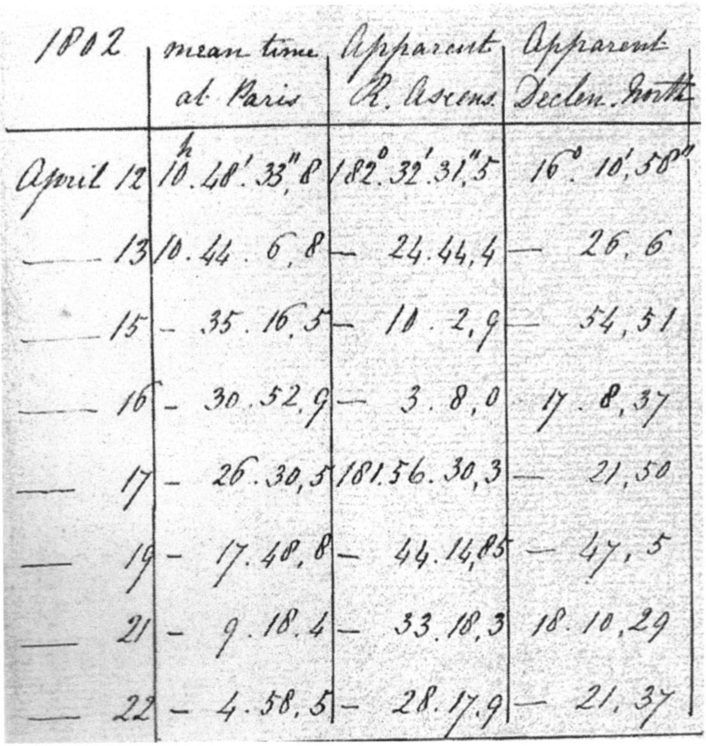

1802	mean time at Paris	Apparent R. Ascens.	Apparent Declin. North
April 12	10h.48'.33",8	182°.32'.31",5	16°.10'.58"
— 13	10.44.6,8	— 24.44,4	— 26,6
— 15	– 35.16,5	— 11.2,9	— 54,51
— 16	– 30.52,9	— 3.8,0	17.8,37
— 17	– 26.30,5	181.56.30,3	— 21,50
— 19	– 17.48,8	— 44.14,85	— 47,5
— 21	– 9.18,4	— 33.18,3	18.10,29
— 22	– 4.58,5	— 28.17,9	— 21,37

Fig. 7.8 Observations of Pallas from Paris in April 1802

declare it to be a comet, as they have not great confidence in the first observations of Dr. Olbers. I hope before this letter must be closed to receive the last observations made here upon this star, and will add them. [Referring to the table, Fig. 7.8] *The declinations are still affected by parallax, aberration and nutation; the right ascensions by aberration and nutation. Mr. Burckhardt thinks his observations of the 10th April, which I formerly sent you, not correct.*

Zach to Gauss Seeberg April 27, 1802

I have to apologize for not having answered two of your dearest letters. But time flies and five or six days appear as one. And nothing makes my life faster than my magazine: hardly have I begun a new issue when it is already finished and thus the year is passing. I am very grateful for the continued comparison of observations of Ceres and you are completely right in waiting and seeing until the entire period of its visibility is over. For a rediscovery your elements are more than sufficient, so why improvement? There will be enough time during summer for calculating the perturbations and here too will be quite some differences and improvements. For example Burckhardt has made an appendix already. Wurm, too, made corrections and Oriani will definitely show something different. You did get Wurm's letter, didn't you? Further below you will find his appendix. Here are the observations of Ceres (Fig. 7.9):

I do not intend to send you Maskelyne's observations, because he informed me he wanted to do it himself. "I have sent several observations of the new planet C. F. to

April	M. Z. Seeberg			AR app. ⚳			Decl. app. ⚳ B.		
	h	m	s	°	′	″	°	′	″
12	10.	32	18,0	178°	26′	34,00″	18°	8′	54,0 :
13	10	27	44.7	178	17	12,30	18	7	42,6
15	10	18	41,8	177	59	23.70	18	5	7,5
18	10	5	17.0	177	35	4.00	17	59	29.1
19	10	0	50,9	177	27	37.20	17	57	8.7
24	9	39	4.8	176	55	46.00	17	42	14,5
25	9	34	47.3	176	50	21.00	17	38	26.4
26	9	30	31,5	176	45	22,60	17	34	46,3
27	9	26	17.5	176	40	49.90	17	30	51,8

Fig. 7.9 Observations of Ceres from Seeberg in April 1802

Dr. Gauss and will send him more. He promises to make the best use of them in his power." In his last letter of April 15th he gave his last observation of Ceres on April 6th as follows: 10h 59′ 52″ RA Ceres 179° 28′ 19″ Decl. Ceres 18° 9′ 10″.2. I can read between the lines that he is not against a comparison with your elements and a publication in the M.C. So if you have a series please communicate it for this purpose. I have published two of your comparisons of Maskelyne's observations. The astonishing thing is that Dr Maskelyne did not say a word about Pallas in his letter of April 15th. And on the same day I got a letter from Sir Joseph Banks of April 16th that Gilpin, Secr. of the Royal Society of Sciences had found Ceres on April 9th and that a very ingenious young astronomer M Lee had observed Olbers' heavenly body on April 13th at 11h 50′ 48″ m. t. RA Pallas 182° 24′ Decl. N 16° 27′. Apparently, the Astronomer Royal has not yet given up but keeps his observations to himself. It irritates him that my observations differ 10″ from his in declination. He believes my spider's threads, which he dislikes, are to blame. But why do my stellar declinations correspond so well to Piazzi's? By the way, me and my 4-foot quadrant are far from competing with 8foot quadrants and entire circles and when I have received mine this summer, as Troughton promises, I will deliver both – excellent declinations and RA. For the month of April Maskelyne's and my declinations correspond better and the difference hardly amounts to a few seconds as you will see. It is my honour to be able to send you an entire series of observations of Pallas (Fig. 7.10).

The RA are very accurate, but friend Buerg finds it hard to observe the declinations at the quadrant. On April 18th and 19th he seems to have mixed it up with a star. On April 24th and 25th he missed it entirely. I am very eager to know what you will find, now that neither ellipse nor circle obey, most likely a very eccentric ellipse, ergo a planeto-comet a new kind of heavenly body. There might be a few of them but since they differ little from small stars – how can we find them? If La Lande had observed his 50,000 stars not only once but several times, he might have found Pallas and other moving stars. [This table of Pallas observations from April 4-27 was published in the May 1802 issue of the *MC*.]

To my knowledge Pallas is being observed in Berlin only. I have not heard anything from Paris, but they must have found it since in all Parisian letters can be detected a ridiculous jalousie. Since they have no part in all those observations and did not lift a finger they now want to show their excellence in calculating, they make a big secret of it "in order not to be preceded" as they write, "we are doing fine without Gauss". I do not need to ask you because it is my firm belief that you will be the first to provide information about Pallas and

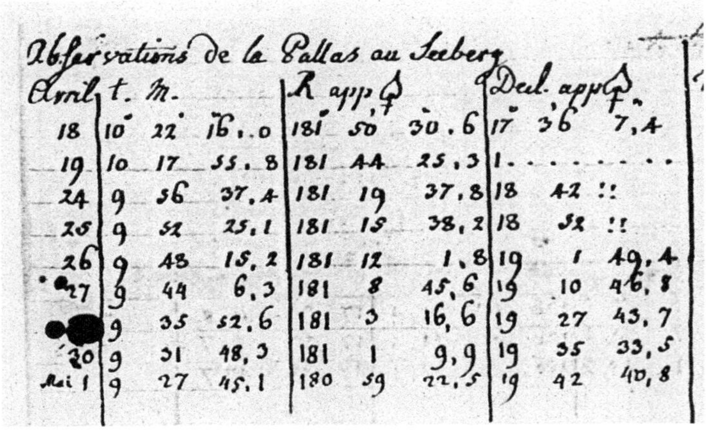

Observations de la Pallas au Seeberg

April	t. m.			AR app.			Decl. app.		
18	10	22	16.0	181	50	30.6	17	36	7,4
19	10	17	55.8	181	44	25.3	1.......		
24	9	56	37.4	181	19	37.8	18	42 !!	
25	9	52	25.1	181	15	38.2	18	52 !!	
26	9	48	15.2	181	12	1.8	19	1	49.4
27	9	44	6.3	181	8	45.6	19	10	46.8
[illegible]	9	35	52.6	181	3	16.6	19	27	43.7
30	9	31	48.3	181	1	9.9	19	35	33.5
Mai 1	9	27	45.1	180	59	22.5	19	42	40.8

Fig. 7.10 Observations of Pallas from Seeberg in April 1802

the first true orbit. Neglect Ceres! And spend all your time and effort on the peculiar Pallas and let us show the high-spirited French, who like it so much to push the Germans into second place, that Germans discovered, observed and calculated Pallas and they did not give one iota. You already mentioned in your last letter that you consider the whole matter a test and you work only for your own and science's sake. Oh please, keep to this fine and noble principle and you will get on well and no-one can and will spoil your life. For you will now soon feel the burden of fame yourself, soon you will know that all that glitters is not gold. But your principles, and what Mr Zimmermann [Eberhard von Zimmermann was Gauss' former teacher at the Caroline College] *has written about you, makes me believe that you will be an Impavidus* [will be intrepid]. *I could not help but mention it in the May issue and I am hoping that you are not unhappy about this because I had more than one reason to do it. Since I know best how people think about certain things and even write openly to me with all their naiveté, I simply had to give them a little avis de lecteur* [notice to the reader], *those who are concerned will understand best. Follow LaLande's example, who made the following calculated mathematico-politico-morale:*

There are a thousand million people living on this earth,
Of these thousand million heads
How many are wicked, foolish, bestial,
But we cannot cure,
We can only pity and serve them.

Wurm has got your elements VII and at once calculated the coefficients of his perturbation equations anew, namely Ceres' distance 2.76996 and eccentricity 0.081406. The smaller coefficients change only slightly. Perturbations of longitude. [These equations were published in the June issue of the *MC*.]

My observations from which Herschel drew such weird conclusions are in the first volume of supplements to the Berlin Astronomical Yearbook. Friend Buerg appears to be mistaken in his calculations for Ceres namely the equat. centri. And they have one minute more than we – your elements give 315° 57′ 16 not 325° 58′ 16″. Here is the calculation. In order not to miss the collection let me close this letter. Soon more observations of Pallas.

[These figures of Pallas differ somewhat from those Zach published in the June issue of the *MC*. Zach also attaches the following: (1) a table of his positions of

Ceres from April 18-May 1, which were published in the *MC;* (2) a table of Méchain's observations of Pallas from April 12-17, and Ceres from Feb. 26 to Apr. 17. These were published in the June issue of the *MC;* (3) a page on equations relating to the perturbation theory of Ceres by Oriani.]

No. 16 Gauss to Olbers Brunswick April 28, 1802

Since you'll certainly welcome everything that to some extent confirms the planet-like nature of Pallas, I'm sending you with utmost speed the calculation of Zach's three observations based on my elements that I've forwarded to you. [These observations from Apr. 15-19 are in the May issue of the *MC.*] *Aberration and parallax are not yet taken into account; the error in RA and Dec. would thereby be substantially decreased.*

Zach has indeed promised to mail me his observations, but I haven't yet received any word. I have only his first three and the ones above kindly forwarded by you. I'm therefore asking you, for security's sake, to also send me the interim ones and those still to come. He wrote me that he saw Pallas on the 12th, but I don't have the observation yet.

No. 17 Olbers to Gauss Bremen April 28, 1802

I'm sending you herewith my two latest observations of Pallas that is still easy to see and appears as an 8th magnitude star. Pallas was compared on both nights with No. 11 Comae Berenices, the position of which I had verified with Mr. de Lalande's two observations in the Hist. Cel. Lalande's two observations differ by 12″ in Dec. [Olbers prints his April 26-27 observations, and the Apr. 10-13 observations from Paris. These are in the June *MC.*] *These observations complete the gap between Zach's observations very well. My letter arrived in Paris only on the 10th at 4 p.m., and in the evening Pallas was already observed.*

Burckhardt is so actively occupied with the calculation of the orbit that he didn't take the time to reply to me but asked Madame de Lalande to do so. Maskelyne was informed of the discovery of Pallas on the 9th; I don't know yet if he observed it. I'm awaiting your next letter very impatiently and will immediately notify you of everything that might possibly arrive from Paris.

Bode to Olbers Berlin April 30, 1802

Upon my return of an absence of seven days I found your letter of the 16th, whose content saddened me and caused the bitter feeling of loss. How is it possible that you can imagine that I wanted to misappropriate the discovery of your comet and even am responsible for the article in Hamburger Zeitung. *You have been set against me by several astronomers and Mr. v Zach and from this and some remarks in the* MC *I learn that I have a secret enemy in him through no fault of mine. He wants to make me smaller, looks out for my mistakes and censures them.*

I only receive Hamb Zeitung *several weeks late and from a friend because of the read articles and according to your hints I too found on April 13, to my annoyance, my article of* Berliner Zeitung *very much altered here. Please find the latter enclosed and judge for yourself whether I gave the* Hamburger *author reason for his announcement; I did not communicate anything of your discovery to Hamburg, but only read your letter to the* Academy *and other scientific societies; how could I have dared to pretend your discovery was mine. I am very sorry that due to my journey this letter can be mailed only today. According to your wish, I mailed today by the riding post the corrected version of the article to the experts of* Hamb Corr *on the condition that it will be published in the next issue and also asked who permitted himself to hand in the article for publication so very much altered.*

Furthermore I am sending you for my legitimation the Berliner Zeitung *of January 14 and 23, wherein I announced your first and my later rediscovery of Ceres. Since you discovered her 14 days earlier, everyone immediately sees that I found Ceres accordingly because I had your information.*

Piazzi seems to have notified Gotha on February 2 and 14 that he had not discovered Ceres again. On March 2 he wrote me: "On the 23rd (without doubt Febr.) I found my planet, without having been able to make any observation before that due to bad weather. I thank you for your felicitations and also congratulate you since you deserve it just as well."

This was his answer to my letter of Jan 12 in which I announced your observation, without doubt he found Ceres again according to your indication, but he does not mention this with one single word –. I would also like to know why this astronomer kept its possible discovery so long to himself, so long that no other could take part in it. I am pleased from the bottom of my heart that you were the first to discover Ceres and Pallas and only your tireless sedulity led you to these discoveries. But I will by no means allow to be robbed of the little honour of being the first, after thirty years of anticipation, who announced, from the first sources, Piazzi's star as the planet between Mars and Jupiter.

Many thanks for the news on your comet, I sent it to Berliner Zeitung *where it will be published in your own words. Bad weather, the fact that the star did not tolerate the illumination at the wall quadrant, that I mistook a different star for it several times, my absence of 7 days, these and other circumstances caused that only two observations at the wall quadrant were successful:*

	RA	Decl.
April 15 10h 37 25 mean time	182 10 31	16 54 30
19 10 17 56	181 44 41	17 46 40

I am curious to know how these observations correspond to yours and others. There will be no shortage of frequent and exact observations of the comet, I do what I can.

On the 29th Ceres culminated at 7′ 12.5 after β Leonis and stood 1° 41′ 38″ higher. I have not been able to observe your comet since that day. On the 29th I believe to have recognised it at RA 180 and decl. 20, did I see correctly? Yesterday it was hazy.

For the Easter fair I printed a brief history of Ceres' discovery, wherein I will also treat your merit in the discovery of Ceres justly. Mr. von Zach again mentions my name in the April issue of MC but in his own way, from which my friends can see that he has something against me. He warns of errors in my star catalogue as if I was entirely to blame for those, although I always indicate my sources: la Lande, Flamsteed etc. Others were wrong, not me, this should be said aloud. I have already hinted at this in a new sic V and VI of the preface to these errors and all the difficulties I encountered during this immense effort. Impartial critics should see this. Recently, Miss Herschel noticed in her work a series of mistakes of Flamsteed's catalogue and his charts (but not while copying the positions of the stars). What is excused in the case of 2,800 stars should be the more so with 17,000. And I even offered to publish all found mistakes in my yearbook 1805f. Usually Mr. von Zach equates La Lande's catalogue with the Augean stables, and it would be a great effort to create order there and now this information seems to favour it and at the expense of me, I know best what I had to do reducing d. l. Lande's inclination. I have already informed you that 111 Virginis is faulty by 15′. I also found D Virginis in Flamsteed and the experts know that his data is no longer exact enough. But why does nobody mention this fact? I said of 171 Virginis myself on p. 56 that Flamsteed mixed it up with 181, I had to list it because of the number. 67 Comae Beren. is correct according to Flamsteed and v. Zach says the declination was too great by 10° … but no star of the Coma can stand 10° farther north. What is this?*

I will hand in my corrections for the January issue, if Mr. von Zach does not take these in, I am forced to publish those in Allgemeine Litteratur Zeitung *with my comments. I have not received any responding letter of Mr. v Zach, who even did not communicate the discovery of Ceres. He told me he wanted to review my celestial chart and no word of this review has been published yet, only the errors of it are mentioned. I am hoping you find my justification right and proper and will not share my words with anyone. Delight me with a more pleasant letter than your last so I can see that our mutual honest friendship has not suffered by this fatal misunderstanding.*

*Flamsteed's 1725 Catalogue appeared in Vol. 3 of his *Historia Coelestis.* The actual star count is 2935, but if duplicates are discarded the total is 2913. Caroline Herschel redid the catalog and submitted it to the Royal Society in 1798, with an additional 560 stars. Bode's Uranographia of 1801 contained 17,000 stars.

[John Flamsteed (1646–1719) was the first Astronomer Royal. An overview of *Historia Coelestis* was given in Appendix 4 of *Discovery of the First Asteroid, Ceres*. For a near contemporary account of *Historia Coelestis*, see Reid (1729).]

No. 18 Olbers to Gauss Bremen May 1, 1802

I'll continue mailing you my future observations of Pallas on a daily basis; its motion in RA and Dec. is diminishing very much. [The RA's will be corrected in Olbers' next letter to Gauss.]

Apr. 28	*11h*	*44m*	*11s*	*181°*	*6'*	*0"*	*19°*	*19'*	*52"*
29	*12*	*3*	*10*	*181*	*3*	*30*	*19*	*27*	*57*
30	*12*	*3*	*25*	*181*	*1*	*25*	*19*	*35*	*37*

The last two positions depend of course on two small stars in the Hist. Cel.; but I always find the observations in the Hist. Cel. to be very accurate. I've taken much effort to improve these last two observations in my own way so that you may be able to use them for your calculations to some extent.

Out of curiosity I calculated the observation of the 29th using your elements. I can't say that I've actually attained an accuracy of one second. I found errors in the longitude and latitude of −2′ 20″ and +0′ 29″ respectively. This agreement of an observation taken 12 days from the basic set of observations (recorded over a 19-day period) appears to me to be good enough, at least generally to accept your orbit determination as being reliable and certain. This all the more so since I have reason to suppose that the error in longitude, using a more accurate calculation, could still turn out to be considerably smaller.

Up to now I haven't informed anyone of your elements, as you requested, and will not do so until you give me permission. However, I now do feel entitled to tell Zach in confidence that your investigations of Pallas indicate an ellipse between Mars and Jupiter.

I'm astounded by your skill, accuracy, and proficiency in the computation of this orbit. The perspicacious method that you discovered must be quite good. It would be a totally futile exercise for me to want to compete with you in computing the elliptical orbit, and I hope you accept the distinction and effort to have determined Pallas' orbit all by yourself. This orbit with its marked inclination and certainly lengthy period of observation could perhaps itself become more reliably known than Ceres' orbit before its rediscovery.

P. S. May 2, morning. Last night I succeeded in getting another observation of Pallas. I consider it to be very good, assuming the comparison star in the Hist. Cel. is sufficiently well determined. [He also includes observations by Zach from Apr. 24-26.]

May 1 12h 27m 15s 180° 59′ 33″ 19° 43′ 31″

The following important letter by Zach about the nature of Ceres and Pallas is in response to a letter from Banks dated April 16:

Zach to Banks Seeberg May 1, 1802

Here I also take the liberty to send you the promised map of the Part of the Heaven, where Ceres Ferdinandea and Olber's Pallas are rambling now. I have described upon, the apparent Path of both. The Ceres from 7 Decb. 1801 where I discovered her first, described till June 29th 1802 a sort of Epicyclus, being in this Period of time just direct, then stationary, and afterward retrograde. She will become stationary again in a few days, about 9 May in AR the 2 May in Longitude. The Path of Pallas marked blue upon the map, is represented in a straight line. She crossed the Track of Ceres the 19 April. This latter heavenly body is *a very remarkable one, and certainly a Middle-Thing between a Planet and a Comet. For it is a Fact now, that the observations and Positions of this star cannot be represented by a circular orbit, a Parabolical orbit has been therefore tried, but with as little success, and we are pretty sure that a Parabola will not do, it remains only the ellipse, and it seems that a very excentric Ellipsis will agree best with the motions of this remarkable body. The Pallas cannot be deemed a comet, if we understand by a comet a hairy blazing star, moving in a parabolical orbit. For she has not all the appearance of a nebula, or a dark gloomy star, not the least Trace of a tail, Bush, or Pencil of spurious Light. She looks rather clearer than Ceres; is about of the same size. She moves not in a parabolical curve. The Pallas cannot be deemed a Planet, if we understand by Planets, heavenly bodies revolving in little excentric ellipses round the Sun, and pursuant to the law of distances, compleated now by the Discovery of Ceres, and extending to the Georgian Planet, & perhaps beyond. The Pallas has no assigned Place as a Planet according to this law in our solar systéme. She moves in a too excentric ellipsis, and has a too great inclination of the orbit, as that she might be ranked amongst our Primary-Planets. This Body gives us therefore the indication of a new species, that we might call PlanetoComet, so we'll have, fixed Stars, Primary-Planets, Secondary-Planets, and Planeto-Comets.* [Compare this discussion about comets with what Zach published in the May 1802 issue of the *MC*.]

The Annals of Astronomy and the Records of time, even since the discovery of the Telescopes, never mentioned a comet, without a Nebula, a hairy bushy Tail, Beams, or Rays of Light. The Denomination of such celestial Bodies itself is taken from Hairs, or Curls, and the greek word κομγ denotes Hairs, or a bush of a Hair, from whence comes the name of κομγτγς.

Mr. De la Lande says in his Astronomie XIX Book, that a Tail, Beams, Hairs, are not a distinctive character of a comet, and in order to support this assertion he quotes several comets, which to his Report had no such visible Distinction. But in this he is greatly mistaken. For instance he says the Comet of 1585 was perfectly round and without a Tail. But the Landgrave of Hesse Cassel who first observed this comet, in a Letter to Tycho de Brahe makes the following description of it: 'Exiquusest, et undquaque crinitus, ut erem etc.' Tycho himself in his Astronom. pg. 752 says of this comet, quedi fibris quibusdam refurbus juxta circumferentiam extitit, minusque illic turebat. Some other authors compared this comet, with the Nebula or Praesepe of Cancer.

LaLande says the comet of 1665 was clear, and had no Tail, and almost no Beams. But Hevelius who observed this Comet in his Cometographia Lib XI p.775 speaks of a Tail. Loci ad quam cauda porrecla fuit, and he gives Angulum Deviationis caudae.

According to LaLande's Astronomy *the Comet 1682 was as round and as clear as Jupiter to the report of Cassini. But in the* Histoire de l'Acad. Roy. *Tome I p. 350 it is plainly spoken of a kernel of the Body, and of a Head. Flamsteed observed the same Comet, and he describes it in his I Vol. of* Hist. coelestis Brittan. *p. 108 with these words: Caput autem tubo pedum 16 considgatum exile apparut, sed spisso inde emanante capillitio, five cauda longa 5 grad* & ... *Hevelius in the* Aolis Eruditorum *1632 pg. 291 gives in Tab. XX a Draught of the appearance of this comet, which is represented with a large and bushy Tail, and he says, 'Initio cauda 12 five grades etc.' I have turned over Lubeniecki* [Stanislaw Lubeniecki, author of Theatrum Cometicum, 1668] *Pingré* [Alexandre-Guy Pingre, author of Cometographie, 1784], *and all cometical chronicles, but I found not one example of a comet reported of an appearance like a fixed star, or like our Pallas, as we behold her now. But why should we not make classifications? To the famous Halleyan Comet of 1531, 1607, 1682, 1759, and which again will become visible in 1835 not such a Middle-Thing? Why should we not distinguish the elliptical, the parabolical, and the hyperbolical comets? What a body is that very remarkable comet 1770, that moves in 5 years round the Sun, and has only once been seen? Where are the comets of 1680, 1729, 1742, 1744, 1763, 1773, 1779 to be ranked? The denomination of comet is too loose, too indefinite, and I hope Mr. Olbers's Pallas will make us acquainted with a new species of heavenly Bodies, and as you observe very well, the Heavens seem to be more peopled than we believed them to be. Perhaps sev-*

eral Stars, which are deemed to be los'd have been such comets as our Pallas. Dr. Herschel and Mr. La Lande have a catalogue of nearly 200 such forlorn stars, certainly the greatest part of 'em arised from errors of observation; or of the Print, or from slips of the Pen. But it is equally possible that Dr. Olbers Pallas, & perhaps one or two such deities more might have been amongst them. Future time, our Progress in practical Astronomy, and the augmentation of observatories will learn (teach) us more. [For a modern examination of this issue, see Wagman (2003).]

But, bless me! What a lucky star that ever shined upon the astronomers! Indeed a great luck! For let us suppose, an astronomer in looking out for Ceres, should first have met with Pallas. Would he not have taken it for the Ceres Ferdinandea? One to hundred thousand he woul'd! Pallas look's exactly as Ceres, Piazzi himself woul'd have been mistaken by the appearance. She moves like her in the Environs and adjacent parts, where the Ceres was exspected. She moves like her very regularly, in an elliptical orbit, certainly every astronomer would have been mistaken, and no wonder! But what would have been the consequence of such a mistake? An unexampled Perplexity and confusion! All astronomers would have pursued the supposed and spurious Ceres. The calculations, the conjectures, and hypotheses would have been endless, and inextricable. For they would have attempted, to adjust, and to make agree Mr. Piazzi's observations of the true Ceres, with our supposed one. The case would have been inexupsable [sic], *and Perhaps a whole century might have passed bevor* [sic] *this intricate case could have been explained and unfolded. For if this misfortune had happened, surely the Perquisition* [sic] *of Ceres would have been neglected; In the mean while the astronomers had put themselves to the Trouble to calculate these paradoxical observations, and the Prophanes* [sic] *had took the pleasure to laught at the credulous astronomers, who believed into the Messiah of a Planet (here without doubt I should have had my good share of Derision, as I was one of the most tenacious astronomers in this Belief since 30 years) the true Ceres might have gone, perhaps forever, for if we had missed or lose the opportunity to observe the true Ceres, in amusing ourselves with the Pallas larvata certainly nobody would have looked for her in 1803, and if we did, would not Pallas have led us astray? No Doubt Ceres was left, and God knows when she might have been discovered again? Let us therefore make an oblation to Urania, that she preserved us from this Danger.*

There is one example in astronomy of a like confusion, which nearly imbroiled the astronomers, but very favourably they escaped the Perplexity. It was in the year 1664 Febr. the 18th that Hevelius mistook a Nebula, or some little cloud for the comet, & produced an observation, which thoroughly did not agree with the precedent observations. Hevelius came into controversy with the french astronomer Auzout, who impugned this observation, but Hevelius supported it with obstinacy, and admitted not the Fault. Very fortunately other astronomers in Italy, Spain and England observed the comet the same day, and so Hevelius was unexpealably [sic] *convinced. Auzout makes therefore the just remark saying 'That what is most disagreeable in this incident is, that if the sky would have been overcast until Feb.18 so that no-one could have observed the comet after that day, he (the comet) would have embarrassed the present and future astronomers by such an odd observation'. The same Thing might have been say'd from Ceres & Pallas, for if Pallas had been discovered first, and before Ceres, Truly this would have perplexed for ever the present and the future astronomers.*

Permit me an other reflexion. It is a long time that the astronomers wished to observe a comet of a so determinate appearance; that it might be observed with great accuracy; For if such a comet should come to pass in a Distance from the Earth, half as this of the Planet Mars, or Venus in its Perihelion, such nice observations of this comet woul'd give us the means to determine the Parallaxe of the Sun, with greater Precision, as the renowned Passages of Venus before the disk of the Sun, which happen so seldom. Indeed this requires a concurrence of accidents very difficult to exspect, for such a comet must be observed in two very different places in the same time in order to get a very great trigonometrical Basis, and Pingré in his Cometographie *Part II p. 151 says himself. 'He assumed that we would have little basis for expecting the combination of these favourable circumstances'. But perhaps Pallas will present the Reunion of such circumstances. If this so well defined Body revolves in an ellipsis whose Period will be determined, its future appearances can be calculated, and if it should take its way between the Earth and Mars, or the Earth and Venus, two different observers in this case will have the time to bespeake to this Purpose corresponding observations, which if all circumstances concurr to be propitious will give us the Knowledge of the Sun's Parallaxe with as great a Precision as Passages of Venus over the Sun. But this is only a conceit fancy. The Elements of the elliptical orbit of Pallas will sole decide, whether such a supposal [sic] can take Place. In the meanwhile our Duty is, to gather materials, and to observe with as much accuracy as possible this remarkable celestial Body, before we can pass our Judgment. This I have done with as much care, and constant application in my Power, and here I have the honor to send you all my observations of Pallas, adding to them those made of the planet Ceres Ferdinandea*

[For an excellent and lengthy study of the state of knowledge about comets in the early nineteenth century, see: *Edinburgh Review* (1835) 61, 82–128.]

Blagden to Banks Paris May 3, 1802

My last letter to you, which was of the 24th of April, contained observations of Olbers's new star down to that time, together with the parabolic elements of its orbit, as calculated by Dr. Burckhardt on the supposition that it is a comet; and, in a letter written to Count Rumford the 27th April, I gave the elliptical elements of its orbit, as calculated by the same gentleman. The astronomers here are still undecided whether to call it a planet or a comet, but I think they are most inclined to the latter opinion. M. Laplace is decidedly so.

No. 19 Gauss to Olbers Brunswick May 4, 1802, 11:45 a. m.

I can finally wish you luck on the ninth planet. I'll advise you of my continuing work with Pallas as it occurs. I had given it up until May 1st. Your penultimate letter with your new observation and the three from Paris was delivered to me late in the evening on April 30 – my birthday – and on May 1 I immediately began a rigorous comparison with the forwarded elements. It happened that none of Zach's six observations and the three from Paris differed by more than 15", though these variations must affect the elements very strongly, partly because of the short time interval (15 days) and your previously noted unfavourable location in the first half of April. In fact, I soon found that a considerable variation would result and would therefore require an iterative method of correction. I therefore considered it preferable to start the calculation from scratch, after correcting the observations of aberration, parallax, and the most likely observational errors, as suggested by the calculation from the comparison of the differences. The very first attempt gave the following new elements, which I haven't corrected any further because, after checking them, they agreed very well with the observations. You can imagine my delight that a much smaller eccentricity resulted and exceeded that of Mercury only very little. The inclination also decreased a few degrees, but the very close similarity to Ceres' orbit seems, after a rough calculation, to disappear and the extraordinary two-ring interlinked configuration is proved correct. ⚴

The elements of Pallas, obtained from the observations at Seeberg and Paris from April 4 to 19, 1802, follow: [These were published in the June issue of *MC*.] *The differences in the nine basic observations amount to only a few seconds and will have little regularity. I haven't yet computed all comparisons but will mail them to you shortly. Your latest communication arrived last night, but, due to a fortuitous delay, I received it only early this morning. You can imagine my impatience while checking the elements against the latest observations. My methods are such that, using the various procedures to determine the geocentric position, one needn't be very attentive during its computation, and even a computational error, which might not be immediately noticeable, is almost impossible. (F. N.: Even though I've solved the task of computing the geocentric position from the elements for*

Ceres and Pallas hundreds of times, I've only rarely encountered an 'error calculi' and then mostly only with the solar positions.) I've just now computed Zach's latest one and yours of May 1 and find, neglecting aberration and parallax:

	RA	error	Decl.	error
Apr. 26	181° 12′ 8″.9	+7″.1	19° 1′ 51″.4	+2″.0
31	180 59 37.3	+4.3	19 43 44.8	+13.8

Aberration and parallax result in a negative error in Dec. with Zach's observations and decrease it very much with yours; aberration slightly increases the error in RA.

It's too risky merely at one's discretion to hazard an opinion on the reliability of the elements. After I've further improved them, I'll carry out an exact calculation to determine how many errors one might expect the observations to have, should the elements be incorrect by a certain amount. Meanwhile, one can certainly no longer doubt that Pallas is truly a planet. I am even inclined to hold that the uncertainty in the eccentricity is barely more than 0.02. Whether the future improvement in the elements will perhaps bring the mean motion closer to that of Ceres or make it identical cannot yet be agreed on with certainty. I am extremely curious whether this is possible. Wouldn't such a unique phenomenon offer wonderful research opportunities? – The movement in the orbit from April 4 to May 1 is 7° 37′; that is, nevertheless, already a respectable arc. If these observations could still be continued throughout this month, I believe that the orbit will be rather reliably determined. If you believe an ephemeris can simplify the observations then I'm prepared to do the computation. The meridian observations are now surely already determined, and hence no one except you will be able to make observations of much greater accuracy.

By the way, you may from now on deal with these elements as you wish. I merely ask that you first compare them with your latest observations made since May 1. Perhaps I myself will publish them in a newspaper in order to satisfy public curiosity. Should Zach send me his observations in tomorrow's mail, then I'll send them to him next Friday which he'll receive on May 11. Thus far, to my surprise, he hasn't forwarded anything to me apart from his first three of April 8. He has completely stopped his almost daily communication since then.

Bode has requested my method for computing the orbits of celestial bodies for the 1805 almanac. I cannot, however, comply on this score because I entirely lack the time to prepare a formal elaboration of it. It would also be much too involved for the almanac, since merely the formulae, without their derivation and just with the scantiest instructions for their use, would fill several pages. I also didn't want to rush anything, but I wanted to elaborate the individual parts with as much elegance and simplicity as I could, and for that reason I wished first to obtain your opinion of it. Perhaps I'll devote a limited amount of work to it, which, considering the general interest shown by the public vis a vis Ceres and Pallas, hopefully won't share the fate of my Disquiss. Arithm., *for which no bookseller wanted to pay the printing costs. As yet I haven't sent anything to anyone except Mr. Zach, and to him just some of the most essential points. I'll forward it to you in its entirety as soon as I can. Kindly excuse the big rush.*

Zach to Maskelyne Seeberg May 4, 1802

I was very much flattered by the few lines you were so kind to send to me, and I return you my sincere thanks for all the favours which I experience from you.

I thank you also for your kind communication of the observations of the new planet Ceres Ferdinandea. I was very happy to see that my observations agree so well with yours. The difference in RA is as much as nothing, allowing for the difference of meridians; but in declination there are some times diversities of 10 seconds, which differences are certainly deficiencies upon my account. For it is impossible to enter into competition with my 4 feet Dollond's quadrant, with your 8 feet mural quadrant. [The next section of the letter deals with instruments and methods of observation. This will be printed in another work by this author.]

I have the honor to send you here annexed a proof sheet of the equations of perturbation of Ceres by the action of Saturn, Mars and Jupiter, which Mr. Oriani from Milan has calculated, and just sent me over. I also take the liberty to send you here my observations of Ceres from the last month, and the whole lot of observations of Pallas. I hope that you have also observed this very remarkable heavenly body. It continued to move in an elliptical orbit very regularly. Her orbit is between Mars and Jupiter but so eccentric that she traverses the orbit of Ceres. Her aphelial distance is twice so great from the Sun, as on her perihelial distance, which is about 2.0. The revolution is between 5 and 6 years. What a various and strange phenomenon! No doubt that several other such Planeto-Comets might exist in the heavens, especially in the great and immense spaces between Jupiter and Saturn, Saturn and the Georgian Planet. But as most of'em must appear as very little stars, hic labor, hoc opus est [this is labor, this is work], *to locate them. Chance can only be the best helper in finding out such telescopic planets.*

I shall now very soon be forced to take leave from Ceres and Pallas, as the meanness of these planets, and the increasing twilight will forbear to confine the observations to the meridian. I must commit now these observations to such astronomers who are provided with equatorial sectors. I shall try my good luck with a Dollond parallactic instrument, but I am much afraid these heavenly bodies will quickly decrease in light, as their distances from the Earth, and the twilight increase rapidly.

Bode to Olbers Berlin May 4, 1802

I am hoping that the reasons for my justification, I gave in my last letter, are sufficient for you. I could only write in the greatest hurry, this is usually the case with my correspondence, and I still have a lot on my mind that I would like to confide to your friendship and noble mind. First, I would like to repeat my fair complaints about Mr. von Zach's behaviour against me, maybe you can show me a way to remedy or you can contribute a friendly share. There is almost no issue of MC without clear expressions of his unfriendly attitude against me. And this is exactly what several of my foreign and my friends here are noticing who ask me personally and in their letters what could be the cause since they cannot find the reason from my part in the MC. What am I to think of his not responding to my letters, maybe six or seven since November, and not reacting to my therein expressed excuses and justifications to the publicly made bitter reproaches? I dearly love peace, especially in a literary respect, and would prefer only to get involved with these accusations against me if necessary. Today I wrote him regarding this matter, do you consider my actions right and proper? Don't you, too, sometimes find the tone of the MC arrogant and presumptuous? What tribunal allowed Mr. von Zach's personal attacks against astronomers? Many of my foreign friends, who were also his friends, complain with good reason about this. Science certainly gains nothing. Many fear the harsh critic, others try to worm themselves into favour by producing the first products.

Furthermore, I cannot understand why Mr. v Zach did not inform you immediately in December about his discovery of Ceres. Only on January 11 he observed its position to parts of seconds. All previous observations were uncertain or not successful. On January 12 he got your news of January 6 on the very same day on which you were so kind as to send me the news and I received your letter on the 12th as well! In the history of Ceres' discovery which I am publishing for pleasure some people will be surprised by the fact that Mr von Zach had sent me letters regularly until December to relay the news and observations on Ceres which he had received and then later no line, no word, I am not to blame.

Yesterday evening your (planet-like) comet had almost the position of 28 Comae Berenice. This star is very faint in the sky; I took it from d. l. Lande, it was supposed to be of 7th magnitude, I found its position only to minutes. The comet did not tolerate any illumination at the wall quadrant and it is also blurred by the dusk. Most likely I will not be able any longer to observe it at the wall quadrant (maybe a couple of times at the circular micrometer), I will only follow its path, it is in good hands now and there will be no shortage of good observations. On Sunday evening it was unusually clear and I could have *observed Pallas and Ceres perfectly well at the wall quadrant, but the Queen Mother, who is a great lover of astronomy, invited me to Monbijou so also this evening was lost for observing.* [Monbijou was a rococo pleasure-house in the heart of Berlin. It was demolished in 1960. The Queen Mother referred to is Frederica (1751–1805), widow of Frederick William II, King of Prussia (1744–1797).] *That there are still smaller planets between the known eight, I do not believe. The appearance of the comets is very different. Previous ones had hair, beards, nebulosities or tails, but now they appear without these ornaments. Maybe the nebulosity around your comet is too thin to be discovered. Our friend Schroeter gives its diameter 4", Herschel, as d. l. Lande writes me, only 1.5. Uranus has 4" too and now appears at its certainly greater distance than that of the comet as 5th magnitude. In order not to confuse the astronomical readership I called your moving star a comet because of its path. By the way, you really discovered a body of a strange appearance and thus something new.*

The fixed stars will not become mobile?! That would be too much! These days Ceres passes o Leonis in western direction very closely. Yesterday evening it culminated 6".5 before o. On the 23rd, 24th, and 25th I was together with Prof. Huth in Remplin to visit Mr. von Hahn. On the 23rd it was nice weather until 10pm and on the 24th until 11pm although the air was hazy. We observed Ceres at a magnification of 240 times at a 20 f. Herschel Reflector. It appeared pale, round and planet-like but no trace of any kind of atmospheric ring. The clouds left no time to observe your comet. On the 25th it was pitch black. On the 26th and 27th we had the fairest weather on our return.

Dr. Gauss sent me already his orbital elements VII according to which I calculated preliminarily the ephemeris for the path of Ceres. In my next letters I am going to send him my further observations at the wall quadrant. But probably he will receive better, as far as I know Mr. v Zach is not able to observe the declination to seconds, but I have no pretensions to decimal seconds.

Olbers wrote Banks on May 4, and this letter was given by Banks to Herschel. It is important as it includes a diagram of interlocking rings, denoting the orbits of Ceres and Pallas. This was drawn first by Gauss in a letter of April 20 that he sent to Olbers.

Olbers to Banks Bremen May 4, 1802

With much pleasure, as you can well believe, I send you full and certain information that our Pallas is a planet; the own sister to Ceres, not inferior to her in dignity and importance, and perhaps on that account another remarkable discovery; as she gives rise to many speculations regarding the origin and history of our Planetary system. With this I send the elements calculated by Dr. Gauss. [Olbers here prints elements of 31 March 1802.] *The orbit* [of Pallas] *is therefore only as eccentric as that of Mercury. But what a great unexpected inclination! And how curious its position with regard to that of Ceres! The two cross each other like the interlocked rings of a chain:* 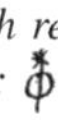 *To my great surprise, and surely to yours too, I hear that Dr. Herschel allows Ceres a diameter of only 1 second.*

Sniadecki to Zach Cracow May 6, 1802

Thank you very much for your letter of April 9 which caused me as much amazement as pleasure. I received it on April 25 together with your journal, i.e. one day after I wrote you on the 24th. I found the star of Mr. Olbers the same evening. According to the observations of April 4, 5, 7, 8 which you had the kindness sending me, on April 25 after Ceres' passage through the meridian I started to observe the passage of small stars between 181 1/2° and 180° RA, between 18 1/2 and 19 1/2 declination north, for your observations made me presume that in this part of the sky the star in question was to be found. After this work I examined with my parallactic telescope that part of the sky that I had just searched in the meridian, since I observe Ceres I map all the small stars where Ceres passes in order not to confuse it with any other stars, I was surprised to find a star that I had not seen earlier and which formed an almost isosceles triangle with the two small stars, which follow immediately 66 Virgo and close to which Ceres was to be found between March 27 and April 3. After having examined all my notebooks and drafts, I convinced myself that indeed this star has not been there before. The next day, April 26, I saw that it had changed its position and the observations at the meridian telescope showed me that the searched star was that which was observed on the 25th sub no. 5-to. Here are my observations of this peculiar star. [Here follow the observations of Pallas of April 25, 26, 27, 28, 29, 30 and May 1, 2, 3 inclusively.] *I hope that you find the results of my observations regarding RA satisfying for I compared Pallas with α, β, η Leo, η Bootis for whose positions I did not consult Bode's catalogue but adopted your positions or those of Maskelyne. I employed your position of 476 Leo of which I calculated the aberration and nutation according to Delambre's formulae, mentioned in* Connaissance des temps, *1788. Thus, I dare to invite you to examine the star's declination, if it is not too great. I cannot talk to you about Pallas' declination with the same satisfaction. This star, which presents itself quite well in my meridian telescope, does not do this in my quadrant, its pale light almost vanishes at the slightest glimmer for illuminating the hairs of the micrometer and it is almost impossible for me to see it precisely. Thus I make every possible effort to determine the positions of 95 Coma Berenices, 14 Bootis, 33 Bootis are Bode's and that of η Bootis is yours. I wanted to compare Pallas to Arcturus but yesterday and today the sky is overcast and I take advantage of the moment and tell you what I have been doing. Without delay I sent your observations of Pallas to Mr. Poczubut and will send him mine tomorrow. He found Ceres on April 3 using my observations sent to him. He observes Ceres with his 8-foot mural quadrant by Ramsden and his excellent meridian telescope. I asked for his observations in order to send them on to you.*

Having written this, I am given this very moment your April issue and your very interesting and instructive letter of April 19. I am infinitely obliged to you for the position of 415 Leo which can be found in a great number of my observations. I agree with you on the error of the star written R4 in Bode, I found the RA of 165 η Leo equally too small by 26″.99. I obtained these results in comparing these stars with α, η, β Leo and as soon as I had a spare moment in the middle of a mass of domestic duties that overwhelmed me at that moment I rectified the RA's of the other small stars of Bode which I used in my observations of Ceres and Pallas. As for the declination, I cannot flatter myself to have obtained the same accuracy. The stars that are of least importance for my observations of the end of April are 372, 222 Leo and those are not strictly correct in Bode. Having corrected 415 Leo and 165 Leo I made the following changes to my observations as I have the honour to present you here. – Some of your improvements perfectly agree with mine, but it is not possible for me to reconcile my declinations of April 2 and 5 with yours, the first was marked dubious in my notebook. Regarding the second I had only one star for comparison for this element and it is not surprising to make a mistake with such a small planet. And another error has crept into the observation of March 2 which is not at all in my calculation, but it occurred while copying that I wrote 31′ instead of 30′ in RA and in the reduction of time.

Since I do not want to delay the departure of this letter and I have to examine the calculation for the following observations according to your improved positions of the stars they will be the subject of my next letter.

Although all my subsequent observations demand revision, I cannot yet send you the continuation of those which can be found in my letter of April 24.

[Here follow the observations of Ceres of April 24, 25, 26, 27, 28, 29, 30, May 1, 2, 3 inclusively as they are inserted in the protocol.]

1802	*Mean time*	*RA Ceres*	*Decl. n.*	*Stars of comparison*
	in Crac ow	*of Ceres*		

1802	T[emps] m[oyen] C[racovien]	A[scensio] R[ecta] ⚳	Décl[inaison] b[oréale] ⚳	Etoiles de comparaison
2 mars	13h 45′ 48″	186° 30′ 27″	15° 36′ 45″	167, 304, 407 ♌
15	12 44 53	184 2 51 ,5	16 58 37	107, 309 ♌
16	12 40 7	183 50 10		269, 107 ♌
17	12 35 21	183 37 40	17 9 10	2, 10, 69, 190 ♌
19	12 25 46 ,4	183 11 54 ,39	17 18 53	α ♌, 125, 190, 476 ♌
20	12 21 1	182 59 15 ,17	17 23 58,8	α, 12, 125, 476, 190 ♌
25	11 57 4	181 54 41 ,3	17 43 36,8	η, 372, 415 ♌ 66 ♍
27	11 47 30	181 29 15	17 50 7,3	η praec[edens]
28	11 42 40	181 16 15	17 53 29	dtto
31	11 28 25	180 39 4 ,74	18 0 51	dtto
2 avril	11 18 58	180 14 56	18 3 48,24	dont [?] 165, 415 ♌ 66 ♍
3	11 14 13	180 3 1 ,25	18 6 9,75	η 665, 415 ♌ 66 ♍
4	11 9 32	179 51 34	18 7 38	η, β, 165, 415 ♌ 66 ♍
5	11 4 49	179 39 58 ,2	18 8 54	β ♌, 66 ♍

Fig. 7.11 Observations of Ceres from Cracow in April and May 1802

I hope that our RA's of Ceres do not differ much for I employed well-determined stars for this element. I cannot assure you an equal accuracy regarding the declinations for the same reason as in the case of Pallas. The planet is hardly visible in my quadrant. In some days' time I will no longer be able to observe it in the meridian.

Schroeter to Olbers Lilienthal May 7, 1802

Our Pallas is a true planet. [Schroeter here prints the 2nd elements of Gauss.]

Orbit a little more eccentric than Mercury's. Its orbit and that of Ceres join like a chain.

Dr. Olbers' obs. May 5 11h 2′ 35″

RA 180° 56′ 28″.2 Decl. 20° 9′ 20″.8

56 6.0 8 59

+22.2 +21.8 error of calculation

which prove at least the chief dimensions of the orbit well determined. Major axis, eccentricity, period and inclination will receive considerable corrections when subjected to continued observations.

Zach to Oriani Seeberg May 8, 1802

Just a few words in a hurry. First thousand thanks for your perturbation calculations. You find the rest printed on the reverse. Your pendulum has left for Zurich and I hope it will arrive in good condition; it has been packed with the greatest care. Since I am about to leave too to accompany my friend Bürg to the frontiers of Bohemia, I am sending you my calculations upon my return. Here are wonders of Pallas. This unique body has a revolution of five to six years between the orbits of Mars and Jupiter. Its orbit is very elliptic, its aphelion distance is twice the perihelion distance and consequently intersects Ceres' orbit. Its perihelion distance is about = 2.0.

Observations of Ceres by Méchain and Bouvard Obs. Nat.

The center lines read:

An error in collimation requires −1.″5 to be subtracted from the first five decl.

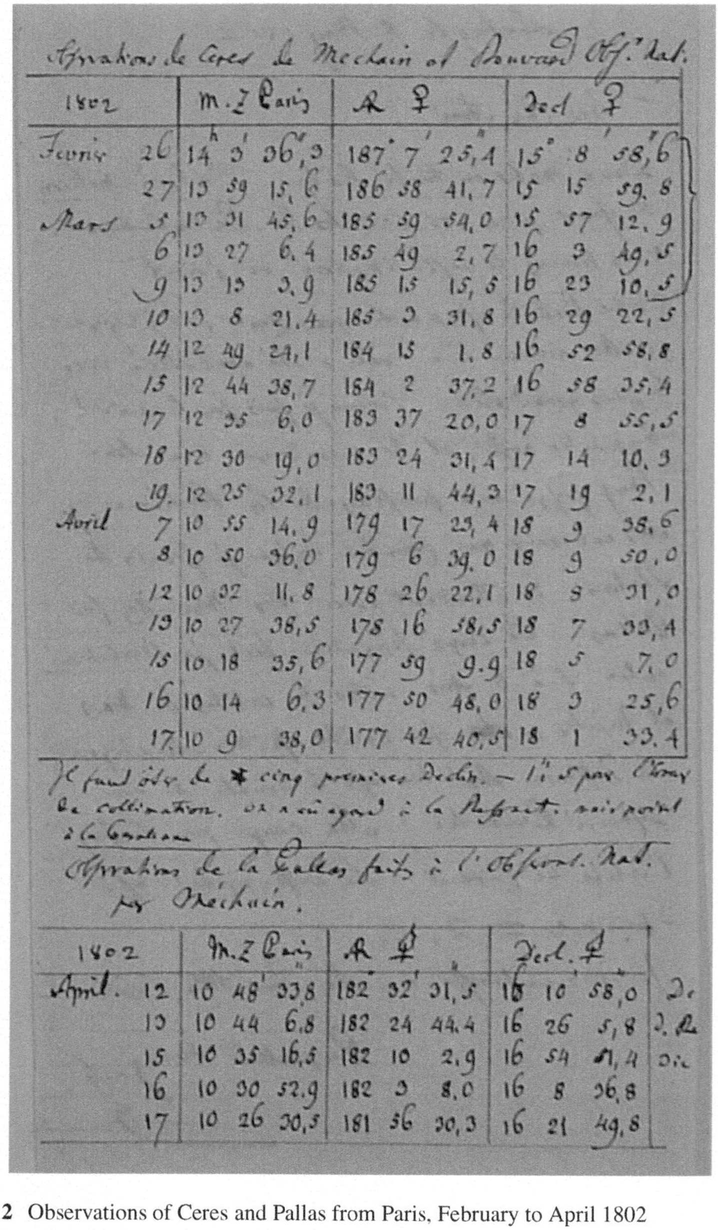

Observations de Ceres de Mechain et Bouvard Obs. Nat.

1802		M. T. Paris	AR ♀	Decl ♀
Fevrier	26	14h 3' 36",3	187° 7' 25",4	15° 8' 58",6
	27	13 59 15,6	186 58 41,7	15 15 59,8
Mars	5	13 31 45,6	185 59 54,0	15 57 12,9
	6	13 27 6,4	185 49 2,7	16 3 49,5
	9	13 13 3,9	185 15 15,5	16 23 10,5
	10	13 8 21,4	185 3 31,8	16 29 22,5
	14	12 49 24,1	184 15 1,8	16 52 58,8
	15	12 44 38,7	184 2 37,2	16 58 35,4
	17	12 35 6,0	183 37 20,0	17 8 55,5
	18	12 30 19,0	183 24 31,4	17 14 10,3
	19	12 25 32,1	183 11 44,3	17 19 2,1
Avril	7	10 55 14,9	179 17 23,4	18 9 38,6
	8	10 50 36,0	179 6 39,0	18 9 50,0
	12	10 32 11,8	178 26 22,1	18 8 31,0
	13	10 27 38,5	178 16 58,5	18 7 33,4
	15	10 18 35,6	177 59 9,9	18 5 7,0
	16	10 14 6,3	177 50 48,0	18 3 25,6
	17	10 9 38,0	177 42 40,5	18 1 33,4

Il faut ôter de * cinq premières Declin. — 1" 5 par l'Erreur de collimation. On a eu égard à la Refract. mais point à la Parallaxe

Observations de la Pallas faits à l'Observat. Nat. par Méchain.

1802		M. T. Paris	AR ⚴	Decl. ⚴	
April.	12	10 48' 33",8	182° 32' 31",5	16 10 58,0	D.
	13	10 44 6,8	182 24 44,4	16 26 5,8	D. R
	15	10 35 16,5	182 10 2,9	16 54 11,4	Dic
	16	10 30 52,9	182 3 8,0	16 8 36,8	
	17	10 26 30,5	181 56 30,3	16 21 49,8	

Fig. 7.12 Observations of Ceres and Pallas from Paris, February to April 1802

The refraction was taken into account.

Letter LIII, Oriani to Piazzi Milan May 8, 1802
[Part of this letter was printed in *The Discovery of the First Asteroid, Ceres.*]
Here are two other observations about the new planet Pallas.

April 25 10h 0m 17s m.t. R.A. 181° 1' 2" Dec. Bor 19° 36' 17"
May 7 9 3 27 180 56 34 20 19 36.

It was stationary in RA on May 5. Let me know if you found it and send me the observations.

[From what he is saying here, we can conclude that Oriani had already sent some Pallas observations to Piazzi. These are not found in the archives.]

No. 20 Olbers to Gauss Bremen May 8, 1802

I certainly needn't tell you how much pleasure your letter of May 4 brought me. My heartfelt thanks for your efforts with my Pallas, or rather with our Pallas. I must say it again: I admire you and your method. In compliance with your request I immediately compared the forwarded elements with my latest observation and found:

May 5 11h 2m 35s	*calc.*	*RA 180° 56' 28".2*	*Dec. 20° 9' 20".8*
	obs.	*RA 180 56 21.0*	*20 8 59*
	diff.	*+7".2*	*+21".8*

The agreement in RA, however, isn't quite as good as it appears to be here. With the same mail I received in a letter from Zach his RA for No. 11 Com. Beren. and of several other neighbouring stars. These differed very noticeably from those that I had obtained from the Hist. Cel. *The difference with No. 11 Comae was 26". Based on Zach's data, the above observed RA must be decreased by 15", and then the difference of the elements is +22".2 in RA. Even this agreement is not good enough, and I have hence sent Zach, Schroeter and Bode your elements with your permission.*

(Do you calculate the RA and Dec. directly from the elements? I mean, without first finding longitude and latitude? This time I still pursued the last method. But if I still had to make many comparisons with the observations which give only RA and Dec., then I would seek the location of Pallas' orbit against one with the equator parallel through the midpoint of the plane containing the Sun and each time bring the position of Pallas and Earth, using three rectangular coordinates, onto this plane and the abscissa line taken to be the equinoctial line. The formulae become apparent easily and effortlessly, and one thus obtains RA and Dec. directly.) [He includes here Zach's observations from Apr. 26-May 1.]

My own observations, all of which from April 26 onwards underwent a considerable correction in RA because of the improved RA of the comparison stars, are presently as follows: (It's a pity I wasn't also able to improve the Dec.'s, but Baron von Zach did not include those stars' Dec.'s. The comparison with Zach's observations indicates that at times mine differ from those by 1'.) Pallas' RA is now again in non-retrograde motion, as will soon be the case with the longitude. It unfortunately still goes through an area totally devoid of well-determined stars.

Most noteworthy, as you rightly mention, is the position of Pallas' and Ceres' orbits relating to each other. To which speculations about the origin and history of our planetary system will Pallas give rise? – The very close approach of both orbits of course disappears using the presently obtained elements. I find the distance of both planetary orbits at the ascending node of Pallas' orbit on Ceres' orbit = 0.22494 and at the descending node = 0.10675. Yet, if the semi-major axis of Pallas' orbit, the longitude of aphelion, and the eccentricity were a little bit larger, then the intersection of both orbits could very well result at the descending node, which I don't find unlikely.

What kind of small planets are Pallas and Ceres? Herschel found an apparent diameter of Ceres, as Zach writes, of only 1", and of Pallas, as Bode informs me from LaLande's letter, of only 1 1/2". In this way, speaking confidentially, irradiation must have interfered with our friend Schroeter's observations. I admit, I have always suspected this; for my very nice

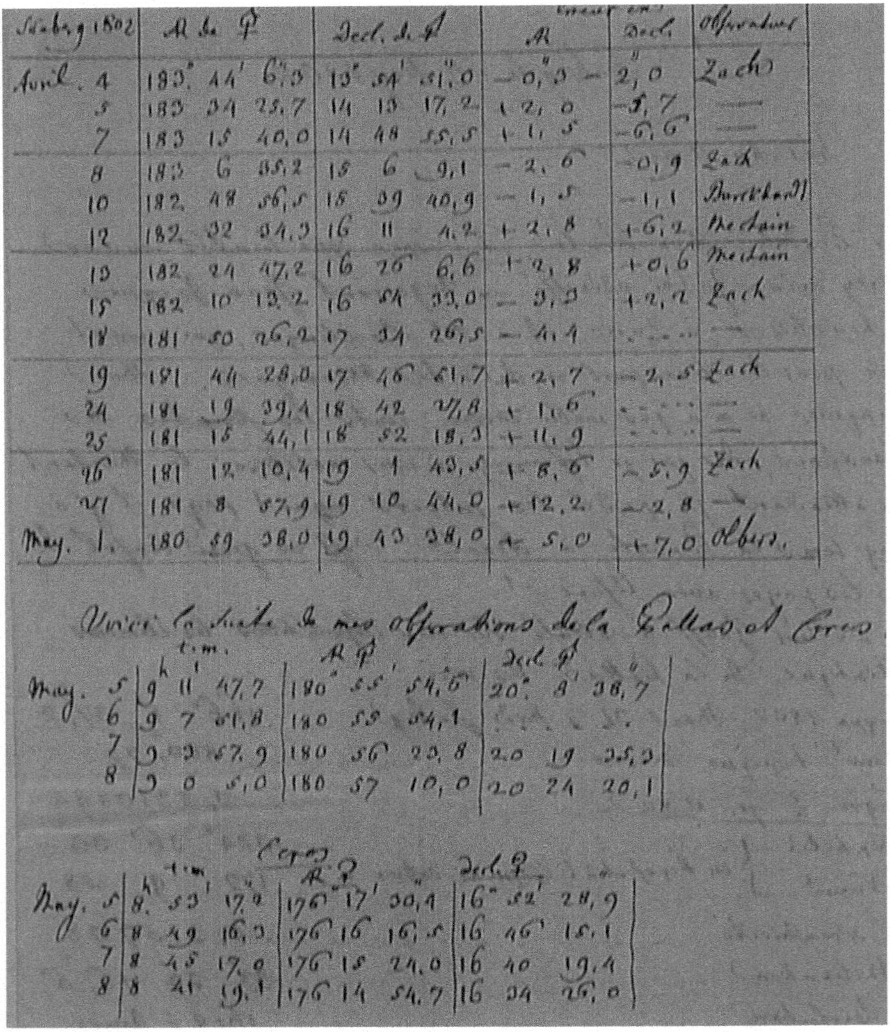

Seeberg 1802	AR de ⚴	Decl. de ⚴	Erreur des AR	Decl.	Observateur
Avril. 4	183° 44' 6",3	13° 54' 51",0	−0",3	−2,0	Zach
5	183 34 25,7	14 13 17,2	+2,0	−5,7	—
7	183 15 40,0	14 48 55,5	+1,5	−6,6	—
8	183 6 35,2	15 6 9,1	−2,6	−0,9	Zach
10	182 48 56,5	15 39 40,9	−1,5	−1,1	Burckhardt
12	182 32 34,3	16 11 4,2	+2,8	+6,2	Méchain
13	182 24 47,2	16 26 6,6	+2,8	+0,6	Méchain
15	182 10 13,2	16 54 33,0	−3,3	+2,2	Zach
18	181 50 26,2	17 34 26,5	−4,4	.....	—
19	181 44 28,0	17 46 51,7	+2,7	−2,5	Zach
24	181 19 39,4	18 42 27,8	+1,6	.....	—
25	181 15 44,1	18 52 18,3	+11,9	.....	
26	181 12 10,4	19 1 43,5	+8,6	−5,9	Zach
27	181 8 57,9	19 10 44,0	+12,2	−2,8	—
May. 1.	180 59 38,0	19 43 38,0	+5,0	+7,0	Olbers.

Voici la suite de mes observations de la Pallas et Cérès

	t.m.	AR ⚴	Decl. ⚴
May. 5	9h 11' 47",7	180° 55' 54",6	20° 8' 36",7
6	9 7 51,8	180 55 54,1	.....
7	9 3 57,9	180 56 23,8	20 19 35,3
8	9 0 5,0	180 57 10,0	20 24 20,1

Ceres

	t.m.	AR ⚳	Decl. ⚳
May. 5	8h 53' 17",0	176° 17' 30",4	16° 52' 28",9
6	8 49 16,3	176 16 16,5	16 46 15,1
7	8 45 17,0	176 15 24,0	16 40 19,4
8	8 41 19,1	176 14 54,7	16 34 26,0

Fig. 7.13 Observations of Ceres and Pallas from Seeberg, April to May 1802

5-foot Dollond, at 240-times magnification, does not even show an appreciable disc for either planet, nor is there a definite difference from a fixed star.

I'm very anxious to receive your method, and ask that you not postpone this for too long. You can certainly depend on my discretion in this matter. I very much approve of your wishing to publish the same as one of your own works.

P. S. Burckhardt must not yet be finished with his calculation, for I haven't received anything from Paris. – Do you wish me to send your elements to Paris? A calculation of ephemerides at 10-day intervals would be useful.

Zach to Oriani Seeberg May 11, 1802

Dr. Gauss has calculated from my observations the following elliptical elements of Pallas: [These are the first elements of Pallas, as published in the *MC*.]

Consequently, Pallas is a planet between Mars and Jupiter, its orbit comes at one point very close to that of Ceres. To a great inclination is added a considerable eccentricity, but not yet great enough to dispute its rank and name of a planet for the rapport between the major and the minor axis is like 1: $\sqrt{}$ *(1- (0.3)²) that is 1 : 0.95 this ellipse is not as oblate as Jupiter's disk. Here follows how these elements represent my observations of Pallas, and those of several others.* [Compare this with the table published in the July 1802 issue of the *MC.*]

center line reads:

Here is the series of my observations of Pallas and Ceres.

Pallas must experience considerable perturbations from Jupiter, but since the eccentricity and inclination of its orbit are great, it will be difficult and the approximations will not be sufficient taking into account the powers and products of these two elements. This will be a new task for geometers but it is still too early to think of it. I hope that we will find Pallas around the end of this year again and four weeks of observing is too little to conclude elements that come close for a body whose perturbations might increase beyond one degree.

Zach to Gauss Seeberg May 11, 1802

I am with one leg in the mail coach, my friend Buerg and I are about to leave, I want to accompany him to the Bohemian border, and I received your letter of May 6th with Pallas' new elements. Here are my latest observations of this peculiar heavenly body. It's getting dark so early and the planet is so small that meridian observations are no longer possible.

I will be back home in eight days, so if it is possible for you to send me your results and calculations until the 20th I could publish them in the June issue. Thus, I hope to find your letter upon my return. I believe to have sent you my observations of Ceres until May 3rd; here is the next installment. [Zach here gives his Ceres positions from May 5-8, and Pallas positions from May 7-11.]

Oriani's calculations of the perturbations for Ceres need to be completed as follows: [See December 1802 *MC.*]

No. 21 Gauss to Olbers Brunswick May 11, 1802

I'm sending you herewith <u>*first*</u> *the list of differences between my most recently sent elements and all of Zach's, and the three Paris observations. I received Zach's observations from Zach himself on the 7th of this month, and I hope he will continue to do so without interruption. There is also one from April 8 among them, which I hadn't yet had. On the 27th he wrote that he hadn't received any Paris observations whatever, which surprises me very much. The French will hardly be grateful to you and Zach, according to what Zach writes me about this, for so obligingly informing me of all the computational data. In the meantime I hope that we won't even need the French intervention and that we'll be able to do very well without both their observations and calculations. The following differences, applied to the observations, give the calculation.*

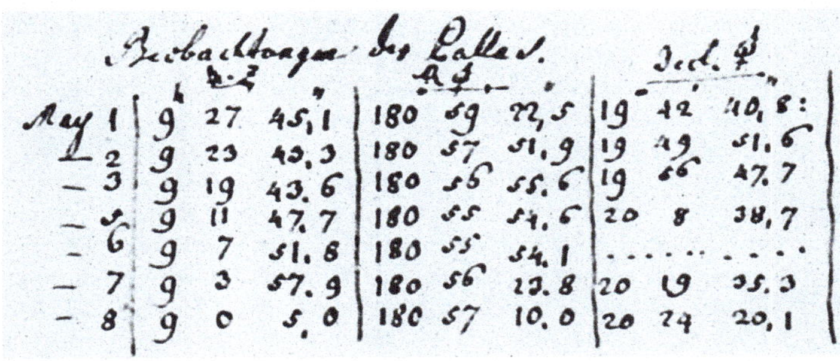

May 1	9	27	45,1	180	59	22,5	19	42	40,8
2	9	23	43,3	180	57	51,9	19	49	51,6
3	9	19	43,6	180	56	55,6	19	56	47,7
5	9	11	47,7	180	55	54,6	20	8	38,7
6	9	7	51,6	180	55	54,1	......		....
7	9	3	57,9	180	56	23,8	20	19	35,3
8	9	0	5,0	180	57	10,0	20	24	20,1

Fig. 7.14 Observations of Pallas from Seeberg, May 1 to 8, 1802

					Pariser				
April	4	5	7	8	10	12	13	15	18
Æ	− 0,3″	+ 2,0″	+ 1,5″	− 2,6″	− 1,6″	+ 2,8″	+ 2,8″	− 3,3″	− 4,4″
Dekl.	− 2,0″	− 5,7″	− 6,6″	− 0,9″	− 1,1″	+ 6,2″	+ 0,6″	+ 2,2″	

April	19	24	25	26	27	29	30	31
Æ	+ 2,7″	+ 1,6″	+ 11,9″	+ 8,6″	+ 12,2″	+ 23,0″	+ 24,0″	+ 25,6″
Dekl.	− 2,5″			− 5,9″	− 2,8″	− 12,5″	− 15,3″	+ 1,7″

Fig. 7.15 First list of differences between Gauss' elements and observational data on Pallas

The considerably increased differences in RA made it immediately probable, or rather certain, to me that my previous elements would suffer a larger change than I initially anticipated from the apparently more accurate agreement with your observations from May 1. The latest Dec's, regretfully, do not particularly agree. I have sought to incorporate, to the best of my ability, all of the observations in the new orbital determination with which I occupied myself the day before yesterday. I wrote Bode yesterday and hope he will immediately send me his observations. His meridian Dec's, which I always found to be quite useful for Ceres, should serve me well. His RA's are often not particularly accurate, as you will have discovered on April 11 and 12, though they are still much better than Seyffer's observations which he sent me and which are also published in the G.G.A. [Learned Announcements of Goettingen]. *With such observations one can hardly do as much in one year as with Piazzi's in one month.* [The elements of Pallas were printed in the June issue of the *MC*.]

I hope these elements will be better than the former that were obtained from 15-day observations. Yesterday and today, however, I've been hindered by various engagements from occupying myself with Pallas and have therefore not yet been able to make any comparisons with the observations. I must hence once again ask you for security's sake to compare them with your latest observations before you share them with anyone else. Should the result turn out well, then you can do with them as you please. I'm still sending the main results to Hamburg today for insertion in the Correspondence. *Zach has already received from me these as well as the former elements. I'll send you the comparison with all of the existing observations in my next communication.*

Based on a hastily-made estimate, the correctness of which I however cannot vouch for, I find for the descending node

Distance from the Sun of Ceres 2.93
Pallas 2.86

and I am now extremely curious if the more accurate future knowledge will reduce the difference or eliminate it completely. I'll let the computation take its course without introducing any hypotheses. It appears to me just as noteworthy that the mean motion has become very similar to that of Ceres, and we don't even know the latter itself accurately to within 1″. As small as the masses of Ceres and Pallas may be, there must be a limit to the difference in the mean motion, within which it cannot fall, without being entirely reduced to zero by the mutual attraction. Perhaps our offspring will, maybe after centuries, be in a position to derive the masses of Pallas and Ceres from the anomalies in the motion of Ceres and Pallas. Moreover, you yourself will have noticed that, based on the existing results, the two orbits do not interlink with each other like rings (as I had said, based on a mere superficial view), but that Pallas' orbit is enclosed by Ceres' orbit. That is to say, at the nodal axis Pallas is, in both cases, nearer the Sun than Ceres.

Tomorrow I'll apply myself to the comparison of Zach's observations with the elements, when I hope to pursue it further. Should it be found that the differences are still not sufficiently regular and determined to make a new improvement necessary or possible, then I'll compute a small ephemeris. This will be facilitated by the solar positions that have already been calculated for use with Ceres; I'll have to use the same moments of time as a basis. I'll utilize my leisure time, which is of course practically non-existent as I still have other work to do, to write down the main points of my method for you. I would consider myself very fortunate if my Disquiss. could arouse your interest in the higher arithmetic, which still remains my favourite among all mathematical entertainment.

Zach to Oriani Seeberg May 12, 1802

Your history of finding Pallas made me laugh and I am obliged to you for sending me your observations of this planet. I sent you mine together with its orbital elements. But the latest observations of Pallas begin already to move away from the ellipse that I sent you, e.g. [These observations from April 29 to May 1 are printed in the June issue of the *MC.*] *Dr. Gauss has improved his elements.* [These are the second elements, as printed in the *MC.*] *What is your opinion of this ellipse? It is so unusual, the orbits of Ceres and Pallas are as interlaced as two chain links. The differences of Pallas and Ceres in the line of Ω is just as peculiar. The distance of Pallas from the Sun is 2.86 and this distance for Ceres = 2.93. What is your opinion of the great accordance of the mean motion of these two planets, maybe they are exactly the same. The discovery of these two celestial bodies will lead us further to other outstanding discoveries in physical astronomy. These two little planets are maybe only fragments of one single enormous one which had its place between Mars and Jupiter. Tell this to your directors who guessed so well the motion of Pallas. Here is what I was able to seize from Pallas.* [Zach here prints his data from May 7-11 for Pallas, and May 6-8 for Ceres.]

No. 22 Olbers to Gauss Bremen May 15, 1802

Please accept my deepest thanks and also my heartfelt congratulations on your much improved determination of Pallas' orbit. Just after the arrival of your letter (May 13) I compared the new elements with my latest observation of May 10, 11h 35m 45s at night. I found: Pallas' true longitude in its orbit 197° 9′ 51″.8, distance from the Sun = 2.46958, consequently heliocentric longitude 6Z 13° 7′ 16″.4, heliocentric lat. 13° 48′ 41″.5. And from this also the solar longitude is 1Z 19° 30′ 24″.2 and the log. of its distance – 0.004587, Pallas' geocentric longitude 5Z 22° 23′ 51″.1, the geocentric latitude 19° 11′ 14″.3. From that resulted the obliquity of the ecliptic = 23° 28′ 7″.

Calc.	*RA 189° 59′ 45″.0*	*Decl. 20° 33′ 44″.5*
Obs.	*RA 189 59 55*	*20 33 26*
diff	*–0 10*	*+0 18.5*

My observation appeared not to be too bad. On this night I had compared Pallas several times with 3 small stars in the Hist. Cel. *of which one preceded and two followed, the first one lying further north and the other two further south. I had improved the RA of the 3 stars, especially that of No. 11 Comae, by using the RA, sent to me by Baron von Zach, of some neighbouring stars. The Dec., however, I derived from the difference in the Dec. of No. 11 Comae.*

You can see, as your elements agree so superbly, that using my observation I really can't attribute any error to it whatever. – I'd like to know which solar tables and obliquity of the ecliptic you use for your calculations, so that there will be better agreement between my yet to be undertaken comparisons and yours.

The weather since the 10th wasn't very suitable for taking observations. Only last night was I able to obtain a few comparisons with the same stars during a storm and through breaks in the cloud-cover; they give:

May 14 12h 3m 43s RA 181° 9′ 37″ Dec. 20° 46′ 29″

This observation may be doubtful due to the mentioned conditions. I've received only two observations from Mr. Bode: [A table of positions from Apr. 15 and 19.] *Zach will have sent you his own observations directly. Though superfluous, I'll note them here:* [A table of 5 positions from May 2-7.] *I've received nothing further from Paris. Your approximate calculation of the distance of Ceres and Pallas at the descending node was very accurate. With your Elements VII for Ceres and Elements II for Pallas I find:*

asc. node of Pallas on Ceres' orbit 0.18725
long. of asc. node 187° 25′
asc. node of Ceres on Pallas orbit 0.07001
long. of asc. node 10° 49′

Whether these distances at Ceres' ascending node or at Pallas' descending node will still decrease could depend on a more accurate knowledge of both orbits. Combining Pallas' orbit with Burckhardt's ellipse for Ceres, I find the distance of both orbits at Pallas' descending node only = 0.06567. I expect, after all, only a close approach of the orbits, not an actual intersection now. But haven't these orbits at one time or other actually intersected? That won't be known with certainty for some years, when we'll accurately know not only both orbits but also all perturbations to which they're subject. The determination of the mutual perturbations of Pallas and Ceres will give rise to completely new and intriguing investigations.

Indeed I still can't wholly abandon the idea that Ceres and Pallas are maybe just fragments of a former planet. The variation in brightness from night to night suggests that both are surely not spherical but rather irregular in shape, and their appearance, due to the varying brightness, depends on whether they present a wider or narrower profile to us. – The whole idea, however, will be confirmed only if we should discover still more similar planetary masses between Jupiter and Mars. I'm now actually missing an 8th magnitude star not far from the position Pallas will arrive at in June. It was, according to the Hist. Cel., *definitely observed on April 10, 1796, and couldn't have been either Ceres or Pallas. This isn't No. 28 Comae Beren., as Bode suggests, which is also missing from the sky and in whose position Pallas was located around May 3. Bode took this star (No. 28) from a manuscript of Lalande; but it was probably never seen in the sky for I can't find its observation anywhere in the Hist. Cel. It was probably another one at the same latitude and still visible in the sky, which could have been assigned to it through a reduction error.*

Bode to Olbers Berlin May 15, 1802

The results of your and Dr. Gauss' calculation on Pallas astonished me and it seemed I heard about things from a different world. Is not actually the observed arc of Pallas much too small to derive its entire orbit from it and do not the results disagree not only with all our experience but also with the entire celestial mechanic. What is de la Place going to say? How can these two planets, Ceres and Pallas, move around the Sun on intertwining orbits in almost the same time, close like sisters – this seems to be as impossible as it is unheard of – what perturbations must they not suffer! All our previous ideas must fail. You assumed the node at 172° and the epoch 1802 on March 31 166°, Pallas must have had still southern latitude then and it stood 13 to 14° north of the ecliptic, its parenthetic longitude was on March 31 178°. It had already passed the opposition thus its heliocentric longitude must have been greater than the geocentric. Here is actually another typo. If the node is at 172°, at which the Earth was on March 13, how could Pallas then appear geocentrically 10 from the ecliptic when it had 183° of longitude? Enlighten me. Sure enough, the correspondence between the observations and your hypothesis is very important but can this small covered arc not be matched with any crooked line? And are the observations really as accurate as this calculation demands? Dr. Gauss sent me his opinion as well which coincides with yours, but you will excuse my incredulity, I admit freely that you discovered a very special *body, which I do not place among the regular comets since it always appears as a 7th or 8th magnitude star without nebulosity. You will earn eternal fame through this fortunate discovery, which sets us astronomers riddles just as Lexell's comet did. If astronomers want to call future celestial bodies that move away from the Sun's orbit more than 60° geocentrically still a planet, I will have to let it happen. How did you get the idea to place Pallas at the exact distance from the Sun as Ceres? Only the position and the shape of the orbit must be different, why not adapt a parabola at a greater distance to the observations? I am inclined to let this strange celestial body, if it only remains parenthetically visible, the name Pallas. Earth is already moving quickly away from it and soon it will become invisible for our telescopes, especially since we will soon have brighter dusks. Will we ever see it again? I experienced the same with Pallas as Mr. von Zach did in the beginning, I observed it partly at the wall quadrant and partly I could not distinguish it at the slightest illumination whereas I was able to see Ceres clearly in full moonlight and full illumination with it. Consequently, I consider Pallas less bright than Ceres although in my 3 ½ f. Dollond at a low magnification, Pallas was brighter. On May 5 I saw Pallas for the last time at the wall quadrant very distinct. According to the well-known clock I determined the right ascension during the culmination 180° 55 37 and decl. 20° 7 47, the latter is very accurate. The first corresponds quite well with your observation and I am hoping to still catch it several times at the circular micrometer. Since the observatories of Paris and Seeberg are chasing it, there will be no shortage of excellent observations.*

I am waiting impatiently for the next issue of MC. What will Mr. von Zach say about this peculiar star? And I might even be presented with a witty joke (i.e., another mean-spirited comment) for my doubts. I am asking you, dearest friend, not to publicise my last remark about Mr. von Zach if you want to contribute something to appease his unjust anger against me. I will happily publish everything you have already communicated or will tell me about Ceres and Pallas in the future in my yearbook 1805 to honour your astuteness and tireless diligence and I will be among the first to publicise your theories, when they have been put straight, about your moving star to honour you. I am humbly asking for your continuing friendship.

PS: Still no answer from Mr. v Zach.

Sniadecki to Zach Cracow May 16, 1802

Here is the continuation of my improved observations according to the rectified position of the stars where we differed slightly and which you were so kind to send me. (Here follow the observations of Ceres recently improved from April 6 until May 11 inclusively, among 20.) There is nothing left to change regarding Ceres' position, the RA's were mostly determined by α, β, η Leo and the results agreed perfectly well, besides I do not trust Bode's positions any longer. I compare the small stars that I was forced to use to known stars, from which I get the improvements so that I do not differ much from those you have sent me but I still prefer yours. I bid Ceres goodbye, I saw it on the 12th for the last time, but I do not send you this observation since I consider it very uncertain because of the moonlight, daylight and the bad weather. Here are my continued observations of this marvellous star of Olbers which is not yet direct again and which I have not been able to see since the 12th. The weather here is awful with lots of rain. [Here follow the observations of Pallas of May 8, 9, 11, 12, 1802.] *I cannot guarantee for the last. Although I missed the declination it seems to me that it was smaller than the previous. I do not hope to see Pallas until the weather is fine again since it is so very small and its light is fading more and more.*

Wilson to Herschel London May 17, 1802

In his [referring to Mr. Vince] *letter to me he expresses a curiosity to know your reasons for a new name to Ceres and Pallas – and seems unwilling to call them any thing but planets – but I think you have already, in the papers to the R.S., fully prepared your response and for his curiosity. If I rightly remember you mentioned to me some other new tract, received from Germany, about the orbit of Ceres – if you think it such as Mr. Tilloch could insert in his* Philosophical Magazine, *or even abridge, I would be much obliged by you sending me the schedule, first time any thing comes from Germany to 19 Bath Street.*

No. 23 Gauss to Olbers Brunswick May 18, 1802

I'm enclosing the result of the comparison of my Elements II with the entire and now closed Seeberg meridian observations of your Pallas. Zach sent me the latest ones from the 2nd to the 8th on the 11th, just as he had received my Elements I and wanted to leave with Buerg. He expected to be back in 8 days. [This table was published in the June *MC.*]

It would be very simple to determine new elements which would agree with the latest RA's a little bit better, but since the Dec. differences themselves are too irregular, one still wouldn't be certain if one thereby obtained better elements. Your RA from the 10th agrees, as you can see from this comparison, very well with those from Seeberg, and I wonder how you can make such sharp observations without sturdy instruments. I'm inclined to think your Dec. of the 10th is a little too small since the calculation among those observed at Seeberg holds so well in the middle, but perhaps the Dec. of No. 11 Comae is also a little uncertain.

If you could only make more such good observations in June, at which time their longer separation will compensate for the somewhat reduced sharpness. Hence we must in any case obtain elements of such accuracy that the rediscovery in 1803 will entail little difficulty. How long do you hope to see your planet this year? Since it is already receding from the Sun it will be visible for a longer time around midnight, and only its increasing distance and hence decreasing brightness will be the main obstacles for obtaining observations. Surely you'll make every possible effort to follow it as long as possible. Today I have currente calamo [with a flying pen] *calculated the following 4 positions. In order to completely assure myself of the accuracy of the calculations I'll still compute the 3 of May 30 and June 11 and 23, and then, using interpolation, again construct an ephemeris for every 3 days. I expect to have them completed by tomorrow noon and will send them immediately to those astronomers with whom I'm corresponding, to Maskelyne among others. This ephemeris will not only help to facilitate the taking of observations. Its accuracy, hopefully to within a few minutes, will also eliminate the doubt and skepticism that some astronomers may still have.*

Midnight in Seeberg RA S. Decl. Distance from Earth

May 24	181°	57′	21°	1′	1.9898
June 5	183	34	20	46	2.1670
17	185	48	20	5	2.3471
29	188	32	19	6	2.5268

Out of curiosity I've computed the position of Pallas on Dec. 31, 1802, midnight, using Elements II, though just quickly: RA 254° 28′, Dec. 3° 53′ north; distance from Earth 3.8082, hence on the left breast of Ophiuchus in the vicinity of stars No. 122, 123 and 131 Bode. It will rise already at 4 o'clock, when the night is pitch black. From the present and future brightness, the latter observed from increasing distance, you'll be able to guess how soon one might have hopes of finding it again. In any case I hope it will be rediscovered before Ceres; the large northerly latitude (on Dec. 31, 26° degrees) will give it a not unfavourable position, especially for the northerly areas. Maybe you'll also be the first one to do so.

For my calculations I'm using Zach's solar tables with Zach's now and then noted alterations to the epoch and apogee. But I'm leaving out not only the Mars equation but also the equations contained in Tables VIII and X, because they nevertheless do not increase the reliability of the Sun's position and the use of tables with double entries with a great number of solar positions is, in the end, an additional difficulty. Thus far I've set the obliquity of the ecliptic to 23° 28′ 5″.3; a few seconds of change, just like any change which affects the result equally or uniformly, doesn't have a noticeable effect on the determination of the elements. [The obliquity of the ecliptic is the angle between Earth's axis of rotation and the pole of its orbit. It changes 0″.47 per annum.] *Only such changes, which noticeably affect the second differential (I'm expressing myself in this manner in order to make myself understood as briefly as possible, for my method contains no differentials.) in the longitude and latitude, strongly influence the elements.*

As the 9th main planet becomes more of a certainty day by day, the more I'm sincerely pleased that you in particular are the lucky one we have to thank for this magnificent and forever peculiar discovery. In a few years' time the outcome may be that either (1) Pallas and Ceres were formerly at the same place simultaneously and hence at one time without doubt constituted one body, or (2) they quietly wander, have wandered, and will wander around the Sun with very similar periods of revolution, while at the formidable descending node Ceres would always be ahead because of the so different central equation. The outcome could also be that these are merely phenomena which, based on our knowledge, are unique and which no one would have dreamt of 1 1/2 years ago. Judging from a human interest perspective, one would certainly not wish the first result. Imagine the shock, the spiritual struggle, the incredulity, the defence of and opposition to Providence we would see develop if the possibility that a planet can be shattered be verified as fact! What will those, who base their framework of knowledge so readily on the unshakeable stability of the planetary system, say if they see that they have built on sand, and that everything is entrusted to the blind and fortuitous play of the forces of nature! I, for my part, think that one should refrain from all such conclusions. It appears to me an almost outrageous presumption that we want to take as the standard for eternal wisdom that which we, through our limited capabilities in our animal existence in the surrounding material world, perceive or believe to perceive, to be perfection or imperfection.

Sniadecki to Jan Albertrand Cracow May 18, 1802

[Albertrand was president of the Society of the Friends of the Sciences in Poland. In this letter (translated from Polish), Sniadecki sends to the society corrected observations of Ceres.]

I do not know what has happened with my work, which I gave to Mr. Czacki, the Starosta of Nowogrod [a political/administrative official], *and which was about the new planet Ceres* [see footnote], *along with certain observations of said planet, as what I was told with regards to that work did not come to pass. If this work reached the Society, whatever use the Society pleases to make of it, I beg you, Sir, to place the hereby attached observations and calculations in the place of the earlier ones, and that the earlier ones be deleted for the following reasons.*

First, in the rapid copying of numbers from my record, I committed two serious mistakes, which mistakes only now did I notice. Second, while using stars from the new work of the Berlin astronomer, Mr. Bode, to compare the planet, and even though I noticed a number of fairly serious and significant mistakes in that work for a few stars, and of which I warned other astronomers, however, during the course of my work, there appeared more mistakes in that [Bode's] *work with regards to other stars, which I was taking as correctly located. Having communed with other foreign astronomers about these mistakes, and having communicated and convinced myself with respect to their accuracy, it was necessary to undertake the calculations of my observations afresh, and to correct and rectify said observations, on the basis of this new information. And so, having sent abroad the corrected results of these calculations, I dare to beg you, Sir, with this request, so that what I sent abroad, match what I communicated to the Society here at home. Mr. Bode, having undertaken a work that is enormously difficult and time-consuming, and not having held it sufficiently long in his consideration, has misled all the astronomers in Europe, who, as did I, must now rework the results of their observations. Just as this* [Bode's] *book appeared, the newly discovered heavenly body disclosed its many defects to the point of indignation. Its correction may employ many astronomers for half a century.*

It would be useful to add to the Society the continuation of my further observations of this planet, as well as news and observations of the new body, discovered on March 28th in Bremen by the astronomer Olbers, and observed here by me from April 25th, which even now we do not know in which class to place – surely further observations will teach us more – but at this time I am so nearly strangled by the press of work and foreign astronomical correspondence, that I barely have time to rest. Therefore, I must leave this work to another occasion, though I will not neglect to share it with the Society.

If at some point it should please the Society to publish these new observations, I ask to please not forget to include the date on which they were communicated, as this sort of work derives much of its value from the time at which it was published.

[F.N.: These observations of the planets Ceres and Pallas were published in *Annals of the Warsaw Society of Friends of the Sciences.*]

Gauss to Maskelyne Brunswick May 19, 1802

As the new star discovered by Dr. Olbers engages at present the attention of all astronomers of Europe, I hope you will receive with kind indulgence the first results of my endeavours about its true orbit, especially as its motion appears to put it already almost beyond doubt, that it is really another planet betwixt Jupiter and Mars whose orbit has the most extraordinary inclination and will, perhaps, give us in a few years, explanations of a very unexpected nature.

As soon as the observations of Pallas were continued through a few days only, it was already decided, that its orbit could not be a circle, the motion being always too fast, whatever distance from the Sun might be supposed. When the series of observations had increased to a fortnight, Dr. Olbers attempted to compute a parabola, but soon found, that it was equally impossible to represent the observations by it; for when he computed again the mean observation, he found an error calcul of 11′ in longitude.

When I had received Dr. Olbers' observations till April 17, for curiosities' sake I attempted to apply to them the same method, which I had made use of in my calculations about Ceres Ferdinandea, and which without any hypothetical supposition yields the true conic section as exactly as the nature of the problem and the projection of the observations will permit. Indeed in the present case I could not exspect [sic], *that the result would be very near the truth, as Dr. Olbers' observation had not the necessary precision, nor was their duration long enough, and besides the fear was accidentally in a very unfavourable position. However the result, which I concluded from those observations agreed generally speaking well enough with that which I afterwards, when all those circumstances had altered for the better, found by the observations of Baron de Zach. The orbit I found was an ellipsis between Mars and Jupiter of an eccentricity greater indeed than that of any other planet (viz 0.3), however not so great as to deprive Pallas of the right of being called a planet.*

The elements which I afterwards have found by Baron Zach's observations from April 4 till May 1, and which differ from this whole series only a few seconds, are the following; I omit the minutes because it is evident they may not yet be depended upon, and I shall soon endeavour to correct them by more observations.

Mean longitude	*March 31, 1802*	*May 31 to Sir J. Banks*
In the meridian of Seeberg	*161° 1/4*	*161° 12′ 43″.2*
Diurnal tropical motion	*12′ 37″*	*757.166*
Excentricity	*0.259*	*.02591096*
Mean distance from the Sun	*2.80*	*2.80068*
Aphelium	*300°*	*log 0.4472636*
Ω	*172° 1/2*	*172° 34′ 35″*
Inclination	*35°*	*35 0 42*
Aphel.	*300 5 4*	

From these elements it is easy to derive two very extraordinary conclusions (supposing they do not differ very much from the truth)

1. that the times of revolution of Pallas and Ceres are nearly or perhaps exactly the same
2. that the orbits of these two planets do nearly coincide in one point at the ascending node of the orbit of Ceres on the orbit of Pallas. This circumstance has yielded an ingenious idea to Dr. Olbers, viz that perhaps Ceres and Pallas might be only fragments of one greater planet, once dashed to pieces by the percussion of a Comet; indeed, the exterious appearance of the two stars and their very variable light seems to countenance such a supposition and to indicate, that their shape is not spherical but considerably irregular; however farther observations and a more exact and certain knowledge of the true orbits will be necessary to decide these highly interesting questions.

Meanwhile I have computed, by the above elements which I design by II, a little ephemeris, which, perhaps, will facilitate the farther observations before Pallas approaches too near the sun. I hope it will differ from the observations only a few minutes, and by the best observations of Mr. Zach from May 2 – till May 11 (from which the greatest difference has been 15") I am inclined to guess, that the calculated RA will be too small about 3 or 4 minutes towards the end of June, yet this estimation may not be depended upon. The declinations also probably will prove somewhat too small, but this presumption is still more precarious and uncertain, as the last declinations observed at Seeberg with a quadrant of 4 feet are not very exact nor harmonizing with each other. [Gauss now encloses an ephemeris for Pallas from May 24 to June 29.]

As the above elements shall not be exact enough to find Pallas again easily, when it emerges again from the rays of the sun in the beginning of 1803 in the constellation of Ophiuchus, it will be of great importance to get still, before its disappearance, as great a series of good observations as shall be possible. I am sure that I shall attain considerably better elements, if you will have the kindness to communicate to me for that use your observations, than I should be able without them. And I have hopes that if you will grant me this favour, I shall have it in my power to determinate the orbit exactly enough, that the planet may be found out again without pain.

Maskelyne to Gauss Greenwich May 21, 1802

On the 21st of April I received your letter of the 3rd of that month, with a further ephemeris of Ceres Ferdinandea from the 21st of April to the 29th of June next, for which accept my best thanks. You was right with respect to my observation of the right ascension of Ceres on the 19th of February; it was wrong reduced; it should have been 12h 31′ 53″02 of time = 187° 58′ 15″.3 of right ascension. I have now the pleasure of sending you all my observations made on the meridian both of this planet, and of the other new planet called Pallas, discovered by your friend Dr. Olbers on the 28th of last March, as he had before he rediscovered Ceres on the 1st day of this year. Having compared together the right ascensions of the Sun found by my transit instrument (assuming that of alpha Aquila by which the others were settled in right ascension) with the right ascensions of the Sun computed from his declinations about the opposite equinoxes, for some years past, I have found the latter right ascensions were greatest, and consequently the right ascensions of my Catalogue required an increase [Astronomical Observations made at the Royal Observatory at Greenwich, 4 vols, London 1787-1800]. *But having taken the observations last year, with particular care not to open the shutters to let in the Sun's rays till the very moment of observing his Zenith distance, I prefer the corrections from them, and have added 3″.8 to the right ascension of the 36 stars of my catalogue. Therefore 3″.8 must be added to the right ascensions of Ceres, sent you before. However I have here sent you all the right ascensions corrected, and the declinations from the first. I have moreover found from my observations of the summer solstitial Zenith distances of the Sun for some years past, the obliquity of the ecliptic no less than 5″ greater than I had computed it by carrying it on as settled in 1786 by a supposed annual diminution of the Obliquity. Hence it is necessary to increase the appt. Obliquity set down at the beginning of the nautical almanac of 1802 by 5″.4. I desire to recommend to you my folio Tables, as likely to be useful to you in your calculations, and desire you will do me the favour to accept a copy of them which I will have the pleasure of sending. I shall now observe Ceres and Pallas with the Equatorial Sector, of evenings, till they shall be lost in the rays of the Sun.*

[He encloses a list of 13 observations, made with his transit instrument between 4 February and 13 May 1802, of the apparent equatorial co-ordinates of Ceres, and 12 between 23 April and 16 May 1802 of Pallas].

Herschel to Gauss Slough May 22, 1802

I have the honour of a letter from you which I ought to have acknowledged sooner, but as you expressed a wish to know some of our results on Ceres, I deferred answering till I could have the pleasure to acquaint you with my observations. The additional discovery of Pallas likewise made me desirous of including this celestial body in my account. On the 6th and 13th of this month a paper of observations was read at the Royal Soc. in which among other particulars I give a set of very accurate measures of the diameters of the new stars. The result of these measures is that Ceres is only about 162 english miles in diameter and Pallas no more than 70. In the calculation of these quantities I have deduced the distance of these two stars from the earth, from your own account of their orbits which certainly must be near enough to convince us that they are extremely small bodies. I enter afterwards into an examination of the nature of these two bodies and compare them with planets and comets. I then define what we call planets and shew that we cannot put these bodies into their class. They are not only out of the zodiac; but Mercury, the smallest of our planets, is more than a hundred thousand times larger in bulk than Pallas. I shew in the next place that they are not comets, and since we can neither call them planets nor comets, it follows that the interesting discoveries of Mr. Piazzi and Dr. Olbers have introduced to our acquaintance a new species of celestial bodies with which hitherto we have not been acquainted.

I enter then into an examination of the principal features in the character of the planets and comets and of these new bodies. Planets are seen to move about in the zodiac. Comets have a visible coma. These new stars are mixed with the small fixed stars of the heavens and resemble them so much that even with a good telescope they cannot be distinguished from them. From this their asteroidal or starlike appearance I take my name, and call these new celestial bodies Asteroids. So that planets, asteroids and comets will make three distinct species of celestial bodies. My definition of this additional species is as follows.

Asteroids are small celestial bodies that revolve around the sun on ellipses more or less eccentric and whose plane can be inclined towards the ecliptic at any angle. Their motion can be direct or retrograde. They may or may not have considerable atmospheres, very small comas, disks or nuclei. I then proceed to shew that this definition will take in future discoveries which from the lately adopted method of observing we have reason to hope will soon be made.

That Pallas is an asteroid might alone be proved from the great inclination of its orbit which according to your elements is not less than 33°. And Ceres which is now actually out of the zodiac, and is so small a body as to have an asteroidal appearance can certainly not with any propriety be separated from its companion. Moreover if we were to call it a planet it would not fill the intermediate space between Mars and Jupiter with the proper dignity required for that station, whereas in the rank of asteroids it stands first on account of the novelty of the discovery reflects double honour on the present age as well as on Mr. Piazzi who detected it. I hope you will occasionally favour me with a further account of the orbits of these asteroids.

P. S. I need hardly add that neither the asteroid Ceres, nor its companion Pallas have any satellites. The small quantity of matter they contain could hardly permit us to expect that they should have any.

Herschel to Méchain Slough May 22, 1802

Regarding the two celestial bodies which were last discovered I am giving you a summary of my observations. In a memorandum, read to the Royal Society in London on the 6th and 13th of this month, I explained in detail my measurements of the diameters of these stars and I believe to have proven that that of Ceres, seen from the earth on April 22, was only 0″.216; and that of Pallas according to an equally good measurement 0″.17; but according to another even more accurate measurement only 0″.13.

Calculating with these as much as our still imperfect knowledge of the orbits of these stars allows, I found that Ceres' diameter is about 162 English miles and that of Pallas only 70.

I explained with the help of all my observations that these bodies cannot be called planets because of their small size and because they are beyond our zodiac. And, as I prove as well, they are not comets either and thus can only be regarded as a species between comets and planets which has been unknown to us and demands a name of its own. Since they resemble small stars and they are difficult to distinguish even with the best telescopes, I called them asteroids.

Here follows the definition of this word: "Asteroids are small celestial bodies that revolve around the sun on ellipses more or less eccentric and whose plane can be inclined towards the ecliptic at any angle. Their motion can be direct or retrograde. They may or may not have considerable atmospheres, very small comas, disks or nuclei."

You see, sir, that this definition leaves us a great deal of space and with tolerating these three species of heavenly bodies, planets, asteroids and comets, we make it much easier to classify future discoveries.

[On this date, Herschel also sent nearly identical letters to Lalande, Laplace, Bode, Zach, Olbers, Seyffer, Schroeter and Piazzi.]

No. 24 Olbers to Gauss Bremen May 23, 1802

I admire, and am very pleased about, the agreement between all of the observations at Seeberg and your elements. If Bode is still in doubt over the planet-like orbit of Pallas, if he still imagines every curve will fit the thus far traced arc, then you and I will have to attribute this solely to his lack of expertise with this type of calculation. The main dimensions of Pallas' orbit are not geometrically established. The major axis, longitude of aphelion, and the inclination of the orbit will always be subject to considerable corrections. However, if one can also regard your specified ellipse as one of the true orbits of the small arc traced from osculating curves, then this osculating ellipse does indeed approach the true orbit very closely. Concerning the larger inclination of Pallas' orbit, I would even maintain that your ellipse would approximate Pallas' orbit even more closely than the ellipse computed for Ceres before its rediscovery, were Zach's observations <u>in Dec.</u> as reliable as Piazzi's, which they still appear to surpass in RA. On the 24th *I received a request from the Prussian captain von Wahl* [C. W. A. von Wahl, 1760–1846] *to send him my hitherto obtained observations in order to compute my comet. I wrote him immediately that it may not be possible to fit just any parabolic orbit to this star's observations. Nevertheless, it would in any case be interesting, if he wished to try, to what extent one could fit a parabola through these observations. I therefore sent him all of my observations at the same time. This week, as I had expected, he informed me that all of his attempts were futile.*

I believe that my following observations are mutually consistent, except that I always had to compare Pallas with a small 9th *magnitude star, the position of which I'm still not quite certain. This 9*th *magnitude star appears in the Hist. Cel. Here are my original observations of Pallas that was compared with this star. My clock, over a period of 24 hours, ran 8s slower than mean solar time.* [These observations from May 17-21 were published the July *MC.*]

All of these observations are as usual a mean of several, but the last one isn't as completely reliable as the preceding ones. I have reduced these observations using the best material thus far at my disposal. However, I still hope to obtain from our friend Zach some better determinations, at least for the 3 stars Comae Beren. My reduction gave the following: [These observations from May 17-21 were published the July *MC.*]

It is unfortunate that with observations off the meridian <u>other stars</u> must be used so often. Precisely that affects the second differential (I understand the expression very well) of the longitude and latitude, if the position of the star itself is in error.

I still can't completely agree with your so confident hopes of Pallas' rediscovery in the coming year. I am much more concerned that we won't find it again before 1805. Out of curiosity I also computed both of Pallas' positions in the years 1803 and 1804. Pallas will be in opposition to the Sun toward the end of June, 1803; concerning Ceres, distance from Earth = 2.61, distance from the Sun = 3.38. How extremely small it will appear then, as it's now already dimmer than a 9th magnitude star? – In the year 1804 it'll be somewhat, but only slightly, better. Nevertheless, dusk won't be so troublesome then. Pallas is in the head of Pegasus which is opposite the Sun at the end of August, distance from Earth = 2.40, distance from the Sun = 3.37. – I doubt very much that I can identify Pallas in both cases with my comet finder, and if not, then the search will become more and more difficult. But if my health and sight don't fail me, then I'll certainly point it out to you again in 1805. Meanwhile, I'll spare no effort in the coming year. It'll be easier to judge on that matter only when we see to what extent we'll still be able to follow it this year.

You are right, that it's best for us to still withhold our always very risky and truly unwarranted speculations concerning the actual nature of the two strange celestial bodies, Ceres and Pallas. Pallas is in any case a very small body. The following extract from a communication of Best to Schroeter will probably interest you.

"Herschel's observations of Ceres and Pallas were read in the Society yesterday (May 6). They go to the 2nd or 4th of May. The majority of the observations were taken with the ten-foot reflector with a 516-times magnification. He denies they have any cometary and planetary characteristics and wants to name them 'planetulas', without thereby detracting from the discovery in the least. The most noteworthy appears to me to be the difference in the measurements, for he finds Pallas so tiny that, if I've heard correctly, 73/m Pallases could be extracted from Mars."

Should this 73/m, as I assume, be 73,000 then Pallas is certainly only a very small planetary fragment! I'm anxiously awaiting your ephemeris. The following summary is an approximate calculation:

May 29	*18h*	*RA 182° 35'*	*Decl. 20° 57*	*Distance to Sun 2.52*	*to Earth 2.07*
June 14	*11*	*185 18*	*20 16*	*2.57*	*2.31*

Here, from Oriani's letter to Zach, are 3 observations from Oriani in Milan. [These observations from Apr. 25-27 were published in the *MC*.]

"The first two observations can have an uncertainty of 8″ to 10″, especially in Dec., because I made them without illuminating the crosshairs of the micrometer." There are, as you can see, no meridian observations.

Please do not publicize the extract from Best's letter which I've sent you, at least don't associate his name with it, since I do not have explicit permission to do so.

Olbers to Lalande* Bremen May 24, 1802

Excuse me, Madam, that it took me so long to thank you for your kind letter of April 14. I have been waiting for the promised results of Mr. Burckhardt's calculations, but apparently he forgot about it and I only know them from the Moniteur. I have not yet calculated the new planet in an ellipse for our famous calculator Mr. Gauss who tackled all those calculations with such surprising ease has forestalled us all. Here are his elements calculated from the meridian observations of Seeberg and Greenwich from April 4 to May 16. He proposes to improve the elements further when the observations are finished which will happen soon in our northern climate where the nights are too illuminated by the twilight which prevents seeing well stars with such a feeble light. Yesterday I again saw the new planet very well. I dare asking you, Madam, to grant your protection to the name Pallas which we gave the new planet. By your intervention your respectable uncle will adopt it and then it will soon be valued everywhere. Herschel claims the very exact diameter of Ceres to be 162 and that of Pallas 70 English miles. I do consider these bodies very small but I cannot imagine how someone can measure such apparent diameters of Pallas in the month of April, when its true size is only 70 English miles. Mr. Herschel wants to distinguish these celestial bodies by calling them asteroids. I take advantage of the occasion to say with pleasure and recognition that we mainly owe the discovery of Pallas to the immense work which your respectable uncle and your husband have undertaken and finished on the 50 thousand stars. Without the help of the precious Histoire celeste francaise *I would maybe not have been able to find Ceres again or discover Pallas. The star of 8th magnitude observed on April 10, 1796 (Hist. Cel. p. 228)*

Passage	*Zenith distance*
12h 20′ 21″	*28.47.9*

is missing in the skies. It was, I believe, another asteroid: for, without doubt, we will certainly find others.

[*This letter was written to the wife of J.-J. Lalande's nephew, so the "uncle" referred to is Lalande himself.]

Olbers to Herschel Bremen May 24, 1802

With pleasure, as you might easily imagine, I am telling you with now utmost certainty that <u>*our Pallas is a planet*</u> *with Ceres' structure and equal in dignity and importance and maybe also a strange thing because it might give us reason for further speculations about the development and history of our planetary system. Here are the second elements calculated by Dr. Gauss for your disposition and further communication* [These are actually the first elements of Pallas by Gauss, as printed in the *MC*.]

it follows:

Distance from the Sun at Aphelion = 3.28004
in the mean distance = 2.69805
at Perihelion = 2.11606

Consequently, the orbit is only as eccentric as that of Mercury. But what an unexpected great inclination and what a weird positioning against Ceres' orbit? Both orbits look like two hooked chain links, like this: 

[This is the same sketch that appears in the Gauss-Olbers letter of April 20.]

It was Dr. Gauss' condition to compare these elements to my last observation before I would communicate them further. I did this and from above elements I calculated:

March 5. 11h 2′ 35 calc. RA 180° 56′ 28″.2 calc. Decl. 20° 9′ 20″.8
observ. -- 180° 56′ 6″.0 observ. -- 20° 8′ 59″
error in RA + 0′ 22″.2 error in decl + 0′ 21″.8.

Errors that are still small enough and at least show that the major dimensions of the orbit have been determined correctly. Otherwise the major axis, eccentricity, orbital period and inclination will see considerable improvements. Pallas is still visible and I hope to observe it until January. You will probably be just as amazed as I am to hear that Dr. Herschel allows Ceres only one second of apparent diameter.

Maskelyne to Herschel Greenwich May 24, 1802

Here are a few of the last observations of Ceres and Pallas made here. But the ephemeris for Ceres is always exact within ½ a minute.

Pallas	*RA*	*Decl. N.*
May 16	*12h 5′ 6″*	*20° 51′ (on the meridian)*
May 17	*5 21*	*20 54 (on the meridian)*
May 21	*6 33*	*20 59 (near midnight)*
May 22	*6 58*	*20 59 (near midnight)*
Ceres		
May 13	*11 45 12*	*16° 1 49″ (on the meridian)*
21	*46 49*	*15 2 18 (near midnight)*
22	*47 8*	*14 55 0 (near midnight)*

Dr. Zach writes by date May 4 that the orbit of Pallas is so excentric it traverses the orbit of Ceres, and the aphelium distance is twice the perihelium distance, which latter is about 2.0, the revolution between 5 and 6 years.

No. 25 Gauss to Olbers Brunswick May 25, 1802

Here, first of all, is the ephemeris of Pallas; the time, as usual, is mean midnight in Seeberg. [The ephemeris from May 29 to June 29, 1802 follows.]

Only the positions of May 24, 30 and June 5, 17, 23, 29 were actually computed using the Elements II. The remaining 7 were derived using interpolation based on a method which is indifferent to the use of the first, second, and third difference, and can differ only a few seconds from the true calculation.

Because of the difference in your position of June 14, I feel obliged to examine yet another one or two positions using a special calculation. Besides, the difference in RA using the Elements II is now becoming larger. Zach must have postponed his departure one more day. On May 11 he got another complete meridian observation that he forwarded to me privately. The errors of the Elements II are RA -16″.2 Dec. -14″.4. Along with this observation he sent 5 from Oriani, the first three are the same ones you're sending me; the last ones are [These observations from Apr 29 & 30 were published in the *MC*.]

It appears, however, that these observations are subject to more error than Oriani had supposed. That Oriani had information of Pallas so soon makes me hopeful that Piazzi also was informed thereof and early enough to still snatch a few meridian observations, especially since the somewhat shorter days in Palermo make it possible to observe it a little longer on that meridian than in our higher latitudes. If my hopes are confirmed, then these observations and particularly the Dec.'s would be invaluable. I don't know if I've previously written you that Piazzi's Dec. for Ceres 1801 agreed inexplicably accurately with the Elements V, and differ only a little over 1″. In the Seeberg Dec.'s the error in the elements appears to be negative whereas your observations make it positive. Méchain's Dec.'s are also good, if they are a close second to Piazzi's. The Greenwich ones, on the other hand, are practically as good as these. I've urgently requested Maskelyne for his observations of Pallas. I've also received Méchain's Paris observations for April 15, 17 and 18 from Zach. Bode however has lost Pallas. On May 21 he wrote that he hopes to locate it again. I regret, therefore, that I've thus far neglected to send him the ephemeris; the next Berlin mail leaves only on the 28th. But he absolutely doesn't want to hear anything about a planet. He believes it to be contrary to all celestial laws, and he imagines a comet at perihelion far beyond Ceres, and that I should just attempt to assume a greater distance and in this way fit the observations to a parabola etc.; Tempelhoff, Trembley, Burja [see note at end of next letter] *and others are of the same opinion. – Since in my method the distance isn't hypothesized but is rather determined from the observations themselves using a systematic 'Calcul', I therefore think I'll not undertake that thankless task. Rather, I'd much sooner utilize my leisure time to refine the elements as soon as I obtain observations that, either because of their sharpness or distance, guarantee a real improvement.*

The prediction you make concerning the rediscovery is of course very sad, though I do hope, should 1803 and 1804 be a disappointment, that 1805 will be auspicious for the believers. I hope that this year's harvest will present us with so many observations that the elements, still to be determined from them, will barely differ by more than 5° even in 1805. The uncertainty in the position of the neighbouring fixed stars will then not be of much consequence, since at the end of this year one can of course determine all of these stars exactly and hence can correct the observations and results. But shouldn't we still have hope, regardless of 1803, to rediscover Pallas? If we were to succeed in determining the orbit so accurately that, for example, in the coming year there were only a 5′ to 10′ error, then shouldn't you be able to discover it, even without your comet finder, merely with your nice Dollond, or Schroeter or Herschel with their powerful telescopes? Generally speaking, the accuracy of the results, other things being equal, is linearly proportional to the quality and quadratically proportional to the duration of the observations. I therefore confidently hope that we'll at least obtain elements of Pallas which are just as accurate as those we obtained for Ceres in 1801, even if the Elements II aren't quite as good yet.

In consideration of the description of my method, I have now successfully overcome a major difficulty which I feared most of all. Just after the Sept. 1801 discovery of the principal formula – which was the only reason for the method's further development and wishing to test it thus brought me into contact with Ceres – I found myself going in a bizarre direction, one in which I wouldn't want to lead you. A few days ago I found another well-defined direction with which I'm satisfied. It may interest you to know beforehand that I didn't sketch a single figure, neither during the development of this method nor during its application to Ceres and Pallas.

P. S. I wouldn't have thought Pallas to be so small that Herschel would require 73,000 to produce one Mars. I don't entirely comprehend on what Herschel could have based his statement, for he knew nothing whatever about the distance of Pallas. To want to distinguish between 'planeta' and 'planetula' seems to me to be almost pedantic. Mercury, Venus, Earth and Mars are also 'planetulae' compared with Jupiter, and perhaps our Sun compared with other fixed stars would just be a tiny 'solculus.'

Bode to Olbers Berlin May 25, 1802

Since there has been fair weather for several days now, I was able to see your Pallas. I have been able to see her very distinctly at the circular micrometer of my 3 ½ f. Dollond northwards between δ and 24 Comae (Pallas did not or only rarely tolerated the illumination of the hairs of our wall quadrant). The observations of one evening correspond quite well but I do not want to label them more accurate as they can be and rather believe that others delivered more accurate results.

		Right Ascension	*Declin.*
May 22	*11h 26′ 41 m.t.*	*181° 46 34*	*20° 59′ 15″*
23	*11 8 40*	*181 52 34*	*20 59 25*
24	*10 37 30*	*181 58 53*	*20 59 36*

I could only compare Pallas with 24 Comae, which is unfortunately only determined by Flamsteed, must consequently assume that this star is correct which is now impossible to determine. Now Pallas is really turning back and moves eastwards with increasing north latitude. It is and remains a very peculiar body, we might call it comet or planet and history has not seen anything similar and you have gained merit for yourself and new views for us by its discovery.

I still see Pallas through a telescope without nebulosity as a 9th *magnitude star. What is to become of it? I am very eager to know whether your and Dr. Gauss' calculation will further improve your elements after a longer series of observations. At the penultimate meeting of the Academy my talk about your calculation of Pallas' orbit caused a sensation and von Tempelhoff, Trembley and Burja negated the closeness and intersection of Pallas' orbit with that of Ceres and labelled together with me Pallas as a very weird comet which passes its perihelion at a great distance maybe far beyond Ceres' orbit. Why don't you try to match a parabola behind Ceres' orbit to the observations. The covered arc of 15 days, about the 108*th *part of the orbit, is barely enough to determine the elements with a certain reliability. Also the outer appearance of Pallas does not decide the matter. Your obtained results are so far from being normal and are a unique success in our solar system.*

Dearest friend, please excuse my doubts, remarks and objections, often truth is discovered by these. The determination of its true orbit is in the best of hands – yours and Dr. Gauss' and nobody could anticipate your work. Please be so kind as to send me your further studies on Pallas' orbit and thus to answer my last doubts. Should not Pallas' north latitude increase with increasing distance from Earth although Pallas is moving heliocentrically farther away from the node?

P. S: I also observed Ceres several times at the circular micrometer, she is considerably brighter than Pallas and of 8th magnitude.

[Gauss' library contained works by Tempelhoff & Burja. Georg Friedrich von Tempelhoff was director of the Artillerie-Akademie zu Berlin; Abel Burja (1752–1816), Professor of Mathematics at the Konigliche Ritterakademie zu Berlin; Jean Trembley (1749–1811) was an expert on probability. The frontispiece of Burja's algebra book began Chap. 1 of this volume.]

Maskelyne to Banks Greenwich May 28, 1802

I thank you for sending Baron Zach's two letters, together with a third, which third I now return to you enclosed. The map of the track of the two new planets in the heavens is a mere curiosity of no real use in astronomy. I will keep it till next week, and then return it to you. Baron Zach wants to class the two new planets by themselves, from the circumstance of these motions, as Dr. Herschel does from their smallness. He seems to think that the orbit of Pallas might be so eccentric that it might in a part of its course come as near to us as Venus does in her transits over the Sun, but the letter you favoured me with yesterday from Dr. Olbers to Mr. Schroeter, shows the contrary: consequently we cannot expect to get at the Sun's parallax this way, not even so well as by the planet Mars in opposition to the Sun.

Memorandum by Maskelyne Greenwich May 29, 1802

Further account of the new planet, Pallas

Dr. Olbers, who discovered it on the 28th of March, after observing it for five weeks, was fully persuaded of its being a true planet. Dr. Gauss has determined its elements as near as could be done from so small a part of its orbit, and its eccentricity is a little greater than that of Mercury, the inclination of its orbit 33° 39′, and its mean distance a little less than that of Ceres. But the most remarkable circumstance concerning it is, that its crosses the orbit of Ceres, approaching the sun nearer than Ceres in its perihelium and receding farther from it in its aphelium than Ceres does. Dr. Herschel has made some curious observations of the apparent diameters both of Pallas and Ceres, from which he infers the real diameter of Pallas to be 95 miles and that of Ceres 162 miles. He considers them as if a different species from the known planets. In their motions and smallness they resemble comets, but in the clearness of their light the other planets.

Zach to Gauss Seeberg May 30, 1802

Back from my journey I find letters from Palermo of Professor Piazzi with observations of Ceres, but unfortunately he does not mention Pallas. The letter is of April 18th. I tarry not a single moment to relay the information in a manner true to the original as postscript A. Postscript B is my reduction to bring them into line with those you used last year for the determinations of Ceres' orbital elements. The calculation of the mean time differs sometimes by a second from Piazzi, but obviously there happened a slip of the pen of 10″ regarding the m. t. in the observation of March 31st. I took Ceres' RA at the passage instrument since it has five threads and the circle only one and on April 9th I had to use the circle for observations of the RA. Professor Piazzi conveys his regards to you. The King is minting a coin on the occasion of the discovery of Ceres and I will see to it that you will get one too.

But apart from that I have not heard anything new regarding Ceres and Pallas from my correspondents. What Sniadecki from Cracow and Derfflinger from Kremsmuenster [in modern-day Austria] *wrote is not of much use for you. You will find these observations in the June issue, which will probably appear eight days later than usual. But because Sniadecki observed mostly according to my determination of stars, I give you his observations of Pallas nonetheless. He promised me (observations from) Vilnius by Poczobut, who possesses an excellent 8-foot wall quadrant by Ramsden, so the declinations will be superb.*

[The observations from Cracow were published in the June issue of the *MC*.]

I think I have sent you already all of Oriani's observations from Milan. They range from April 25th to May 7th. Since he observes at the equatorial sector he will be able to send us some more. Did you check whether his observations correspond to meridian observations? To be on the safe side here are his observations of Pallas of May. [These were published in the *MC;* Zach then goes into a lengthy mathematical digression about Oriani's perturbations of Ceres.]

All your letters (the last one of May 21st) came in time and I would like to express my thanks. Everything, even the ephemerides for Pallas has its place in the June issue. Very soon you will receive the advance sheet, right now it is being set. The register of the June issue is also a little late. Your suggestion to derive the ascending node, heliocentric longitude and true distances from the geocentric longitude and latitude will be in that issue. Friend Burckhardt was not successful with Pallas. He also found that circular and parabolic orbits are not correct. La Place and Méchain searched in vain too. La Lande wrote "the distance between Mars and Piazzi has been found to be 2.1, but this is not confirmed". Burckhardt is working now on an ellipse of twelve years and is convinced it will represent the observations. In order to tease these gentlemen I want to send only your comparison of my observations of 27 days without your elements II, but I am afraid that Dr Olbers has sent those already to Paris.

[Zach attaches a table of Ceres observations made by Piazzi. This was published in the June issue of the *MC*. Zach also attaches a table of Ceres observations by Poczobut.]

No. 26 Olbers to Gauss Bremen May 30, 1802

Many thanks for Pallas' ephemeris. I don't see that my data for the position of June 14, 23h (maybe I had written 11h) differs much from your ephemeris. But on the whole my computation was carried out carelessly. The observations of Pallas follow, first the original and then after my reduction. Mr von Zach will, however, send you his much more accurate reduction at my request. I'm indicating the two compared stars as No. I and No. II. They appear, as I have already informed you with the first ones, on p. 68 in the Hist. Cel. *as follows:*

1794 April 28	*2 Comae*	*11h 53m 0.3s*	*26° 13′ 6″*
	5 Comae	*12 0 57.5*	*27 8 4*
	No. I	*12 1 12*	*27 52 13*
	No. II	*12 7 43*	*27 41 30*

Pallas was compared with No. I until May 21; thereafter with No. II. You've already received the final few observations.

May 22	*11h 17m 47s*	*Pallas advances No. II*	*1m 50″.0 5′ 51″ south*
23	*11 13 46*		*1 27.3 5 26*
24	*10 50 52*		*1 2.0 5 18*
25	*11 7 38*		*0 35.5 5 27*
26	*10 45 56*		*0 7.8 5 59*
28	*10 50 7*	*Pallas precedes No. II*	*0 50.5 7 3*

From these observations and for the times indicated I've derived: [Observations from May 22 to 28] *It appears that your elements still agree well in RA but result in a one minute increase in Dec.*

Mr. Bode has sent me 3 observations using a circle micrometer. Pallas was compared with No. 24 Comae, whose position only Flamsteed has determined. [These observations from May 22-24 were published in the *MC*.] *Flamsteed's RA of No. 24 will probably be too large by 1′. In the meantime I wouldn't advise improving your elements for Pallas again until the observations are concluded. Their accuracy exceeds that necessary for its discovery.*

Zach to Banks Seeberg May 31, 1802

Having prosecuted Dr. Olber's Pallas from April 4th till May 11th in the meridian, Dr. Gauss upon this set of my observations calculated the elements of an elliptical orbit of this very remarkable heavenly body, which represent with great accuracy all Seeberg observations. It appears in general by these calculations, that: Pallas is a planetary heavenly body, that moves between the orbits of Mars & Jupiter, with a very great eccentricity and inclination, and whose orbit comes very near, to the orbit of the planet Ceres, perhaps touches it, perhaps even cuts it, like two links of a chain in this way 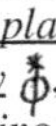*, which cannot yet be attested with certainty, the observed arc run over by this planet being too small. Notwithstanding, it appears already that the distances of Pallas and Ceres in the line of nodes of their orbits, in very nearly equal. In the descending node, the distance of Pallas from the Sun is = 2.86 and the same distance of Ceres = 2.93. In the ascending node, these distances are of a greater inequality. Another very remarkable circumstance is, that the mean motions of Pallas and Ceres are very near the same, perhaps absolutely the same, which cannot yet be attested; the period of the observations of both planets being by far too short. But as much appears already, that these mean motions will not differ very much, and in this case, as little as the masses of Pallas and Ceres may be, they will, nevertheless exact a very sensible action one upon the other, and give therefore occasion to very curious and interesting investigations in the mechanics of the heavens. The new planet Pallas will also call forth the utmost exertion of our analytical powers.*

Hitherto the two elements of a planetary orbit, viz, the eccentricity, and the inclination, had been considered as an infinite little quantity, and so it could be, as these two elements in all our old planets are very small; so then, the higher powers of them could be neglected without danger, as they produced no sensible term in the approximating series. But this is not the case made by Pallas, where the eccentricity of the orbit and the inclination are so very great.

Here are the elements of the orbits of the planet Pallas calculated by Dr. Gauss. [These are the elements given by Gauss to Maskelyne in his May 19 letter.] *By these elements the whole series of my observations are represented thus.* [Zach prints his data from April 4 to May 11.]

Pallas and Ceres are now come too near to the sun, and the twilight permits no meridian observation. But the astronomers who are provided with equatorials of great perfection, as for instance are, in Greenwich, Oxford, Richmond, and of Sir George Schuckburgh [1751–1804] *will be able to prosecute these two planets a longer time. The observations of Pallas will chiefly be of a very great value; as the series of meridian observations is not above 5 weeks. If more observations are not procured, it will be with some difficulty we shall find again Pallas next year, for the elements of an orbit calculated upon so little an arc as 7½°, can be in fault of some degrees till January 1803. You will do, most honoured Sir a great benefit to science in general, and to astronomy in particular, if you engage the English astronomers, who have so very excellent, and fixed equatorial sectors, to prosecute Pallas out of the meridian as far as they can, to this purpose I take the liberty to send you here an ephemeris of this planet's motion calculated by Mr. Gauss, this will enable and facilitate to the astronomers the research, and the observations of Pallas.* [Zach prints an ephemeris from May 24 to June 29.]

Maskelyne to Herschel Greenwich June 2, 1802

I have just received your favor dated May 1, for June 1, yesterday. I was going to write to you by this post, Dr. Gauss of Brunswick by date May 19 has sent me fresh elements of Pallas and an ephemeris of its place till 29th this month, as follows. He thinks the calculated RA may turn out 3′ or 4′ too small toward the end of June, and the declination also too small, but both these are uncertain. He derives some extraordinary conclusions from these elements, supposing they do not differ much from the truth.

1. The periodic times of Pallas and Ceres are nearly, or perhaps exactly, the same.

2. Their orbits nearly coincide in one point at the ascending node of the orbit of Ceres on the orbit of Pallas. Hence, Dr. Olbers conjectures, "they were once but one planet, which was divided into two by the percussion of a comet; indeed the exterior appearance of the two stars and their very variable light seems to countenance such a supposition and to indicate that their shape is not spherical, but considerably irregular."

[The 3rd elements of Pallas, and an ephemeris from May 24 to June 27.]

Lalande to Herschel Paris June 3, 1802

I have found with pleasure your letter about the planets of Piazzi and Olbers. I wish that we could interest others for acceptance of your ideas. But it is not possible to believe that the planet of Olbers has a diameter of only 70 miles. We could not see it.

Zach to Oriani Seeberg June 4, 1802

How you made me happy by the sheer quantity and interestingness of your letters of May 1, 5, 8, 18 which I have all received. Do not be impatient at all about your improvements of the calculations of the perturbations of Ceres. I have inserted everything and in my next issue of June you will find everything at the correct place. The sheet I had sent you previously was only a proof but what I am sending you here is actually what will be published; and as you can see, your corrections do not appear. I am asking you to always do the same and to send me your calculations as fast as possible to arrange a date and actually this time you anticipated the positions which have not been calculated yet. Regarding the formulae in another hypothesis of the mean distance and for another eccentricity, I only received them yesterday and can only publish them in my July issue, but I communicated them to Dr. Gauss who will redo his VII elements according your calculations.

I am endlessly grateful for your Pallas observations but would be even more so if you could send them now that we can further observe this planet at the meridian. Your observations with the equatorial sector are priceless now and I am asking you to observe this little planet attentively, it may be visible until the end of this month. I stopped observing it at the meridian on May 11 and since I do not possess an equatorial I am unable to follow it to the western horizon. I am sending you here a small ephemeris of Pallas which will facilitate its search in case of a long interruption due to bad weather and you have lost it. I am asking you with insistence to send me those observations for you can easily imagine how interesting this extraordinary body is which might be the most remarkable star in the entire universe as you will see in a second. You will find the orbital elements I and II of Pallas first, based on my observations and you will also see how much the observations agree. But the wonderful thing is that most likely both, Pallas and Ceres, have the same revolution, the same mean motion, that their orbits touch and maybe intersect or are one in the other like two rings of a barrel. The observed difference of 7° 1/2 is still too small to decide it but it can already be seen that it will come close to one or the other hypothesis. From this the conjecture can be drawn that Ceres and Pallas are fragments of one single planet who had its place between Mars and Jupiter. Maybe the impact of a comet, maybe expansive forces in the planet itself made the body burst and Ceres and Pallas are scattered samples, maybe there are several of these widespread samples in this region of which some have escaped on parabolas, hyperbolas and others were restrained on ellipses. I will even tell you, but in private, that we are tracing a third fragment but do not talk about it for we must not, as you know, divide the skin of the bear before it is caught, you will be the first to know if our hopes and suspicions have come true.

As small as the masses of these two planets are, there can be no doubt that there must be a very sensitive interaction between these small bodies. Your perturbation calculations of Ceres will be all the more difficult to calculate since the eccentricities and the inclinations are considerable. These two little planets will bring a new era in physical astronomy and give enough brain food for other geometers. Until we are there, we need to try to collect every piece of information, the more so since we will not at all see Pallas at its opposition in June 1803 – the planet will almost be in Cerberus at 9″ 8′ of long. and 48 degrees of latit. north. But the distance from the earth will be 2.6 and when it was discovered in March 1802 it was only 1.38. In June dusk in our latitudes is troublesome, which makes Pallas' visibility doubtful. It will be a wee bit better in the opposition of 1804 which will occur around the end of August at 19° latit. north in the head of Pegasus. The distance of the planet from the earth will still be 2.4 but the nights will be more favourable. It is only in 1805 that we will see Pallas <u>for certain</u> and it will be closer to the earth. From this you can see what diligence it takes to get as much observations of Pallas as possible before it leaves us for such a long time. Herschel finds the size of Pallas and Ceres extremely small and gives especially Pallas not a second in diameter, he says that it takes 73,680 Pallas-sized bodies to equal Mars. Here is an ephemeris for Pallas, midnight in Seeberg. Ask Cesaris and Carlini to send me their observations of Ceres and Pallas, please ask them with compliments.

Méchain to Herschel Paris June 4, 1802

The observations of the new star discovered by Dr. Olbers and of Piazzi's star which you kindly gave me are endlessly curious and interesting. I admire your ingenious means of determining such small diameters and I would be delighted to be able to understand them, though I do not at all doubt their accuracy. Mr. Schroeter in Lilienthal found with his large telescopes Ceres' diameter by 2″ smaller, as you have certainly heard, around the end of last January. And he even suspects two satellites. Mr. Zach sent me at the end of March a series of diameters of this planet measured by the same. Mr. Schroeter finds it to be 529 geographical miles. This is quite different from those you found. Thus we await impatiently the result of your observations for this planet and for Pallas as well.

Ceres' orbit has been quite well determined and even all the perturbations she experiences from the other planets which are considerable. Oriani in Milan and Burckhardt here worked successfully on them. The eccentricity and inclination of her orbit do not appear to be too considerable to place this planet not in the same category as the other planets; even the physical considerations that you oppose to those according to your observations merit the same. Pallas is more rebellious and still difficult to grasp. Until now the orbit has not been determined; in the elliptical hypothesis the eccentricity has to be enormous which has not yet been determined at all. The parabolic hypothesis, which Laplace favours most, is just as unsuccessful. I found a parabola after a period of twenty days of observing that corresponds well for that time but which differs by more than 30′ for the geocentric longitude. I will try to rectify it according to the latest observations; but I have not much hope to be successful. We need to believe to find it with time. I observed this planet at the meridian from April 12 until May 14, every time the sky was favourable; now I am observing it beyond the meridian and I believe to be able to follow it for another month. I also observed Ceres at the meridian after her rediscovery on January 27 in the morning. I will also observe her from time to time beyond the meridian until she is lost in the rays of the Sun. I have sent Dr. Maskelyne my first observations of this planet and asked him to communicate them to you.

Letter LIV, Piazzi to Oriani Palermo June 4, 1802

On May 17 I received from Zach the news of the new discovery; on the 19th the same was confirmed by you to me. Based on the 2nd obs. by Olbers, on the 3rd by Zach, and on the last one by you I calculated the circular elements. Here is what I found.

Longitude 183° 47′
Ascending node 170 12
Inclination 27 1

[The orbital radius is missing.]

With this data I looked for the new star at the meridian passage on the 20th, 21st and 22nd of the month, but no luck. For us there is a difference of 1 degree long. and 3 degrees inclination, and that made me doubt the planet's nature. [The first circular elements of Pallas calculated by Oriani were in the last letter. They were published in the *MC* June 1802, vol. 5, pg. 604.]

Bode to Olbers Berlin June 5, 1802

While I am impatiently waiting for the crucial results of your analytical calculation of Pallas' orbit, I made an attempt to preliminarily determine this orbit according to Lambert's construction method. I have been sketching several cometary orbits according to it in order to find the elements and to show the entire parabolic path as far as the accuracy of this method allows. I am asking you to do the same for Pallas and I do not know why I have not done it yet. Enough, the result of my calculation and drawing seem to indicate that Pallas is a distant comet which passed already its perihelion and is now moving away again from Sun and Earth. I am telling you, my dearest friend, these elements only in private since they will need further considerable improvement which I will find through repeated sketches and calculations which will prove my assumption of Pallas right. I cannot see where and how I could have gone wrong since everything is in harmony and even the observed geocentric longitude and latitude that are derived from the construction and the calculation are as accurate as one can expect from this method. Sketch and your observation of March 29 and April 29 and mine of June 3 form the basis. The parabolic elements of Pallas' orbit are the following:

Time of perihelion 1801 26 Sept. 15 h

Distance of perihelion 1.9298

Position of perihelion 33 24 55

Node 5 24 56

Inclination 47° 53'

Log. of diurnal mean motion 9.53788

True anomaly at the 1st observation of March 29 71° 1'

------------------------ 2nd June 3 83 45

Distance of Comet from Earth on March 29 1.93 from the Sun 2.92

---------------- ------------ June 3 2.97 3.47

Helioc. Longitude on March 29 9h 6z 2° 21' 8° 9' N

------------------------------ June 3 12h 6 11 21 17 8 N

Please be so kind as to send me your opinion of these elements as soon as possible. Until now I have not received any calculations of Pallas' orbit from foreign astronomers, although there should be no shortage of observations. I have just received from Abbe Poczobut of Vilna his observations of Ceres; he says of Pallas only that he had received Mr. von Zach's observations and that the bad weather in May impeded his search for her.*

[*Marcin Poczobut was a Jesuit. The term "Abbe" was in widespread and ambiguous usage. It could mean the abbot in charge of a monastery, a priest who had income from a monastery, or a person with nominal religious position but no actual duties. Poczobut was likely in the latter category.]

No. 27 Gauss to Olbers Brunswick June 8, 1802

The news that Dr. Maskelyne was able to observe our Pallas on May 16 on the meridian will certainly please you as much as it pleased me. He was good enough to send me all of his observations. There are 12, the first one on April 23. They are, except for a few, quite superb. He will hence still observe Pallas on the equatorial telescope until it becomes obscured by sunlight. His letter of May 21 was written on the same day as mine in which I sent him the ephemeris that he will have long since received.

I have just finished the task of computing new elements based on these observations but haven't been able to conduct any test whatever. I'm therefore still not quite sure if they're free from errors. The arc traced out from April 4 to May 16 is 11° 24′. I have hope that the elements are already almost exact. I'm now also convinced, based on geometrical calculations, that they are in general correct. It is most strange that the period of revolution has of itself, without any action on my part, become almost identical with that of Ceres. The daily mean motion of 5″ is hardly doubtful. I'm asking that you personally check the elements using your most recent observations. [These interim elements do appear in the *MC*.]

Asc. Node	*172° 28′ 18″*
Aphelion	*300 58 48*
Incl.	*34 39 11*
Eccen.	*0.2476402*
Log. semi-major axis	*0.4425664*
Tropic	*769.5414″*
Epoch	*162° 25′ 45″.9*

Zach writes me that Mssrs. Burckhardt, de La Place, Méchain etc., all were unsuccessful in their attempts. Burckhardt now supports an ellipse having a 12-year period with which he hoped to describe the observations. Seyffer had informed me of a parabolic orbit, computed by Burckhardt.

Incl.	*54° 58′ 30″*
Asc. Node	*176 45 34*
Perih.	*113 52 3.50*
Dist. Perih.	*0.8432*
Date of perih.	*1801 Sept. 29 16h 48m*

I've already made a start on a draft of my method. In the June issue of the M.C. you'll find a small but very unimportant attempt, which I had already sent to Zach a fairly long time ago, to obtain a number of solutions to the problem of finding the heliocentric locus from the position of the orbit and geocentric locus. At that time I hadn't yet possessed your determination of the cometary orbit, otherwise I wouldn't have taken on the effort to look for solutions to a problem that you have long ago already solved so nicely. Nevertheless, a feature of mine is that one can easily completely guard against computational errors. Besides, I've found it more convenient not to make use of this method at all when determining the planetary orbits by hand, but rather to use other methods instead.

I've now also obtained Piazzi's observations of Ceres up to mid-April and the entire set of Greenwich observations. Oriani's perturbation formulae are already in the individual copies of the May issue of M.C. Schubert [Friedrich Theodor Schubert, 1758–1825] *has sent me his, which almost agree with them. I'm thinking of shortly beginning the first improvement of Ceres' orbit, but only when I'm a bit further on with Pallas. For the final touch to the correction equations I think I'll make use of a specific procedure.*

Bode to Olbers Berlin June 11, 1802

I had just finished my letter of June 5 when I had the honour to receive your friendly and nice letter of the 1st. You are too kind to give me your further observations on Pallas and to answer my doubts about this weird star, to advise me, to reassure me, and to share your assumptions. Now that also you, a master of comet calculation, have left the calculation of Pallas' orbit to the astute and learned Dr. Gauss, I have abandoned any kind of research in this respect. Consequently, I am hoping, dearest friend, that you have not publicised the results of my preliminary attempt to find Pallas' orbit according to Lambert's construction method which I confided to you in my last letter. I am only embarrassed to have caused you new calculations since I dared to ask for your opinion.

I based the first attempt on the first and last observations that are 66 days apart. As is generally known, Lambert's preassumption that the mean rad. vect. divides the found chart in the ratio of the intervening time is no longer correct, but Pallas, regarded as a comet, is already considerably distant from its perihelion, which decreases this incorrectness slightly.

You have again immortalized your name – whatever might become of Pallas. I finished my treatise on Ceres already four weeks ago but the engraver delayed us. You, dearest friend, will not find anything new in it. It only contains (as the one on Uranus) the history of Ceres' discovery together with some remarks for my astronomical friends who keep asking about it orally and in their letters.

PS: The treatise also deals randomly with Ceres. Since I communicated your first result to several friends and the Academy, that the orbits of Ceres and Pallas intersect and come very close around the node, and an accident is to fear when both would come close there at some time, they confused it with our Earth and rumour has it in Berlin that I announced a future comet which I believed would threaten the existence of our Earth in September.

No. 28 Olbers to Gauss Bremen June 13, 1802

Here are my further observations of Pallas that I'm again sending you in their original form because I can't trust my reduction since I must still make do with small stars from the Hist Cel. *Apart from those stars which you've marked No. I and No. II, I've now used another, No. III, which appears twice in the Hist. Cel.*

				Zenith Distance
5 Comae Beren	12h	0m	57.5s	27° 8' 4"
No. III	12	12	14	28 1 0
11 Comae Beren	12	10	54	29 53 21
No. III	12	14	1.6	28 1 6

(footnote by Olbers: With regard to the RA of the last line: At the 3rd crosshair, hence at the middle crosshair 12h 13m 32.6s)

My observations, which I consider more or less correct (those of June 3 the best, those of the 2nd and 10th somewhat less certain), are as follows:

May 30	11h	40m	46s	Pallas advances	No. II	1m	54".5	8'	56"	south
30	12	14	20		No. I	8	24.0	1	37	north
Jun 2	11	13	4	Pallas precedes	No. III	0	53.2	5	48	north
3	10	46	7			0	18.0	3	38	
6	12	53	45	Pallas advances	No. III	1	43.0	3	57	south
8	11	9	10			3	2.0	9	32	
10	12	33	50			4	31.2	15	52	

There results, based on my reduction:

May 30	*182° 41' 31"*	*20° 56' 20"*
30	*182 41 43*	*20 56 10*
Jun 2	*183 7 18*	*20 51 19*
3	*183 16 13*	*20 49 9*
6	*183 46 33*	*20 41 34*
8	*184 6 21*	*20 35 59*
10	*184 28 43*	*20 29 39*

May 30 (FN by Olbers: The times in the original were recorded hastily.)

I'm most thankful for your new elements of Pallas that I take to be very accurate. I, as you, am happy that Maskelyne was able to observe Pallas for so long. I can't tell you enough, dear friend, how much I admire you for all of these calculations.

Herschel was good enough to write me. He found the diameter of Ceres and Pallas to be 162 and 70 English miles respectively: "by a set of very accurate measures". – Pallas' apparent diameter in April must have been only 1/10". I don't understand how it's possible to measure this. He derived the distances from your first ellipse. Herschel wants to distinguish Ceres and Pallas not only from planets but also from comets and names them 'asteroids.'

How curious and extraordinary that the major axes and periods of revolution of Ceres and Pallas are the same. Both indisputably belong together.

Zach to Oriani Seeberg June 15, 1802

Dr. Herschel makes the diameter of Ceres 162 English miles and that of Pallas 70 miles. He also proposes to give the appellation asteroids to them both.

We have reason to fear we will only see Pallas again in 1805 as you can see for yourself by its elements and its positions in 1803 and 1804. The opposition will take place in the summer with the great dusks and the distance of this planet, or asteroid (as Herschel says), to the earth will still be very great thus we have every reason to believe that our telescopes will not reach it. Mr. Schubert in Petersburg calculated the perturbations of Ceres and is mostly in agreement with you. I will send you his formulae in my next letter as well as the observations from Vilnius and Greenwich of Ceres and Pallas. The French citizens are still a mystery. Friend Burckhardt found an ellipse of 12 years for Pallas, hopefully less.

[Zach appends observations of Pallas made by Olbers, and the third elements of Pallas by Gauss.]

Maskelyne to Herschel Soho Square June 18, 1802

It has occurred to me that as the elements of the orbit of Pallas found by Dr. Gauss give the mean distance 2.8 whereas I believe you took it considerably less, I believe 2.1 according to what was reported to you from a communication I received from France, in computing the real diameter from the apparent diameter you had observed, that the real diameter would require to be augmented. I thought you would like to make their alteration before it is printed. I spoke to Dr. Gray about it, yesterday, who will stop the printing your paper, till he receives your directions. From Dr. Gauss' elements the distance in parts of the orbit may be readily computed, and thence the real magnitude inferred. I shall have the authority to compute it myself as soon as I get home tomorrow, and send you the result which you may compare with your own calculations. I shall compute for the 22nd of April for the most exact apparent diameter of Pallas which you mentioned to me 0".13.

Maskelyne to Herschel Greenwich June 23, 1802

I now fulfill my promise in the letter of June 19th. I have calculated the distance of Pallas from the earth on April 22nd at midnight by Dr. Gauss' elements which I sent you, and find it to be 1.976 whence as you measured the app. diameter of the planet 0".13 its appt. diam. at the mean distance of the earth from the sun should be 0".2568 and taking the mean distance of the sun at 93606000 English miles the true diameter of Pallas will be 116.56 or say 116 English miles, which approaches nearer to that of Pallas [he means Ceres] *which I think you said was 162 miles. I hope this will save you some trouble as it has given me a good deal, principally owing to the greatness of the excentricity of the orbit. You will be pleased to send what directions you think proper to Dr. Gray about altering the number for real diameter of Pallas or convey me your directions, and I will impart them to him. On the 20th I observed Ceres and Pallas, the 1st is reduced to a star of 10th magnitude and in the last observations of the evening between 11 and 12, the first was in RA 12h 4' 9" Decl. 10° 26'N. The 2nd in RA 12h 25' 2" Declin. 19° 49'N. The difference from Mr. Gauss' elements is exactly what he supposed they would be. The obs. were made about midnight. I have this day received a letter from M. Méchain who had received the account I gave him of your observations of the two new planets. He writes that they are impatient to see your paper upon the subject.*

No. 29 Olbers to Gauss Bremen June 24, 1802

I'm delighted to be able to send you herewith 3 good observations of Pallas taken on June 19, 20 and 21. I hope these might be of use for further corrections to the elements.

I've already mentioned the observations of June 10 previously. The weather here was adverse from the 10th to the 19th. On the 14th it was a bit clear, but Pallas couldn't be reliably compared with No. III because of bright moonlight and a somewhat hazy atmosphere. At 11h 56m 14s Pallas trailed the star by 7m 30.9s and was 30' 53" further south. There results:

RA 185° 13' 47" and Dec. 20° 14' 38"

But on the 19th, 20th and 21st it was very clear. It was possible to compare Pallas with the binary star No. 24 Comae Ber. from which Zach determined the RA. I've unfortunately had to retain Flamsteed's Dec. Of course the position determination in Bode's catalogue is attributed to LaLande; but this is only a spelling or printing error. The RA and Dec. specified therein are Flamsteed's. Here are my observations, always the mean of many, among which there is excellent agreement.

Mean time			
June 19 11h 13m 55s	Pallas preceeded	0m 7.6s	north 25' 57"
20 11 7 51	Pallas followed	0 45.0	21 37
21 11 8 54		1 38.6	16 55

Using the above, I've derived the following positions of Pallas taking due account of aberration and nutation:

	RA	north Decl.		
June 19	186 16' 28"	19° 53' 51"	−96"	+26"
20	186 29 39	19 49 31	−17	+8
21	186 43 4	19 44 49	−25	+6

(FN by Olbers: The corrections have been added by Gauss himself.)

I wonder if I'll still be able to make such good observations again. Pallas will be too faint and dusk at our latitude is too light. At midnight, when dusk is least troublesome, Pallas is already too low and I'm therefore afraid that I'll soon have to discontinue the observations since I won't be able to recognise the emergence and disappearance of the faint planet with certainty. In more southerly latitudes, e.g. in France, one will be able to continue these much longer.

Dr. Burckhardt has also finally finished with his calculations. As I don't know whether you read the Moniteur, I'm sending you herewith the complete article from No. 203, Saturday, 23 Praireal:

The planet discovered by Olbers on March 28 has been calculated by Citizen Burckhardt, who after some lengthy and laborious calculations has found the elements of this planet in the following manner.

Long. March 31	*162° 51′ 14″.2*
Asc. Node	*172° 28′ 57″*
Aphelion	*302 3 2*
Incl.	*34 50 40*
Eccen.	*0.2463*
Distance	*2.791*
Diurnal sid. movement	*760.84″*
Sidereal revolution	*1703.7 days*

Citizen Burckhardt was obliged to calculate the perturbations that this planet experiences due to the attraction of Jupiter and which were causing some very perceptible differences in the observed positions. But these calculations are very complicated due to the large inclination and large eccentricity of this planet. Lalande." [End French text.]

The computation must have been much more difficult for Dr. Burckhardt than for you. Nevertheless he was already acquainted with your improved elements No. II, through which, in my opinion, the main difficulty of the determination would have already been eliminated if he hadn't wanted to include the perturbations in the calculation. To improve elements is, after all, always much easier than to find them in the first place.

Have Burckhardt's elements, by including perturbations, really become more accurate and more useful for future findings than yours? – Both, after all, agree by and large very well. The largest difference is only in respect of the major axis and the period of revolution, which is dependent on the major axis, and the daily motion. Burckhardt's period of revolution is noticeably different from that of Ceres, with which yours agrees so perfectly.

I can see immediately that Burckhardt's ellipse for Pallas would give a very good approximation at the descending node with Ceres' orbit. In fact I find at this node, if I simultaneously also apply Burckhardt's ellipse to Ceres, the distance of Pallas from the Sun = 2.88223, the distance of Ceres = 2.92327, difference = 0.04104. – Your elements for Pallas No. III and Ceres No. VII give the distance of Pallas = 2.845797, of Ceres = 2.930747, difference=0.084950. – Actually we still can't say with certainty how closely both orbits perhaps intersect each other here.

Zach to Gauss Seeberg June 24, 1802

I would like to thank you for your important letters of June 10th and 20th. I continue to send you my collected materials concerning Ceres and Pallas. This time you are getting the Milan observations of the two planets; Oriani promises observations of Pallas at the equatorial sector, which are important, because such are rarely made outside of Greenwich. Also included are Olbers' observations and my reductions of those. Dr. Olbers sent me his original observations of Pallas and asked me to determine those as accurately as possible with the new aids. I did this – but successfully? You will know best. Everything depended on *three small stars, observed only once by La Lande in his Histoire célèste. But unfortunately La Lande's determinations are not always the most accurate. In the meantime I did everything to determine these three stars as well as possible. My reductions differ slightly in RA from those of Olbers, especially around June 6th, 8th and 10th. Dr. Olbers has to decide whose reductions are to be trusted. You will see it also while comparing the elements.*

I stopped here. Dr. Herschel wrote and tried to suggest his term asteroids to me. He wants to introduce three distinct species. Planets, asteroids and comets. He wrote: "I hope this classification will meet with your approbation and that you will do me the honour to adopt it." But I have no inclination to do so because his definition of asteroids is not convincing. I rather stick to the name planet, together with you and Olbers. Only if there are several small planetulus between the older ones we can talk about a new classification, but today smallness, inclination and eccentricity do not decide on planetism or not planetism. Thus, Herschel's definition of asteroids is arbitrary. Thank you 1000 times for the relayed Greenwich observations – they complete the collection in the M.C. I am planning to publish the continued observations in the July issue. So I am waiting eagerly and impatiently for what you might be so kind and give me; I would very much like to publish in this issue the comparison of the Greenwich and Dr. Olbers' observations with the elements III of Pallas.

According to Dr Maskelyne 3".8 can be added to all of my observations of Ceres and Pallas since they are based on his 36 fixed stars catalogue. Whether this applies to other observations as well remains to be seen because other observations are based on Bradley, La Caille or Ferdinand Mayer. [James Bradley, Astronomer Royal, 1673–1762; Nicholas La Caille, 1713–1762] *But this correction is so small, smaller than the ability of most astronomers. This correction can also be applied to my solar tables, since they are based on Maskelyne's stars as well. Maskelyne's hint is hopefully different from the one he gave in the January issue of 1802.* [Zach adds tables of observations of Ceres and Pallas made in Milan and Bremen. These were published in the *MC*, July 1802.]

PS: I was about to close this letter when I received Dr. Burckhardt's latest elements of Pallas' orbit from Paris. After having produced in different attempts ellipses of 25 yr and 9 yr he found at last the ascending node. [These elements were published in the Sept. *MC*.]

Vossische Zeitung Berlin June 24, 1802

Article by Bode:

In the night I was observing the new main planet Ceres at the observatory, not far from 20 Virgo and the moving star called Pallas, discovered by Dr. Olbers in Bremen in March

1802	AR	Dekl.	1802	AR	Dekl.	1802	AR	Dekl.
April 4	− 1,0"	+ 1,5"	April 23*	− 0,1"	− 0,6"	Mai 3	+ 1,3"	− *21,8"*
„ 5	+ 0,7"	− 2,7"	„ 24	− 5,4"		„ 4*	+ 2,1"	− 0,3"
„ 7	0	− 2,8"	„ 25	+ 3,1"		„ 5	+ 2,3"	− *2,5"*
„ 8	− 3,9"	+ 2,8"	„ 25*	+ 0,3"	+ 3,4"	„ 6	+ 6,2"	
„ 10†	− 1,7"	+ 1,5"	„ 26	+ 2,4"	− *7,8"*	„ 7	+ 0,9"	− *14,9"*
„ 12†	+ 2,2"	+ 8,4"	„ 26*	+ 3,1"	− 4,5"	„ 7*	+ 39,3"	− 0,5"
„ 13†	+ 2,4"	+ 2,3"	„ 27	+ 0,8"	− *4,1"*	„ 8	+ 3,0"	− *9,3"*
„ 15	− 2,4"	+ 2,9"	„ 29	+ 4,3"	− *12,2"*	„ 9*	+ 2,2"	+ 5"
„ 15†	+ 1,3"	+ 1,9"	„ 30	+ 2,2"	− *13,9"*	„ 11	+ 2,9"	− *31,3"*
„ 16†	+ 1,7"	− 0,5"	Mai 1	+ 2,1"	+ *4,0"*	„ 11*	+ 3,4"	+ 0,3"
„ 17†	+ 3,4"	+ 3,0"	„ 1*	+ 2,8"	+ 3,3"	„ 13*	+ 4,6"	− 1,7"
„ 18	− 5,6"		„ 2	+ 7,8"	− *5,5"*	„ 14*	+ 4,0"	− 1,5"
„ 19	+ 1,1"	− 4,4"	„ 2*	+ 0,2"	+ 2,0"	„ 16*	− 1,7"	+ 0,4"

Fig. 7.16 Comprehensive list of differences between Gauss' elements and observational data on Pallas

28. Dr. Gauss' observations, based on previous observations, and the discoverer's observations show that the latter is a peculiar planet roaming between Mars and Jupiter, close to and in Ceres' orbit. Should these weird and unheard of facts be confirmed by a rediscovery of Pallas, as in the case of Ceres, after its return from the Sun, Olbers' discovery is one of the most important.

No. 30 Gauss to Olbers Brunswick June 25, 1802

*Sincere thanks for your two letters of June 13 and 24. I would have answered sooner if a small unpleasant task hadn't interfered: namely, to push on, as actively as I could, the comparison of Pallas' third Elements with the complete set of 39 meridian observations which were made available to me – a task in itself not intellectually appealing, but one which is necessary for obtaining more accurate future improvements; I wanted to share the results with you. Here now is this result which nevertheless will require some small improvements here and there, for I calculated several observations, e.g. the Greenwich ones on May 9, not yet using the elements but rather initially using only interpolation. The symbol ± indicates that the computation gives more/less than the observation. I've indicated the Paris and Greenwich observations with a + and * respectively. I'm writing the differences for the Seeberg Dec.'s which appear after the start of the Greenwich data, in red, because they, as the comparison shows, don't compare at all well with the latter, so that one might see the true course of the errors better. The Greenwich RA of May 7 will be 2.5s too small.*

From this it appears that the Greenwich observations in Dec. as well as in RA are exceptionally good. The meridian observations by themselves won't allow the derivation of more reliable elements; I would have to change them in such a way, if I didn't consider it better to wait for the complete set made off the meridian, that for example ceteris manetibus (remained with others) the computed Dec. at the start of the row and the RA at the end would be reduced by about 1" to 2" and 3".5 to 4" respectively. One can easily see that the agreement in RA would as a result be significantly better. Meanwhile the elements would be changed only trivially.

Based on a rough calculation, which may easily be off by 5" or maybe even more, I arrive at the following difference in your 3 observations due to the 3rd Elements:

June 19	– 9" in RA	+26" in Decl.
20	–17	+8
21	–25	+5

You can see that I can't actually make any improvements in the elements now. Because the Dec., which affects the elements more than the RA (that is, the effect of latitude on some elements is more than six times that of longitude; sapienti sat [enough wisdom]*), can easily be flawed by 20" in Flamsteed's stars; I'm cautious, therefore, not to deviate from the truth yet again because of the Dec. I even believe that the 3*rd *Elements are perfectly adequate for the 1803 discovery if only the planet is bright enough, and that they'll hardly be off by 1/2° at the time of opposition. To take perturbations into account before Pallas has traversed a much larger arc, appears to me almost superfluous. Ignoring these perturbations, an ellipse that exactly fits the observations already takes these into account and must include them a fairly long time. (F.N. by Gauss: Thus far the observations of Ceres do not yet indicate the slightest effect of the perturbations; I had to expect that the 7*th *Elements would be off by 40" in May, even if Ceres' orbit were a true ellipse, because I only had such inadequate observations for the determination of the 7*th *Elements. I'm certain that all of the hitherto existing observations of Ceres can be fit quite superbly, and without being forced to do so, to an ellipse all the while ignoring perturbations.) It appears to me therefore not just an unnecessary, but also a difficult, task to first subtract from quantities – whose small differences and* *still smaller differences of the differences which must be the source of our knowledge –, and very large quantities – (like the perturbation equations of Pallas most certainly would be) which, furthermore, can in no way be determined exactly based on the present state of perturbation theory –, in order to have to add them again later. Should I have the inclination and time this year to investigate the effect of Jupiter, then I'd adopt a different method and regard the elements themselves as variable. I even expect that this will also generally be preferred with Pallas in future; for if one, as is usual, will want to improve the positions computed with the mean elements using equations, then I think these will be so large and numerous that my patience would hardly be sufficient to calculate a large number of positions this way. I almost think it will always be easier to calculate a position using elements without tables, than to calculate with tables perhaps 30 or 40 equations for longitude, latitude, and radius vector. Hence it appears to me that it may well be best for the time being to indicate the changing osculating true elliptic elements at least about every three months within the tables. This itself might be useful in Ceres' case – I also believe that Pallas after some revolutions will be the best means of determining Jupiter's mass.*

You are doing Mr. Burckhardt an injustice if you interpret his work to be the easier one. If I ever have the pleasure to communicate to you my method of determining the preliminary orbit in extenso, then you'll probably agree with me that it's by far not as difficult to make the first orbital determination as to improve it later. If the reports, which Seyffer sends me concerning Burckhardt's method used in his first unsuccessful investigation of the orbit, are substantiated, – namely, that he had used assumed values of all eccentricities between 0 and 1 and all distances from Mars to Jupiter, not having found any suitable ones –, then he has done more than I from Oct. 1801 to the present, and my patience would hardly have been up to such a task.

Mr. Herschel also gave me information on his 'Asteroids.' What surprises me is (1) that he doesn't announce it as being a modest proposal, but rather says simply 'I call them', and (2) that his reason in Ceres' case consists in that it now 'is out of the Zodiac.' That shows a very biased and, it seems to me, unphilosophical outlook. It is likewise strange that he withholds his measured apparent diameter. Should it be correct, then, as it seems to me, a smaller mass is hardly important to be able to distinguish Pallas or Ceres from the remaining planets.

Zach to Oriani Seeberg June 26, 1802

Thousand thanks for Mr. Cesaris' observations of Ceres and Pallas, but it is yours I am waiting for with utmost impatience. It is only you and Dr. Maskelyne of Greenwich who could make them with an equatorial sector consequently, my friend Dr. Gauss and I wait for them like the famished wolf for the ewe. Did Piazzi not observe Pallas at all? We suspect this from his declinations. In the meantime I am sending you my observations. I presume and if I am mistaken, I ask you to calculate the perturbations for Pallas, here are the elements III improved by Gauss derived from 42 observations at Seeberg and Greenwich. [These elements are printed in the July *MC*.]

Please notice as quite singular that the mean motions of Ceres and Pallas are approaching each other always to that extent we improve the elements, I believe that we will find them in the end to be equal; and Ceres and Palls still represent a very remarkable celestial game of billiards in its kind. Burckhardt for his part found elements that come close to those of Gauss, but he employed the perturbations that Jupiter has had on Pallas, only using the first powers of the eccentricities, but that is not enough for the latitudes. He found: [These tables are to be found in the Sept. *MC*.]

Mr. Schubert of St. Petersburg also calculated the perturbations of Ceres according to the method of the c. Senateur, who becomes nobler and more unbearable in his arrogance each day in what he is assuring and writing. Schubert is mostly in good accordance with you. In my next letter I will send you the copy. The braggarts of Europe did not at all tackle this task, they wanted to beat everyone else but like in the gospel, Signor Abbate, the first will be the last and the last the first. So be the first for Pallas. Mr. Herschel just wrote me that he had observed Ceres and Pallas with his large telescopes, neither has any satellites at all, he found the diameters much smaller than Schroeter. For Ceres 162, for Pallas 70 English miles. Furthermore he does not want these corpuscles to be called planets, he invents a class of its own and calls it asteroids. But this is nonsense. For in this case Mercury is an asteroid as well in comparison to Jupiter. So this nomenclature means nothing, it is arbitrary and offensive.

Did I send you the observations of Ceres by Mr. Poczobut from Vilnius made at the mural by Ramsden? In any way, you will find them printed in my next issue.

No. 31 Olbers to Gauss Bremen June 30, 1802

It pleases me very much that you haven't allowed yourself to be tempted into carrying out new calculations using my observations of June 19, 20 and 21. Flamsteed's Decl. for 24 Comae is actually 18″ too small. I have now received the June issue of the M.C. and find two Dec.'s in excellent agreement, for No. 11 Comae therein. Since in the Hist. Cel. 24 Comae is observed twice together with 11 Comae, the Dec. of 24 Comae could therefore be derived much more accurately. In order that you yourself realise how reliable this Dec. now is, I'll set down the calculation. [These complex calculations can be found in Schilling (1900) vol. 2, pg. 58.]

The mean difference in Dec. is –5″. That merely indicates your Elements no. III still agree so well in Dec., that using <u>my</u> observations no variation whatever is detectable. To what extent you can rely on the difference in RA is best decided by a fourth observation. We have very bad weather here and almost always cloudy sky. On the 25th it was a bit clear, but a strong storm blew all the clouds away and I could make an accurate observation. But on the 26th the evening was very clear, the atmosphere totally calm, and I was very satisfied with my observations. Based on the mean of 4 very well matching differences in RA and 3 for Dec., Pallas at 11h 18m 34s trailed 24 Comae by 6m 14.9s and was 8′ 40″ further south. Hence for June 26 at 11h 18m 34s, apparent RA 187° 52′ 20″, apparent Dec. 19° 19′ 32″.

I'm very anxious to discover how this observation will agree with the previous ones. However, I very much approve of your wanting to wait until the observations have ceased before making any changes to the orbit. In particular it will still be necessary to have several observations with which Pallas can be compared with <u>other</u> stars; for a possibly incorrect position of 24 Comae would still subject all of these observations to a constant error dependent on that position. As soon as it becomes clearer in the first days of July, I'll be able to compare Pallas with No. 32 and 33 Comae.

You tell me I'd be doing Mr. Burckhardt an injustice if I believed his work to be the easier one. I would, after having received your method, satisfy myself that it is <u>by far</u> easier to make the first orbit determination than to improve it later. I can't of course give an opinion on whether this is easier using your method because I'm not acquainted with your method yet. But Burckhardt wasn't acquainted with it either and Seyffer's letter shows how wasted his efforts were to find this <u>first</u> orbit determination. I wouldn't have been off by so much. I had outlined a method for myself to find the first approximate elliptic elements using 5 observations, but you anticipated this before all of us in such a brilliant way! Of course this method, requiring so many observations, isn't elegant and I also would have had to perform many calculations. Compared with <u>these</u> computations, those which necessitate the correction of the orbit appear to me to be easier, at least when finding an ellipse which is entirely satisfied by 3 complete observations from the known approximate orbit. I believed myself capable of accomplishing this after having seen the account of the elements that you published in the Hamb. Corr., and particularly as I already knew from my preliminary calculations concerning the parabolic and circle-hypothesis, that in the case of Pallas small changes in the longitude of the ascending node correspond with very large changes in the inclination. – On these grounds I concluded Burckhardt's achievement to be the easier one. But as he included the perturbations of Jupiter, his calculation may indeed have turned out to be extremely extensive and difficult.

You have completely convinced me that for now it would still be better, in the case of the small arc thus far traced out by Pallas, simply to stay with the osculating ellipse. I agree with you that maybe in the future it would be easiest always to calculate the planet's posi-

tion directly from each newly obtained set of elements instead of from so many tables of equations. Your excellent idea, to consider the elements themselves as variable rather than the perturbation equations, pleased me very much. You will, of course, as you led me to believe in one of your letters, present us with one of your own works on Ceres and Pallas. Now much we'll learn from it! — I'm truly looking forward to receiving your method.

No. 32 Gauss to Olbers Brunswick July 6, 1802

It pleased me very much that you were able to observe Pallas so long and were hopeful of being able to continue the observations still well into this month. This renews my hope that the planet, in the coming years, will surely not elude the combined investigations of astronomers, partly because we thereby obtain beforehand the means to determine the position for the coming year with adequate accuracy, and partly because Pallas, although at that time further from us and the Sun, at least won't be in the haze of the horizon as much. Your last observation of June 26 appears to agree very well with your previous ones. After a rough calculation, which might very well be off a few seconds, I find

Calculated	*RA 187° 52' 2"*		*-20"*
	Dec 19° 19' 32"	*difference*	*0"*

Parallax is neglected here; its inclusion would increase the calculated RA a bit and decrease the Dec. The elements therefore appear to truly underestimate the RA; the Dec., however, still agrees well enough so that nothing can be decided. I still wonder over the accuracy of your observations which even appear to improve with the planet's decreasing brightness. Bode has sent me some observations also taken with a circular micrometer (the last one from June 19), which agree much less among themselves.

A few days ago I received Oriani's meridian observations of Pallas from Zach totaling 15, the last one from May 22. Although only whole seconds are given, they do agree among themselves very well, except that all of the RA are too small compared with those from Seeberg and Greenwich by a few seconds, perhaps the consequence of a difference in the comparison stars. Here is the result of the comparison with the Elements III;

	RA	*Decl.*		*RA*	*Decl.*
May 4	*+15″.7*	*-5″.9*	*May 12*	*+7″.8*	*–1″.4*
5	*+19.8*	*+5.1*	*17*	*+11.1*	*–4.0*
6	*+16.3*	*+6.4*	*18*	*+15.5*	*+0.6*
7	*+10.8*	*+2.1*	*19*	*+16.4*	*+0.2*
8	*+ 8.3*	*+0.4*	*20*	*+14.8*	*–2.0*
9	*+ 6.9*	*+0.1*	*21*	*+10.6*	*–1.5*
10	*+ 7.8*	*–2.0*	*22*	*+6.9*	*–1.4*
11	*+ 9.2*	*-3.6*			

When I compile the observations of various astronomers in this way, I almost fear, and fully realise, that I perhaps may not be able to make any improvement in the elements. I admit that of the observations made off the meridian I trust yours the most, inasmuch as you are able to use reliable stars. I'll hardly be able to reliably count on an accuracy of 15″ from those made with an equatorial telescope and which I'm expecting from Greenwich and Milan. The interpretation of Oriani's RA shows that, in this connexion, the agreement of the observations among themselves is still not a fully valid proof. All observations, good or bad, will have to be kept in their original form and the questionable stars will have to be determined with utmost care in the coming year. Fortunately this doesn't apply ad faciendum [for doing], *but rather ad melius faciendum* [for doing better]; *at least I think that the III Elements of the coming year would be good enough if one could only see Pallas, and should one not be able to see it, then even the best elements can't help. It's another matter if one were to be forced to extend the elements to 1804 or even 1805. Then the best accuracy would of course have greater importance.*

According to Oriani's latest meridian observations of Ceres (until May 21) the error using the VII Elements in RA and Dec. is +38″ and +29″ respectively, the latter not having changed at all in almost half a year. From this one can easily conclude that all of the hitherto existing observations can be very well represented by an ellipse. I think it will always be interesting and instructive if I were to determine in addition to a new orbit, in which I were thinking of including the perturbations as accurately as possible, another one in which they would be removed altogether. I'm curious if in 1803 the observations themselves will already show a <u>*definite trace*</u> *of an unknown influence? Only when Ceres and Pallas will have made several revolutions will they be the best means of determining the mass of Jupiter more accurately than has been possible using hitherto available data.*

When I've set aside the work involving Ceres and Pallas (I'm just waiting for the end of the meridian observations at Palermo before starting on the former), it'll be my first concern to prepare an outline of my method for you; I've already made a good start. I myself very much long for such an intellectually stimulating task, and look forward to it as a respite from the numerical calculations. Once the method which one wishes to employ has been set out, then these calculations become a mere repetitive task. If I subsequently decide to make the method publicly known in condensed form, then the applications and illustrations therein couldn't be better selected than from Pallas and Ceres. Indeed, the variety of specific circumstances enables me to give nothing but genuine illustrations for all of the various methods that I've used, one being more suitable here and another there.

[Enclosed is an ephemeris of Pallas from June 28 to August 28, 1802.]

No. 33 Olbers to Gauss Bremen July 14, 1802

I must finally report to you the complete results of my observations of Pallas taken this year. Of course the weather has allowed only a few observations, but I hope that two of them are good.

From June 26 to July 3 it was completely overcast. In the area where Pallas should have been, I found three small telescopic stars in a line curved slightly to the north. Of these stars, the leading one was by far the smallest and most insignificant, and this one, as revealed by the observation, which, however, was very uncertain due to cloud cover, was Pallas.

On July 4 it was very clear. Pallas, small and dim, was compared with No. 33 Comae. The observations agreed well among themselves. At 11h 28m 14s it led the star by 3m 2.5s and was 22′ 19″ further north. I obtained: RA 189° 52′ 31″, Dec 18° 33′ 42″.

On July 8 the atmosphere was extremely clear and calm. Today I diligently devoted myself to an accurate determination of Pallas' position. Fortunately it was practically on the same parallel of latitude of No. 32 Comae, allowing the precise determination of the difference between your Dec. and that of this star to be obtained using a circular micrometer. It was compared 7 times with No. 33 and 6 times with No. 32 in RA and 3 times with No. 33 and 5 times with No. 32 in Dec. The observations agreed very well among themselves and there resulted:

11h 5m 16s difference in RA from No. 33+1m 11.6s;		*Pallas southward 2′ 25.4″*
11 13 20	*32 +1 22.1*	*0 18.5*

Based on my reduction I concluded:

July 8 11h 5m 16s	*RA 190° 56′ 9″.0*	*Decl. 18° 9′ 0″.7*
11 13 20	*190 56 16.3*	*18 8 58.6*

(F.N. by Olbers: Should both Dec.'s on closer inspection not agree well with each other, then the second one should be considered the more trustworthy; on the contrary, the first RA should be considered more trustworthy.)

I've retained the decimals because they resulted from the calculation and not because I considered my observations accurate to a decimal place in the seconds figure. If I'm not too mistaken, this is the most accurate of all observations of Pallas that I have made. You must still wait for the reduction that Zach will undertake at my request because I'm still not quite certain about the Dec. of the two comparison stars.

The weather cleared up very late on July 9. The very dim planet was compared twice with No. 32. At 11h 53m 47s Pallas trailed No. 32 by 2m 29.5s and was 6′ 42″ southward. This gives: RA 191° 13′ 10″, Dec. 18° 2′ 35″. The observation isn't at all as reliable as on the previous day.

Since Pallas now appeared so dim, was increasingly lower in the sky at dusk, and in addition was situated near the full moon, I considered it advisable to stop the observations. The resulting observations must hence be uncertain, since the appearances and disappearances could no longer be discerned with certainty. In France and especially Italy one may still be able to pursue them until the end of July.

You have nevertheless revived my hope to actually locate the planet at its next opposition. It will, I believe, surely approach Earth more closely at that time just as it did on July 9 of this year, and its weaker illumination by the more distant Sun will be more than compensated for by its greater altitude above the hazy horizon.

I've obtained some more recent information about Herschel's measurements from the Nicholson Journal *No. 7, p. 221.* [A journal of natural philosophy, chemistry and the arts, edited by William Nicholson.] *Herschel used 'the lucid disc micrometer', that is, he compared an illuminated ring, which he saw with one eye, with the planet's greatly enlarged disc seen with his other eye in the telescope. Hence his procedure was the same as that used by Schroeter, who found the apparent diameter to be 20 to 30 times greater!! He concluded from his observations that the apparent diameter of Ceres at a distance of 1.634 was 0.22″ and of Pallas at a distance of 1.187 was 0.17″ or 0.13″. It appears evident from his letter to Méchain, printed in the Moniteur, that these measurements pertain to April 22. There he gives the apparent diameter of Ceres still more accurately as 0.216″. In any case his adopted distance of Pallas is much too small. He concludes: diameter at a distance of 1 for Ceres = 0.35″, for Pallas = 0.21″ or 0.16″. The true diameter is 163 and 95 or 71 English miles.*

Ceres' colour is redder than Pallas'. They have, more or less, a vapour or 'coma', though occasionally, if the atmosphere is very clear, this haze scarcely surpassed the diffused light which surrounds every small star. There is no likelihood here that either of the two heavenly bodies has satellites.–"Herschel added some comments at the end of his lecture which indicate that the apparent 'comae', which envelop Ceres and Pallas are nothing more than the effect produced by the aberration of light around the image of every small star."

The contrast between Schroeter's and Herschel's measurements is most surprising. Just between us, I trust neither of them. I believe Schroeter has included too much spurious light in his measurements, and he would have perhaps found a fixed star to be just as large. – And Herschel? – I mean, the eye could easily be misled in comparing such small dimensions. Even if he enlarged Pallas 500 times it would have appeared to him (according to his stated diameter) only as a 1′ 5″–diameter disc appears to the naked eye. With such a diameter a disc actually still appears as a point, and whether one of two such small disks appears larger than the other depends only on the brightness of these small disks. The light from Pallas must certainly have become very feeble in the telescope after a 500-times magnification, and hence a probably brighter, though much smaller, disc could still appear as large as Pallas to the naked eye. – Nevertheless, I am convinced that Herschel is much nearer the truth than Schroeter.

There may be something at work here that Hume touched upon in his philosophical work *A Treatise of Human Nature*:

But as there is a certain degree of an emotion, which commonly attends every magnitude of an object; when the emotion increases, we naturally imagine that the object has likewise increas'd. (Hume, 1739: 2.2.8.6)

Here he uses the word magnitude in the sense of size, not brightness. Schroeter was certainly prone to flights of fancy, so perhaps his emotions got the better of him when studying the new and exciting objects Ceres and Pallas. For an examination of visual acuity in astronomy in the seventeenth and eighteenth centuries, see George (2006).

Zach to Oriani Seeberg July 16, 1802

Here are the promised perturbations of Ceres by Schubert. And your observations compared with the elements III of Pallas by Mr. Gauss, which you will find in the attached proofs and their comparison with my observations and those of Mr. Maskelyne. It is with impatience that I am awaiting your observations of Pallas to improve these elements for the fourth time. I am sending you the observations of Mr. Olbers who continued until July 9. [Compare this table with the one published in the August 1802 issue of the *MC*.]:

[The sixth lines reads: 4 July 11 28 14 Pallas precedes 3′ 2″.5 in time 33 Com. Ber. and 22′ 19″ north.]

La Lande wrote me: 'La Place always wanted Olbers' planet treated as parabolic since the great inclination does not allow in physical hypotheses that it is among the planets. But

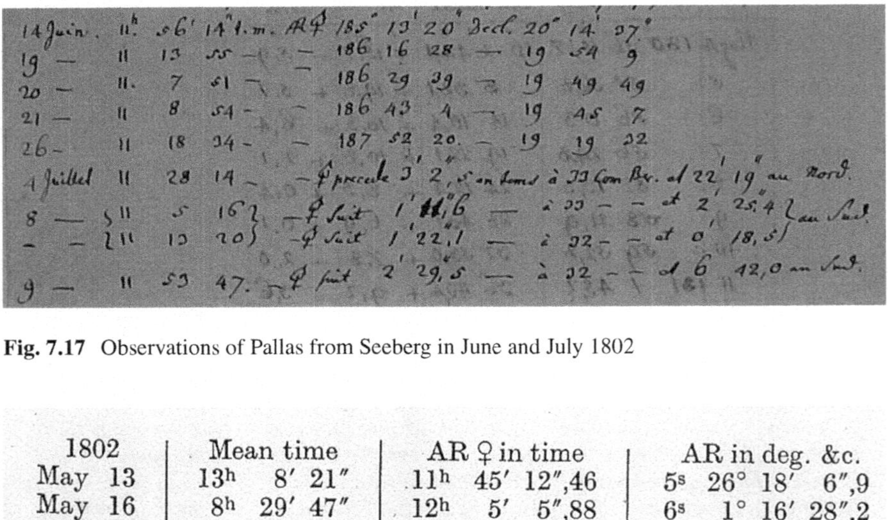

14 Juin	11h	56′	14″ t.m.	AR♀ 185° 13′ 20″	Decl. 20° 14′ 27″		
19 —	11	13	55 —	— 186 16 28	— 19 54 9		
20 —	11.	7	51 —	— 186 29 39	— 19 49 49		
21 —	11	8	54 —	— 186 43 4	— 19 45 7		
26 —	11	18	34 —	— 187 52 20.	— 19 19 32		
4 Juillet	11	28	14 —	— ♀ precede 3′ 2″,5 en temps	à 33 Com. Ber.	et 22′ 19″ au Nord.	
8 —	11	5	16	— ♀ suit 1′ 11″,6	— à 33 —	— et 2′ 25″,4	au Sud
— —	11	13	20	— ♀ suit 1′ 22″,1	— à 32 —	— et 0′ 18″,5	
9 —	11	53	47.	— ♀ suit 2′ 29″,5	— à 32 —	— et 6′ 42,0	au Sud.

Fig. 7.17 Observations of Pallas from Seeberg in June and July 1802

1802	Mean time	AR ♀ in time	AR in deg. &c.
May 13	13h 8′ 21″	11h 45′ 12″,46	5s 26° 18′ 6″,9
May 16	8h 29′ 47″	12h 5′ 5″,88	6s 1° 16′ 28″,2
June 28	11h 35′ 18″	12h 33′ 28″,34	6s 8° 22′ 12″,6

1802	Decl. N	Geoc. Long.	Geoc. Lat. N
May 13	16° 1′ 48″,6	5s 20° 7′ 55″,5	13° 12′ 55″,6
May 16	20° 51′ 18″,9	5s 22° 31′	19° 33′ 45″,6
June 28	19° 8′ 11″	5s 29° 43′ 48″,6	20° 49′ 18″,9

Fig. 7.18 Observations by Maskelyne

there can be no doubt.' What is your opinion, dear friend, of this reasoning? And what do you think of Dr. Olbers' idea that Ceres and Pallas are fragments of one planet? It is very peculiar that the revolutions of these two fragments are the same! Bonaparte has doubled the Prix de la Lune for Bürg and wants him to come to France and I have been charged with the negotiations. But Buerg will not do this foolish thing, at least I will not encourage him and on the contrary – I want my dear Burckhardt back.

What is your opinion of Herschel's asteroids? Do you believe in the measurements of the diameters of Ceres = 0. "216 and of Pallas 0. "13. I do not believe it, with these pretensions one makes a fool of oneself.

There is fear that we might not find Pallas again next year. But I do not despair with my 8-foot telescope, aperture of 4 inches. Adieu, my dear and estimable friend, please delight me with some lines from your hand, compliments to your colleagues and thanks for their observations. Piazzi wrote me that he could not find and observe Pallas. What a pity!

Maskelyne to Gauss Greenwich July 20, 1802

On May 31st I received your favor of May 19, with your elements of the orbit of Pallas, and an ephemeris of its right ascension and declination from May 24 to June 29th for which I return you my best thanks. I observed Ceres on this meridian till May 13th and Pallas till May 16th. I have since observed them with an equatorial Sector of 5 feet length of telescope, having an aperture of 4 1/10 inches.

I should have been glad to have received some further ephemeris of the two planets from you, which might have been the means of my making more observations, as I have not always had time to make these computations from the elements. If you have any such by you for any time to come be pleased to send them to me directly by the post.

On the 20th of this month (2 days ago) at 10h 37′ 38″ mean time I observed a star of 11 magnitude in the RA 193.46.5 and declination 17.2.55 N it followed 38 Comae Berenices by 3′.41″.62 of time, and being 1° 9′ South of it. I take the mean place of the star from Wollaston's Catalogue in RA 12h 51′ 22″.73 time and 18° 11′ 55″ north declination. I take this to be the planet Pallas, but I am not sure; but shall know on searching the heavens another night. There were other small stars near it.

When I can no longer see the planet Pallas, I will send you the whole of my observations. I found the right ascensions of Pallas in your ephemeris too little towards the end of June as you supposed, but the declinations rather greater.

Maskelyne to Gauss Greenwich July 30, 1802

On the 26th I received your letter dated July but without the day of the month. It contained your elements of Pallas's orbit No. III. The weather has been very bad here, else I think I might have seen the planet longer. However I observed Ceres on the 3rd instant for the last time, and Pallas on the 18th of June. Having now finished my observations of these two planets for this season, I have the pleasure to send you them. I must apprize you that the observations made out of the meridian, viz. with the equatorial sector, are much inferior to the others; but may be of use to improve the elements to help finding the planets when they shall emerge from the rays of the Sun. If you shall take the trouble to calculate ephemerides of them, be pleased to send them to me, and I will endeavor to catch an early sight of them. You will judge from these observations what my telescope will do in seeing the planets in a faint state. It is a very powerful one, the aperture being 4.1 inches.

[Maskelyne encloses a second copy of the transit observations of Ceres and Pallas sent in his letter of May 21, 1802, together with two further observations of Ceres (on June 20 and July 3) and six of Pallas made out of the meridian with the equatorial sector.]

The right ascensions & declinations of the stars compared with the planets with the equatorial Sector are as reduced immediately from Wollaston's Astronomical Catalogue without any allowance for aberration & deviation etc. Therefore, as the planet was pretty near the stars in these observations, the places of the planet set down are the places without the Bradleian aberration. [The phenomenon of stellar aberration discovered by James Bradley, 1693–1762.]

On February 3 Ceres appeared to me 8M; March 4 9M or less; April 22 9M. May 17 9M. June 20 10M. July 3 10M.

Your calculation of the place on May 7th points out an error in my observations that day of 2″1/2 of time in the RA=37″1/2 of the Equator owing to a mistake of 10″ at one of the wires, which divided by 4, as the middle wire was missed, to observe the Zenith distance, produces the 2″1/2. The corrected RA = 12h 3′ 45″.60 = 6s 0° 56′ 24″.0 from which your computation differs +1″.8, which agrees very well with the other observations. I addressed a line to you on the 20th by the post, but since find reason to think that it was a fixt star I had observed on the 18th and that the planet could not be seen. However if you have computed its place on the 18th I should be glad if you would send it to me.

Pallas on April 21 appeared of 9M. April 22 9M May 17th 9M June 14th 9M June 18 11M. June 20 11th to 12M. June 28 10M. I began a new set of magnitudes with my Equatorial sector, calling the 1st or brightest but less than a star of 6th Magnitude to be of 7th M and calling the smallest that is visible with the telescope of the 12 Magnitude the following one and brightest of the two stars (it being a double star?).

On May 4 I found Ceres with	*aperture,*	*4 inch*
Pallas with	*″*	*, 5*
Uranus with	*″*	*,15*
95 or o Leonis 6M with	*″*	*, 2*

[In each case] *just visible or barely visible with great difficulty. From this you may judge of the relative brightness of Ceres and Pallas & of their visibility in future, considering also their altitude as on Bouguer's Table in his* traite optique. [Published 1760.]

Gauss to Maskelyne Brunswick July 30, 1802

Your favour, dated Jul 20 I received on the 26 of this month. I regret very much, that I had been induced, by the early conclusion of the German observations of the New Planets, to think a continuation of their ephemeris unnecessary. As I now see, that your superior instruments permit a considerably longer series of observations, I instantly have done, what was still in my power to do: I have constructed by the IIId elements of the planet Pallas, sent you in a letter of July 9th, a new ephemeris, which I hasten to send you. I am afraid, that, after the arrival of this letter, the planet will be exceedingly faint, this ephemeris perhaps will serve at least, so decide, whether the planet at any time has been confounded with a fixt star or no. You have sent me an observation, which you say was made "the 20th of this month (two days ago)". It is evident, that an error is committed either in the date of your letter or in the day of the observation. I cannot but make the latter supposition, as your letter could hardly come from Greenwich to Brunswik in three days (it being arrived here already July 23 in the evening). According to the IIId elements the computed place of Pallas was July 18. 10h 37′ 38″ m.t. Greenwich RA 193° 44′ 59″ D. 17° 2′ 1″ which tolerably agrees with your observation, which, perhaps, still may suffer some change after a more accurate reduction, as the position of 38 Comae Ber., you made use of, seems to be only that settled by Flamsteed. But perhaps this reduction cannot be made, before the star shall culminate again by night. In a letter I wrote yesterday to Mr. Zach I have enquired if he is in possession of any later determination of 38 Comae Ber.

By passing to me Taylor's Logarithms, *you will make me greatly your debtor. They will rightly abridge all my calculations and spare me much time and labor, which I may employ otherwise. As soon as I shall have corrected once more the elements of Ceres and Pallas by the last observations I am to compute an ephemeris for the year to come for both planets. Before hand I found by the IIId elements of Pallas*

1803 June 28 midnight RA= 275° 45'
at Seeberg, about one Decl.= 23° 12' North
day before the Longit. = 9^s 7° 42'
opposition Latit. = 46° 31' North
Distance from the earth = 2.56

[Gauss appends here an ephemeris for Pallas from June 29 to August 31, 1802.]

The point of chronology raised in this letter was examined by Forbes (1971:228). Despite his incredulity, the letter did indeed take only five days to travel from Greenwich to Brunswick; Gauss actually received it on the morning of the sixth day. Forbes confused the issue further when he stated in this note 'Another reason for doubting Gauss' own words...' when he actually meant 'Another reason for doubting Maskelyne's own words...' It was Maskelyne, not Gauss, who made the date error in the July 20, 1802, letter from Maskelyne to Gauss.

Bode to Herschel Berlin August 5, 1802

Let me thank you for your kind letter of May 22 and your results of your studies regarding Ceres and Pallas. I read them not only with pleasure but also with a kind of astonishment because it remains inexplicable to me how you can obtain such small true diameters of these two celestial bodies since you agree with Dr. Gauss regarding their true distances, as I believe. I cannot imagine how extremely small you must find according to this the apparent diameters and how you are able to determine such small diameters. How can planetary bodies reflect their light if they themselves have only apparent diameters of parts of seconds. In March Ceres appeared as a 7th magnitude star and was visible to some of my friends with the naked eye. I am eager to read and study your treatise read to the Royal Society. I believe or am still convinced that Ceres is the eighth main planet of our solar system and that Pallas is a neighbouring extraordinary planet (or rather comet) revolving around the sun. Thus there would be two planets between Mars and Jupiter where I have been expecting since 1772 only one and the known beautiful progressive order of the distances of the planets from the sun is only completely proven by this discovery since there was a gap at the distance 4 + 24 = 28. I have just published a treatise on Ceres. I have read your interesting researches on Ceres and Pallas, not only with pleasure but with a certain astonishment; as it remains to me inexplicable how you can suggest such small real diameters for both these bodies, when you agree, as I think you do, with Dr. Gauss about their real distances. How excessively small you must find the apparent diameters, of that I have no conception, nor how you were able to measure such minute diameters. How could planetary bodies send back reflected light if they (are so small). Ceres was seen in March by several of my friends with the naked eye.

No. 34 Gauss to Olbers Brunswick August 6, 1802

I have indeed resolved to take on, after the completion of my computations for the two new planets, the pleasant task of preparing for you a rather detailed outline of my methods used in the orbit determinations. Yet I fear that it could still take a fairly long time before I completely finish with those calculations.

I haven't received anything about the two planets for a long time, other than a letter from Maskelyne who still observed Pallas on the 18th, although he was uncertain whether or not it was Pallas. It was "of the 11th magnitude." He's promised me his complete set of observations as soon as he can no longer see the planets. I've also sent him, at his request, an ongoing ephemeris of Pallas, a copy of which is enclosed. [This ephemeris is no longer extant.] *What a nice prospect for the rediscovery in the coming year when it will come noticeably nearer than on July 18. Maskelyne's eyes and instruments are unrivalled. His equatorial telescope is 5 feet and has a 4 1/10-inch aperture.*

I still have a small task in mind for the future concerning your idea of the cosmogony of Ceres and Pallas. As soon as I've made the final determination of both orbits, thereby enabling me to indicate more accurately the point at which the paths will next appear, I'll compute a small ephemeris where this point will appear at various times of the year. From this one can prepare a small chart. One will need something of this nature as a guide in the pursuit of other planetary debris. – Do write me some time when convenient whether your comet finder, with which you discovered Ceres and Pallas, is the same as the one you used for your fine time determination. I'm pleasantly occupying myself on occasion with more practical astronomy; I'm getting excellent time determinations, although I'm only using a very small telescope with attached sextant kindly loaned to me by Zach. Is that comet finder a local or English instrument, and about how much is it? It would be a delight to procure one for myself. I hear that Tiedemann in Stuttgart is also constructing a similar one.

Zach to Gauss Seeberg August 10, 1802

I owe you three replies and thank yous for three of your precious letters, dear Doctor. But you know my situation. And only today can I get rid of my guilt, after having used all your notes and calculations as you can see from the enclosed galley proof. And at the same time you will learn to your and my regret that it had not been possible for Piazzi to observe Pallas. We have to content ourselves with Oriani's and Olbers' observations. That the latter match up to those made by Oriani at the equatorial sector can be seen from my reductions on page 187. [He is referring here to the upcoming August issue of the *MC*.] *You will even notice that Olbers' deserve to be preferred. Compare only his observations of July 8th of a twofold comparison to those by Oriani of the same day of a threefold comparison. Olbers' distances are accurate and Oriani's only differ by ½ minute in RA! The first observations on p. 188 are very accurately reduced and will – hopefully – correspond to your elements III. Oriani promises later observations. I determined accurately only 38 Coma Berenices of Dr. Maskelyne's observations. You will find this star on page 187. And I reduced Dr. Maskelyne's observations of July 18th and found RA = 193° 46′ 6″.64, thus only different by 1 1/2 sec. But I found the declin. 17° 2′ 35″.8 and thus different by 192 from Dr. M's.*

But I believe the doctor stated in his letter the dist. decl. 1° 9′ only à peu près [about] *in minutes and that is the reason why I did not publish it in the August issue. I just mentioned it as the latest date for an observation of the planet. I will soon send you the galley proof of the following page with further news about Pallas, right now it is in press. I will not publish Schubert's formulae. I would like to quote some parts of your letters, give some of your observations with the sextant and a comparison of the watches etc in the next number in extracts.*

No. 35 Olbers to Gauss Bremen August 18, 1802

I'm happy that you occasionally occupy yourself with astronomical observations, of course only for relaxation, since you can certainly leave the observing to us ordinary people and apply your time much more to the sciences. My two comet finders are German products. The achromatic one, from the deceased Hofmann in Leipzig, costs 13 Thlr., has a 6° field of view and gives very sharp images, so that with a magnification of 8⅔ times I can still see the Jovian satellites if they're not too close to the planet. This is one of the best ones made by Hofmann (I've used many of his), and I use it for my time determinations. However, the aperture is not large and hence doesn't admit sufficient light to be considered a very good comet finder. – The other one is not achromatic, has a 2-inch aperture constructed by Weickardt in Leipzig based on the same design as a Ramsden, admits an extraordinary amount of light and costs 15 Thlr. I searched for and found Ceres with this one. – By July 8 and 9 Pallas had already been invisible to both comet finders for a long time and was still visible only in my 3-inch aperture Dollond.

The fact that Maskelyne still saw Pallas on July 18 pleases me but wasn't unexpected. I had to discontinue the observations mainly because of dusk. Greenwich already lies 1 3/5° further south, and this must be taken into account at dusk in July. I wouldn't be surprised if Pallas were visible in Milan until the beginning of August.

You will have received the reduction of my observations, which was kindly taken over by Zach, in the August issue. I'm anxious to know how the observation of July 8, which I par-

ticularly bring to your attention, agrees with your elements. – Zach has incorrectly derived the RA of June 26 from my data.

It seems to me, based on what you've written, that the symbol for Pallas □, or even ♀ which was suggested by Zach, is just as irksome to you as it is to me. I at least must always take care not to confuse it with that for Venus. – Why don't we replace the leaf-shaped top of Pallas by a three-pointed one? i.e. 🜍. It is then easy to draw and not easily mistaken. – The fact that in chemistry 🜍 is the symbol for sulphur certainly won't cause offense. It should be easy and clear to record and identify.

It appears to me more and more likely that we'll see Pallas in the coming year even though it seems to me that you, dear friend, do not sufficiently take into account that Pallas will be illuminated only half as much next June as in July of this year.

I wasn't totally idle in Rehburg vis a vis Pallas. Rather, I've investigated whether it might have been observed together with the 50,000 stars in the Hist. Cel. Fr. I am truly hopeful of finding it among them. The section from 5:30 to 9 o'clock RA and from 60° to 62° zenith distance was observed on March 1, 1797. I now find according to your elements on March 1, 1797 RA of Pallas 107° 24′, that is 7h 10m in time and the southern Dec. 13° 26′, hence zenith distance from Paris 62° 17′, thus only a little bit out of the section. Burckhardt's elements, disregarding perturbations, give the RA 118° 59′, southern Dec. 12° 14′, or RA in time 7h 56m, zenith distance 61° 5′. At least it is not unlikely that it was located within the section and was then certainly observed along with the stars, since it appeared at least as a 7th magnitude star at that time. Should you still calculate improved elements for Pallas, then I ask you to investigate immediately its position for March 1, 1797. – I'll then see whether an observed star is missing as soon as this area of the sky again appears at night. I'm requesting beforehand that you keep my hunch, which I haven't confided to anyone except our friend Zach, to yourself. It would be an important discovery if we had such an old observation of Pallas. Your idea, regarding the calculation of an ephemeris for the descending node of Pallas' orbit on Ceres' orbit, has also occurred to me.

I'm now starting a preliminary calculation of a rough ephemeris for Pallas' movement in the coming year in order to better familiarize myself, perhaps still in this year, with the small stars among which I'll have to search for it in the coming year. As soon as you undertake another improvement of your elements, please forward them to me immediately.

Oriani to Zach Milan August 28, 1802

You have received for one month my observations of Ceres and Pallas until July 8, here follow the latest. I saw Pallas again on the 17th and 18th of this month, but it was so tiny that I could not successfully observe it. The latest observations prove at least that we will see this planet next year at its opposition. It is very easy to find the perturbations of Pallas based on those of Ceres supposing that these two planets have the same mean distance to the Sun. Only the name Pallas has to be replaced by Ceres.

Piazzi has sent you his last memoir on Ceres. Since May 23 he has not observed this planet due to the lack of equatorial instruments. He told me that the last name Ferdinandea, regarded as useless by other astronomers, has won him an increase of 100 ounces/ about 50 louis/ to his salary. He sent me recently the copy of a letter from Herschel, in which the latter intends to call the two new planets asteroids. Here is a part of this letter:

"Moreover if we were to call it (Ceres) a planet it would not fill the intermediate space between Mars and Jupiter with the proper dignity required for that station." Presto vedremo dei Duchi, Conti e Marchesi anche in Cielo. The Italian words are from Piazzi. You have already successfully proven the planetism of Ceres and Pallas; consequently, it is useless to ponder Herschel's new dynasty.

Zach to Oriani Seeberg August 28, 1802

I am endlessly grateful, my dear friend, for your observations of Ceres and Pallas. I reduced them in my way and I wished you had sent me all the originals of your observa-

tions. Do it in the future. I am sending you here my printed reductions. Look at your journal if there is not a mistake in the m. t. of Ceres on June 28. You have sent me 10h 3′ 10″. But I believe it must be 9h 45′48″. Same thing with Pallas on July 3: instead of 10h 9′ 54" it must be 10h 7′ 54″. In the register which you sent me there is app. Decl. C. B. 18° 8′ 39″ same thing for Pallas. This certainly is a mistake of the copyist. I made a Emendationem ex Ingenio and I read for 135 C. B. 18° 10′ 39″. We are waiting impatiently for the continuation of your observations if there is one. But always send me your original observations. [A section follows here on Ceres perturbation calculations.]. *Adieu, dear friend. I will send you soon the ephemerides of Ceres and Pallas.*

Letter LVI, Oriani to Piazzi Milan September 1, 1802

I observed Pallas till Aug. 8; I saw Pallas again Aug. 17 and 18, but it was so small and dim that I could not make any observations. Now I am very busy in reorganizing the formula to calculate the perturbations of Pallas; the very large inclination and the not so small eccentricity require new and complicated formula not even available in Laplace's Mechanics.

Ende to Olbers Celle September 7, 1802

I am very grateful, dearest friend! for the announcement of the discovery of a comet. I saw it at almost the same time as you did, namely on September 2 10h 15'. But what havoc did you wreak with this discovery! Are you not aware that according to Sicilian on the 10th and according to the local mob reckoning (and I mean the mob in frock coats as well as the one in smocks) on the 29th is to hit us, set fire, flood us, or better, destroy us! Some have already given your Pallas this role of a strangling angel. And now you are frightening the humans for a second time! Yesterday I informed Zach of your discovery and admonished him to do penance for the comet which is to destroy Erfurt and Gotha on the 10th is here!

No. 36 Gauss to Olbers Brunswick September 10, 1802

Since I increasingly realise that neither the Italian nor English off-the-meridian observations compare with yours, I have derived a new orbit from your observation of July 8 along with Zach's first and Maskelyne's last. I admit nevertheless that I'm not really satisfied with the result. After checking, I found that these new elements (given below), which accurately describe the three observations taken as the basis, nevertheless differed by 12″ in longitude and latitude near the middle of the meridian observations (April 25). I'm still undecided if I should let it go at that or if I (2) should determine another orbit wherein the differences would be apportioned among the observations as much as possible, or if I (3) should take the trouble to take into account, in the calculation, the influence of Jupiter on the elements treated as variable. Should you have missed one of the stars of the Hist. Cel. *Observed on March 1, 1797, and do not wish to undertake the work resulting from that omission, then I would certainly choose alternative (3). In this case I would be very obliged to you if, as soon as you're sure of the omission of an observed star, you would immediately inform me of it and send me the observed position.*

The elements of Pallas referred to above are as follows:
epoch March 31, 1802, Seeberg, noon March 31

Longitude	*162° 55′ 6″.8*
Anomaly	*221 16 25.3*
Daily mean tropical move.	*769″.7263*
Log semimajor axis	*0.4424992*
Asc. Node	*172° 26′ 30.7″*
Aphelion	*301 38 41.5*
Incl.	*34 36 58.8*
Eccen.	*0.243888 = sin 14° 6′ 58″.09*

I have, in the meantime, computed an ephemeris based on these elements, which follows herewith. (F.N.: This ephemeris follows in the next letter.) The III Elements give:

	smaller RA	greater Decl.
1803 Febr. 4	13'	3'
June 28	32	2

Taking the brightness on July 8, 1802 to be unity, then based on this standard it was:

1802	*April 4 = 4.23*
	May 4 = 2.75
	July 16 = 0.90
	July 26 = 0.80
	Aug. 7 = 0.70
	14 = 0.653
	24 = 0.60

I'm now very curious if and how early, in your opinion, the higher altitude will compensate for this weaker light. If only the owners of the most powerful telescope had your skill, then I believe there would certainly be no doubt concerning the rediscovery. That skill is the most relevant here, makes me conclude that even Piazzi in 1801 could not locate Ceres off the meridian even though he knew its position accurately.

Huth to Herschel Frankfurt on the Oder September 10, 1802

Although I enjoyed reading your nice observations and ways of measuring the diameters of planets and could convince myself of the correctness of both, I still doubt the conclusion you are drawing "that Ceres and Pallas cannot be planets" and I believe that the smallness of these celestial bodies does not exclude them from being planets, if I may say so.

I believe planets have always been defined as celestial bodies revolving around the Sun in the same direction on slightly elliptical orbits. This kind of motion can be found in the case of Mercury, Venus, Earth, Mars, Jupiter and the George's planet, and furthermore that their orbits intersect at small angles. In the case of Ceres and Pallas we have the described motion but not the position. Their orbits intersect the orbits of those celestial bodies moving with them around the sun at greater angles. Consequently, Ceres and Pallas give us the following new experience: There are celestial bodies moving around the Sun on unchangeable and slightly elliptical orbits with a common direction but with a greater inclination of their orbits than those of the others or Ceres and Pallas teach us that there are planets beyond our zodiac, in which the known planets are visible.

But why should stars with planetary motion (and it is the motion that is characteristic for planets and not the size, not the positioning of their orbits, not the existence of satellites, rings or atmospheres, not the differences of distances from the Sun, all these things are minor facts and allow a great variety) outside the zodiac, I am asking you, not be called planets? I think it unwise to introduce new names especially a general one, if we can avoid it. New categories entice us to see differences where there are none. The fact that Ceres and Pallas are smaller than the other planets cannot hinder us to expel them from the set of planets, especially in regards of their consistent nature and planetary motion. That their orbits are this close cannot deceive us because their size is similar and most likely their density as well, that is why their momenta come almost to a balance with the gravitation against the Sun. They cannot become each other's satellites because they are partly similar in mutual gravitation and partly because they are too small and too far apart to provoke a mutual approximation, at least for the time being. Maybe the characteristic of our planetary system is the following: that there are many such small celestial bodies of similar size and density. Maybe when the planets formed, there also developed from a certain quantity of planetary matter a certain amount of planetary grains of similar size, but where Earth, Venus, Jupiter, Saturn and George's planet developed, the matter formed larger spheres that attracted neighbouring smaller ones thus building satellites. It does not seem unlikely to me that in a very short time between Mars and Jupiter, but closer to Mars, there will be found more bodies like Ceres and Pallas. And more of those might be seen in our zodiac. But all of them will be planets if they show planetary motion just as Ceres and Pallas.

No. 37 Olbers to Gauss Bremen September 11, 1802

I'm very anxious to learn how long Maskelyne, the Parisians, and Oriani have seen Pallas. I've already calculated for myself a sort of ephemeris, at about 20-day intervals, for Pallas in the coming year. I'll hence be able to prepare a chart in advance based on the ephemeris. I'll start looking for it in February but don't expect to find it before April. In the area through which it will pass in the coming year are an unbelievable number of small telescopic stars which will complicate its discovery very much. Whether Pallas has been observed in Paris on March 1, 1797 I'll be able to know only in the coming month.

Bode to Olbers Berlin September 14, 1802

I would like to express my gratitude for the information on the discovery of a new comet and its observations. Bad weather, full moon and hazy air hindered me to find it despite my efforts made. Yesterday evening the moon arose later and the air remained clear until 8 o'clock, then fog and mist, and by 10 o'clock it was completely overcast. If the brightness of the comet will decrease as you described even further I will hardly observe it exactly but can only determine its position parenthetically with the circular micrometer. But it is also already being observed in Paris since I received a letter yesterday from Mechain, dated Sept 1, with the news that he discovered the comet already on August 28 below the two nebulae at the western hip of Oph., just when he was done with observing Pallas. You certainly know already about Mechain's discovery of the comet for I read about it together with yours in the papers, but maybe not that Mechain observed Pallas until the end of August. So I take the liberty, dearest friend, to relay this observation in a hurry. Too bad, Mechain did not send me the ascension and declination of Pallas of the 28th because he had not calculated it because of the comet. But I received earlier ones from July 12 through August 7. [This is the same table as published in the Nov. 1802 issue of the *MC*.]

I am asking you not to disclose these observations to anyone, if Mr. Mechain does not send them completed, they will be published in the Yearbook 1805, which will leave the press at the beginning of the next month.

	MT		RA		Dec		
1802 Aug 28	9 h	24	249°	18	6°	11 ¾	S
30	9	47	249	55 ½	1	41 ½	S
Sept 1	9	18	250	35	2	29	N

Mr. Mechain has expressly asked not to publicise these observations or to communicate them to others because he has not yet determined the stars compared properly.

PS: Dr. Gauss has sent me his latest elements of Pallas' orbit and an eph. for next year. If you write him very soon, please give him for his calculations in my name one or the other observations of Pallas by Mechain.

No. 38 Gauss to Olbers Brunswick September 14, 1802

I had to finish my last letter in a hurry due to a disturbance so I forgot to enclose the mentioned ephemeris of Pallas for next year. I am hurrying to send it. It may be used together with yours for a mutual verification and for assessing the degree of uncertainty that we have to expect. I take advantage of this opportunity to add currente calamo *several things. Several days ago I determined another purely elliptical orbit of Ceres.*

Epoch Seeberg		*Aphelion*	*Node*
1801	*77° 22′ 42.″3*	*326° 14′ 52″*	*80° 59′ 37″*
1802	*155 34 4.2*		
1803	*233 45 26.0*		

Eccentricity	*0.0804785*
Log. Semi-major axis	*0.4419513*
Diurn. Trop. Motion	*771.1832″*
Inclination	*10° 37′ 18″*

This ellipse will hopefully still match all previous observations quite well (I have not determined any positions according to it yet) although I do not consider it for the non plus ultra of correspondence. I obtain according to it and the latest elements of Pallas by a rough calculation the distance from the sun in the average line Ceres 2.92644 Pallas 2.85551 and the line itself 7° 32′ longitude 10° 11′ south. latitude helioc. According to these elements is the geoc. longitude on Jan. 28, 1803, midnight Seeberg latitude 280° 26′; latitude 5° 4′ south.

Zach to Gauss Volckershausen September 15, 1802

At the moment I am in the countryside close to Wanfried/Werra, visiting His Highness the Landgrave of Hessen Philipsthal [Wilhelm, 1726–1810] *when your letter of the 9th reached me. I am hurrying to express my sincerest thanks for Pallas' ephemerides and the treatise by Mr. Lichtenstein. Ceres' ephemerides is already included in your* Vienna Ephemerides *pro anno 1803 where Triesnecker had calculated them according to your elements. There can be no doubt that we will see Pallas again next year because Oriani followed her until August 8th. He saw the planet on August 17th and 18th but was not able to observe it. He writes: "The last observations prove at least that we will see that planet next year at its opposition." Here are the observations reduced by Oriani; again he has sent me his complete diary, I will take on a more accurate reduction first thing upon my return (Jan 20). And Oriani calculated Pallas' perturbations, which is easily done if you assume that Ceres and Pallas have an equal mean distance. Furthermore he finds:*

Annual trop. motion of the aphelion		*106″.1*
------------------	*ascending node*	*−7.2*
Annual change of eccentricity in seconds		*−1″.36*
--------------------	*inclination*	*+0.81*

Oriani and Piazzi make fun of the asteroids. To be precise of that part of Herschel's circular letter which he also sent to Italy "Moreover if we were to call Ceres a planet, it would not fill the intermediate space between Mars & Jupiter with the proper Dignity required for that station" and Piazzi's reply to this is "Soon we will see dukes, counts and marchesi in the sky as well." Oriani wrote: "You have already successfully proven the planetism of Ceres and Pallas and consequently it is useless to spend our time on the new Herschelian dynasty."

[Zach appends tables of Ceres and Pallas observations from Milan.]

Zach to Sniadecki Volckershausen September 17, 1802

Oriani followed Pallas until August 8 and even saw it on the 18th. He no longer doubts that we will see this planet again next year. It is my pleasure to send you on the reverse an ephemeris, that of Ceres you will find in the Ephem. of Vienna of 1803, which Mr. Triesnecker will send you without doubt. What do you think of Herschel's new dynasty that he wants to introduce? The Italians make fun of the asteroids. Piazzi wrote: Presto vedremo dei Duchi, Conti, e Marchesi anche in cielo. Oriani wrote me: 'You have already proved successfully Ceres' and Pallas' planetism and consequently it is useless to spend a long time on the new dynasty of Herschel.' But in my September issue I bring more decisive proof for the planetism of Ceres and Pallas. My friend Buerg will not leave for Paris. I have sent my Burckhardt there, but the Frenchmen of previous times are a different story, at least that is what I believe, but at present I won't let Buerg travel into the country of thieves and boasters – Alteri tempi, alteri cure! [Different times, different cures!] *Right now I would like to recover my Burckhardt. Buerg's lunar tables are being printed. Piazzi was awarded money by the King in order to coin a medal on the occasion of Ceres' discovery, but instead he will purchase an English equatorial sector of six feet. That is worth far more.*

Zach to Oriani Volkershausen/Hessen September 17, 1802

Thousand thanks for your precious observations on Ceres and Pallas, you topped it all, illustrious colleague, for no other astronomer has observed them so long as you. If we find Pallas again it is you who we owe it. Your observations delighted Dr. Gauss and Olbers and I enjoyed them as much as they surprised me. Dr. Olbers with his excellent Dollond of 5 feet had already given up on July 8.

And also thousand thanks for your remarks regarding the perturbations of Pallas, but I would have been even more delighted if you had sent me your analytical formulae, I had printed them in my journal where I try to collect everything that has been found out about and done regarding these two new planets. Until today I have been quite successful to collect everything.

Piazzi's remark about the celestial deities made me laugh, this bon mot is brilliant. You said I had successfully proven that the two stars were planets but you will be even more content to hear what I said about this matter in my September issue. Herschel's dynasty is not popular in Germany either.

Dr. Gauss improved his elements III of Pallas according to observations until July 8. Here are his elements IV. [These can be found in the October *MC*.] *He improved them also according to your observations, in the meantime the elements III are sufficient to search for and find Pallas. Here are the Ephemerides for 1803 calculated from the elements III. The difference between III and IV is not great. Consequently, both elements are sufficient to find Pallas. We will see what the elements that are derived from your observations will give us. You will find the ephemerides for Ceres in the Vienna ephemerides for 1803 that Triesnecker has sent you without doubt. I believe we are tracking Pallas, in de la Lande's Histoire céleste, but the matter is not certain yet and in my next issue you will know the truth.*

No. 39 Olbers to Gauss Bremen September 19, 1802

Many thanks for your informative letter of September 14 to which however I won't reply today because I have important observations to share with you concerning our Pallas. Méchain wrote to Bode on September 1 that he had discovered the comet on August 28 just as he had finished the observation of Pallas. Subsequently, Méchain observed Pallas until the end of August! He did not communicate his latest observations to Bode, but did do so with his 8 observations up to August 7 that Bode sent to me. Bode writes: "I ask that you communicate these observations to no one in extenso unless Mr. Méchain sends them in their entirety: they will appear in the 1805 yearbook which will be completed by the beginning of next month." – The following postscript is added at the end: "Dr. Gauss has kindly sent me his latest elements of Pallas' orbit and its ephemeris for the coming year. Should you write him soon, I request that you send him, in my name, one or the other of Méchain's observations of Pallas to be used for his calculations, and give him my regards."

That ban to share his observations probably refers only to Zach. This is due partly because, as you know, Zach and Bode are now not friends, and partly because Bode wants to reserve these observations for his yearbook and does not want them made known beforehand in the M.C. I now certainly expect that Méchain will also have sent the same observations to Zach (although, for some time now, Zach does not appear to be receiving as many reports as usual from Paris). Nevertheless I'll naturally comply with his wish to the letter. Only I don't understand why I shouldn't send you all 8 of Méchain's observations, since I know for certain that you won't use any of them in a way contrary to Bode's intention. Here are all 8: [These were published in the Nov. *MC*.] *"These observations are made off the meridian and the result of the observations repeated 6, 9 and 12 times daily with near-by stars is from Le Francois and Zach."*

Méchain's exemplary observational accuracy is as well known to you as it is to me, and therefore you will probably give preference to these observations rather than to those of Oriani, which extend to August 8, and to those which our friend Zach will have forwarded to you. These observations will thus serve as a splendid verification of your latest elements, the result of which I'm naturally curious. – I'm very annoyed that I didn't again seek out Pallas after dusk and subsequent to my return from Rehburg. I would certainly still have been able to observe it in August.

Zach to Méchain Seeberg September 20, 1802

Oriani followed Pallas till August 18. [The remainder of the letter deals with a cometary discovery.]

No. 40 Gauss to Olbers Brunswick September 21, 1802

Many thanks for your kindly communicated obser. by Méchain of Pallas. I am hurrying to send you the comparison with the elements III, which might be wrong by several seconds. [Compare these numbers with those that appeared in the November, 1802 issue of the *MC*.]

To judge how the new elements of Pallas match these observations I recalculated the position for midnight of August 4. It is (in both cases without aberration which I apply later according to my own way):

acc. to elem. III 198° 59′ 52.″2	*14° 57′ 47.″7*
IV 199 0 58.3	*14 57 48.1*

		Æ Calc.	Decl. Calc.	Differenz			
Juli	12.	192⁰ 2′ 1″	17⁰ 42′ 45″	Æ	— 9″	Dekl.	+ 2″
„	14.	192⁰ 35′ 34″	17⁰ 29′ 31″	„	— 12″	„	— 25″
„	16.	193⁰ 9′ 29″	17⁰ 16′ 5″	„	— 5″	„	0″
„	17.	193⁰ 27′ 8″	17⁰ 9′ 4″	„	— 3″	„	+ 3″
„	18.	193⁰ 44′ 41″	17⁰ 2′ 7″	„	— 15″	„	+ 15″
Aug.	4.	198⁰ 58′ 3″	14⁰ 58′ 30″	„	— 49″	„	+ 1″
„	5.	199⁰ 17′ 24″	14⁰ 50′ 58″	„	— 47″	„	— 8″
„	7.	199⁰ 56′ 15″	14⁰ 35′ 57″	„	— 52″	„	— 16″

Fig. 7.19 Comparison of Gauss' III elements of Pallas with observation

The new elements indicated for Aug. 4 66" more for the RA than the elements III, on July 8 those indicated 24" more. The decl. for both times the decl. was up to 1" identical. Around August the error of the elements IV in RA was about +18"; in the first half of July it was according to your observation of July 9 = 0, according to Méchain's obs. about +¼ min. An amalgam of 3 or 2 parts of the new elements with 1 part of the new [should read "old" not "new"] *would be a proper mean of all these obs. I definitely cannot improve the elements according to these Parisian obs. The above comparison shows that the decl. are not accurate to 10" and 10" decl. has as much influence on the elements as 18" RA. The latest obs. of the end of August would justify a reliable correction. I appreciate very much having received the first obs; now I can hope that the ephemeris will not differ much from the skies; most likely the true motion will fall between the calculated III and IV elements and closer to the latter. I am very delighted that Méchain was able to observe Pallas that long, now we can no longer doubt its rediscovery next year. On August 28 the brightness was = 0.578. I would be very curious to know how long Schroeter would have been able to see Pallas with the 13ft. telescope. How? Should it not now still be possible? The position of Pallas will hopefully be calculated to 1' accurately and this position is quite high in the evenings after dusk.*

I am afraid that Dr. Maskelyne shares your fate, dear friend, having doubted too early the possibility to observe Pallas further. In his letter of July 20 he sent me an observation of which he was not sure it really was Pallas. (It really was). RA = 193° 46' 5", decl. 17° 2' 55" N. mean time 10h 37m 38 s, calculated RA = 193° 44' 59", calculated decl. = 17° 2' 1". But Maskelyne indicated the difference in decl. to 38 Com Beren. only in min (–1° 9'). He asked for a continued ephemeris which I sent him already on July 29. But in a second letter of July 30 he wrote: 'I since find reason to think it was a fixed star and that the planet could not be seen.' And he sent his complete observations. But there are only four of them of June 11, 18, 20, 28. On July 4 he observed two stars, none of those was Pallas. Consequently, I did not send them on to Zach. In the last half of July the weather was very bad in Greenwich and I cannot say whether Maskelyne observed after that and upon the reception of the ephemeris. I rather doubt it because he probably would have informed me otherwise. I do not have Oriani's observations yet, but hopefully receive them today from Zach.

Bode to Olbers Berlin October 9, 1802

I saw your comet several times and tried to observe it at the circular micrometer but its utterly pale light and blurred boundary and the fact that it did not show a nucleus made the determination of its position – at least for me – very uncertain.

No. 41 Olbers to Gauss Bremen October 10, 1802

The comparison of Méchain's obs. of Pallas and your elements IV which you sent me delighted me very much. It seems to be unnecessary that you work on finding more exact elements for the time being. For mere elliptical elements I consider these accurate enough and the considerable effort to determine variable elements can be postponed to that point when a greater arc will be able to give these variable elements more certainty. The elliptical elements, which you calculated so accurately, are in my opinion sufficient to find the small planet again next year. For since you indicate the position up to ½° accurately, we can always bring the telescope into such a position that the planet will certainly be somewhere in the field of it; and then it is only a matter of recognizing it among all the other small stars. – This is not an easy task because in the region where we have to search for Pallas next year there is an uncountable amount of small stars.

No. 42 Gauss to Olbers Brunswick October 12, 1802

Since my last letter I haven't carried out any further important computations concerning Pallas. I also believe as you do that the present results will be quite adequate for its discovery, especially since the latitudes, or what here in the vicinity of the Coluri Solstitt. is one and the same thing, the Dec.'s, will probably still be considerably more reliable than the

RA. I'd be surprised if those in the first half of this year were off by more than 5 minutes. Should you, however, still find Pallas in the Hist. Cel., *then I'll hardly be able to withstand the urge to make use of it. Yet I'm still uncertain in this case if I would view the elements as variable or compute the perturbations, applied to the positions calculated from a true ellipse, using quadratures. Mr. Burckhardt, as I see from the M.C., has calculated the perturbations only to the first power of the eccentricity and inclination. This procedure, which presently in the case of Ceres will be entirely adequate for several years, appears to me, in the case of Pallas, not to be of much more help than if the perturbations are completely removed. Actually, the calculation of the perturbations of Pallas will in future be a real worry for the analysts. To assist oneself by using quadratures from year to year, has of course all of the required accuracy. Yet one would nevertheless wish, and the dignity of science demands, that in future the position calculated for all more remote times should be determined only when one has sufficient observations. But if one wants to express the perturbations by the usual method using series, then this will converge extremely slowly, and I believe that when one wants to compute the position to within 1", perhaps several hundred equations for longitude, latitude, and radius vector will be required. I've begun occupying myself with an improvement in Ceres' orbit.*

No. 45 Gauss to Olbers Brunswick October 26, 1802

[In this letter, Gauss gives Olbers his V elements for Pallas; see *MC* Dec. issue.]

Gauss to Maskelyne Brunswick November 18, 1802

I owe to you still my best thanks for the kind communication of all your observations of the two new planets and for your welcome and useful present of your folio tables, which I received by favour of Mr. Tatter in the month of August. I must beg your kind forgiveness for paying my due acknowledgements so lately. My last letter written July 30th*, which contained the calculated positions of Pallas for the months of July and August, I hope you have received. Since that time I have calculated, twice more, elements of Pallas. The first (IV) after the last observations of Dr. Olbers are printed in the October issue of Baron Zach's Journal; the latter after the last observations of Mr. Oriani conducted August 8 at Milan, which are V*th *in order, I have the honour to communicate here following as they are not yet published but will appear in the December issue of the* Monthly Correspondence. [See *MC* Dec. issue.]

These elements I had already computed, when I was informed that Mr. Méchain has been so fortunate as to observe Pallas till the end of August; but till now I have not yet received any of his observations posterious to the last of Mr. Oriani's, which agree very well with the cotemporeal obser. of Mr. Méchain. To my very great surprise Mr. Messier at Paris has observed Pallas till 21 Sept.; his last observ. is

Sept 21 7h 58' RA Pallas 215° 48' 46"
Decl. 8 59 28

Supposing the time to be mean time, the place of the planet calculated after my Vth *elements is*

215. 49. 14 Difference +28" (Distance from earth 3.518)
9. 0. 16 +48" (Altitude above the horizon 13° 10')

There are already extant two ephemerides for the course of Pallas in the year to come, one constructed by Dr. Olbers after my III^{d} *elements* [Zach's *M.C.* September] *the other by myself after the IV*th *elements (ibidem October). I have now the honour to send you a new one computed after my* <u>*last*</u> *above elements (V), continued above a month further than the two preceding ones, which I hope will be near enough to the truth to find the planet out again, though I am afraid, that the faintness of its light and the innumerable little stars which lye in its way in the galaxy will be very untimely circumstances. The last column of the table expresses the value of the quantity*

$$\frac{1}{(\text{distance of Pallas from the sun x dist. of Pallas from earth})^2}$$

and may be considered as the measure of the planet's apparent quantity of light, neglecting its elevation above the horizon. The same expression yields for this year

1802	*April 4*	*0.08997*
	May 16	*0.04740*
	August 10	*0.01455*
	Sept. 25	*0.01030*

I have also corrected my VIIth elements of Ceres after the observations of the last season and introduced the perturbations caused by the action of Jupiter. The result of these researches is exhibited in the Monthly Corr. *November, but till now I have not calculated any ephemeris for 1803 by the new elements, because the course of Ceres 1803 after my VII elements is already calculated by Mr. Triesnecker and Bode in their respective astronomical ephemerids, which I hope will be sufficient to find the planet out. The difference between the VII elements and the new ones (VIII) towards the opposition is by a calculation made somewhat hastily.*

Longitude 10′) both being greater after the new elements
Latitude 3 1/2′)

If I should undertake the calculation of a more correct and extensive ephemerid, I shall not fail to send it you as soon as it is finished.

No. 47 Olbers to Gauss Bremen November 24, 1802

The Moniteur *No. 53 again has a paragraph by Burckhardt on Pallas. Messier has observed it up to September 24 and still tried to observe it on October 10. But the resulting observation is questionable because of the nearness to the horizon and smoke from chimneys. Burckhardt is occupied with perturbation calculations that he still hasn't completed. He has, however, calculated the following pure elliptic elements, irrespective of perturbations, using the most recent observations.*

Asc. Node	*172° 27′ 35″*
Aphelion	*302 3 2*
Incl.	*34 38 0*
Long of perihelion	*121° 12′ 19″*
Solar movement	*+0.3″ per day*
Anomaly April 4	*10h 51m 17s = 42° 21′ 9″*
Semimajor axis	*2.769915*
Eccen.	*0.2463*
Sidereal revolution	*1683 days 20 hours*

These elements represent the five observations in the following way:

Date	April 4	May 20	July 3	Aug. 5	Sept. 20
Helio in long.	+1".4	+1".0	−1".6	-0".6	−6".0
Helio in lat.	0.0	-2.7	−3.5	+13.5	−18.3"
Geocent. in lat.				+11	−12"
Observers	Zach & Burckhardt	Le Francais	Mechain	Mechain	Messier

(I don't quite know why and how the heliocentric errors are given here). With these elements one finds that the planet will be at the 15 Pluviose year 11 [February 4, 1803] *at midnight 267° 41′ RA and 5° 38′ Dec. North. This agrees very well with your results.*

Gauss to Bolyai* Brunswick December 3, 1802

My dear Bolyai, it's been quite some time [since I have written] and all that I could use as an excuse for my negligence would not be convincing enough for a severe judge and consequently I'd rather admit I am hoping for your forgiveness.

Since then I have mainly been working on the two new planets Ceres and Pallas. I assume you know the history of these two discoveries. If not, I'd like to refer you to von Zach's Monatliche Correspondenz *(which is at least read in Hungary) where you can find the first news about those since the June issue of 1801 and my complete works regarding them from the December issue 1801 until today. But you can also find the most important in Bode's Astronomischen Jahrbuch for 1805, but less complete. If you cannot get hold of any of these and wish nevertheless first hand information please write and I will do my best to enlighten you as it is possible per literas. If you have read von Zach's MC you might already know that my work on Ceres has already improved my situation: our generous sovereign has granted me a pension of 400 Rthl. which has put me in an independent and carefree situation. But it is uncertain whether or not I am to enjoy this situation for long.*

*Farkas Bolyai (1775–1856), Hungarian mathematician and friend of Gauss.

Chapter 8
Bode's Asteroid Book

C. Cunningham, *Early Investigations of Ceres and the Discovery of Pallas*,
DOI 10.1007/978-3-319-28815-4_8

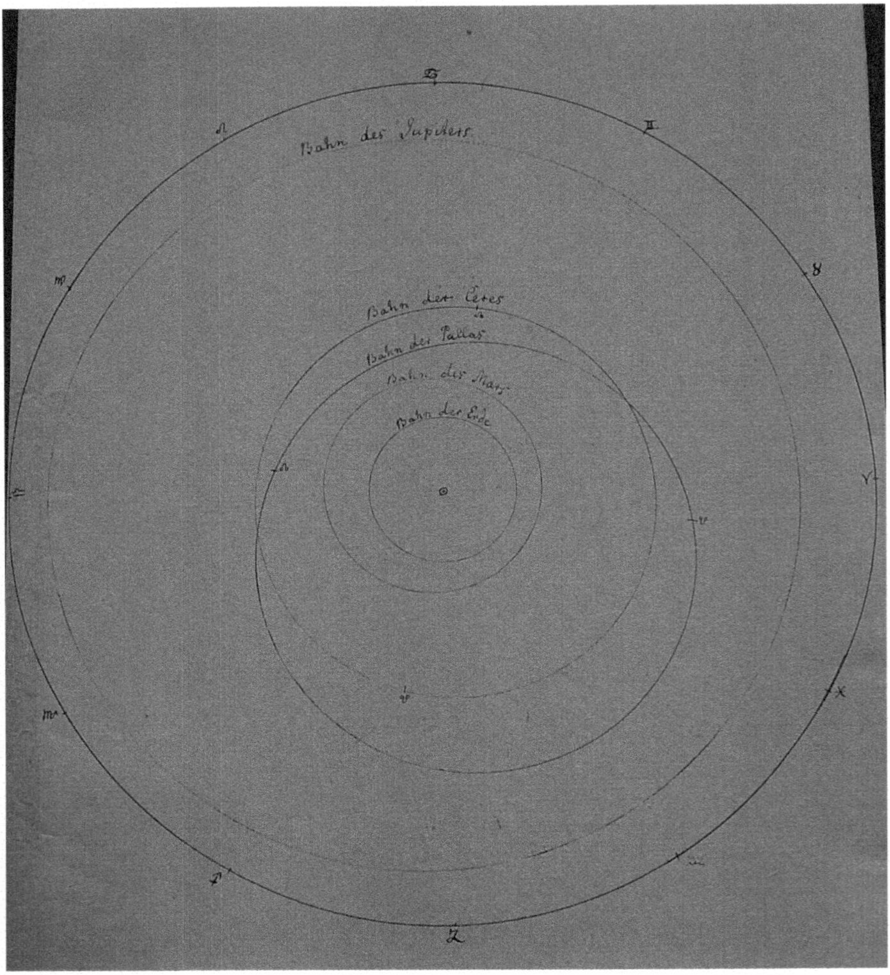

Fig. 8.1 Map drawn by Olbers showing the orbits of Ceres and Pallas (Staatsbibliothek, Bremen)

This chapter contains the section about Pallas in Bode's asteroid book of (1802b). The first part of the book was published in *Discovery of the First Asteroid, Ceres.* Bode dated his book May 17, 1802, just two months after the discovery of Pallas. A survey of the discovery of Ceres and Pallas was also published in Bode (1803). Here is Bode's text, beginning on page 95 of *The new planet between Mars and Jupiter*:

> *They* [Ceres and Uranus] *were then joined unexpectedly by a strange new 7th or 8th magnitude variable star without noticeable nebulosity which also the commendable and tireless Mr Olbers in Bremen, had the luck to discover in the northern wing of Virgo, nearly at exactly the same position where he first had rediscovered Ceres on January 1st. He kindly communicated this important news on March 30th and gave me the following three observations, the first of this star:* [table of positions of Pallas on March 28, 29 and 30]
>
> *"What shall I think of this new star,?" Mr Olbers asks. "Is it a strange comet or a new planet? I do not dare to judge it yet. It is certain, that it does not resemble a comet in the telescope, no trace of nebulosity or atmosphere around it can be seen."*
>
> *I received this letter on April 5th; and because it was fine in the evening I searched for it and soon found it with the searcher at the position marked on my map, southeast of D Virginis, a 7th magnitude star without nebulosity which I could consider the one of Olbers; at the wall quadrant the observation did not succeed because the air became hazy. This was the case on the 7th, too (on the 6th it was completely overcast).*
>
> *On the 15th I could, at last, observe it at the wall quadrant and found it at 10h 37′ 25″ mean time and compare it with β and o Leonis; the stars' apparent right ascension 182° 10′ 31″ and declination 16° 54′ 30″. I have observed it several times at the wall quadrant and the circular micrometer and saw it now and then in the 2-foot searcher and 3.5-foot Dollond. Mr Olbers had the kindness to inform me in his later letters about his further observations of this strange star, which he names Pallas; among other things in a letter, dated April 6th as follows:*
>
> *"I continue to send you news of the strange mysterious star, which I discovered on March 28th. It continues to move steadily with slightly decreasing right ascension and declination, and its appearance is still that of a small 7th magnitude fixed star, slightly fainter than Ceres*.*
>
> ** I have found this faintness completely confirmed. A body that, according to Mr Schroeter, has a diameter of 4 seconds and still appears as an 7th or 8th magnitude star, must be illuminated extremely faintly.*
>
> *These three observations are of April:*

	Mean time	*apparent RA*	*apparent dec.n.*
April 1	*8h 0′ 40″*	*184° 15′ 38″*	*12h 54′ 25″*
2	*7 56 55*	*184 5 8*	*13 14 28*
3	*8 0 37*	*183 54 32*	*13 34 16*

> *With impatience I am now waiting for your message whether you could find my star and what you think about it. I do not dare any judgment of this curious stranger." –*
>
> *I consider it a very distant comet, maybe to be found beyond Ceres' orbit and announced it publicly as such. It probably is now near its perihelion and therefore is describing a fairly concentric circle regarding the earth's orbit. – There can be comets without phosphorescent tail and coma and this one either lacks it completely or its misty shroud – its phosphorescence – is only faint and thin and therefore not visible at its great distance.*
>
> *Thus Pallas appears as a planet-like faint 7th or 8th magnitude star, whose apparent diameter Mr Schroeter determines, according his letter of April 21st, to be 4″.6; nearly that of Ceres and who saw it with his large telescopes blurred, diffuse and nebula-like (traces of*

its coma). It has made its way in the sky since the day of its discovery on March 28th until now (end of April) with slow and decreasing steps with a high inclination against the equator and the ecliptic, from southeast to northwest in a strange way coinciding straight through Ceres' apparent (here knot-like) winding orbit. Therefore the observation of this small planet-like comet interrupted from time to time that of Ceres and Uranus. I expected it several times at the wall quadrant in vain because of its faintness. Mr. Olbers says in a letter, dated April 16th, about the nature of his variable star the following:

*"I could see from my observations so far that this strange star is not a planet, if you consider planets those bodies that change their distance to the Sun only slightly. No circular orbit corresponds to this star, its orbit is considerably elliptical; whether it resembles a parabola time will tell, I do not know yet.**

** Its described arc from March 28th to April 16th can only be of a few degrees, from which nothing certain can be said about the shape and position of its true orbit.*

"You only have to observe this mysterious stranger in your telescope to be convinced this is a totally different body than those appearing normally as comets. If you would like to call it nevertheless a comet, although it has no coma at all, I have no objections." –

Mr Schroeter writes further on April 21: "We have observed eagerly Mr Olbers' Pallas since March 30th; Mr Harding is determining its position, I do the measuring and everything else. In these local powerful instruments it is nebula-like and diffuse like Ceres and resembles it like one sister to the other. With such a nebula-like imitation I found its apparent diameter on March 30th 4″.635. On the 31st it had changed its light very much, appearing brighter without blurred outlines, like a fixed star, and could not be measured; under such changed circumstances it appeared with a diameter of 3″.244. On April 2nd it was strangely blurred and diffuse again and had a diameter of 4″.753. On April 3rd Pallas showed a diameter of 4″.611 and Ceres only 4″.309; on the 4th Ceres 4″.680, the 9th 4″.063; on the 15th, and on the 30th alike, Pallas appeared brighter and with a diameter of 3″.562. Undoubtedly Pallas does not describe a circular but parabolic orbit. In the strict sense it cannot be called a planet. And in the broader sense comets are planets; Ceres and Pallas are alike and resemble the hitherto known planets more than a comet. Nonetheless the former describes a circle-like orbit, the latter does not. The fact that recently in such a small celestial space (i.e. the northern wing of Virgo whose smallest stars have been observed continuously together with Ceres, passing through, for six months so accurately that led Mr Olbers to the discovery of his variable star.) two of these wandering celestial bodies, not visible to the unaided eye, were discovered, leads one to suppose there are more of their kind, which might show us in the end the transition, how elliptical comet orbits were formed from circle-like more or less eccentric planetary orbits. I consider these two discoveries very important for physical astronomy and accordingly I would call Pallas and Ceres alike, disregarding the earlier characterization, a comet-like planet, rather than a planet-like comet. Forgive my imagination, which seems to have more than one reason."

We know from history no such comet, which appeared as a small star without nebulosity and tail; and before the invention of the telescope one only knew the large comets, visible to the naked eye. Most of the recently observed ones we only discovered by telescopes and followed them in their paths; some of them appeared as small stars shrouded in mist without tail; but never has a comet been observed without this misty shroud.

Nevertheless it is possible that comets might be completely free of it and may appear at a great distance as planet-like stars. Does Mr Olbers' Pallas belong to this kind of comet? Who knows, whether there had been some comets, among those observed and then never found again, stars which had not been observed continuously and noticed maybe only in their short path, and which were therefore considered fixed stars. It really seems to be this way.

One of those might have been the variable star in Leo observed by Prof. Huth in December last year. Comets have been considered different from planets until now because of two facts: their extreme eccentric orbit and the great angle of inclination of their orbits to the ecliptic. If one disregards these distinguishing features and assumes these bodies as stable and solid masses they are planets as well and especially those deserve that name whose orbit we completely know and which return regularly to the Sun, like the comet of 1759 whose orbital period is 75 years, but which is the only one of this kind to be known. The mysterious one of 1770, whose orbital period Mr Lexell calculated = 5½ yr, and several others whose orbital periods have been calculated but which are not relevant here.

Thus Mr Olbers really made an important discovery in this extraordinary heavenly body and provided the astronomers with a wealth of material for further research on the true nature and its true orbit. In the astronomical yearbook of 1805 I will give further details of it.

Chapter 9
Schroeter's Asteroid Book

C. Cunningham, *Early Investigations of Ceres and the Discovery of Pallas*,
DOI 10.1007/978-3-319-28815-4_9

Lilienthaliſche Beobachtungen

der

neu entdeckten Planeten

Ceres, Pallas und Juno,

zur

genauen und richtigen Kenntniſs ihrer wahren Gröſſen,
Atmoſphären und übrigen merkwürdigen Naturverhältniſſe im Sonnengebiete,

von

Dr. Johann Hieronymus Schröter,

Kön. Groſsbr. und Churf. Braunſchw. Lüneb. Juſtitzrathe und Oberamtmann, mehrerer Academien und Societäten der Wiſſenſchaften Mitgliede.

Mit zwey Kupfertafeln.

Göttingen,

in Commiſſion der Vandenhöck - Ruprechtſchen Buchhandlung. 1805.

Fig. 9.1 Title page of Schroeter's 1805 book on Ceres, Pallas and Juno

The first part of Johann Schroeter's asteroid book entitled *Lilienthal Beobachtungen der neu entdeckten Planeten Ceres, Pallas und Juno*, published in 1805, was printed in *Discovery of the First Asteroid, Ceres*. The second part, on Pallas, appears here. The concluding section, on Juno, will be published in a later volume. The original section numbers are retained here.

In much of the book Schroeter appears to be making a case against possible detractors, believing that his observations are beyond question. These observations include detailed measurements of the atmosphere of Pallas, which does not actually exist. Schroeter was quite chuffed to have found a dense atmosphere around both Ceres and Pallas, but his colleagues across Europe were unconvinced. Some believed his eyes were deceiving him. In a discourse on human understanding that was written in 1690 but posthumously published, Peter Daniel Huet (1630–1721); Fig. 9.2, bishop of Avranches in France, neatly summed up the underlying and age-old cause of the dispute:

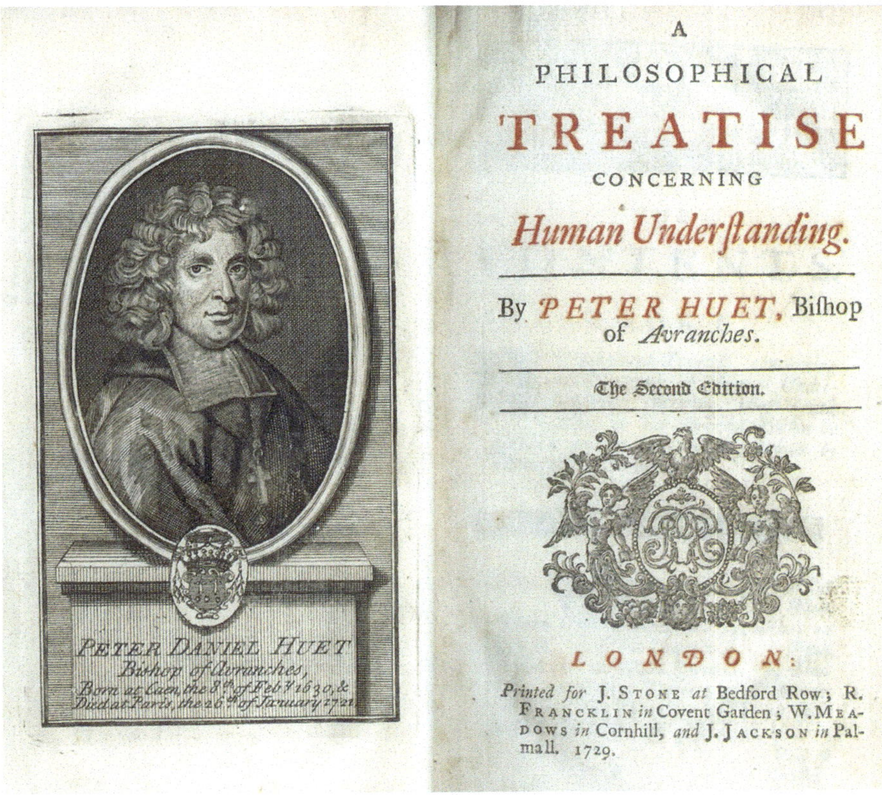

Fig. 9.2 Peter Daniel Huet and the title page of the 1729 edition of his book.

Our senses don't feel outward Objects, but only the Impression of the Images or Forms which flow from without them, and that this Impression, which comes from without, has not the same Effect in every Man, but is differently felt according to the Diversity of the Organs of Sense. This is very elegantly expressed by the Satyrist in these words: Our Eyes deceive us, and the Uncertainty of our Senses imposes upon our Reason.

The following translation covers pages 176–241 of Schroeter's book:

69

On Mar. 30, 1802 Dr. Olbers notified us of the most peculiar discovery of Pallas. It was still dusk when we were looking for this new wandering star which we found at 7h 44'; I with the 13-foot reflector at a magnification of 136 times and Mr. Harding with the 3-foot achromatic telescope. It stood south of three 8th magnitude stars)*

**) These stars are mentioned in the* Hist. cel. Franc. *vol. I, page 149 and observed*

in April, 1795, at 12h 17'20".5
12 18 0.5
12 18 19

at RA 184° 36' and 12° 14' n. declination.

It had a dim light and appeared in the 3-foot achromatic telescope as bright as a 7th magnitude star. At a magnification of 136 and 288 times of the 13-foot telescope it appeared noticeably larger and more planet-like than Ceres; and even if it, too, did not appear absolutely clear, but comet-like with a rather nebulous boundary, its boundary was nevertheless much clearer than such a planet's.

70

While Mr. Harding was determining its position, I immediately started to calculate its apparent diameter at a magnification of 288 times at my 13-foot favourite.

Accidentally the disc micrometer was still at the distance at which I found Ceres' [angular] diameter on March 28 at exactly the same magnification with a projected disc of 3.333 lines. At this distance, however, Pallas appeared larger. Therefore I had to bring the projection micrometer slowly closer until Pallas' image was the same size as the imagined disc and I found the unknown distance from the eye to be only 515 lines, instead of 594 lines in the case of Ceres.

In comparison to Ceres Pallas' light was equally pale at a magnification of 288 times; but still slightly stronger, because Pallas' image, when projected right between two illuminated projected disks, retained its disc like appearance and lost less of its pale nebulous boundary through the light of the disks, and then still was like a projected disc of 2.800 lines. How pale Pallas' blurred nebulous shroud must be to cause such an image, follows from

1) the fact that as soon as there was a little wind its image appeared considerably smaller and more blurred and

2) the fact that I was as always able to see Uranus with the naked eye, whereas of Pallas, when it appeared with its nebulous shroud, like Ceres, the naked eye could not discern any trace.

71

According to this imaginary measurement the calculation for Pallas' diameter including its nebulosity showed:

L. 3.333 Lin. = 3.5228353
– L 515.000 Lin. = 5.7118072
7.8110281 = tang. 22' 15"

$$= \frac{1335''.000}{288} = 4''.635;$$

But if Pallas' image was projected between two illuminated disks of the projecting apparatus, thus the outer fainter part of its atmospheric envelope would become invisible, its entire diameter was only 3".893. Mr. Harding determined with the circular micrometer for March 30 8h 26' 39" mean time this new wanderers' apparent RA 184° 35' 52" and its northern declination 12° 14' 59". Pallas was accompanied by a very tiny very faint star east of it, which could only be seen in the 13-foot reflector, following the planet at 8h 30' in 5".4 and at 11h 7' in 11".7.

72

All circumstances of this first observation and measurement, the similar motion, size and the very similar light atmospheric cover showed already that Pallas was a sister of Ceres, of the same origin and with identical particular planetary characteristics.

Furthermore the fact that it appeared considerably smaller as soon as a little wind was blowing, seemed to hint at a quite considerable core or planetary sphere, because fixed stars of a certain size appear due to the motion of our hazy air and the thereby caused scintillating (rather larger than smaller), but small faint fixed stars for this very reason disappear and become invisible.

73

Thus, I had good reason to hope for good observations and continuous measurements like in the case of Ceres. And the success showed very soon that Pallas is subject to a peculiar and nearly more striking change than Ceres: Its light had changed totally, when we could observe it again, due to bad weather only on April 1, after sunset when the sky finally cleared up. It appeared in the 13-foot reflector at a magnification of 136 and 288 times MUCH BRIGHTER, WITHOUT ANY NEBULOUS BOUNDARY LET ALONE A PLANET-LIKE APPEARANCE just like a fixed star and thus could not be distinguished from a fixed star. From time to time, however, it appeared as a little disc, but only by its sight it appeared considerably smaller than 48 hours ago, when I had found it with its nebulous atmospheric shroud. Because of this obviously smaller diameter it was very convenient for me to have the projection micrometer still at the same distance of 515 lines, at which I had found Pallas, at a magnification of 288 times equally large as a projected disc of 3.333 lines, because I thus could make a direct and reliable comparison.

What could be determined by sight was confirmed by the micrometer: when Pallas appeared from time to time in strong dew as planetary disc, it was just as big as a little disc of 2.333 lines and by no means larger.

And again, Pallas was accompanied by a tiny star as it was 48 hours ago, which one could have considered a satellite, unlikely though it might be. I determined its position against the planet from 8h 21' to 10h 50', and in the meantime the planet was moving considerably west, but the tiny co-star did not seem to have a motion of its own.

While I was occupied with this, Mr. Harding determined the planet s position using the circular micrometer and at 8h 44' 3" mean time Pallas' RA was 184° 15' 3" its apparent n. declination 12° 55' 27". I used the three stars mentioned in section 69 of the Hist. cel. Fr. *vol. I and Bode's No. 225 Virgo as reference points.*

74

If this observation and this measurement are compared with the previous ones of March 30 and the following ones, they seem to be of special peculiarity to the physical astronomy, because:

1) accidentally during the observation it favourably happened to be dewy. Every practical observer knows that dew like any other atmospheric modification causes a shimmering or glimmering of the major planets' borders and their discernable features of the surface, eg at the Moon's crater rims or mountains, so that a small crater becomes clearly visible, but more often cannot be distinguished due to the glimmering.

The reason Pallas appeared only from time to time as a planetary disc was undoubtedly caused by falling dew. But that it was more luminous as a little disc, brighter than 48 hrs. ago, WITHOUT ANY NEBULOSITY AND REMARKABLY SMALLER, could not possibly have had its natural cause in any accidental operation of our atmosphere; because such tiny objects, due to atmospheric fluctuation, appear as blurred and dim images and by no means as a DISTINCT AND BRIGHT image.

It follows from this that

a) on March 30 Pallas appeared at the very same magnification and telescope in a pale light and shrouded; this pale colour and nebulous boundary and an obviously larger diameter could not at all be an optical illusion. And a pale light, blurred image, and a larger diameter match each other perfectly.

b) the pale light caused by the shroud, which was measured too, vanished after 48 hours and Pallas now appeared as a strikingly brighter and at the same time smaller disc, without nebulosity, the physical cause of all this could either be a rotational period of 48 hours or a clearing up of Pallas' atmosphere. If one tries to find the cause in Pallas' rotation and assumes that after 48 hours the hemisphere, which may reflect the sunlight stronger, was turned towards the eye, then, however, the also measured, now vanished nebulous shroud could not have ceased to exist altogether; unless one could and would assume that this brighter hemisphere had as well a lighter atmosphere, the other one a rather dim and saturated one. But this seems to be highly unlikely and moreover the following observations contradict that assumption: if it was a consequence of a rotation of 48 hours, thus on April 3 after 48 hours the same alternating appearance must have been observed, which was, according to section 80, not the case.

75

Undoubtedly the cause was an accidental exceptional clearing of Pallas' atmosphere which reflected on March 30, while saturated, the sunlight and formed a discernible boundary or shroud, which was measured as well and which gave the better-reflecting sphere a pale appearance by its fog-saturated character, but which cleared up from March 30 to April 1. If you assume this, which could be seen clearly from these two observations and the following one from April 3, thus on April 1, since the cause of its visible shroud had disappeared, the effect of it had to vanish, too. The sphere must have appeared without nebulosity considerably lighter in colour and smaller in diameter without twice its atmosphere, measured on March 30. Everything is so consistent, that one cannot judge it differently according to such exceptional appearances' nature. Furthermore, according to the measurement of April 1,

2) Pallas' apparent diameter and the apparent vertical expansion of its considerable atmosphere can be determined.

a) For the apparent diameter of the sphere the calculation shows

L. 2.333 Lin. = 3.3679147

- L 515.000 Lin. = 5.7118072

7.6561075 = tang. 15' 35"

$$= \frac{935''.000}{288} = 3''.243;$$

b) On March 30 Pallas' apparent diameter, including its then visible, but on April 1 vanished nebulous envelope: 4".635. If you subtract from this the diameter, found on April 1, of the mere disc without atmosphere = 3".243, thus for the double size of the atmosphere on both sides, or rather round about remains 1".392; and this number divided by 2 is 0".696 for the apparent vertical height of Pallas' atmosphere, seen on March 30, 1802, with the 13-foot reflector.

76

On April 2 around 9h it was most peculiar to see only 24 hrs. later that Pallas appeared just as she did on March 30, at all magnifications of the 13-foot telescope, as a clearly discernible planetary disc with nebulous boundaries; and that its light was EXACTLY as DIM, PALE and FAINT and did not reach by far the brightness of 24 hours ago, although the night was equally clear. In order to measure the diameter including the nebulous boundary or envelope accurately, I had locked the disc micrometer at an unknown distance from the eye and compared the image of the planetary disc, at a magnification of 288 times, to the projected disc of 3.333 lines, by projecting the image of the planet as usual close to the micrometer disc. The former appeared considerably larger than the latter, so I gradually had to draw the micrometer closer and closer to the eye until both disks appeared equally large next to each other. I then found the unknown distance 502 lines. According to this Pallas' diameter including the atmosphere is

L. 3.333 Lin. = 3.5228353

– L 502.000 Lin. = 5.7007037

7.8221316 = tang. 22′ 49″

$$= \frac{1369''.000}{288} = 4''.753;$$

77

At that time the results of 4″.635, 3″.243 and 4″.753 of the three measurements did not make much sense to me, since I did not know then how these contrasting measurements would automatically fit together. After so many observations and measurements of that kind which I had made for other heavenly bodies, I had to be convinced of a modification of Pallas' atmosphere that took place from March 30 to April 1. This was precisely the reason why observations and measurements of Pallas were not as enjoyable as the ones I had made for Ceres. Thus, the following measurements were made rather indifferently and I could not explain the contrasts I found satisfactorily. If you now compare the third measurement with the two previous ones, I am truly pleased to see how strikingly it confirms the results of the first two observations: for that

1) Pallas' atmosphere returned to its previous state within 24 hours, because the nebulous envelope around the core was visible again and at the same time the light of it had changed, too, to a dim, pale and faint light.

2) If 24 hours ago only the diameter of the bright sphere had been measured, this measurement confirmed it strongly. Because according to this measurement, made with the again visible atmosphere, the diameter, in correspondence to the diameter on p. 75, for the sphere was 3″.243 + twice the atmosphere = $\frac{1''.392}{4''.635}$ *And without then even thinking of it, this measurement showed it impartially. The difference is only =* $\frac{4''.753}{0''.118}$.

If the insignificant variation of Pallas' distance from earth is not taken into account, because on March 30 it was only 1.577 and on April 2 only 1.395 thus amounting to + $\frac{18}{1000}$ *this is only a minor error in measurement, which should be allocated to the results of all three measurements of March 30, April 1 and 2 for correction, which would be for each of it* $\frac{39}{1000}$ *of an arc second.*

I think someone who has an impartial closer look at these perfectly corresponding measurements and regarding Ceres' excellent progressive series of measurements and the following vivid discussions, might feel that with such accuracy an optical illusion is impossible.

78

While observing on April 2 we compared Pallas to Ceres and Uranus at a magnification of 136 times at the 13-foot telescope. Ceres and Pallas appeared veiled and without great difference, just like two sisters of the same origin.

At a magnification of 288 times we were able to discern each planet's nebulous, planet-like and ill-defined disc. A BRIGHTER star, however, standing close to Ceres appeared without disc as a fixed star.

Uranus appeared much more defined and luminous; and for this reason it appeared deceptively larger and compared to Ceres and Pallas like a small version of Jupiter, without us noticing however any oblateness. Nevertheless, this time it did not appear as clearly defined as we had seen it during previous comparisons. But today its appearance resembled that of Ceres and Pallas, albeit less blurred.

Maybe and most likely this exceptional blurred image was the reason why its diameter was found slightly larger than it should have been in proportion to the other measurements. With the greatest attentiveness possible I moved the micrometer until I could obtain Uranus' image equally large to a projected disc of 3.333 lines. I then found the distance of the micrometer from the eye 530 lines, thus only 28 lines more distant than in Pallas' case. Uranus' apparent diameter

$$\frac{3.333}{530.0} = \textit{tang.}\ \frac{21\,37''}{288} = 4''.503$$

slightly more than Herschel's or our measurements showed. The measurement was as accurate as possible. Maybe our normally clear atmosphere, but more likely an exceptional modification of Uranus' atmosphere at the time of measurement, was the cause why the planet was not as distinctly defined as usual so that its disc appeared exceptionally large. By the way, Mr. Herschel also noticed appearances in Uranus that seemed to portend atmospheric changes.

79

By the way, the observation of April 2 is to be commented as follows:

1) At 11 o'clock, when the previous comparison of those three planets was made, this planet had ascended too high to measure Ceres conveniently and accurately.

2) Again, like on March 30 and April 1, Pallas was accompanied by a tiny very faint star, positioned northeast, which followed the planet at 8h 44' for 1''.5. And as well a second, even fainter little star, twice as distant and positioned northeast of the planet, followed the planet for several seconds. Both little stars were, according to my observations of 8h 44', 9h 0' and 10h 30' at the circular micrometer, no satellites.

3) Mr. Harding found at 8h 19' 11'' Pallas' apparent right ascension 184° 4' 33'' and the apparent n. declination 13° 14' 14''.

80

On April 3 at 9 o'clock, Pallas appeared in the 13-foot reflector at a magnification of 288 times just the same as at our first observation on March 30: with pale light and blurred, similar in shape to the planet-like fixed nebula at ν Aquarius, as a round not clearly defined but nebulous planetary disc. And I could see at one glance from the fixed micrometer that it was, including its nebulous shroud, slightly larger than Uranus on the previous evening.

I moved the micrometer to an unknown distance from the eye and after some examination and moving back and forth, Pallas and its nebulosity appeared at a magnification of 288 times equally large as a projected disc of 3.333 lines. The unknown distance of the micrometer from the eye proved to be 517.5 lines. According to this the calculation for the apparent diameter is:

$$\frac{3.333}{517.5} = \textit{tang.}\ \frac{22\,8''}{288} = 4''.611$$

81

To convince the reader how unfamiliar I was 2.5 years ago with the natural course of things and how unbiased these measurements and observations then were, regarding to what I said in section 74 and 75 about the only now concluded diameter of Pallas' sphere and about the vertical height of the visible expansion of its atmosphere, I am quoting from my diary of April 3 what I was writing about the calculation of the diameter:

"According to exact measurement certainly not smaller with nebulosity. Even between the disks it appeared considerably larger than a disc of 2.800 lines. Even when windy it kept its disc and the air was hazy. And certainly its appearance as a fixed star of the first (when it appeared from time to time only as disc and considerably brighter) had a different cause in itself. Maybe, after further appearances of this kind, its rotation can be assumed."

It was my idea then, as correct as it might be in general, to think naively of its possible rotation, disregarding the facts that were already in the observations of March 30, April 1 and 2, 1802, according to p. 74 and 75. Thus, the more impartial the observation and measurement of April 3, 1802, confirms:

1) what has been deduced from the observation of April 1 about the diameter of Pallas' sphere and the vertical dimension of its atmosphere. Now, not like on April 1 only the sphere appears in much brighter light without nebulosity, but the disc is observable again in a pale dim light with nebulous envelope and the actual measurement, repeated under these circumstances, corresponds exactly, because this one, without neither taking into account the slight difference in Pallas' distance from the earth of April 1 =1.389 and on April 3 only +0.011 = 1.400 nor the inevitable errors in measurement, must have shown for the measured diameter of the sphere alone on April 1 3".243 and for the double expansion of the atmosphere $\frac{1''.392}{4''.635}$ *and for both really was 4".611, thus the difference is just* $\frac{24}{1000}$ *of an arc second.*

82

At the same time this observation shows:

2) with certainty that regarding a possible rotation I was not mistaken and that the physical cause of the peculiar appearance of April 1 could not be caused by an alleged rotation, because otherwise the part of the sphere, reflecting the sunlight so much brighter and which must have been at least, as it is in general, the entire hemisphere facing the earth, would be now facing the earth again. From March 30, 9 o'clock, when Pallas was observed and measured in all its nebulous cover to April 1, 9 o'clock, when the mere but much brighter sphere without all nebulosity was observed and measured, 48 hours had passed. Accordingly, every possible rotation must correspond to this period and therefore the brighter hemisphere must have faced the earth again, on April 3 at 9 o'clock, the time of the actual observation; which was not the case. In fact it appears to me, because of the conspicuous, huge atmosphere of Pallas and its density most unlikely, yes, not even imaginable, to hope for the discovery of a rotation at a time when the atmosphere might clear up for a considerable time, like on April 1.

83

During this observation of April 3 Ceres and Uranus were compared to Pallas.

1) Ceres had in the 13-foot reflector at a magnification of 136 times a slightly reddish light and of 288 times it apparently had a stronger light than Pallas, but was still equally pale and not as clearly defined as Pallas and appeared as a disc, as usual. It formed together with S and D Virgo a triangular, north above the bright 6th magnitude star N 3 Virgo, which appeared very small and not as little disc.

This already showed how far we had always been from an optical illusion. More remarkable is, however, which I expressed explicitly in my diary, that this time Pallas' nebulous boundary appeared JUST FOR ONCE to be slightly larger than usual. It was very helpful to measure Ceres' diameter for comparison. But such measurements of Ceres and Uranus are not part of the main objective of observations on Pallas and thus must be regarded as random and cannot be used for any conclusions on the main objective, because measurements with such a large reflector always require efforts and the main effort should go on the account of Pallas, which alone is quite tiresome. After this more general remark I am quoting from my diary's contents:

"I found that its diameter was only slightly smaller than that of Pallas (because after having measured Pallas, the micrometer stayed fixed at a magnification of 288 times for comparison) and moved the micrometer until Ceres was now smaller than larger, but rather a wee bit larger than smaller, on average appeared to be the same size as the disc of 3.333 lines and I found the distance from the eye 554 lines. Thus:

L. 3.333 Lin. = 3.5228353

- L 554.000 Lin. = 5.7435098
7.7793255

$$= tang. \frac{20\ 41''}{288} = 4''.309;$$

If we would like to assume this occasional measurement highly accurate, which might be justified by the above words "but rather a wee bit larger than smaller", the remark that this time the nebulous cover appeared a little bit larger or to be of larger size would appear to me normal and thus real and truly justified. That this might be the case for atmospheres of considerable size, which are capable of different densities and degrees of saturation and thus are now more visible then less visible, may be proven by the comet of 1799 and my articles about the latest astronomical discoveries in the second part of the third volume Beyspiele (Examples). *And in this respect this remark might be valuable for those observing Ceres and Pallas in the future with large and powerful telescopes*).*

**) To a certain extent what I said about an apparent exceptional, irregular and only partial oblateness of the planets Jupiter and Venus in the miscellany of the second volume of my* Astr. Beytraege (Astronomical Memoirs) *No.1, p. 7-15 can be associated with it.*

2) Uranus, observed at 10h 45' with the very same magnification of 288 times at the 13-foot telescope, appeared despite the hazy air comparably more distinct than Ceres and Pallas and although equally faint with more luminosity than both of these recently discovered planets. I saw it again with the unaided eye, very faint though, and the result of all previous comparisons was that it really possesses more true physical size than both other sister-planets covered in their atmospheric haze.

Because Uranus' diameter had been determined very accurately by Mr. Herschel I regarded a further comparative measurement of it redundant, especially since the previous showed that the result would be the same to some decimal seconds. By the way, for April 3 at 8h 21' 53" Mr. Harding found the apparent right ascension of Pallas 183° 54' 21" and the apparent n. declination 13° 35' 31" using the circular micrometer.

84

On April 4, at 9 o'clock, I tried to observe and measure Pallas again at the 13-foot reflector, because the light of the new Moon, since April 2 at 4h 9', had started to impair the observations.

At a magnification of 288 times, I moved the disc micrometer back and forth since Pallas' image equaled a disc of 3.333 lines and found the distance of the micrometer from the eye only 510 lines. Accordingly, the calculation for the apparent diameter resulted in

$$\frac{3.333}{510.0}\ Lin. = tang. \frac{22\ 28''}{288} = 1348.000$$

=4''.680;

which corresponded to the previous product except +0''.069 and to the result of 4''.635 of section 71 and of the observation of April for the sphere's diameter + twice the expansion of the atmosphere, disregarding the change of Pallas' distance on April 4, quite perfectly.

Because the true existence of the visible atmospheric cover mattered, I tried only a 6.5-inch opening for the 13-foot telescope; but still Pallas appeared as nebulous as before and its nebulous envelope appeared as pale as on March 30, April 2 and 3; but this had to be exactly so, if the now measured apparent diameter should equal that of imagined days under the same circumstances as shown above. Pallas had a pale light even at a lesser magnification.

85

The observation of April 4, 1802, is annotated as follows:

1) 24 hours earlier Pallas was not accompanied by a co-star, but now there was one east following north, which was so faint, it was only to be guessed in such a powerful reflector. It is remarkable anyway that this very faint co-star was at the very same position, at which *according to section 71 during the first observation on March 30 also such a very tiny and faint star was discerned, only visible in the 13-foot reflector; it was in the east, following the planet then at 8h 30′ for 5″.4 and at 11h 7′ for 11″.7. Since all other little co-stars, observed in the meantime, would not match this one, it is unlikely that this co-star is a satellite. It still is and remains possible, if you assume that this very tiny, faint, hardly visible little star, which is comparably less brighter than the 2nd old satellite of Saturn in its easterly elongation, but has in the easterly elongation its brightest light*)*

**) See Fragments for a better knowledge of Jupiter's satellites, their characteristics, true proportions of size, rotations and atmospheres, p. 233, Vol. II of the* astr. Beytraege.

and that, just like the considerably brighter 2nd satellite of Saturn, according to our local observations, is partly invisible in its westerly elongation in the 13-foot reflector, this much fainter satellite is invisible during the rest of its orbit. If this is so, an orbital period of 5 entire days would follow. It is at least worth the effort to pay attention to it during future observations.

2) Ceres was compared to Pallas during this observation and it has to be said that in the 13-foot reflector at a magnification of 136 times, Ceres did not have its usual white-reddish, but pale light; through this the often observed change in light of these two planets was proved which is to be found unmistakably everywhere throughout the creation.

86

The weather and other circumstances hindered our further observations until April 9, 8h 30′. At a magnification of 136 times of the 13-foot telescope Pallas had a slightly reddish and rather strong light, but at a magnification of 288 times a white and quite bright light. Because this observation was made only 5 hours prior to the first quarter of the Moon which occurred at 13h 18′, the already quite bright moonlight must have the same influence on the disappearance of the less dense and visible outer layers of its nebulosity, if Pallas' nebulosity equalled that of Ceres.

Due to strong wind an accurate measurement was impossible. In some windless moments, however, Pallas' disc, when projected onto a dark background, appeared slightly smaller than a projected disc of 3.333 lines. I gradually moved it away and soon Pallas appeared now slightly larger then slightly smaller than such a projected disc, what is normally the case when measured and the atmosphere is unstable and one image equals the other without being able to correct further. I found the unknown distance of the projected disc from the eye to be 587.5 lines.

According to the successful experiences in the case of Ceres where the brighter light of the background absorbed the fainter of the atmospheric nebulosity completely – or most of it except for the densest layer surrounding the sphere – it is remarkable that if Pallas' image was projected onto white illuminated background the brighter disc or sphere at a certain distance of the micrometer from the eye equalled a projected disc of 2.80 lines.

87

According to these measurements it follows that

1) the apparent diameter of Pallas, including its denser and brighter nebulosity, which the faint moonlight could neither darken nor make invisible was:

$$\frac{3.333}{510.0}\ Lin. = tang.\ \frac{19\,30''}{288} = 1170.000$$

$= 4''.062;$

and

2) the sphere, including the densest layer of mist directly covering it:

$$\frac{2.800}{587.5}\ Lin. = tang.\ 16'\,23'' = \frac{983''.000}{288}$$

$= 3''.413;$

I had never expected both results to correspond so well to the formerly found rates and the similar course of all measurements of Ceres: because

1) according to page 77, disregarding Pallas' inconsiderable distance of 1.441 on April 9 from the Earth, the already quite bright moonlight made 0".573, and thus 0".286 of the whole or a little more than ⅓ of the tiny outer layers of its nebulous atmospheric boundary invisible; and

2) the measured diameter, projected onto a bright illuminated background, of the sphere equals that found according to section 77 except for 0".170.

As comparatively strong as Pallas' light was under the above mentioned circumstances, similarly striking was the appearance of the compared Uranus in its white pale light but in perfect planetary shape in comparison to Ceres and Pallas, so that the peculiarity of the nebulous boundaries of these two planets could not be ignored.

88

On April 15, 1802, after 9 o'clock we made another observation and measurement of Pallas that I think is equally informative.

Since on April 18 at 3h 28' in the morning there was a full Moon and the observation occurred only 2 days and 6 hours prior to the full Moon, the intensity of the moonlight differed only slightly from that of the full Moon. Pallas was almost at conjunction with the Moon and thus this planet was close to the Moon, so that the strong moonlight had to render the nebulous atmospheric cover completely invisible or except for a very small part directly covering the sphere - just like the illuminated paper of the projection micrometer mentioned in section 86 did, when the planet's image was projected onto it and as it was the case with Ceres at full moonlight. Notwithstanding this, Pallas showed again much luminosity. I proceeded as usual measuring its diameter at a magnification of 288 times at the 13-foot reflector and approached and withdrew the micrometer until the image, projected next to a projected disc of 2.800 lines, appeared first a thread larger, then a thread smaller; but in comparison to the next larger disc it appeared larger by 1/6 = 0.089 lines; I found the unknown distance of the projected disc from the eye = 580.5 lines. According to this the calculation for the apparent diameter

$$\frac{2.889}{580.5}\ \textit{Lin.} = \textit{tang.}\ 17'6'' = \frac{1026''.000}{288}$$

$$= 3''.562;$$

89

On comparing this result to that of 3".413, which according to section 87 was gained from the measurement of Pallas' image projected onto the illuminated white background of the projection screen, when the bright background like the light of the nearly full and very close Moon rendered the nebulous boundary just as invisible, again both results are equal except for 0".149. And if I had refrained from making the correction of 0.089 lines, which was made only because of cautiousness, thus the actual result, obtained from a measurement in strongest moonlight, would only be 3".452 and differ from that of the measurement on brightly illuminated background only by 0".039. Everybody can see clearly anyway that the strongest moonlight because of our illuminated atmosphere can never possess such a darkening power as a solid illuminated object, onto which, like in our case, is projected a nebulously defined celestial body.

Both measurements were made under different but similar circumstances that without doubt made the atmospheric haze invisible except for the densest layer surrounding the solid body. If both results are compared to that one obtained from the strange measurement of Pallas' bare uncovered sphere of April 1 which was 3".243 for its diameter, they differ, not taking into account Pallas' different, but regarding the moonlight inconsiderable distance of 1.492 from the earth on April 15, only by 0".170 and 0".319 in very few and thus inconsiderable decimals of an arc second from it. Everywhere you look in these measurements of Pallas there is true coherence and congruence so that their reliability and accu-

racy cannot be doubted. With this I had to conclude these observations for the time being, because in addition to other business my time was consumed by establishing the new Goettingen Observatory on the instructions of the Royal Electorate Ministerii [he is using the Latin form] *and a journey to Goettingen. After these observations' congruence and accuracy, without any further discussions, I can move on to the physical astronomy – equally important considerations and conclusions on the equally peculiar nature of the planet Pallas, namely:*

I) The determination of the diameter seen from the mean distance of the earth from the Sun and the true diameter of the solid sphere itself.

90

According to section 75 the apparent diameter of Pallas' solid sphere on the first day of April 1802, when it appeared without any nebulosity in a pure, bright and clear light, was 3.243 seconds and later tests and measurements from the 9th and 15th of the same month, when the sphere was only covered by the densest layers of its atmospheric nebulosity, correspond to the result equally well and show its accuracy, as in the case of Ceres, so that undoubtedly the spheres' diameter of both planets can be assumed as certainly determined as the diameter of Uranus and Jupiter's satellites. If you assume the earth's mean distance from the Sun=1, the size of Pallas' diameter is inversely proportional to the distance of the planet from earth; and since according to Herrn Dr. Gauss' elements Pallas' distance from earth was = 1.389 on April 1, 1802, it follows that:

distance	diameter	distance
1.000 :	3".243 = 1.389 :	4".504

and thus the diameter of Pallas' sphere seen from the mean distance of the earth from the Sun is 4.504 SECONDS; and consequently according to the well-known calculation its true diameter is 455.43 or rounded 455 GEOGRAPHICAL MILES.

91

It is particularly peculiar that in the vast solar system this newly discovered planet, according to Herr Dr. Gauss' elements, moves just like all other main planets, at the same great distance as Ceres, between Mars and Jupiter around the Sun, at exactly the same distance from the Sun where we had since long suspected one (but not several) planets and tried to find it; and that Pallas' and Ceres' orbits are inclined towards each other and intersect like two intertwined rings; and that Pallas can be rightfully called a sister of Ceres of the same origin; and that according to its solid sphere its true diameter is greater than Ceres' by 1/4 but is almost equal to that of our Moon but that it is so much smaller than the smallest of the known 7 main planets, yes, its size is according to page 401 of volume II of my Beytraege *even slightly surpassed by the smallest 2nd satellite of Jupiter, whose true diameter has been determined = 456 geographical miles by TOTALLY DIFFERENT ways of determination, cp. p. 390.*

92

If Pallas' true diameter

1) is compared to that of the earth = 1719 geographical miles, consequently 1719/455=3.77 so that Pallas' diameter goes 3.77 times into that of the earth.

2) is compared to that of Mercury = 608 geogr. miles, consequently 608/455=1.33 and Mercury's diameter is 1 1/3 times Pallas'.

3) It equals the Moon's diameter of 468 geogr. miles except for 1/36; but it comes closest to that of the smallest 2nd satellite of Jupiter = 464 miles, except for 1/46.

93

But even more peculiar is, compared to the relatively small size of its sphere,

II) the Nature of Pallas' Atmosphere.

Both Herschel's and the local observations show that this planet truly has in proportion to all other major and minor planets a disproportionately large and dense atmosphere and that such an atmosphere is similar to that of comets, the atmosphere is densest close to the sphere and fades into outer less dense layers, as Herr Dr. Herschel predicated, using the same expressions, which is mentioned in section 46 of his treatise of both planets Ceres and Pallas.

94

If you compare the local measurements of Pallas' diameter they are very useful to determine the expansion of the atmosphere, visible with the local 13-foot reflector, to a tenth of a second or to 1/63 of the entire diameter.

According to p. 71, 76, 80 and 84 the observations were made close together, namely in a space of six days, when Pallas' entire diameter, including its atmospheric nebulosity, was measured on clear dark nights when no moonlight could influence the visibility of the outer thin layers, and the results of the measurements are:

March 30, 1802	*4".635*
April 2	*4".753*
3	*4".611*
4	*4".680*

the mean value is = 4".670.

This mean value is as accurate as possible, because the results correspond among each other so well and three of them have the same decimal and the greatest deviation of April 2 and 3 is only 0".142 and the deviation of such a mean value is only = 0".083. According to Gauss' elements the differences on the following distances of Pallas from the earth are equally insignificant:

March 30, 1802	*1.377*
April 2	*1.395*
3	*1.400*
4	*1.406*

since they daily were only 6/1000 of the mean distance of the earth from the Sun.

95

Accordingly, the above mentioned mean value of 4".670 of the obtained results corresponds to that of April 2, 1802, when Pallas' distance from earth = 1.395 and is useful to determine accurately the size or the vertical height of Pallas' atmosphere and to give general remarks on its density. The distances are inversely proportionate for the entire diameter of the atmospheric nebulosity surrounding the sphere as seen from the mean distance of the earth from the Sun:

distance	*diameter*	*distance*	*diameter*
1.000 :	*4".670 =*	*1.395*	*: 6".514*

so that accordingly the entire diameter seen from such a mean distance is = 6.514 SECONDS,

and it follows further that the nebulosity's true entire diameter including the sphere is 658.68 geographical miles.

If from these 658.68 miles the sphere's diameter of 455.43 miles is subtracted, consequently for the double expansion or vertical height remains 203.25 and thus for the single vertical height, from the surface of the sphere to the outermost tiniest layer of the atmosphere 101.62 geogr. miles.

96

So this is the vertical height of Pallas' atmosphere, as far as it can be discerned with the local 13-foot reflector or an equally powerful instrument, at a distance of approximately 1.400 and under similar atmospheric conditions. Without doubt Pallas and Ceres might appear proportionately smaller with a less powerful visual instrument and with an even less powerful one or at a greater distance might be hardly discernable or not be seen at all, according to what has been said in section 17. And it is just as certain that this heavenly body's atmosphere would be better visible at a smaller distance from earth but under the same or similar circumstances. If you compare both tables of the distances and results of the measurements for Ceres, given in sections 18 and 43, regarding the progressive visibility of the atmospheric nebulosity, to the ones for Pallas and keeping in mind that according to the ratio of Pallas' elliptical orbit and the position in the solar system the distances cannot decrease considerably, I very much doubt that its atmospheric nebulosity will become visible to a considerably larger extent. Everything is limited; and consequently the visibility of atmospheric nebulosities. When Ceres, according to section 19, from Jan. 25 until Feb. 12, 1802, at a considerably greater distance than Pallas, namely from 1.903 to 1.736 within only 18 days was hurrying towards earth by 0.166 of the distance, the peculiar visibility of its atmospheric nebulosity grew by 0".867, disregarding the very different distances. But now the visibility of the outermost layers, growing through this approach, was starting to fade, so that I assume, according to p. 25, it will soon have reached its end. This really happened, for according to sections 43 and 44 the visibility increased from Feb. 12 until March 16, the time of inferior conjunction, within 31 DAYS only by 0".153 and at the time of inferior conjunction the apparent differences were just minor errors in measurement. Nevertheless Ceres' distance from earth still decreased by 0.134, almost smaller by the alterum tantum (as much more).

All this obviously shows that the progressively increasing visibility of Ceres' atmosphere had already reached its limit by the end of February 1802 and thus justifies in the case of Pallas' atmosphere my assumption that any larger expansion would not become visible, all the more since the visibility of the expansion of Ceres' atmosphere had already reached its limit around March 7 at a great distance of 1.613. Without Pallas being observed at a far smaller distance from earth, namely only 1.377 and thus by 0.236 or 1/6 to 1/7 of the entire distance.

If that, what I have uttered at several opportunities about the expansion of atmospheres of heavenly bodies in general remains true, at least according to my opinion, I am thus justified to assume the following sentence:

that the expansion of Pallas' atmosphere, as far as it is visible and perceivable in its modifications and effects to us, inhabitants of the earth, just as it is the case with all other planets, is considerably smaller than the atmosphere of Ceres and is approximately in a ratio of 101 : 146 or of 2 : 3.

97

Nevertheless, it is and remains a peculiar exception that such a small planet whose size is barely the size of our Moon, possesses such a visible and expanded atmosphere which itself is something between a planet's and comet's atmosphere and taking all this into account it appears worth the effort, at least to me,

2) to compare the proportion of density, refraction of rays, dawn and dusk of Pallas' atmosphere to the already known atmospheres of the Earth and the Moon.

If we assume, as we have to do according to all facts and analogies, that Pallas' atmosphere just like that of Ceres and all other planets does not have any considerable inherent light and that it only reflects the sunlight, then it is certain, according to our observations that except for a vertical height of 101 geographical miles, it must be so dense that it is capable of refracting the sunrays and to spread dusk and dawn across the planetary ball because it really reflects the sunrays to such a height; and according to the remarks made in the case of Ceres' atmosphere and the proportion of the vertical height, except for which the atmospheres of our earth and Moon because of their density are able to reflect the twilight in dark night, that the atmosphere of the planet Pallas, reflecting the sunlight high up to 384971 Toise [the old French measure of length: 1 Toise = 1.949 m] *in general is approximately*

a) 10.13 times as high and dense as our earth's atmosphere equally dense only to 38000 Toise and

b) 293.20 times as high and dense as the atmosphere of our Moon might be except for 1313 Toise.

98

What I am predicating here about the density of Pallas' atmosphere, as was done on Ceres' atmosphere in section 66, is only generally relevant for the proportion of the vertical height for which this atmosphere is not yet dense enough; but is by no means relevant for the density near Pallas' surface and thus what has been said about the Melanderhielm Theorem in section 67 is here also true.

99

At least such contemplation serves also here to understand – most probably a natural cause – why Pallas' atmosphere, besides its extremely high density, equal to that of Ceres, is subject to such vital and strange changes, which are not restricted to smaller areas or zones like in other planetary heavenly bodies, but are extensive and thus striking. That such great changes and alternating modifications affect a considerable part of the surface of the sphere really occur in the atmospheres of Pallas and Ceres, cannot only be observed at Slough, Remplin and here with excellent powerful telescopes, but also by the deserving discoverer of Pallas and at several other places with excellent achromatic telescopes, so that there truly is a congruent experience.

Outstanding among the present observations is the fact that on April 1, 1802, Pallas' sphere was unveiled from its atmospheric envelope and appeared stripped of all nebulosity in clear naked shape, but already 24 hours later, on April 2 it was shrouded in mist, pale and dim, just like it had been the case 72 hours before, on March 30, and it continuously appeared with this nebulous face until April 3 and 4, 1802. In section 82 it has already been proven that this was not due to a possible rotation. This was doubtless a fast clearing of Pallas' atmosphere that probably must have covered a very considerable part of the sphere's surface, because the atmospheric nebulosity had become invisible and the bare pure disc appeared without any nebulosity in bright and shining brilliance.

100

Opinions like these will probably lead to further contemplations and studies of that kind; and I would like to express at this opportunity the following sketchy thoughts. If you compare Pallas' and Ceres' atmospheres according to their extremely great vertical height and the thus connected density in proportion to their exceptionally small spheres to other larger planets, and because of their observed atmospheric appearances and modifications (I am considering Mercury and Venus circling the Sun, where maybe the closeness of the Sun had had an exceptional impact on the natural structure and probably the extreme light intensity renders many of the atmospheric changes invisible to us), then, the nature of the atmosphere's height and density of all other planets seem to be expressed in the different sizes and qualities of the atmospheres themselves. According to the above explanation

a) Ceres and Pallas have the highest and densest atmospheres among the planets in proportion to their small spheres. And their obvious atmospheric change is among all others the most striking and vital because their <u>entire</u> spheres appear now pale and dim and then bright and shiny without this being due to rotation; which is not the case with the other planets.

b) The next densest atmosphere is that of Jupiter which in proportion to the earth and the approximately threefold gravity of falling objects, is several times denser than the earth's atmosphere. Into this picture fits perfectly the fact that the sphere in general appears at times pale and dim, and then bright. On the other hand compared to all other planets it is subject to the greatest atmospheric changes so that entire equatorial and polar zones, being the largest part of the surface, sometimes appear in excellent brightness and at other times form in continuous gray light whole stripes partly similar to the monsoons of our earth. Also in accordance with this is that according to volume I of my Astronomische Beytraege *p. 1-137, and in particular sections 90-117, the described observations, the peculiar rotation-independent atmospheric motion of Jupiter's clouds or winds are partly of higher speed than the most violent of our earth; and in perfect harmony with this is on the other hand that bodies on Jupiter's surface are about 3 times as heavy as they are on our earth and that if winds should have the same force and impacts on them, it must be connected to the several times denser atmosphere.*

It follows then:

c) the density of our telluric atmosphere, which we, however, cannot fully and correctly judge in its effects, due to a lack of meteorological data across the entire surface. Nevertheless, according to all observations it is certain that it resembles the strongest of Jupiter's atmosphere in its effects.

d) The atmospheric changes and peculiar movements or winds on Mars' surface, which are quite similar in speed to our earth winds, are described in detail in my Areographischen Fragmenten *which are almost ready to go to press and*

e) our later observations, made with the 10-foot Dollond in daylight, on Mercury and some of its discerned atmospheric spots and stripes will result in a similar corresponding proportion.

f) My observations of the Moon's atmosphere prove that according to many of the measurements of the lunar twilight, it is 28.94 times less high and accordingly less dense than that of our earth. According to this small ratio of density there are never any appreciable atmospheric spots, let alone stripes, visible. Minor variable atmospheric phenomena are primarily to be seen in the valleys and craters where the atmosphere is most dense and usually they stretch over very small areas of several seconds in diameter.

Everywhere in these natural conditions there is a strange harmony; and thus both Ceres' and Pallas' atmosphere deserve to be compared in order to see and admire nature's great analogy and diversity.

Chapter 10
The Scientific Papers on Ceres

C. Cunningham, *Early Investigations of Ceres and the Discovery of Pallas*,
DOI 10.1007/978-3-319-28815-4_10

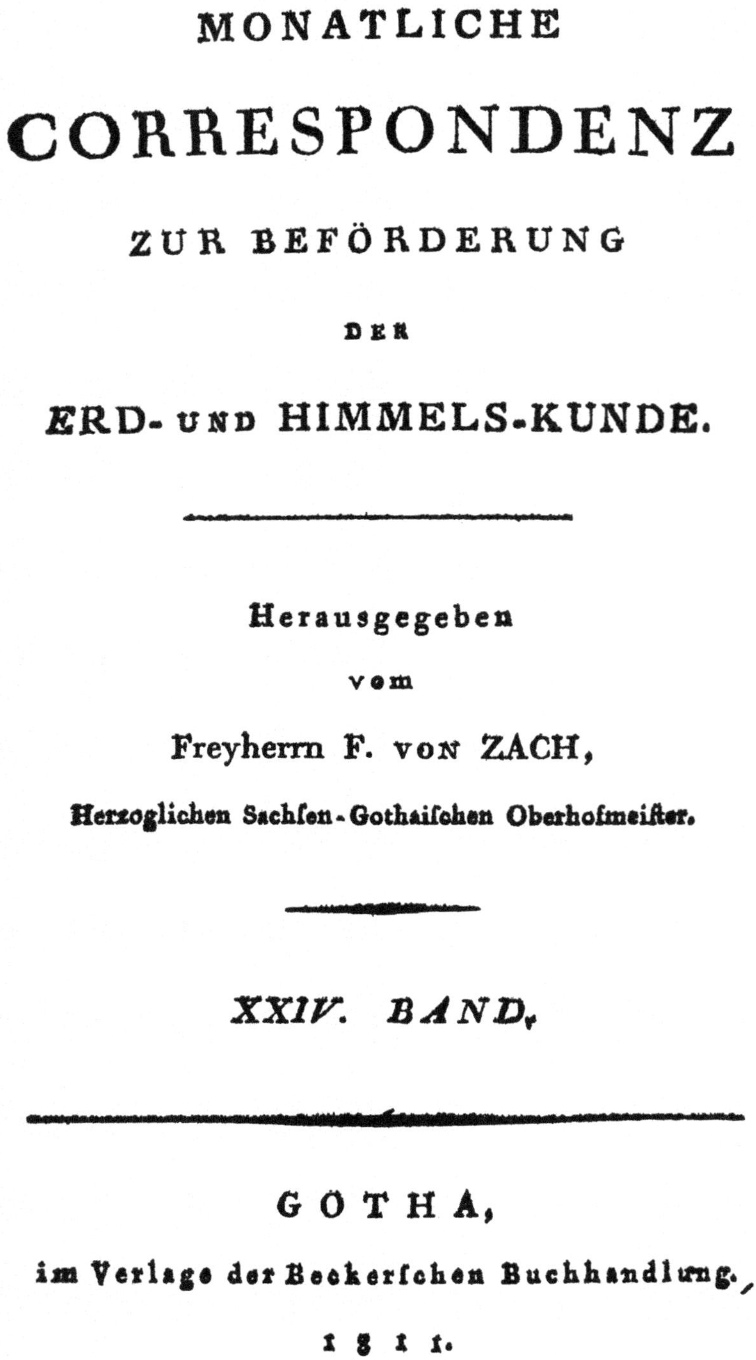

MONATLICHE

CORRESPONDENZ

ZUR BEFÖRDERUNG

DER

ERD- UND HIMMELS-KUNDE.

Herausgegeben

vom

Freyherrn F. VON ZACH,

Herzoglichen Sachsen-Gothaischen Oberhofmeister.

XXIV. BAND.

GOTHA,

im Verlage der Beckerschen Buchhandlung.

1811.

Fig. 10.1 Title page of Baron Franz von Zach's journal *The Monatlichte Correspondenz*

From 1802 to 1807, William Herschel submitted five papers (in whole or in part) about the asteroids to the Royal Society. Together, they comprise the bedrock of the English response to the discovery of the newly found celestial objects. His first report, read before the Royal Society on February 18, 1802, was not published until 1912, when his *Collected Scientific Papers* were edited by John Louis Emil Dreyer (1852–1926). His other papers were published in the *Philosophical Transactions of the Royal Society*. The second study, dealing with both Ceres and Pallas, was read on May 6, 1802. The third, from 1803, gave a description of Ceres and Pallas. The fourth paper, written in late 1804 and presented in early 1805, deals with the third asteroid, Juno. The final paper, from 1807, presents his observations of Vesta, the fourth asteroid. In the first two studies, he examined the color and apparent sizes of the objects, looked for possible satellites following them or atmospheres surrounding them, and estimated their magnitudes. The final two studies were more restricted in nature, concentrating almost entirely on the apparent sizes of Juno and Vesta. The complete papers will be published in subsequent volumes of this series.

It may fairly be observed that Herschel applied his long-held beliefs about observational astronomy to the study of these four new objects. In what might be termed his philosophy of astronomical study, Herschel wrote this about his study of supposed lunar volcanoes:

> *... the phenomena of Nature, especially those that fall under the inspection of the astronomer are to be viewed, not only with the usual attention to facts as they occur, but with the eye of reason and experience. In this we are, however, not allowed to depart from plain appearances; though their origin and signification shoud be indicated by the most characterising features.* (Herschel, 1787: 229)

The Paper of February 1802 About Ceres

Herschel's fine observational skills were well known, a fact that prompted Carl Gauss to urge Herschel to study Ceres and Pallas (Forbes, 1971):

> *I am hoping to give you a large degree of precision with an already existing as well as a (yet to be made) observation and it will need to be decided whether the Ceres Ferdinandea had the same fate as Uranus Georgius, but was not recognised earlier on.*
>
> *I am very curious whether your observations will enrich us this time with Ceres' satellites or other physical characteristics [oddities] of this planet and I would be extremely honoured if in this case you would send word to me regarding this matter.* (Gauss, 1802a).

Herschel needed no prodding, as his curiosity had already been piqued for months. Once he had a reliable ephemeris in hand, he began observing Ceres on February 7. He first saw it in a 10-foot reflector with a 9-inch mirror at a magnifying power of 600x: "I immediately perceived a star which appeared sufficiently different from another [star] at no great distance, to occasion a surmise that it was the planet." (Herschel, 1802b: cix) Herschel did not say at this point why it appeared to be different from a nearby star.

He then changed magnification to 1200 but "... found a doubt still remaining that there might be a mistake." (Herschel, 1802b: cix)

Switching to the 20-foot telescope, which had a mirror of 18.7 inches, he again looked at what he supposed to be Ceres (Fig. 10.2), using magnifications of 300× and 600×. While this still did not resolve his doubt, he was able to secure additional observations through partly cloudy skies over the next few days. This decided the matter: "... the star I had examined must be the new planet." (Herschel, 1802b: cx)

On the matter of magnification, it must be noted that with his 20-foot reflector, a standard magnification was 157×, which gave a field of view of 15′4″. It was with this telescope he discovered 2500 deep-sky objects from 1783 to 1802. He used a 7-foot reflector, with a mirror of 6.3 inches and a magnification of 227×, to discover the planet Uranus in 1781.

For a modern telescope with a well-figured mirror, the maximum usable power is 60× per inch (Clark, 1990: 30). Therefore, Herschel's 9-inch instrument had a theoretical maximum of 540×. Light, detail and field of view would have been seriously compromised at a magnification of 600×. The image must have been quite poor at 1200×, so it is not surprising using such an extreme magnification did nothing to allay his doubts. With the 18.7-inch reflector, a magnification of 900× is theoretically allowable, so the magnifications he used (300× and 600×) were within reasonable limits. On a more practical level, the quality of his optics, stability of his mounting (which was outdoors and subject to wind), and atmospheric stability (which was certainly not spectacular compared to a mountaintop site) certainly put a major constraint on his ability to see small planetary disks. A magnification of 600× can certainly be considered an upper usable limit.

The specific issue of observing Ceres at high magnification was addressed by Baron Franz von Zach in a letter to Gauss:

Fig. 10.2 William Herschel used this 20-foot telescope to study Ceres and Pallas

The high magnifications impede a clear image of the planet. These high magnifications are the reason why Ceres appears so faint and dim, especially at the quadrant, overkill is possible! I saw the planet with my comet searcher and small telescope much better and more distinct than with any of my large and excellent instruments and unfortunately I only have high magnifications and smaller are not easily or quickly obtained. (Zach, 1802f)

On February 13 at 5 a.m., Herschel (Herschel, 1802b: cx) finally saw Ceres with "... great distinctness." He claims to have seen a "...disk, though very minute, being perfectly well defined all round" (Herschel, 1802b). He further estimated the disk of Ceres to be one second in size. Since the actual diameter that Ceres subtends at opposition is only 0.8 arc-seconds, Herschel was clearly seeing more through the eyes of hope than reality. Determined to make a diameter estimate, he concluded, after comparing Ceres to Uranus which was also visible that morning, that Ceres "... is less than five-eighths of the diameter of our Moon" (Herschel, 1802b: cxi). Given that the Moon has a diameter of 2159 miles, this puts an upper limit on Ceres' diameter of 1350 miles. For a first approximation, this is quite good, the real diameter being 578 miles.

Herschel's friend Patrick Wilson (1743–1811) was quite anxious to know the status of the paper about Ceres: "Does your paper on the new Planet come in this week?" (Wilson, 1802a; his underlining). The paper on Ceres was read to the Royal Society the day following this letter, February 18, 1802.

The Paper of May 1802 About Ceres and Pallas

There was great advance interest in what Herschel was going to report about the newly recovered Ceres, as we learn in this letter from Patrick Wilson, who was Regius Professor of Practical Astronomy at Glasgow University from 1784 to 1799. Here he is reporting a conversation he had with Sir Joseph Banks:

Last Sunday's evening I was at Sir Joseph's rooms, when he enquired kindly about you, and expressed some hopes of hearing farther from you as to the new planet by Thursday, in consequence of our having lately some intervals of a clear starry heavens.

P.S. I have received back Piazzi's and Baron Zach's schedules upon the new planet – pray shall I send them out to you in a parcel by coach? Mr. Tilloch has been greatly obliged by them. (Wilson, 1802b)

The Mr. Tilloch referred to was Alexander Tilloch (1759–1825), a Scotsman who established the *Philosophical Magazine* in 1797. He published numerous articles about the asteroids.

In his second paper, which was read before the Royal Society on May 6, 1802, Herschel (1802d) reported on observations he made of both Ceres and Pallas; the former from February 25 to May 4; the latter from April 21 to May 4.

This paper created a sensation across Europe, and became one of the most discussed and derided papers ever printed in a scientific journal. This was caused by two main factors, one observational, the other an attempt at a synthesis of knowledge, which is what we expect of a great scientific advance.

On the observational side Herschel arranged the presentation not in chronological order but under various subject headings. First he was concerned with their magnitude, which in early nineteenth century terminology meant not their brightness but their apparent size.

Next he gave a list of dates on which a search was made for possible satellites of Ceres and Pallas. This is followed by a single paragraph dealing with color. The next two sections deal with their appearance in terms of their disk, and whether or not they are surrounded by an atmosphere or coma.

His controversial conclusion is contained in a section "On the Nature of the New Stars." Going beyond the observational work, Herschel became most famous (or notorious) for the introduction of the term 'asteroid.'

The paper concluded with a brief account of observations made on May 4, 1802, after the main paper had been written. This dealt entirely with the appearance of Ceres and Pallas—whether or not they showed a coma.

Herschel tried to continue his observations of Ceres and Pallas, but the weather did not cooperate, as we learn in a letter from Patrick Wilson:

> *I'm sorry the weather has continued so long unfavourably for your seeing and examining further the new planet. The circumstances of the two lucid points seemed not much to be realized on by the company at the Royal Society on the 18th current – and amidst the conversation of the members, as passing out, that night, I overheard once or twice some wishes expressed for your soon having two or three hours more of serene atmosphere, in order to settle finally the matter of the lucid points.* (Wilson, 1802c, his underlining)

The Paper of June 1803 Giving a Description of Ceres and Pallas

Herschel's third paper that treats the subject of the asteroids was read before the Royal Society on June 9, 1803. This lengthy paper actually concerns itself almost entirely with double stars, but he begins the paper by addressing the issue of nomenclature. Without mentioning Ceres and Pallas by name, he says they can just as easily be called planetoids as asteroids, but gives his reasons why they should not be lumped together with the large planets.

The Journals

Most of the early papers on asteroids were published in *Monatliche Correspondenz* (*The Monthly Correspondence*), edited by Baron Franz von Zach. This is not surprising. In the early nineteenth century the *MC* was the *only* monthly astronomical journal in the world. This contrasts dramatically with the number of German periodicals dealing with the law. Between 1781 and 1810, some 26 were published (Klippel, 1999). Even the medical community had an earlier presence than the oldest science. *The Journal of Pharmacy* was founded in Germany in 1794. Another source of early papers was the *Berlin Astronomical Yearbook (BAJ)*, but this was an annual publication in contrast to the monthly publication of Zach's journal. Papers in the *BAJ* typically appeared a few years after they were written. In French, several extracts are included here from *The Moniteur*.

Also included here from 1802 are papers in the English publications *Journal of Natural Philosophy*, first published by William Nicholson in 1797, and *The Gentleman's Diary*.

We begin with the first paper in *The Monthly Correspondence* about the discovery of Ceres. Since Zach mused about nature itself in this paper, the following lines from an earlier age are an appropriate place to begin:

There is no art delivered to mankind
that hath not the works of Nature
for his principal object. So doth the
astronomer look upon the stars, and,
by that he seeth, setteth down what
order Nature hath taken therein. Sir Philip Sidney (1554–1586)

Regarding a New Primary planet of our Solar System long suspected between Mars and Jupiter and now likely discovered. By Baron Franz von Zach. (*MC*, June 1801, p. 592)

That there must have been a particular primary planet of our Solar System located between Mars and Jupiter which had been impossible to locate until now because of its dim light and small size had, to our knowledge, first been suspected or at least first publicly expressed by the immortal Lambert. In his Cosmological Letters Regarding the Constitution of Planets (which were published in Augsburg in 1761), there appears the following remarkable passage on page 7, right at the end of the first letter: "And who knows, whether or not there are lacking planets which have progressed out of the wide space between existing between Mars and Jupiter." Lambert was probably led to this idea upon comparison of the different distances of planets from one another, and he must have found at that time, that the distance between Mars and Jupiter suddenly becomes entirely disproportionate. To fill this space, he placed a primary planet, and since it hadn't been visible for two hundred years – since the invention of the telescope – had it being torn from the mighty power of the Sun by a marauding comet to continue through immeasurable space as a satellite. He concludes with the statement: "Does the same apply to celestial bodies as it does on earth, that the strong wear out the weak, and are Jupiter and Mars only designated to capture booty?"

What the astronomers of this opinion could confirm regarding the existence of such a planet was a certain relation that the six known primary planets maintained in their distances from the Sun; the relation which was confirmed in an unexpected way by Dr. Herschel in 1781 by a seventh newly found primary planet, Uranus, on the other side of Saturn's path. This remarkable relation was first made known in Professor Bode's second edition of his Introduction to the Knowledge of the Starry Heavens, published in 1772. To represent this in an approximate way and with small numbers, which are easily overlooked, the distance of the Sun from Saturn is divided into 100 equal parts; it follows:

1. *Mercury...........4 such parts distant from the Sun*
2. *Venus............* $4 + 3 = 7$
3. *Earth...........* ..$4 + 2 \cdot 3 = 10$
4. *Mars...........* ..$4 + 2 \cdot 2 \cdot 3 = 16$

5. *Hera or Juno*. .$4 + 2 \cdot 2 \cdot 2 \cdot 3 = 28$
6. *Jupiter*.......... .$4 + 2 \cdot 2 \cdot 2 \cdot 2 \cdot 3 = 52$
7. *Saturn*.......... ..$4 + 2 \cdot 2 \cdot 2 \cdot 2 \cdot 2 \cdot 3 = 100$
8. *Uranus*............$4 + 2 \cdot 2 \cdot 2 \cdot 2 \cdot 2 \cdot 2 \cdot 3 = 196$ *etc.*

Or, expressed more simply, the nth planet calculated from the Sun is distanced $4 + (2^{n-2} \cdot 3)$ *from it. Or one represents, as Professor Wurm has done, the mean distance of the first planet by 'a', the difference of distance between the first and second by 'b', the mean distance of Earth from the Sun = 1: therefore, the mean distance of the nth planet from the Sun is* $= a+(2^{n-2} b)$*. This law is founded on no theory known to us, at least, one hasn't been able to prove it mathematically, and it was concluded empirically out of analogous conclusions. In no science has the human spirit brought forth, solely through mathematical logic and keenness of geometrical reflection, more, more certain and purer truths as in the science of astronomy. If one observes the greatness and loftiness of objects, with which science is concerned, and the smallness of Man and his domicile; when one considers the immeasurable variety and interrelation of heavenly phenomena, which all follow out of one single, very simple natural law of gravity throughout all of creation; when one thinks about the profound mathematical methods and types of calculation which had to be discovered in order to depict the diversely combined phenomena, to manage a certain perpetual correspondence of these calculations with the actual occurrences of the heavens, then the layman as well as the initiate must certainly acknowledge that no science does more honour to the spirit of mankind, that no science has made more a priori discoveries, and that no science is so founded on more solid evidence as the lofty science of astronomy.*

Mathematical astronomers (since there are some who are not) do not accept something that cannot be mathematically proven. However great possibility that the announced relationship of planetary distances, or at least as an approximation of it, could occur in nature, there still were astronomers who doubted the conclusion of this unproved law, and consequently also doubted the existence of an invisible planet to be found supposedly between Mars and Jupiter. It remains characteristic and significant that, to our knowledge, there were no astronomers of any nation other than those of Germany, who tackled this conjecture in their textbooks or who wrote about this subject. How is this to be explained? Was the mind of a great German man, the mind of a Kepler, to rest solely on the Germans! Not that the Germans believed 'implicitly' in the existence of such a planet, or regarded it as proven. Professor Bode made mention of this planet in his valuable textbooks and in all their numerous editions since 1772, but he speaks of its existence as 'suspected' and as an 'analogy', but not as a proven truth.

As early as sixteen years earlier, I had been occupied with the calculation of analogous elements of this elliptically-orbiting [Ger. – *latirenden*] *planet, as one can see in a letter which I wrote to Professor Bode in September of 1785, printed in the* Berlin Astronomical Yearbook *of 1789 (p. 162-3). But I myself explained these investigations as idle daydreams, and dubbed my calculations as chimerical. Jokingly, I compared them with the attempts of the initiates who searched for gold. When I had the great pleasure of once again seeing my valued friend in 1798 in the presence of Lalande in Gotha, we came to speak about this subject. The venerable chief of astronomers had no firm belief in this planet, indeed, he never said a word of it in the three editions of his Astronomie, and I was in the habit of calling those who had a strong belief in this 'astronomical alchemists'.*

Professor Wurm was occupied in 1787 with similar thoughts regarding possible planets and comets in our Solar System, the central idea of which he put forward on the Berlin Astronomical Yearbook *of 1790 (p. 167) and undertook further in the following year's issue (p. 188), which extended into the formation of the system of satellites. But he also remarked that he was far from wanting to impose his astronomical 'raptures'–he expressly described his investigations as such – on anyone as true.*

One has cause to be all the more on one's guard with analogous conclusions, since the significant example of our great master must serve us as an example that one should not place too much trust in one's over-zealous powers of imagination. One may put their play down on paper, however, it must be held it against certain types of natural scientists who propose to catch Nature on the thin ice of mysticism, to replace a generally understandable language with an incomprehensible invented language, and to want to explain 'obscurum per obscurius'. These people, who show their opposition to this, in their opinion, unimaginative and unproductive math, ought really to heed the above cited reference to the likes of Laplace, unless of course they rank this great man with these unpoetic, unaesthetic and unphilosophical minds which they regard with pity.

Kepler, who often took delight in similar astronomical reveries and allowed his fiery imagination free reign, believed to have made a very important discovery that the five regular bodies fit into the spaces of six then-known planetary paths, and indeed, after new observations, their distances from one another did agree with this rule very well. But alas! (as Professor Wurm remarks) Euclid and Nature leave no regular bodies for Uranus, and I add, also for Hera; and with this, Kepler's 'useful ideal' suddenly falls completely to pieces.

One could very well raise the same question here which was posed upon the occasion of the discovery of Uranus – why had these planets not been discovered long ago – and give the same answer as was given by Court Counsellor Lichtenberg [Georg Christof Lichtenberg, 1742–1799], *who found this question not very much more reasonable than that of Lelio's servant in Lessing's oeuvre, who would have liked to have known why his master's father was to return today, and not a year earlier or later, which would have been much more understandable to him.* [Gotthold Lessing, 1729–1781, German dramatist and writer on philosophy; he believed the search for truth is more valuable than the certainty gained by clinging to doctrinaire orthodoxy. Lelio was a spendthrift character in his 1749 three-act farce *The Old Maid.*]

It is therefore the most natural to suspect, as Professor Bode did in his Commentary on Astronomy *that this planet is smaller than Mars, and emits too little light from its surface due to its considerable distance from this planet, which is why it has escaped our keen eyes until now. Who knows the composition of its surface? We know of planets which shimmer with different nuances of colour, like red and green as with Mars and the double star γ in Andromeda, which wax and wane and sometimes even disappear entirely from our view.*

Kant and [Matthias] *Wünsch, in their cosmological writings, are of the opinion that this planet does not exist in and of itself, but that it is incorporated in Jupiter, which is then correspondingly that much bigger than it should be according to the probable rule, because it, so to speak, takes the place of two planets. Kant attributes the smallness of Mars and its numerous satellites to the same cause. But this hypothesis is not required to explain the previous invisibility of this planet. This can be done much more naturally, according to appropriate natural laws. How long was Uranus hidden from our eyes? But it not only stood in the sky, but was also seen and observed, as we now know, by French, German and English astronomers 20, 30 and 90 years prior to Herschel's discovery of it. How could Professor* [Christian Ernst] *Wünsch* [1744–1828] *then make the following remark in the second edition of his* Cosmological Conversations [Kosmologische Unterhaltungen für junge Freunde der Naturkenntniss] *of 1791 'ten' years after the discovery of Uranus: "What kind of body must it have been, which one could not find in such proximity with even the best telescopes, irregardless that one could, with the greatest diligence, peer almost every night at every small spot in the heavens." But should this planet one day be discovered, or be discovered already, it will certainly be very understandable, why this planet, appearing as a telescopic star, could remain hidden so long among an innumerable army of these stars. Professor Wünsch is of the opinion that since, through the best telescopes, the satellites of Saturn and*

Uranus are visible, but are revealed only by a pale reflection, the hidden planets must also have been similarly visible. But Professor Wünsch did not take into consideration that it is entirely different to find one particular, indistinguishable, extremely small moving point of light 'somewhere' in immeasurable space than it is to look for a satellite which must not only be found "always" in the vicinity of its primary planet, but also in the field of the observer's telescope. Since Professor Wurm is a good arithmetician, he estimated the degree of probability and possibility of discovering a satellite or such a planet. It is possible that this planet has been seen several times, like Uranus, but it is also possible that this planet is 'not always' visible. If it reveals itself only as a telescopic star in its greatest perigee to Earth, it could very well be beyond the range of our most powerful telescopes in its greatest distance from the Earth and then disappear entirely, which increases the difficulty and the element of chance in its discovery.

All these obstacles were partly foreseeable, and it was only possible through 'chance', or through a 'systematic order', to find this planet among innumerable telescopic stars. When I undertook a new revision of the stars in Goettingen in 1787, I had the intention at that time to search for this planet, in which endeavor the most Serene Founder of the Uranian Temple in Gotha [Duke Ernst II] *encouraged me. I limited myself only to the stars of the zodiac, and produced my zodiacal star catalogue in right ascension, with the conviction that only in this methodical way was only possible to come upon this hidden planet.*

When, in the Autumn of the previous year, I had the pleasure of making an astronomical voyage to Celle, Bremen and Lilienthal and to spend several happy weeks in the most instructive company of the most commendable and learned German astronomers, it was the opinion of these reasonable men that in order to get on the trail of this long hidden planet, it could not be merely the affair of one or two astronomers to peer at the entire zodiac right down to telescopic stars. Six astronomers gathered in Lilienthal at that time, thus founding on September 21, 1800 an exclusive society of 24 practical astronomers throughout Europe to systematically search for the planet suspected between Mars and Jupiter. They elected the Senior Civil Servant Schroeter as their president, and I was granted the honour and trust to be nominated permanent secretary of this astronomical society. The plan of this society was, among other suggestions, to divide the entire zodiac among the 24 members. Through a draw, each member received a zone of 15° in longitude and 7 – 8° in northern and southern latitude for inspection, and each was entrusted with very watchful supervision. Each member was to draw up a very exact star chart including the smallest telescopic stars of his section, and through repeated revisions was to ascertain the unchanging state of his district or every wandering celestial body. Through such a strictly organised policing of the heavens, divided into 24 sections, we hoped eventually to find a trace of this planet, which had so long escaped our scrutiny, if it did exist and make itself seen. By order of the society, I had invitations sent out to several of the most famous practical astronomers in Europe in the name of the society, and asked them join in this mutual astronomical purpose. Almost all accepted with pleasure. Some members of this society are already at work and have already sent in interesting reports regarding their inspections. Shouldthe honor of the first discovery of this planet be denied our embryonic society, not only will the presumable discoverer of it rankamong the members of our society, the delivery of our invitations being hindered by the disquiet of war, of postal delivery and of ocean travel, but this society has already contributed much and will continue to amend our star catalogues in future. Since this is not the sole purpose in the large area with which astronomy is concerned, the continuance of these will still lend their helpful services.

In February of the present year, 1801, Lalande wrote to me from Paris. Piazzi, the astronomer in Palermo, had discovered, on January 1, a very small comet in the shoulder blade of Taurus. It appeared as a magnitude eight or nine star, without the tail and without any nebula. But since no further details regarding its position and course were reported, such a small star was not to be found. In anticipation of more exact reports, I paid no more attention to it.

In April, I received a letter from Professor Bode from Berlin dated the fourteenth of the same month, in which he had the kindness of informing me that he had received a letter dated January 24 from Piazzi from Palermo, in which he reported to him that he had discovered a small comet on January 1 in 51° 47′ right ascension and 16° 8′ northern deviation. On January 11, it changed from a retrograde to a direct rotation, and on January 23, its right ascension had been 51° 46′ and its northern deviation, 17° 8′. He hoped to be able to observe it for the entire month of February; it was very small, a magnitude eight star, without any nebula. Professor Bode continues in his letter: "The appearance and movement of this alleged comet suddenly struck me upon reading Piazzi's letter; I therefore immediately wrote to him and requested the results [sequence] of his observations. In the meantime, until these are affected, I can't help but to report to him, that I have found through a well-known and simple calculation that both observations from January 1 and 23, as well as the consequent standstill of the 11th, applies well to the assumption that this is no comet, but rather the planet between Mars and Jupiter, which has remained unknown until now, assuming its distance from 2.75 to 2.80. What do you think about this? Admittedly, it's unfortunate that the third observation is lacking. But since the standstill matches the indicated positions so well, the matter has been of great interest to me. Please write and give me your opinion with the next post; I may be mistaken and desire instruction – agreement alone is so very strange. Have you perhaps already further observations of this curious comet?"

Immediately after reading this letter through, I quickly looked up my old books of calculations from 1784 and 1785, found them, and immediately showed Professor Pasquich (who had been present upon the letter's arrival) that the elements of this planet's path, provisionally and analogously calculated as early as 1785 and referred to in the Berlin Astronomical Yearbook *of 1789, had indicated as its distance from the Sun 2.82 and its period as 4.74 years or 4 years and 9 months.* [Johann Pasquich, 1754–1829, Hungarian astronomer, head of the Observatory in Buda from 1806–1824] *I immediately answered Professor Bode that my two provisional elements of this long hidden planet, calculated by means of my 'dreamed' analogies 16 years before, which had been deposited to him in a sealed note in October 1785 when I had the pleasure of becoming personally acquainted with this worthy scholar and friend in Berlin, fully coincided with his and therefore also with Piazzi's observations. I was therefore not alone in the opinion that this alleged comet could actually be the missing and long searched for planet. I also wrote to Oriani, in Milan, from whom I received a letter a few days later, and found that he was entirely of the same opinion that this star was a planet located between Mars and Jupiter. Piazzi himself had informed Oriani (Piazzi's letter to Oriani was written on the same day as the one to Bode, January 24) that he regarded the planet, which was at first regarded as a comet, as an actual planet. The honour, therefore, of not only first discovering this planet, but also first recognizing it as a planet, can accordingly not be disputed by Professor Bode. One ought almost think that he wanted to also reserve the honour for himself (which one should not hold against the first discoverer of this planet) of having first calculated the elements of its path, all the while remaining meagre if not incomplete in the announcement of his observations of this remarkable body. Professor Bode immediately reported his discovery and suspicion to the* Royal Prussian Academy of Science, *had the news of it printed in the* Berlin Newspaper *no. 57 of May 12, in the Intelligensia page of the* General Literary Newspaper *of Jena no. 90 of May 6, and in the* Hamburg Impartial Correspondence *no. 76 of May 13. Out of this, it came to the general public through several other political newspapers.*

A few days after receiving the letter from Professor Bode, and before I could answer him, I received the letter from Oriani of Milan, dated April 7, in which he reports to me: "I just received a letter from Piazzi in Palermo which contains some news which deserves your undivided attention as well as that of other astronomers. He writes that he had observed a magnitude 8 or 9 star on the shoulder of Taurus on January 1, 1801. On the 2nd of January, he found this star approximately 3′ 30″ more to the north, and approximately 4 minutes moved forward against o Aries. On both following days, the 3rd and 4th, he found approximately the same movements. Because of the overcast sky, he could not observe it on January 5, 6, 7, 8 and 9. He saw the star again on January 10 and 11, and later on January 13, 14, 17, 18, 19, 21, 22 and 23. From the 10th to the 11th, its retrograde movement became direct again. He also adds that on the first day of observation (January 1), its right ascension was 51° 47' and its deviation was 16° 8' northerly. On January 23, he found RA 51° 46' and in northerly declination 17° 8'. He continues that he first announced this planet only as a comet, solely because he continuously observed it without a nebula and with a very slow movement. He therefore came upon the thought and the suspicion several times, that this could very well be a planet. Unfortunately, this letter, written on January 24, was in transit for 71 days. It was therefore difficult to guess the position of this new planet from the two positions given by Piazzi after such a long period of time. I have tried in the meantime to use the fact that this heavenly body switched from retrograde to direct on January 10, and have, in assuming an orbit, found that its distance from the Sun must be three radii of our planet's orbit, so that this star could very well be a new planet, the path of which must fall between Mars and Jupiter. It is presumable that the path of this planet, as all others, will have a notable eccentricity, and consequently, the hypothesis of a circular path, which I have assumed, must be insufficient to correctly represent its movement and geocentric position after such a long time. We must therefore wait for the later observations by Piazzi, which he will have continued. The sky was always overcast for us since receiving Piazzi's letter; perhaps you have a more favourable sky for astronomical observations than we do. With this conviction, I now send you my elements of the path calculated from, as you have undoubtedly seen, incomplete observations, through which you will be able to calculate the approximate position of the planet. Heliocentric longitude of the body on December 1800 2Z 6° 54', heliocentric longitudinal movement in 100 days 18° 9', longitude of the ascending node 3Z 8° 32'. Inclination of the path 3" 50'. [The letter Z is an abbreviation of a zodiacal sign. One Z is 30 degrees.] *But as said, these results are subject to grave doubts, since they are based only on two incomplete observations and on the extremely inadequate hypothesis of orbital movement. In the meantime, I flatter myself with the thought that this letter will soon come to you, and before this planet loses itself in the rays of the Sun; perhaps you will be fortunate enough to find it with your superior instruments, and then share with me more precise news..."*

Immediately after receiving Oriani's letter, I calculated the position of the planet with his elements and searched for it in the heavens during various clear evenings, but unfortunately, the news came too late; the small planet had moved too close to the Sun, so that it soon sunk entirely in its rays and in the haze of the horizon. I later discovered that Oriani, in his great haste and desire to share his news with me as quickly as possible, must have been mistaken in the calculation of these provisional elements (as I later found), especially in the ascending node and the inclination of the path. But even without this error, I would not have been able to find this small planet, since when I received the letter, the dusk must have already been too strong and the star too close to the horizon to make such an uncertain search possible. Professor Bode had also, as he told me in a letter dated May 12, searched for it several evenings in vain.

Since there remained no hope at this time to observe this strange planet until its return from the Sun next August or September, I got down to the calculation of its path as well as I could in the meantime, given the incomplete and imperfect nature of those few observations reported. I did this not with the intention of locating this planet again within 2 or 3 months when it would return from the Sun, since I hoped to receive better and more precise information from Piazzi and his continued observations from February, March and April before then; I did this rather because I wanted to be able to hazard some conjectures beforehand, with somewhat more confidence, regarding the actual existence of a planet to be found between Mars and Jupiter.

Piazzi's reported observations are for the calculation of a path partly incomplete and partly inadequate. 1) His two known observations are reported only in minutes and are therefore only approximate. 2) At least three observations are required to calculate the path of a comet or a planet. 3) The times of the observations are not given. With the first one, one may assume that at least the 'next' minute of the observation is correctly given. As far as the second difficulty is concerned, as we have already mentioned above, Piazzi perhaps intentionally withheld the third observation, perhaps because he wanted to first calculate the path of the planet himself (since he did regard it as such before January 24). But if this was the case, then he did give away his third observation in a way, in that he referred to the circumstance of the planet's standstill from the 10th to the 11th of January. Griani and Bode knew how to use this circumstance – and I likewise used it – to calculate a more precise path of this planet from these sparse observations. The third difficulty was to be treated by means of the following conjectures: As we all know, Piazzi is concerned with drawing up a great star catalogue. He was favoured with great fortune, which I had hoped for upon completion of my star catalogue, and for which the Lilienthaler Astronomical Society *had methodically calculated, and the planet would certainly not have escaped them had not Piazzi beat them to it. Piazzi was provided in his observatory with a splendid transit-instrument and an entire meridian from* [Ramsden], *with which he undoubtedly came upon this small planet. This planet must have reached its culminating point on January 1, the date of the discovery, at approximately 9 o'clock in the evening; in this season and at this hour, Palermo is completely benighted, and because of this, Piazzi could very well observe this small magnitude 8 or 9 star even at midday. There was a full Moon, and it stood at the horizon of Palermo for already three hours; therefore, it must have been a very bright night. But the Moon stood over four hours or approaching 68° distant from the star. This circumstance alone gives rise to my suspicion to the point of being almost a certainty, that he did not discover this supposed comet in a free handed way, with a comet finder for instance, but must rather have found it with a meridian instrument. The right ascensions of the planet cited above converted into time were therefore converted from its culmination time to its sidereal time; in turn, I have converted these into mean solar time, thus bringing out the actual moments of observation of this planet. With this, and with the inclination of the ecliptic 23° 28' 10", I have obtained the following data in the calculation of the path:*

1801	*Mean Time in Palermo*	*Geoc. longitude*	*Geoc. ll latitude south*	*Place of the Sun +20"*	*Log. dist. Sun-Earth*
January 1	*8 43' 15"*	*1 23° 29' 40"*	*2° 37' 5"*	*9 11° 1' 40"*	*9.9926158*
23	*7 16 41*	*1 23 43 40*	*1 38 50*	*10 3 22 28*	*9.9932351*

Since the heavenly body came to a standstill on January 10, I found the elongation of the planet for this moment to be 7 Z 26° 41' 41", and by means of Keill's [John Keill, 1671–1721] *theorem found that the tangent of elongation equals the radius of the path divided by the square root of this radius + 1, this radius itself = 3.071, and with this, by dint of Kepler's theorem, the orbit (3.071)1.5 = 5.3817 years. According to well-known methods, I further found the following more precise orbital elements:*

Epoch mean heliocentric longitude in 1801	*2Z 6°*	*55'*	*40"*
Longitude of the ascending node	*2 22*	*0*	*50*
Mean annual movement	*2 6*	*54*	*25*
Inclination of path	*7*	*47*	*40*
Orbital radius	*3.071*		
Synodic period	*5.3806 years*		

If one looks more closely at these elements and compares them, it is most striking and curious that the distance and orbital time of Piazzi's star are exactly the same as those of the famous comet of 1770, the calculation of which gave astronomers so much to busy themselves with, and the course of which Lexell could represent as nothing other than an ellipse of 5.5 years. Pingré [Alexander Pingre, 1711–1796, French astronomer] *also found its distance as 3.09 and the orbit as 5.4 years. Burckhardt, who just recently took the prize at the Parisian National Institute for this repeated investigation could similarly do nothing but submit to an elliptical path of 5.5 years following newly conducted observations of this strange heavenly body. Was this comet from 1770 the long-hidden planet? Or is Piazzi's star perhaps the comet from 1770? In both cases, whether the body be a planet or a comet, why hasn't it been seen observed more often and have been discovered long ago? The causes, a few of which we have cited above, may be various. But, to remain with only one possible and adequate type of explanation, one need only read the observations of this comet for 1799 carried out with physical consideration by Schroeter, published in volume 3 of his* Articles Concerning the Newest Astronomical Discoveries. *In this, the strange appearances of the unpredictable, changeable modifications of the nebula surrounding the comet as well as the known atmospheres of the planets, including that of the Sun, are explained with regard to the evidence. In the nebula surrounding the comet from1799, Schroeter observed not only unpredictable changes, but also rapid and quite disproportionately great changes both in the extension as in the light of this photosphere. Dr. Herschel recently presented the* Royal Society *in London with a treatise regarding the constitution of the Sun, in which this great astronomer explains Sun spots as an elastic non-radiant gas which produces on the surface of this opaque solar body the radiant fluid which surrounds it, or divides this photosphere, through which we may then see parts of the dark body, which we perceive as spots. Could it not be the case that we see a body at one time which is lost to us in another time? We know several of these which have a periodical light change of varied duration, as, for example, the many variable stars.*

What ever happened to the famous Tycho star in Cassiopeia from 1572, which suddenly shone as brightly as Sirius and even exceeded Venus and Jupiter in brightness in their perigee, so that they could be seen with the naked eye on the brightest day, and then completely vanish after two years, after which time not even the slightest trace of it could be found. In old chronicles, it is reported that during the time of King Otto I (in approximately 945 and also in 1264), a new and unmoving star had revealed itself between Cassiopeia and Cepheus. Afterwards, some astronomers came to the suspicion that it had been the star from 1572, which appears approximately only every 300 years. What similarity does this have with the equally famous discovery of a star on the foot of Serpens in 1604, which suddenly became so bright that the star's brightness exceeded that of a magnitude1 star and then completely disappeared the following year and hasn't been seen since? What are Herschel's planetary nebula spots? Didn't Schroeter find that very large layers of nebula had disappeared from the sky! What are our great fiery and radiant balls which exceed the speed of a cannon ball's flight many hundred times, which streak past, shine, burst and disappear? It seems to be agreed that they are bodies from space and not from the atmosphere. Montanari [Geminiano Montanari, Professor of Mathematics at Bologna, later Prof. of Astronomy at Padua] *believed this as early as 1676. In his Fiama volante, he calculated the height of this fireball, seen throughout Italy in that particular year, as 40 Italian miles. The great fireball, seen and relatively well observed throughout all of Europe is probably not fresh in the memories of our readers. This giant fireball was calculated at a height of 60 English miles by English astronomers and natural scientists, and 1.16 miles in greatest diameter. It must have run through the diameter of the Earth in 7 minutes.*

Why shouldn't comets also be radiant at certain times and dark at others? The comet of 1770 could therefore after all be opaque at one moment and in a phosphorescent state in the next. Perhaps the rarity of their return could be explained in this way, as well as by the power of disturbance [perturbation] *of the larger and denser bodies. They return, we do not see them; they are there, we do not recognize them. Determining the identity of a body from the identity of the elements of a body's path alone is, as all astronomers know, only possible with a high measure of certainty, in only one case among 91 calculated comets, through one confirmed hypothesis which can only be valid when applied ' directly', but not 'the other way around.' "Two comets, which do not have identical elements of their paths, could very well be the same body." Admittedly, to my knowledge, no one has explicitly asserted this, but Lexell has already said this implicitly; when regarding the 5 year elliptical path of the comet of 1770, the question was raised as to why this comet, which has such a short orbiting time, hadn't been seen long ago and more frequently. Lexell was of the opinion as was likely his great teacher, Euler, under whose supervision Lexell worked that the influence and perturbation of enormous Jupiter, very close to which this comet passed on May 27, 1767 and August 23, 1779, "could have completely altered its path." Burckhardt is also of the same opinion in his prize article about this strange body* [a reference to a prize offered by the Paris Academy, which he won in 1801. The topic of Academy prizes and the orbit of Pallas is examined in the next volume in this series].

But how frightening is the mere thought of wanting to calculate the perturbations of such a body! Wouldn't such an undertaking exceed our powers of analysis? The so very difficult theory of the Moon would seem like elementary calculus against that of a body having such a changeable path. It is to be hoped that such an obvious appearance would bring about the need to give our perturbation calculations a new direction, so that the difficult theory of approximations could become more complete, and so that the influence of successive integrations would be better and more precisely determined on the neglected quantities.

But the geometrician, who could represent 'all' coordinates of the movement of every body in the quickly approximated series of the sine and cosine of the angle, which depend on its true movement, is perhaps not of this earth.

That the comet of 1770 could not be a planet – for this, some will want to find proof in the nebula which this body has had, resulting from Messier's detailed observations. But is it therefore proven that planets may not and cannot have nebulae? We have planets without satellites and some without this retinue. We have planets with two and several rings. Why should there not also be a planet with a nebula? in order to prove that comets are nothing other than planets. This difference in the designation comes out of the times of ignorance and must exist today only in the use of language to differentiate between those bodies whose appearance is of short duration and who do not remain visible throughout their entire path and those bodies which are always visible provided they are not situated next to the Sun. Even the state which otherwise seems unique only to comets, that some among them move retrogradely while all planets move directly, is only illusory. One may refer to the explanation which Laplace and Lalande have given in my A. G. E. *II B. S. p. 170 and p. 259 upon the occasion of the retrograde satellites of Uranus. Lalande states: "The word retrograde impresses through its expression; in reality, it is nothing." In 1755, Kant conjectured in his* Universal Nature and Theory of the Heavens, *p. 58 of the original edition, that the retrograde movement of some comets could be only an optical illusion like that of the geocentric movement of planets.*

Every newly discovered object must also receive a new name. A name in itself is obviously nothing of great importance. In the meantime, we have seen with Uranus, how solidly it held and still holds to unite all voices in the commonly accepted designation, from the Thames to the Neva. If the new body discovered by Piazzi really is the suspected primary planet found between Mars and Jupiter, then it has already been given, I think, a fitting and also happy name 15 years ago by a great and serene protector and patron of astronomy, the founder of the Uranian Temple in Seeberg. Uranus has once given us the increased right, for the sake of uniformity, to take the name of this new planet, as all the old ones, from mythology. The Duke of Gotha suggested the name Hera, the goddess the Romans gave the name Juno, daughter of Saturn, sister and wife [consort] *of Jupiter. Jupiter would therefore have his parents and ancestors above him, and his wife and children below. The Greek name Hera is preferred to the Latin name Juno, because*

1) the latter has been ascribed to the planet Venus. Pliny, in his Natural History *Lib. II C. VI writes: Below the Sun walks the great star some call Venus...others call it, however, Juno, others Isis, others the mother of God. St. Augustine in* The City of God *Lib. VII Ch. XV names Venus the 'Star of Juno.'*

L. Apuleius, in Of the World *p. 252 (Edit. Bipont.) says Juno, on the other hand, esteems (to be) the Star of Venus. Hera is always shrouded in clouds; our planet also remained long hidden. The name will still remain fitting even if this new body turns out not to be the suspected planet; we will have included the stars instead of the goddess.*

2) Hera is at the same time the name of a city in Sicily, through which the memory of the discoveries made on this island and the glorious name of the discoverer of this eighth primary planet will be contained and immortalized, as long as there exists tradition and history on our world. This city of Hera, located as Palermo is on the coast of Sicily, is otherwise also called Hybla minor, and it is of this that Pausanias speaks of in Eliacis *Lib. VI Ch. VI and Cicero speaks of in his letters* ad Atticum *II 1, and which comes up in the Itinerario Antonini. Finally, Hera is also the mother of Vulkan, who has his workshop in the fire-spewing mountain of Etna on this island.*

The objection that had been raised upon the naming of the planet Uranus that all planets bear Roman and not Greek names, can occur all the less in this case, since with the discovery of the new planet by Herschel, the Greek name Οὐρανός *was maintained since it would have been commensurate with the then unbroken analogy to name it according to the Latin,* Coelus. *This may have occurred as a good prognostic. All older planets, the discovery of which are lost in the darkness of time would therefore maintain their Latin names. All newer planets, the genesis of which will be put down for the latest posterity with the name of the discoverer, should bear Greek names to make the distinction. What seemed a break of analogy now becomes harmony. If ever a planet be discovered beyond Uranus (the area of creation has, like its creators, no limits), the Greek would then be its hieroglyphic designation.*

A fitting designation of this new planet will now have to be taken into consideration. A newly discovered planet can reasonably be given the sign of a newly discovered metal. This idea was followed with Uranus, but through this one perpetuates an error, or rather sets our former ignorance regarding the components of platinum as a memorial. Therefore, the suggestion to designate this planet as such is more fitting: the lower part indicates a planet, the upper, a fixed star; it should namely indicate a fixed star which has become a planet. But since this sign has already been introduced and is being used in the Viennese Astronomical Ephemeris, *one need only reverse it to avoid misinterpretation. Earth and Venus likewise have turned symbols. Mars and Uranus have likewise very similar hieroglyphs.*

The Continued Reports Regarding the New Primary Planet of our Solar System Long-Suspected between Mars and Jupiter and now likely discovered (*MC*, July 1801, p. 53)

On the 16th of May, I [Zach] *received a letter from my highly respected friend Dr. Olbers from Bremen, in which he enquired about the unexpected, great newspaper report that recognised the common comet discovered on the first day of this century, previously announced as a moving star, as the eighth primary planet of our Solar System which was missing until now; he requested some more detailed information regarding this great astronomical event to satisfy his thirst for knowledge. I had anticipated Dr. Olbers's request, since I myself knew well what a gain it would be for science to forward these observations to such a learned and astute astronomer as Dr. Olbers as soon as possible. And I actually received an answer from him on May 30 in which he had calculated and communicated to us new elements of an orbit for this planet from the two observations by Piazzi on the 1st and 23rd of January, of which he had been informed.*

"From such a small star," writes Dr. Olbers, "without any nebula, slowly moving so closely to the ecliptic, it was easy to guess at a planet. However, the credit remains Piazzi's to have not only discovered this new planet, but also to have recognised it as such. Would Piazzi have robbed our budding society of the honour of having discovered a new planet? Because our society certainly would have found it, had it begun working according to our plan, since a moving, magnitude 8 star would not have easily escaped notice."

From the two observations sent to him and known by him alone, Dr. Olbers calculated, assuming an orbit, the following elements of a path, but naturally remarks that these are not to be determined with any reliability, since the observations are only separated by 22 days and are only supplied in minutes. The visual lines are also not favourable. However, he found the following data under these probable conditions:

Orbital radius	*2.947465*
Longitude of ascending node	*2Z 21° 55′ 10″*
Inclination of path	*7 54 38*
Heliocentric longitude on January 1, 1801	*2Z 7° 40′ 36″*
Sidereal period	*5.04096 years*
Daily heliocentric movement	*11′ 43″.87*
Annual movement	*71° 24′ 57″.6*

"With these elements," continues Dr. Olbers, "one will be able, with difficulty, to calculate the position of this planet in advance, so that it will be possible to find it upon its reappearance in the morning in August, if it really doesn't differ from a magnitude 8 star in merely looking at it, since it probably has a not undistinguishable eccentricity. In opposition, it can probably grow in light-intensity to a magnitude 6 star. I hardly doubt that it will be found and that it has been observed among Lalande's stars. I am therefore very hungry for any further observation which Piazzi will make known."

On May 16, I received an answer from Professor Bode, in which he informs me: "It was most agreeable to discover from your letter that you are of the same opinion as I regarding Piazzi's comet, and that Oriani also, and even Piazzi himself, support this. How many times have I not wished that I could experience this discovery.... I have often been laughed at regarding the certain harmonious progression in the distances of planets from others.... Assuming the distance of 2.75, I find that the heliocentric difference in longitude between the 1st and 23rd of January corresponds to the observations very well; the planet reaches its node, which I set in Taurus, its inclination must have exceeded 6°, and in this I also find a reason as to why this planet hasn't been found. If the existence of this new planet is confirmed as a consequence of several observations which I impatiently await from Piazzi, I will not hesitate to immediately share them with you..."

Until the end of May, I received no further reports of this body. In the meantime, I had, at any rate, given my Parisian friends reports about it and shared our elements of the path with them, and since I certainly suspected that Lalande, who had received the first report of the comet from Piazzi, had received the continued observations and had also shared his suspicions of a planet with him. I therefore requested the observations of this planet from Lalande, which would have come to his knowledge. But not without a little displeasure did I receive, at the beginning of June, several letters from Paris, from Senator La Place dated May 29, from Lalande and Burckhardt dated May 26, from De Lambre dated May 24, from Méchain dated May 26, from Henry dated May 28, and not one of these six astronomers, who had informed us of several important observations and new discoveries, ever utters a syllable regarding this new planet! Only Méchain mentions Piazzi's comet in his letter, which proved to me beyond a doubt that nothing yet was known in Paris of this new, suspected planet at the end of May, whereas we in Germany already had knowledge of it in

March through Professor Bode. Méchain merely writes: [trans. from French]: *"Have you already seen the comet, which the journals have announced as having been discovered in Palermo last January? No one here has seen it. Our astronomers have found nothing since the one of December 1799. Sometimes I search for it, but without success."*

On June 10, I once again received a letter from Professor Bode, in which he had the kindness of telling me the following: "I received Piazzi's first letter on March 20, and I answered it on the next post. However, he did not wait for the response, but – imagine my pleasure and at the same time my frustration: I received a letter from Piazzi, broke it open with anticipation, and only found the following regarding the new star, which I now faithfully offer you: [tr. from French] *'I wrote to you in January, announcing a comet which I discovered in Taurus and which I followed until February 11 until I was attacked by a serious illness, from which I am not yet entirely recovered. If I can restore myself to health, I will calculate the elements which I will then send you. In the meantime, I am informing Lalande of my observations.' He therefore simply announces that he observed the star, which he still called a comet as in his first letter to me, until February 11, and then became ill, without communicating the observations.*

On June 18, I received the following letter from Dr. Burckhardt in Paris: "I hasten to inform you of what I have learned regarding Piazzi's comet until now, as incomplete as this information is: however, I hope to send you the continuation of my investigations in the next mail. Lalande received Piazzi's observations on the evening of May 31. I immediately began to calculate its path. Two days later we received your letter with your and Oriani's investigations, which led to the hope of having found a planet in this body. My investigations had already revealed to me that the described arc was not considerable and I therefore believed that a parabola must necessarily have met with your satisfaction. The slight geocentric and heliocentric movement of this comet made me no end of trouble when determining the orbit. I first chose the observations from January 14, 21 and 28 and found it necessary in this circumstance to choose the most distant observations, namely those of January 1 and 21 and February 11. During these 41 days, the comet changed in its geocentric longitude by only 3° and in its heliocentric only by 10½°. When I wanted to improve the parabola found through my method by means of Laplace's method, I found that the equations of condition provided no means to do this. I attempted Laplace's method of approximation, but with just as little success, as I could have foreseen, since the unavoidable errors in the observations have a great influence on the differences of heliocentric longitude and latitude. I then tested 8 hypotheses by means of Laplace's method of improvement without ever coming closer to the truth. I then calculated the following orbit which meets all three observations to ± 2½ minutes:

Orbital radius 2.74
Epoch 1801............................2Z 8° 16′ 20″
Ascending node.........................2Z 20° 15′
Inclination of path..........................11° 21′
Period....................................4½ years

"As varied as the experiments which were tried until now were, they did not prove that there is no possible parabola for these observations. I decided to use a method which has often succeeded where all other methods of interpolation have failed. As soon as the equations of condition are so that they cannot be made equal to zero without giving both variables extremely improbable values, one must therefore be satisfied to change one of the variables until a hypothesis has been formed in which both errors are equal and opposed, using whichever error is then the least possible which can be obtained while maintaining the variables. The latter value is then changed and the former value is determined anew through attempts of making both errors equal and opposed. The change in the absolute values of the least possible error in both cases indicates which change has to be made so that the value of the least error becomes zero. For instance, in the case of Piazzi's comet,

I made the logarithm of the distance from the Sun = 0.438; the least error was ± 4′. I therefore realised that I had to further decrease the distance; I formulated the following parabola after 20 hypotheses:

Position of the ascending node..........2Z 20° 50′
Inclination of path................................ 9° 41′
Position of perihelion...................... 4Z 8° 38′ 25″
Smallest distance from the Sun...........2.21883 its log. 0.3461250
Log. of diurnal movement.................9.4409408
Time of transit in proximity to Sun..1801 June 30, 19:00 1′

"This parabola satisfies the three observed longitudes. However, it is not possible to represent the three latitudes through them. The errors in longitude are −1′ 47″ and +38″ on the 14th and 28th of January.

"I believe to be able to ensure that there is no parabola which satisfies these observations better. Piazzi wrote nothing about the accuracy with which he was able to observe this comet. My inclination differs greatly from your and Oriani's determination. This stems from the first observation, in which you and probably also Oriani were sent the declination approximately 30 minutes too great. For this and other reasons, I have requested a new, entirely reliable copy of his observations from Piazzi. It will then be shown whether something more exact can be found regarding this peculiar body, which will still be very uncertain however, since the described arc is only 10°."

On June 21, the letter promised by Dr. Burckhardt arrived with the following contents: " I am sending the continuation of my investigations of Piazzi's star as promised. I have spared no effort in searching for an ellipse, although the arc described is still too slight to be able to hope for great accuracy; but I believe to have promoted and simplified the finding of this body through this.

Position of ascending node.................2Z 20° 58′ 30″
Inclination of path..................................10 47 0
Position of aphelion............................2 8 59 37
Time of transit through the aphelion January 1801....1.3328
Eccentricity............ 0.0364
Log. of semi major axis....................0.4106586

"This ellipse represents the longitude and latitude of five observations to a few seconds: greater accuracy could easily be achieved, but this is superfluous since the described arc is too slight. To get an idea of the components belonging to the changes, I have decreased the position of the aphelion to 45°, or set the true anomaly on January 1 to exactly 45°, then the eccentricity is 0.0344 and the logarithm of the axis 0.41544. The period, 4.20 years. I have made some attempts to decrease the position of the aphelion to 90° to 100°, but without success.

"This ellipse gave me the following determination: and I sincerely hope that all friends and enthusiasts of astronomy will want to concern themselves with finding this body, though we in Paris will neglect nothing to discover it." The subject is too important to not earn the united efforts of all astronomers. It would have been much better had Piazzi announced his observations earlier; it would then have been discovered earlier and have been observed longer. Sickness, namely, forced Piazzi to give up his observations on February 11. [Table: positions of the newly discovered body by Piazzi, see *MC*, p. 62].

In order to make finding this small body even easier for the enthusiasts of astronomy, we have drawn up the following instructions here, for those who are provided with no instruments other than a telescope, as well as for those who can place their telescope on a parallactic tripod. First, you will find enclosed at the end of the book a hurriedly drawn map which sketches the entire course of this body from July 17 to September 18 (Fig. 10.3).

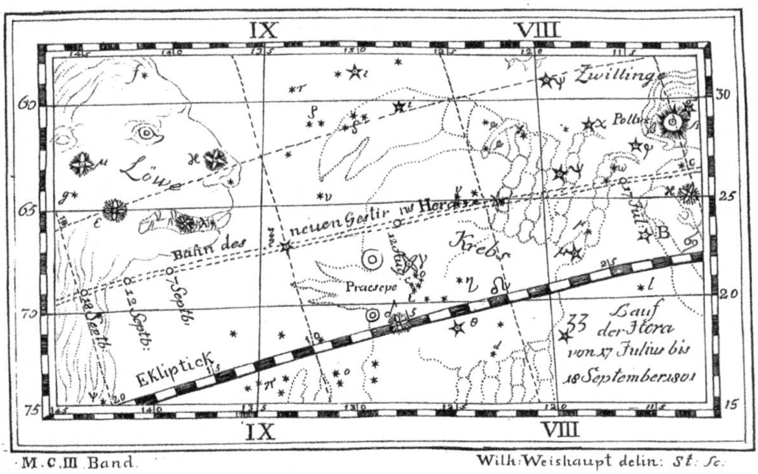

Fig. 10.3 The path of Ceres, as depicted in a chart published in the July 1801 issue of the *Monthly Correspondence*. Ceres is denoted on the chart as Hera

Hardly would such a small star, as the suspected planet reveals itself, be found any earlier, since it will rise on July 17 in our northern regions of Germany at around 3½ o'clock, and the rising of the Sun follows at 4 o'clock. It follows that such a small and indistinguishable star, so close to the horizon and during the light of dawn will not be easily discovered. From July 17 to August 12, this body will pass through the constellation of Cancer. Until August 12, the planet will be easier to discover. It will rise approximately one hour before the Sun and will come to be situated one and a half degrees over the northern donkey γCancer [the 4.7-magnitude Gamma Cancri, or Asellus Borealis is known as the Northern Donkey], *and will be almost parallel between the stars λ Cancer and λ Leo. On August 25 and 26, it will pass very close to the star ξ Cancer. On September 7, it will come parallel to γ Cancer and γ Leo, and will have entered the constellation of Leo; on September 23, it will come 2½° under x Leo, and will come to stand 1½° under γ Leo on October 5. By this time, this body will probably have long been discovered and will have been observed by astronomers with better instruments. But on October 5, the planet will rise at 1:30, therefore 5 hours before sunrise, in pitch-black skies. It will therefore be exceptionally observable during the new Moon. It is unnecessary to say more about it since the next August issue of our Monthly Correspondence will probably anticipate this and offer more detailed information.*

For those who have the possibility of parallactic movement with their telescopes, we will supply the following calculated right ascensions and declinations for easier and quicker location of this new body. Those astronomers equipped with meridian telescopes will be able to observe this small body only with difficulty in the meridian before the end of November. [a table of data is included here, see MC p. 64.]

That a new planet would be conferred several new names was to be expected. In the Leipziger Allgem. Literar. Anzeiger [May 11, 1801] *No. 72, an unnamed source* [who wrote he was "not astronomer by profession, but a zealous friend of astronomy"] *suggested the name ' Vulkan.' He believed that it would not be improper to give the god who forged the weapons of Achilles a place in the sky next to the god of war, the husband of Venus next to her lover. Vulkan would also not be able to complain that the honour was paid to him too late and that such an inconspicuous planet had been given his name, since he himself, due to a small mistake on the foot, is not fleet of foot or otherwise of splendid form. Vulkan, as the son of Jupiter, belongs to the family and has, in this respect, a well-founded claim to the honour intended for him. Doctor [of medicine] and Professor [Johann Albert Heinrich] Reimarus* (1729–1814) *in Hamburg is of the opinion that it should be called Cupido. Because it was once established that planets be named according the gods of antiquity, he would therefore be (counting from Venus downward) the next from Mars, a lover of Venus. Others believe that the name Cupido is also fitting because the name is associated with the idea of blindness. The new planet appears only as a magnitude eight star and cannot be seen with the naked eye. But should the planet be confirmed, the question of a name will be decided by the majority, and perhaps even by chance. It is also possible that a general consensus will never come to be, as was the case with Uranus. In Italy, the name Ferdinandeum Sidus is maintained, in France, 'la planète Piazzi', and in the rest of Europe, some kind of mythological name, until time and circumstances will completely decide. Fortunately, the name doesn't matter if this important victory only remains in space – a victory which, as one of our most worthy and ingenious scholars states, does not cost blood and tears as is the case in political conquests in this small Earth, but rather only human diligence, the spirit of observation and keen perception.*

In the most ancient times, to designate the order of planets in their distances from the Sun, they were incorporated in Latin verses of commemoration. Thus, we have, for instance, the ancient, popular, but mistaken verse: Saturni atque Jovis sidus, Mars, Sol, Venus alma, Mercurius, claudit utima Luna chorum.

When Herschel discovered the new planet beyond Saturn, Poinsinet de Sivry [1733–1804] *wanted to have it named Cybelle, according to the spouse of Saturn and represented the order of the seven planets in the following verse:* [refer to table, *MC*, p. 66]

One of my friends expresses the order of the now eight planets in the following not unsuccessful verses, which, according to the custom of usual memorial verses, expresses a further thought: [This Latin verse was published in *Discovery of the First Asteroid, Ceres.*]

Continued Reports Regarding the Long-Suspected Ninth Primary Planet of our Solar System (*MC*, August 1801, p. 155)

Since Professor Piazzi tracked his newly discovered star only until February 11 and since he hadn't informed other astronomers about his discovery earlier, nothing was expected of this beyond his own observations. But even in informing people of these, he seems not to have been particularly liberal. At first, he sent only a couple of observations to Oriani and Bode, and even these were incorrect. Through these, the two astronomers, as well as Dr. Olbers and myself (as our readers will have discovered from the previous book) were misled and inevitably brought to somewhat incorrect elements. Later, he sent his complete observations to Lalande in Paris, but under the condition that they would not be publicly announced. From these, Dr. Burckhardt calculated the elliptical elements of a path, which were immediately communicated, but he harbored some doubts about the correctness and exactness of these observations. He therefore believed to be able to conclude from the differences of the current observations, that the right ascension of this body must have been decreased by 2 minutes 30 seconds on January 1, and that Piazzi had been mistaken or had written the observations of the day incorrectly by 10 seconds. The differences between January 11 and 13, and 14 and 17 also do not fit particularly well.

Later, Piazzi sent his observations with the same conditions to Professor Bode, and the latter had the kindness to report us the following about it: "…When I came recently back from the country, I found a third letter from Piazzi, dated May 1, and behold! the long-awaited observations of his new star, 21 in number, from January 1 to February 11, had finally arrived, but with the express wish to make nothing public before he had done so. Due to his friendship, I am compelled to keep my word, and since I am to hope that you will see it the same way, I will share the following observations in confidence…" Although we had received the three positions sent along with Piazzi's observations, it was everywhere with the request not to make them publicly known. This is why we cannot presently announce them to the readers of Monthly Correspondence. *So much we can say to reassure those to whom these observations have also been imparted, that the three copies we have received from different sources are all identical, except for the fact that the observation of February 1 in the copy Lalande received from Piazzi is regarded as doubtful. An error in transcription can be ruled out, and the copies can be assumed as correct and genuine since two copies of this transcript from Palermo themselves stem from the original.*

Professor Bode also immediately noticed that the deviation of the first observation was given as a half degree less than Piazzi had written in his previous letter. Through this, the inclination of the path grows to almost 12°. "This till now unheard-of planetary inclination (writes Bode) is nearly enough to make me doubt its existence. But for the sake of my hypothesis, I'll imagine that its consequent great geocentric latitude and its occasional exceeding of the borders of the zodiac is why it has remained hidden from astronomers for so long, who, until now, only had the opportunity of comparing planets with fixed stars only in the proximity of the ecliptic. Piazzi writes in his letter [tr. from French]: *'I was also struck by the apparition of this comet, but it hardly seems that it could be a planet. Upon inspecting my observations, you will perhaps share my opinion. In the meantime, I beg you not to publish the results before I do.' What do you say? How could Piazzi have declared the new star a planet in his first letter of January 24 to Oriani? I have asked myself the same thing"....*

Mind you, Piazzi did regard this new star as a possible planet even before January 24. He supports this opinion with reasons which our readers will have discovered in the June issue of M C. *p. 608. He expressly wrote to Oriani that he had announced this star only as a comet at first on the basis of always observing it without a nebula and with a very slow movement. Several times, he had come upon the thought and suspicion that it could very well be a planet. Apparently, Piazzi's opinion changed afterwards and he returned to the idea that it was a comet. We cannot understand how he could make conclusions of the nature of this planet from the mere inspection of the observations, since even Laplace did not hazard a decision even after Dr. Burckhardt's calculation of the elliptical elements, and is thus of the opinion that further observations are needed. In the meantime, Oriani, Bode, Olbers, Burckhardt, Prosperin* [Erik Prosperin, 1739–1803] *and Fuss* [Nicolaus Fuss, mathematician, 1755–1826] *have all come to "suspect" that this new body could be a planet, and all observations till now also suggest an elliptical path.*

In a later letter from Senator Laplace from July 19, which we received during the correction of the present arc, this great geometrician makes a more exact account of this body and assures us that he is not averse to regarding it as a planet; the objection of a few astronomers regarding the overlarge inclination of the path seems but a minor objection to him. Since the opinion of this scholar is of the greatest weight, we will therefore entirely reproduce this part of his letter here: [tr. from French] *"You're correct in what you write me regarding Piazzi – it's a pity that he did not inform the astronomers in time; for we will have quite a bit of trouble in trying to relocate this star. That its inclination is greater than other planets can prove a slight problem for people of the opinion that it is a planet but that its eccentricity is less than that of Mercury. I am therefore not far from believing that this star is a planet and I urge you to examine it well once it escapes the rays of the Sun." And later in the same letter, in which the senator informs me of the 3rd volume of his immortal work* Méchanique cèléste, *which is now ready to be printed, he takes this opportunity to again give an explanation regarding this curious star:* [tr. from French] *"The new star observed by Piazzi could not have a great influence on planetary movements because of its extreme smallness, but if it is a planet, I hope that its orbit will be sufficiently known so that I can include it in my work on the perturbations." Only Professor Klügel, as Professor Bode informs us, wants to have nothing to do with this planet – his reasons are unknown to us. At this time, no astronomer regards Piazzi's new star as proven planet, at least to our knowl-*

edge. Everything that has been said, discussed and calculated about it are the 'conjectures of a possibility'; 'all' have hazarded only possible 'hypotheses' about it and have reserved doubts; 'all' were of the opinion that they had to await further observations of this star after its return from the Sun and that time alone would give us a 'certain' explanation about it. If the good fortune were to be Professor Bode's, as had happened with Uranus, that this body would also be found in some star chart, all our doubts would then be removed and a great light shed on this subject. He writes, at least, that he will therefore make the effort. Lalande also does not doubt, if this new body is otherwise a permanently visible star, that he can find it in his immense star catalogue of 50 thousand stars.

So much is certain that Piazzi came upon this star, as we had correctly surmised in our first essay (June, p. 612), upon completing his star catalogue and while searching for and determining very small stars in the meridian. Yes, an error in writing or even of printing seems to have given rise to this important discovery, for Piazzi stumbled upon this star, as Oriani reports from Milan on June 17, while wanting to search for the 87 Mayer's star according to Wollaston's General Astronomical Catalogue (1789), and did not find it in Mayer's star catalogue. The error came from Wollaston, who had wrongly attributed the determination of this star to Tobias Mayer [1723–1762, astronomer at the Univ. of Goettingen] *instead of La Caille, who deserved it. This star is also mentioned in De La Caille's* Zodiacal Star Catalogue *(Ephemer. des mouvemens cèléstes 1765–1775, p. XVII), in Bode's complete star catalogue in his edition of the* Flamsteed Celestial Atlas *(Berlin 1782) p. 18 no. 243, and also in his new, magnificent celestial maps, sheet XII.* [Ed: modern photometry gives this star a visual magnitude of 6.2; it is better known by the catalog designations BD +16.484, HD 22695 or HR 1110. Piazzi entered it in his own catalog as '3 hours 103'.] *Piazzi, in wanting to observe this star, hit upon this new star, which was located only 14¼ minutes west and 16 minutes south of it, and went through the meridian only 57 temporal seconds before the other.*

Dr. Burckhardt's elliptical path comes very close to the orbit; he is also of the opinion that no parabola other than his could be found to correspond to the observations. But Soldner [Johann Soldner (1776–1833) director of the observatory at Bogenhausen near Munich] *in Berlin calculated a parabolic path, the elements of which deviate considerably from those of Burckhardt. But because of our promise, we are not able to communicate them here. We can only report that Professor Bode outlined these two differing parabolas, then the orbit, and Burckhardt's ellipse, and that 'all' represent 'several' observations very well.*

Such a small arc which this planet has described until now should cause no displeasure. When Uranus was discovered, several different experiments were conducted in the same way. Boscovich [Ruggero Boscovich, 1711–1787, director of Brera Observatory in Milan; he was the first to compute the orbit of a planet from three observations of its position.] *showed in a short essay that there were four parabolas which satisfied a three month long observation of this planet. Lexell showed that there could be several parabolas from 14 to 18 in perihelial distance, which a long sequence of observations of several months could represent very well. Further observations had to be made before something certain could be known about the entire path. Portions of it could be represented in several parabolas, but subsequent observations ruled out one after the other until Lexell calculated the orbit, and finally, Laplace the true ellipse. Dr. Olbers therefore quite correctly states that, quite likely, no elliptical path of Piazzi's star could be calculated at this time with any measure of reliability. This thorough astronomer therefore writes on July 4: "Dr. Burckhardt obviously assumed that this planet was in its aphelion at the time of its first observation. Actually, different combinations, each of three observations, will have to be made to see to what extent these different combinations give the same elements for the elliptical path." Dr. Burckhardt himself expects no great exactness of his ellipses and he is far from passing them off as the 'true' ones. He remembers that the described arc is too slight for this and says that he endeavoured to represent the few available observations in one ellipse simply because he believed to be able to more easily promote and simplify the location of this star following its return from the Sun in this way rather than through parabolic elements. Nothing more can be done than what the conditions of the problem and Piazzi's 'few' observations allow.*

With these reservations, all the efforts and calculations of those astronomers who have concerned themselves with this puzzling body until now have been presented to the readers of the Monthly Correspondence. *Its appearance in the planetary system is too significant than to do anything other than employ all insight and diligence, calculate all possibilities and probabilities, through which the location of this strange body can be made easier. This has become so much more necessary since a second chance can hardly be expected to present itself to our eyes without any effort when dealing with such an extremely small, inconspicuous body which has no typical characteristic and easily loses itself in an infinite army of similar bodies.*

Admittedly, it is an eternal pity, and all astronomers without exception lament this, that Piazzi hadn't observed this stranger longer than to February 11. Had he not have taken ill, or had it pleased him to report his discovery earlier, other astronomers would have tracked this star in February, March and April, and if these observations would have led to no complete conclusion regarding the nature of this body, they would still have instructed us to the extent of being able to relocate it with some measure of certainty. But several astronomers now fear, not without reason, that it will be difficult to so easily look for this stranger now. Professor Bode is of the opinion that, since it shows itself as a magnitude 8 star, it should only be found just before the first light of dawn and in a considerable height above the horizon. For this reason, he believes that there is little hope of discovering it before the beginning of September, and then only when all other conditions are favourable. The longer the epoch of its discovery has to be set back, the more the true movement of this body will deviate from our provisionally calculated path, and the more difficult it will therefore be to find this plain wanderer in space.

We agree with the opinion of Professor Bode when there is talk of finding this body in a general way. But we believe that well-directed equatorial or parallactic instruments could more easily succeed in getting on the track of this stranger earlier, even if the previously calculated positions were to be off by a few degrees. In the suspected area, in a zone of several degrees, differential observations in right ascension would be made of all small stars, along with the well-defined fixed stars known at the moment – in this way, one would be able to differentiate the determined and undetermined stars from a planet through the star catalogues and by means of repeated observations, without first having to conduct differential observations in declination beforehand or having to wait for a dark night to be able to have an overall view of the respective position of the body at one time. But since the daily movement of the body is very great and amounts to nearly 1½ minutes in time at the beginning of September, the movement alone is therefore adequate to recognise the changeability of this body in one hour. In this period of time, it will take from between 3 or 4 temporal seconds. Admittedly, everything depends on the quality of the telescope and on the condition of the atmosphere. But since the atmosphere seems innately purer in the northern areas of Germany during this time of the year, those astronomers who are equipped with better parallactic instruments should spare no effort in hunting for this star as soon as goodness and fortune allows.

Since apparently so many astronomers have wanted to find a planetary body in Piazzi's star, others have therefore expressed doubt about it. The great inclination of the paths usual for planets until now seems to have given Professor Bode some misgivings. It has been calculated as 11° to date, and Soldner finds it to be as high as 18° in his parabola. But there can be no objection out of physical reasons, since, after all, the determination of the width of our constellation 'a posteriori' has been derived, and out of a mere empirical theorem. Professor Piazzi finds a reason in this, as Oriani informs us, to doubt the planet because the observed arc of its retrograde seems to stand in no proper relationship with its daily movement. On January 1, the day of the discovery of this body, he found it already in the process of being retrograde; he had then observed the retrogradation as only 9 or at most 10 days, but the duration of it must have been at least 100 days with this planet and the arc amount to between 9 and 10°. But since neither the aphelion can be ascertained with any reliability at this time, nor the different distances of the body from the Sun and from the earth, the dissimilar speed of the elliptical movement, the inclination of the planes be exactly assumed, it is also difficult to determine the arc of retrogradation through these elements, and the margin for error can be very great. Lalande showed (astr. art. 1190) that when, for example, the orbital position of Mars is calculated, an error of 2½° can occur in the angle of commutation in certain circumstances. In comparison, Professor Prosperin also finds that the observations by Piazzi of this body fit almost exactly within a cycle, that is, in a planetary path which returns on itself, and that its observed station accepts this very well. Here follows what this celebrated astronomer, who is intimately entrusted with similar calculations, writes on June 30 from Upsala: "Since only two observations are needed to calculate its radius and all remaining elements in assuming an orbit, and since the observations of the 1st and 23rd of January are known to us, nothing remains other than to explore whether the planet's stand-still agrees with this orbit, if not, then this orbit cannot be circular, or nearly circular."

Continued Reports Regarding the Long-Suspected Ninth Main Planet of our Solar System (*MC*, September 1801, p. 279.)

We are finally permitted to share with our readers the observations by Piazzi which have so long been kept hidden and secret. After Piazzi had forwarded several erroneous copies, which, according to his testimony, arose from an incorrect reduction of the right ascensions by his assistants, so finally, the present authenticated copy, which we will communicate faithfully and correctly here, has come into being. Only the second and fourth columns, namely, the right ascension of the body in time, and the northern declination have been sent in from Palermo; all other columns have been calculated by myself. I have used the inclination of the ecliptic 23° 28′ 12″ in calculating the geocentric longitude and latitude, and used my improved solar tables in calculating the mean solar times and solar positions and distances (Fig. 10.4).

At this time (August 26), neither the circumstances nor the weather have favoured the location of this body; nor have we heard of the more fortunate successes from other areas. Perhaps the coming month will see the eagerly longed-for and desired discovery. We are still working over the opinion of a great northern astronomer regarding this curious body.

The permanent secretary of the Royal Swedish Academy of Science, Sir Melanderhjelm [Daniel Melanderhjelm, 1726–1810], *states the following regarding this subject in a letter from Stockholm from July 22: "It also seems more likely to me that Piazzi's new star is a planet situated between Mars and Jupiter, than being Lexell's comet from 1770. I and Lexell, who was my friend and also my and Prosperin's student in Upsala in 1763, often corresponded about the nature of this comet at that time, and wondered whether it could be a planet. But out of the present calculations and elements, I believe the conclusion that Piazzi's star could more likely be the missing planet. Whether it is equally indifferent to the law of attraction and Kepler's Law, in which distance from the Sun the planets move, the harmonious progression of these planetary distances is one more reason among others to believe in the existence of this planet. I sincerely admit that this is the case with me at least. Beyond this, I find in the smallness of this new planet a certain economy of Nature, if I may express myself so. This small planet occupies along with Mars, which also numbers among the small planets, such a position in our planetary Solar System that it will and cannot cause any great perturbation in this system.*

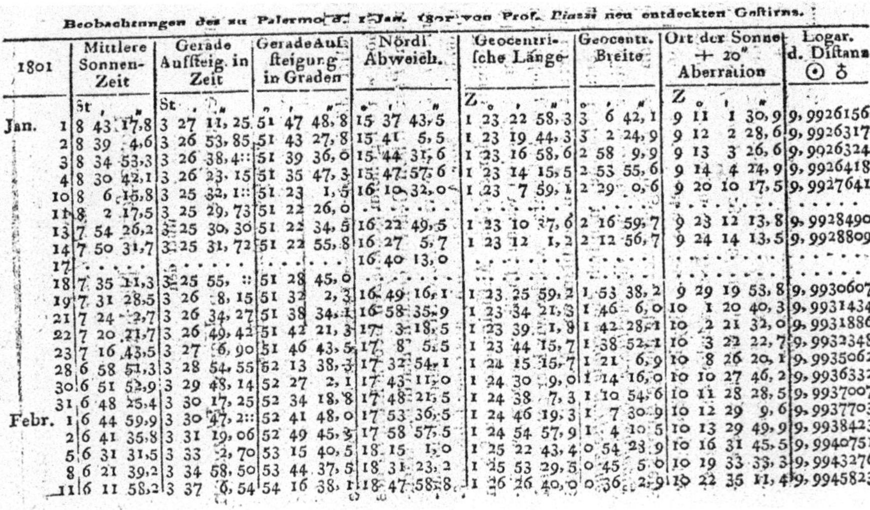

Beobachtungen des zu Palermo d. 1. Jan. 1801 von Prof. Piazzi neu entdeckten Gestirns.

1801		Mittlere Sonnen-Zeit	Gerade Auffteig. in Zeit	Gerade Auffteigung in Graden	Nördl. Abweich.	Geocentrifche Länge	Geocentr. Breite	Ort der Sonne + 20″ Aberration	Logar. d. Diftanz ☉ ♂
		St ′ ″	St ′ ″	° ′ ″	° ′ ″	Z ° ′ ″	° ′ ″	Z ° ′ ″	
Jan.	1	8 43 17,8	3 27 11,25	51 47 48,8	15 37 43,5	1 23 22 58,3	3 6 42,1	9 11 1 30,9	9,9926156
	2	8 39 4,6	3 26 53,85	51 43 27,8	15 41 5,5	1 23 19 44,3	3 2 24,9	9 12 2 28,6	9,9926317
	3	8 34 53,3	3 26 38,4::	51 39 36,0	15 44 31,6	1 23 16 58,6	2 58 9,9	9 13 3 26,6	9,9926324
	4	8 30 42,1	3 26 23,15	51 35 47,3	15 47 57,6	1 23 14 15,5	2 53 55,6	9 14 4 24,9	9,9926418
	10	8 6 15,8	3 25 32,1::	51 23 1,5	16 10 32,0	1 23 7 59,1	2 29 0,6	9 20 10 17,5	9,9927641
	11	8 2 17,5	3 25 29,73	51 22 26,0	. . . . .	. . . . .	. . . . .	. . . . .	. . . . .
	13	7 54 26,2	3 25 30,30	51 22 34,5	16 22 49,5	1 23 10 27,6	2 16 59,7	9 23 12 13,8	9,9928490
	14	7 50 31,7	3 25 31,72	51 22 55,8	16 27 5,7	1 23 12 1,2	2 12 56,7	9 24 14 13,5	9,9928809
	17	. . . . .	. . . . .	. . . . .	16 40 13,0	. . . . .	. . . . .	. . . . .	. . . . .
	18	7 35 11,3	3 25 55, ::	51 28 45,0	. . . . .	. . . . .	. . . . .	. . . . .	. . . . .
	19	7 31 28,5	3 26 8,15	51 32 2,3	16 49 16,1	1 23 25 59,2	1 53 38,2	9 29 19 53,8	9,9930607
	21	7 24 2,7	3 26 34,27	51 38 34,1	16 58 35,9	1 23 34 21,3	1 46 6,0	10 1 20 40,3	9,9931434
	22	7 20 21,7	3 26 49,42	51 42 21,3	17 3 18,5	1 23 39 1,8	1 42 28,1	10 2 21 32,0	9,9931886
	23	7 16 43,5	3 27 6,90	51 46 43,5	17 8 5,5	1 23 44 15,7	1 38 52,1	10 3 22 22,7	9,9932348
	28	6 58 51,3	3 28 54,55	52 13 38,3	17 32 54,1	1 24 15 15,7	1 21 6,9	10 8 26 20,1	9,9935062
	30	6 51 52,9	3 29 48,14	52 27 2,1	17 43 11,0	1 24 30 9,0	1 14 16,0	10 10 27 46,2	9,9936332
	31	6 48 25,4	3 30 17,25	52 34 18,8	17 48 21,5	1 24 38 7,3	1 10 54,6	10 11 28 28,5	9,9937007
Febr.	1	6 44 59,9	3 30 47,2::	52 41 48,0	17 53 36,5	1 24 46 19,3	1 7 30,9	10 12 29 9,6	9,9937703
	2	6 41 35,8	3 31 19,06	52 49 45,3	17 58 57,5	1 24 54 57,9	1 4 10,5	10 13 29 49,9	9,9938423
	5	6 31 31,5	3 33 2,70	53 15 40,5	18 15 1,0	1 25 22 43,4	0 54 23,9	10 16 31 45,5	9,9940751
	8	6 21 39,2	3 34 58,50	53 44 37,5	18 31 23,2	1 25 53 29,5	0 45 5,0	10 19 33 33,3	9,9943276
	11	6 11 58,2	3 37 6,54	54 16 38,1	18 47 58,8	1 26 26 40,0	0 36 2,9	10 22 35 11,4	9,9945823

Fig. 10.4 Piazzi's 1801 observations of Ceres

At the moment, it is admittedly difficult to hazard certain conclusion and remark about this new heavenly body given the available data – one must wait for further observations. I would also not want to decide whether Lexell's comet could indeed not be this supposed planet. The difference between a couple of elements of the path is not enough to make two different bodies out of it with any certainty, as you quite correctly remark. They could very well belong to one body and the perturbations of Jupiter could have been great enough to cause such a difference and change. Even the changes in the form of this body, its visibility and its invisibility can decide nothing here, since we still know too little of the physical constitution of this body, as you remark, to go in about the reasons for them. Banned to a northern corner of the world, it is not our lot to discover the Mirabilia Coeli [wonder of the heavens] – *the luck of rediscovering this planet will also scarcely be granted us. Our Nicander is also of a very weak health and is sickly; our two other astronomers, Svandberg and Osverbom, are now in Lapland engaged in graduating. I have also sent the best instruments of our observatory there."*

[In 1767, Jupiter's powerful gravity changed the orbit of Comet Lexell to one that would bring it close to Earth. The comet then streaked within 1.40 million miles (2.26 million km) of Earth – just under six times the Earth-Moon distance – on July 1, 1770. Even now, this remains the closest known cometary approach to Earth. So close was this passage that Earth's gravity actually shortened the comet's period around the Sun by three days. When the comet didn't return as expected in 1776 or 1781, astronomers Anders Johann Lexell and Pierre Simon Laplace painstakingly calculated the comet's motion over time and discovered that another close approach to Jupiter in 1779 had again radically changed the comet's orbit. This time, it may have been sent out of the Solar System altogether.]

Regarding the comet (moving star) discovered by Herr Joseph Piazzi, Royal Astronomer of Palermo, in Taurus on the first of January, 1801, which can be regarded, with great probability, as the long-suspected 'primary planet of our Solar System located between Mars and Jupiter

By Bode, written in September 1801. BAJ (1804), p. 251. This German paper is the same as the French language paper in Memoirs of the Royal Academy (1801), p. 132.

On March 20 of this year (1801), I received a letter from Herr Piazzi from Palermo dated January 24, in which he informed me of the following: "On January 1, I discovered a 'comet' in Taurus at 51° 47' right ascension and 16° 8' northern declination. On the 11th, it changed its retrograde movement to direct, and on the 23rd, it had 51° 46' right ascension and 17° 8' northern declination. I will continue observing it and hope to be able to follow it throughout the entire month of February. It is very small and is similar to a magnitude 8 star without any perceivable nebula. I beg of you to let me know whether it has already been observed by other astronomers; in this case I should save myself the trouble of computing its orbit. In the meantime, I request of you not to make your published results known before I have done so myself." [Positions of the comet observed from Palermo from January 1, 1801 to February 11 are given.] *It was discovered when searching for the star indicated as #87 'in Mayer' by Wollaston in his catalogue, but which does not appear in the zodiacal catalogue of the former. It constantly appeared as a magnitude 8 star and was not recognizable with the naked eye. The observations were conducted completely on the meridian.*

Actually, these are the more exactly reduced observations sent to me in a later letter, dated July 30, by Piazzi. He states in the same letter that his assistant made a few errors in calculating the prior one; at that time, he was so weak that he could undertake nothing. Now he wants to contemplate calculating his observations.

On July 2, I made a report to the Academy regarding the continued research and calculation of the true path of this planet according to Piazzi's observations. I had already tried to predetermine the heliocentric position of the star within the assumed orbit from the very first details, but since Herr Piazzi had set the declination on the day of its discovery, January 1, as 15° 38' instead of 16° 8', in other words, half a degree smaller, the inclination of the path (approximately 6° from the first observation) grows almost twice because of this, and this inclination of the path, hitherto unheard of with a planet, could very well upset the suspicions regarding Piazzi's star if one hadn't had a reason in this significant inclination as to why this planet hadn't been discovered earlier; for, since it often exceeds the boundaries of our constellation and since, normally, only planets in proximity of the ecliptic are compared with fixed stars, it could have all the more easily escaped the attention of astronomers. In order to mention only a little concerning the results of my repeated calculations, I took the observations of January 1 and 23 and February 11 as a basis and found the following (Fig. 10.5):

Assuming the distance of the planet from the Sun to be 2.95, which must then amount to a period of 5.067 years according to Kepler's Law, it led to the following in a circular path:

	Helioc. longitude			*Latitude*		
January 1	*2Z 7°*	*38'*	*22"*	*2°*	*19'*	*17"S*
23	*2 11*	*58*	*12*	*1*	*22*	*38*
February 11	*2 15*	*57*	*15*	*0*	*33*	*25*

W. Z.	geoc. Länge.	Breite.	elong. öſtl.	☉	diſt. ☉.
1. Jan. 8U. 40'	12. 23°22'58"	3° 6'58" S.	132°21'32"	9.11. 1.26.	0,98315
23. — 7 4	1 23 44 12	1 39 7	110 22 2	10. 3.22.10.	0,98457
11.Febr. 5 58	1 26 26 34	0 36 19	93 51 35	10.22.34.59.	0,98762

Fig. 10.5 Calculation of Ceres' distance from the Sun in early 1801

From this, it follows that the average period corresponds to 4.83 ±2 months, agreeing with Kepler's Law. The differences of the heliocentric latitude give the inclination of the path to be approximately 11° 56′ and the position of the Ω 2 Z 18° 44′. A planet in this distance from the Sun must come to a standstill in an eastern elongation of 123°, and Piazzi's star had this elongation precisely on January 10 when it became stationary, which was extremely favourable to the theory that it was indeed the planet suspected between Mars and Jupiter. In the same way, I calculated the observations of January 2 and 16 and February 19 in an orbit with a distance of 2.80, and, according to this, the

January 2 hel. long. = 2 Z 8° 37′ Lat = 2° 19′ S.
February 5 " " = 2 15 50 " = 0 56
and out of this, the period is 4.64 years.

Kluegel's method of obtaining an approximate determination of the orbit of a more distant upper planet from two observations, when applied to this new star, indicates nothing other than the expected planet.

In the meantime, with obliging helpfulness, Herr Colonel Lieutenant Baron von Zach in Gotha often shared reports, investigations and calculations regarding this planet conducted by Oriani, Laplace, Burckhardt and Olbers enclosed in letters he had received, as well as his own research and calculations. They are complete in the June, July and August issues of his monthly correspondence, along with many interesting comments. Space does not allow to cite them all here; I only remark that attempts were made by Dr. Burckhardt, Dr. Olbers and Herr Soldner to represent the true course of this body as a comet during the 41 day period of Piazzi's observations through parabolas or through ellipses or circles, or represent it as a planet. However, though heliocentric, since it only describes the small arc of approximately eight degrees which suits almost every conic section or orbit [circle] in that area, no great exactness of the results was expected. But that which resulted from an ellipse or a circle corresponded much more easily and exactly with the observations than that from a parabola. Herr Dr Burckhardt in Paris found the following elements of an elliptical path after many calculations, which represent the longitude and latitude of five observations to within a few seconds. The Ω is 2 Z 20° 58′ 30″. Inclination of the path is 10° 47′. Position of the aphelion is 2 Z 8° 59′ 37″. Time of the intersection through the aphelion 1801 January 1.3328. Eccentricity is 0.0364. Semi major axis or mean distance from the Sun is 2.5743. Sidereal orbit is 4.13 years.

According to this, Herr. Burckhardt had determined the geocentric positions where the star was to be found after its return from the Sun in the early hours in the following way (Fig. 10.6):

	U. M.	Länge. 4 Z.	Breite N.		U. M.	Länge. 4 Z.	Breite N.
12. Aug.	10 54	4° 21′	4° 51′	8. Oct.	22 0	28° 12′	6° 53′
						5 Z.	
7. Sept.	16 19	15 28	5 41	14. —	3 0	0 12	7 8
12. —	22 0	17 40	5 52	19. —	7 0	2 11	7 22
18. —	3 0	19 50	6 3	24. —	11 0	4 8	7 37
23. —	8 0	21 58	6 15	29. —	14 45	6 3	7 53
28. —	13 0	24 5	6 27	3. Nov.	18 0	7 56	8 9
3. Oct.	17 41	26 9	6 40	8. —	22 0	9 48	8 26

Fig. 10.6 Burckhardt's predicted positions of Ceres in 1801

Herr Doctor Olbers had the friendly helpfulness of sharing his thoughts, investigations and calculations of the true path of this new body with me in various letters. Among other things, he gave me examples that Piazzi's observations don't fit at all within a parabola, but rather that the star must move within an ellipse which doesn't differ very much from a circle. Further, he also believes that the great inclination of the path against the ecliptic is no reason to doubt the planetary nature of the body. With the assumption of an orbit, Herr Doctor Olbers calculated the following elements from the observations of January 1 and February 11: Ω 2 Z 20° 22′ 45″. Inclination of the path 11° 3′ 36″. Heliocentric elongation of Ω in the path in the first observation 11° 46′ 53″.5. Radius of the circle 2.730185. Sidereal period 1647.75 days. Daily heliocentric movement 13′ 6″.528. The interim observations correspond with this in the following way:

	Observed Longitude	Observed latitude	Calculated longitude	Calculated latitude
January 13	1Z 23° 10′ 37″.6	2° 16′ 59″.7	+1′ 46″.7	−0′ 30″.9
19	1 23 25 59.2	1 53 38.2	+2 17.0	−0 26.9
31	1 24 38 7.3	1 10 54.6	+1 56 1	−0 18.2

The observations themselves are not any more exact; from this, Herr Doctor Olbers believes to be able to conclude that 1) Piazzi's star really does move within an ellipse which doesn't deviate very much from a circle, and is in reality, as I suspected, a planet. 2) Even in this not very eccentric path, it must have been very close to the apsis line during the observations, because otherwise, the observations could not correspond so 'exactly' with a circle-hypothesis. 3) It can't be determined from this with any certainty which portion of the apsis line, whether perihelium or aphelion, it was close to in January and February. 4) This uncertainty has a considerable influence on the positions of the star calculated in advance; one would then have to know whether its movement is increasing or decreasing. If one therefore takes the locations produced from the orbit as a basis, the actual ones can't deviate too much from them. According to this, he finds the geocentric longitude to be some 2″ smaller than Herr Burckhardt and the latitude agrees to within a couple of minutes. Moreover, the planet could show itself before September mornings, and even then, it will appear very small. On January 1, it resembles a magnitude eight star, its distance was 1.968; on August 19, however, this distance will amount to 3.645 and on September 4 3.536.

At this point, it also depended on whether this small moving star could be located in the early hours after its opposition with the Sun, and the united efforts of all astronomers were directed to this goal. Then, on August 19, 20 and 21, while I was visiting Herr Marshall von Hahn in Remplin [Count Friedrich von Hahn, 1742–1805], *I looked for the star, which was then expected in Cancer, in the early morning, but it only came over the horizon with the bright dawn. At this time, in the early morning, we had moonlight. On September 3, 4 and 5, the Moon traversed Cancer. On the 6th, it stood illuminated and extremely small in Leo. With the rising of the Moon, we had 2¼ hours of clear air. At the observatory, I set the 2-foot Dollond star-finder and the 3½-foot Dollond on the head of Leo, where the star was to be located at 15½° Leo and 5.75 N. latitude according to Herr Dr. Burckhardt's ellipse. In the course of searching, I found this star, but while being busy examining it more carefully with the Dollond, a rising wind quickly blew clouds in the way, which then covered the easterly sky. Then, a six day-long period of constantly very dark, foggy and rainy weather descended. The morning of the 14th was clear again. From the observatory with the aforementioned telescopes, I most carefully examined all the small stars in the area of this star between 3 and 4 o'clock and produced a drawing of their position. Early on the 15th, I believed I found the star, since its position seemed to have moved a third of a degree east from the previous day's investigation (corresponding to the star's current movement); but toward 4 o'clock of the 16th, I found, though only through breaks in the clouds, that I had been deceived in a curious way by confusing two stars. The morning of the 17th saw fully overcast skies. On the 18th and 19th, I could once again see, and I diligently investigated all the stars in the area with my 3½ foot Dollond as soon as they had risen above the haze of the horizon from approximately 3 o'clock to the break of dawn, but I didn't find any which characterized itself as a planet in form or which had changed its position against the neighbouring stars. It was hazy on the 20th. I will continue this investigation until the light of the Moon prevents it.*

For the moment, I've received no news regarding other astronomers' investigations of Piazzi's star. It must be considerably removed from that location indicated by the calculated circular or elliptical path, or must have such a weak light or small size that our usual finders and Dollonds are no longer sufficient. Trying to locate it becomes ever more difficult and uncertain the longer it takes to discover it, and only a repeated fortunate coincidence could lead to finding it. Should the assumption that it really is the planet suspected between Mars and Jupiter be proven true through its discovery and planetary movement, as all calculations of the apparent course observed solely by Piazzi seem to indicate, and should talk then move to the naming of it, I would like to suggest the name Juno (Hera, in Greek), as I already informed Baron von Zach in Gotha in May. We must remain with mythology for the sake of analogy and to avoid flattery, and because the planets found over Jupiter carry the name of his ancestors and those standing closer to the Sun the names of his spouse and children.

The 7th nebula from the 1st class of Herschel's catalogue belongs to the [radiant ones]. *He observed it on January 23, 1784, 31° 41′ 15″ in right ascension east and 40′ south of 49 Leo. The longitude 6 Z 3° 20′ 23″ and latitude 11° 22′ 0″ N. follows from this. After a period of time, he couldn't find it, and therefore explained it as a telescopic comet. If I assume that it was our new planet and that its distance from the Sun = 2.′ 73, then its*

helio. long.	*January 23 1784*	*15 [St.]=*	*5Z 15° 12′*	*lat. 8° 45′ N*
	January 23 1801	*7 [St.]=*	*2 13 16*	*1 21′ S*
			8Z 28° 4′	

If I now calculate that since this time, thus, in 17 years, it has made 3 orbits + 268° 4′, then this equals 1658 per orbit. Herr Doctor Olbers found it to be 1648 days. This close agreement is remarkable. Only the latitude cannot be brought into line with the inclination of the path 11° and the Ω in 2 Z 20°, unless Herr Herschel wrote 40′ 'north' instead of 40′ 'south' of 49 Leo; I have written to him about this matter.

Today (September 25), I finally received a letter from Herr Piazzi dated August 1, which had the following contents: "I believe that you are correct in regarding my star as an actual planet. The later observations, during which it had lost much in the way of light, led me astray. I have tried to represent its course in a parabola, but to no avail. The course it has covered is still too small for an exact calculation within an ellipse. An arc corresponds the best and is sufficient to locate it again. I have calculated the following elements of the path from the later observations: radius 2.6862. Ω 2 Z 20° 46′ 48″. Movement in the path from January 1 to February 11 9° 2′ 29″.7. Inclination 10° 51′ 12″. Epoch 1801. 2 Z 8° 46′ 41″. Movement in 100 years 22° 6′ 33″.7. Sidereal orbit 1628.27 days. Diameter in the distance from object to the Sun 19″. Size 1.33 the diameter of the Earth. These elements are taken from a treatise which will appear shortly and which I will send to you through Herr Doctor Triesnecker.

The observations were done with a telescope which magnifies 50 times and has a three inch aperture. I estimated the diameter of the star to be 7″. In the first days, I tried to observe it with a night telescope and with another achromatic objective with a four inch aperture, but it was impossible to distinguish it in this way.

I embrace you heartily that you have first announced my new planet, to which I would like bestowed the name Ceres Ferdinandia. Before the month of November, I scarcely believe that it will be located."

Continued Reports about the Long-Suspected New Primary Planet of our Solar System

MC (October 1801), p. 363.

Towards the end of August until the middle of September, we made several and various attempts at finding the eagerly awaited new arrival in the morning hours. The bad weather that generally prevailed did not particularly favour us in these endeavors. At first the bright Sunrise, then the moonlight, and finally the rain, fog and haze, made all attempts at finding it impossible. All our non-local astronomical friends and correspondents, our most famous French and German comet lookouts, Messier [Charles Messier, 1730–1817, compiled the famous Messier catalog of nebula], *Méchain and Bouvard* [Alexis Bouvard, 1767–1843, Director of the Paris Observatory], *Herschel, Olbers, Bode and Schroeter were no more successful in their pursuits. All of our reports agree in that all attempts remained without success because of the generally prevailing bad weather.*

In the meantime, the well-founded opinion that this newly found body is of a planetary nature is not only maintained by most astronomers, but rather, further calculations and investigations confirm this suspicion more and more.

For instance, Dr. Olbers attempted to see whether three observations would be satisfied by parabolic elements. The results of his calculations turned out to be 'negative.' He was not in a position to represent more than 'three' longitudes and 'two' latitudes, or 'two' longitudes and 'three' latitudes in a parabola of the 'three' observations used as a basis for calculation. He duplicated his calculations and so that our readers could see how little a parabola is suited, we will produce "one" result of his calculations here for examination.

Parabolic Elements for Piazzi's star as calculated by Dr. Olbers

	Ascending 3 longitude and 2 latitude	*Ascending 2 longitude and 3 latitude*
Longitude of the node Ω	*2 Z 19° 50′*	*2 Z 21° 7′*
Inclination of path	*10 38*	*9 48*
Longitude of the perihelion	*3 25 4*	*4 10 6*
Time of solar transit	*1801 Jun 8.16h 16′*	*1801 Jun 25.7h 38′*
Distance of perihelion	*2.53510*	*2.13268*

These parabolas also approach those calculated by Dr. Burckhardt and which we have cited in volume IV of Monat. Corr. *p. 60. Dr. Burckhardt already assured at that time, and repeated his assurance upon the occasion of Soldner's path, that he hardly believed that another parabola would satisfy the observations better than his. On the whole, the fairly certain conclusion follows out of this " that Piazzi's observation can be by no means tolerably represented by a parabola, and that they could befit only a planetary or an elliptical movement."*

Consequently, Dr. Olbers was really on the point of calculating a new elliptical path of this planetary body out of the complete, above-mentioned improved series of Piazzi's observations, since he felt a very great confidence in the exactitude of these observations, which led him to not only statements in decimals and hundredths of seconds, but rather to Piazzi's name, his great exactitude otherwise referred to, and, as is well known, his splendid instruments.

But Dr. Olbers soon saw that the cited right ascensions must have had considerable errors here and there. He partly found the same errors in the routine discrepancies which we criticised and revealed in Vol. IV of the Monat. Corr., *p. 156. In February, he even suspected errors of nearly 20 temporal seconds, which could not very well be improved through some probable conjecture for such thankless as well as superfluous work, since our Burckhardt has already done in this all that could be done. He therefore contented himself with seeing how much Piazzi's new observations deviated from an orbital hypothesis. He first tried to find an orbit in the observations of January 1 and February 11, and after we had given him, in handwriting, the table of Piazzi's observations calculated in the previous issue, p. 280, he improved the orbital elements in the following way:*

Rad. of orbit....................................2.730185
Long. of ascending Ω...................2Z 20° 23' 45"
Inclination of path.............................11 3 36
Heliocentric distance from Ω in the
path in the 1st observation................... 11 46 53.5
Time of orbit...................................1647.75 days
Daily helio. movement.....................13' 6".528

The observations lying between these correspond with these orbital elements in the following way:

	Observed Longitude	Observed latitude	Calculated longitude	Calculated latitude
January 18	*1Z 23° 12' 24".3*	*2° 16' 28".8*	*+1' 46".7*	*-0' 30".9*
19	*1 23 28 16.2*	*1 53 11.3*	*+2 17.0*	*-0 26.9*
31	*1 24 40 3.4*	*1 10 36.4*	*+1 56.1*	*-0 13.2*

From these slight differences/deviations of the orbital hypothesis, Dr. Olbers draws the following conclusions:

1) *According to the observations, Piazzi's star is 'really a planet' and moves in an ellipse which is not very eccentric.*
2) *It is likely that this planet was not far from the apsis-line during the observations, that is, close to either the perihelion or the aphelion.*
3) *It seems hardly likely that anything certain can be fixed regarding the measurements of the true ellipse from the observations which are so similar among themselves and deviate little from the orbital hypothesis. And if also, as Burckhardt found, and as all parabolic elements confirm, the planet increased its heliocentric velocity and decreased its distance from the Sun somewhat during the observations, it will therefore be very difficult to determine with certainty for such a slight arc (the orbital hypothesis cites it as 8° 57') and such a minor eccentric ellipse whether Piazzi's star went through its aphelion shortly before January 1 or whether it went through its perihelion shortly after February 11. Dr. Olbers admits that Burckhardt found an ellipse which harmonizes very well with the observations, with which the aphelion falls on January 1, but it seems to him that an ellipse which doesn't coincide that much worse ought also be found if the perihelion is set a few days before or after February 11.*
4) *The uncertainty of whether Piazzi observed his star in the proximity of the aphelion or perihelion has an influence on the future pre-calculated positions given to locate it. If the new planet had gone through its aphelion before January 1, its heliocentric velocity will have increased, and its geocentric longitude must also be greater in August and September, as suggested by the orbital hypothesis. However, if it went through its perihelion in February, its heliocentric velocity will have subsequently decreased, and its geocentric longitudes must be 'smaller' in August and September, as per the orbital hypothesis. Because we can't know at this time which of the two cases applies, the future search for this body is more certain to use as a basis the positions concluded from the orbital hypothesis, which cannot deviate from the real ones too much and which in both cases constitutes the mean.*

In the case of Burckhardt, the positions deviate longitudinally by no more than two degrees, and latitudinally only by a couple of minutes. This new planet will most certainly be located again with the points determined by Olbers through the orbital hypothesis as a starting point, and by searching these latitudes a few degrees backwards and forwards and noticing all the small stars located therein.

Regarding the objection of a few astronomers that the great inclination of the path of Piazzi's star casts a serious doubt as to whether this body could really be a planet, Dr. Olbers explains: "This great inclination given by the elliptical elements of Piazzi's star which is so unusual for a planet seems to me no basis to disbelieve that this body is a planet. We can offer absolutely no reason why planets must have such a slight inclination. Even the hypothesis of the great Laplace, who states that the planets were deposited from the gradually contracting solar atmosphere, is not only highly improbable, but rather, I trust myself to say, obviously wrong, because the movements, namely the projectile speed of the planets do not coincide with it, and this is, as far as I know, with the exception of Buffon's [Georges-Louis Buffon, 1707–1788, a French naturalist who was the first to propose the theory that the planets had been created in a collision between the Sun and a comet] *reveries, the only hypothesis through which it has been attempted to give a physical cause from the slight inclinations of planetary paths. It is known that Newton found convincing proof for the arbitrary arrangement of an omnipotent creator in these slight inclinations. It is therefore still not proven that no planet could have an inclination of 11 to 12°."*

Our readers have already seen in the August issue (p. 159) that even Laplace calls this objection slight. But more than this, and with greater reason, the opinions as to the suspicion whether Piazzi's star and the comet of 1770 are one and the same are divided. Dr. Olbers finds it completely improbable. He writes us about this: "In itself, the Comet of 1770, surrounded by such a tremendous atmosphere, could never reveal itself as a magnitude 8 star without any nebula. But the path of the comet of 1770 could have been moved by Jupiter, and therefore, there must be one point in its path when it approaches very close to Jupiter. Consequently, it can assume absolutely no dimensions which correspond with those concluded through Piazzi's observations." Professor Bode is of the same opinion. This astronomer expresses his doubts with the following reasons: "Is such a tremendous transformation in the shape and location of the path possible through the attraction of Jupiter? This comet was seen in Hamburg on July 1 in Lyra with a nebula surrounding it, with the apparent size nearly that of the full Moon and completely round. Could this comet appear 3 times further away than the Sun, without any nebula? The comet of 1729, which was observed over four times further than the Sun, appeared through a sixteen-foot telescope with a nebula as great as Jupiter's, as seen through the above mentioned telescope. I would rather accept that the comet of 1770 appeared only with such an approach to Earth, than appearing at that time with such a strongly radiant nebula; it was, according to Lambert's calculation on July 1, only 7 times the Moon's distance from us."

As the case may be, future observations will soon and certainly decide the questions surrounding Piazzi's star. But what concerns the comet of 1770, it appears that this puzzling body will long be shrouded in inpenetrable darkness until a lucky chance, brought about through time and circumstance, allows this body once again to be seen. Professor Wurm therefore asks: "What, ultimately, is to become of this body? A 'planeto-comet' or something in between?" In vain, we lose ourselves in this type of conjecture. But it is the duty of this historiographer is to gather all opinions and voices and to report faithfully. And since we have undertaken this duty, we will want to fulfill it conscientiously.

**** Shortly before the printing of the last sheet, a very bright morning was granted us on September 23. I had the pleasure on this night of searching through the area a few degrees east and west of the expected location neighbouring the present position of the suspected planet with a parallactic instrument (a 3½-foot Dollond) and a good comet-finder until the break of dawn in the company of the royal astronomer and chairman of the*

Prague observatory, David [Martin Alois David, 1757–1836], *and Professor Buerg* [Johann Buerg, 1766–1835, Austrian astronomer whose chief work was his set of Moon Tables published in 1806] *But until now, we have been unable to notice a change in position of any star against others, and have also been equally unable of finding one displaying planetary qualities. I'm afraid that finding this body will cause us much of trouble and cost many a sleepless night, since subsequent observations seem to rob us of all hope of finding this inconspicuous arrival prior to the complete absence of any daylight or moonlight.*

Earlier, we gave our readers new elements of an orbit which the tireless astronomer, Dr. Olbers, calculated from the entire series of observations by Piazzi. As uncertain as these may be in some of their components of determination, given the nature of the matter, so much is certain: this new wanderer 'must appear extremely small now in September.' On January 1, 1801, on the day of its discovery, it was similar, according to Piazzi's estimation, to a magnitude 8 or 9 star. Its distance from Earth at that time was 1.968. Now, according to Olbers' elements, the distance on August 19 was 3.645, and on September 7, 3.536 – that is, nearly twice as great as on the time of its discovery. Since the light, or rather, the visible brightness decreases all the more, as the inverse quadratic relationship to the distance – it is easy to calculate the present radiance, or rather, the present 'invisibility' of this planet. The contemplation of this did not escape us even upon the first announcement of this body.

Piazzi's Account of the Discovery of Ceres

MC (November 1801), pp. 558-582.

On October 15, I [Zach] *received, via Vienna, a letter from Palermo, dated September 1 of this year, from the director of the Royal Sicilian Observatory, Giuseppe Piazzi, in which this famous astronomer had the goodness of enclosing a short discourse on the new body discovered by him. We know how to satisfy the thirst for knowledge of our astronomers and the curiosity of our non-astronomical readers no better and effectively than to give them a short, synoptical extract from this short Italian pamphlet, which will likely not find its way into German bookstores either easily or quickly. This discourse of two sheets bears the title:* Risultati delle Osservazioni della nuova Stella scoperta il di' 1 Gennajo all' Osservatorio Reale di Palermo. Da Giuseppe Piazzi Ch. Reg. [Chierico Regolare] Direttore del medesimo. Presentati alla supreme generale Diputazione degli Studj. in Palermo 1801. Nella Reale Stamperi.

[Zach then prints a summary of Piazzi's monograph on Ceres; the text below resumes from p. 571 in the Monthly Correspondence.]

A brief appendix follows Piazzi's discourse in which he reports that Oriani communicated his and other astronomers' calculations to him regarding this body. He accordingly cites Oriani's parabola as well as Burckhardt's parabolic, orbital and elliptical paths, exactly those which appear in the July issue of Monat. Corr. *pp. 59, 60, 61. At first glance, it seems strange that Piazzi names Dr. Burckhardt as the calculator of his elliptical, but not his elliptical and orbital path – he seems to attribute these to another astronomer unknown to him. But this can be explained in the following way: Piazzi excuses himself in that he could not discover the name of the astronomer who calculated the parabola and orbit from the two German papers sent to him by Oriani. These two German papers were none other than pp. 59, 60, 61 and 62 of the July issue of our* Monat Corr. *On the preceding p. 58, which Oriani did not send, Dr. Burckhardt's name appeared; on those which Piazzi received, the name was mentioned in reference to the elliptical path. This is the reason that Piazzi could not know from whom the elements of the parabola and orbit originated, which is why he indicated the name of this astronomer by dots in his discourse.*

From the same pages, Piazzi saw that Dr. Burckhardt harbored some doubts regarding the correctness of the copy of his observations, through which he suspected that some errors had snuck in. He recognizes that this was the case with his first copy, and as a result sent corrected copies to Lalande, Oriani and Bode that were exactly identical with the first, upon which he based all his calculations and which we have inserted in Volume IV of Monat. Corr., *p. 280. To anticipate the wishes of those astronomers who showed a lively interest in this discovery and to dispel even the slightest doubts regarding his observations as much as was in his power, he carried out their reduction again from scratch. Instead of a few less precisely determined reference stars, which he had used at the beginning, he now used better determined ones, and took into account their own movement and the deviation [declination] of his instrument and used such care as is required to attain the highest level of exactness. In spite of this, this yielded only minor differences from the previous in right ascension, which have almost no, or at most, a minor influence on the calculation of paths, and this is why he initially regarded great exactness as superfluous in the reduction of observations. As a consequence of this last rigorous examination, to maintain the greatest keenness, 1″.5 was removed from the first four right ascensions, and that same amount from those of January 10, 11, 14, 19, 21, 23, 28, 31 and February 1; 3″ had to be removed from the right ascensions of February 3 and 8. Piazzi had observed this body mainly through his two instruments and his meridian telescope at the meridian circle, but he always preferred the meridian instruments for the right ascensions if he could conduct his observations completely with this instrument. When this was not the case, he took the mean of the observations from both instruments. Meanwhile, the difference was never more than 0″.2 in time, except on January 19, on which day he was in the area of one temporal second more than the transit instrument showed. As far as the observed declinations are concerned, he found nothing to be improved. Incidentally, if anyone wishes to take a look at his original observations, he is willing to communicate them with the greatest pleasure. These same observations will appear in print in volume IV of the Palermo Observatory along with the remaining observations of 1794.*

This is the only true and authentic story of discovery of this long-suspected and now likely discovered primary planet of our Solar System, which we have had sent to us by the discoverer himself from Palermo and which we have communicated to our readers here in faithful extracts. Contradictory and misleading reports have appeared in the political newspapers which have led to doubt and misunderstanding, through which the publisher of Monat. Corr. *has attracted widespread verbal and written inquiry. Since it is becoming impossible to answer every enquirer in writing, we see ourselves doubly compelled to choose the present public means to dispel the unfounded rumors which have reached the general public. In these newspapers, it was supposedly stated that Professor Seyffer in Goettingen received a letter from Piazzi in Palermo, in which he informs him that he now declares the star, newly discovered by him, to be a comet. But whoever has read the above story of this planet's discovery with some attention will see that this rests entirely on an error in view of time and of a confusion regarding the date. Since Piazzi probably could have written Professor Seyffer of his opinion at the time when he still had it, the letter, as is quite possible, could have been delayed in the mail, arriving at a time when Piazzi had again changed his mind, having come to a better realization after concluding his calculations, as he himself explains in his discourse. Piazzi's letter to Professor Seyffer proves at most that the date of this letter (August 4) was either incorrectly written or falsely cited, or that this entire letter was misunderstood. Since no German astronomer doubted the planetary nature of this body at this time, the calculation of parabolic elements had long been given up by astronomers, who now concerned themselves only with orbits – Burckhardt had already calculated an elliptical path. The report from Seyffer's letter therefore obviously came to the political newspapers at an inopportune moment, and the sender was therefore probably not especially familiar with what goes on in astronomy and with the history of this remarkable body. Otherwise, he would not have allowed this report of a limping messenger appear in the newspaper, through which he only led the public into confusion.*

In another newspaper, this report is contradicted, with the entire letter from Piazzi to Seyffer called into question and declared apocryphal. We in our position confess candidly that we do not find the least cause to doubt this letter. Why shouldn't Piazzi write to the successor of one of our most famous German astronomers, Tobias Mayer, the professor of astronomy in such a world famous university as Goettingen, as well to the celebrated astronomers of the Royal Berlin Observatory! This is much more likely since Piazzi speaks in his discourse of Tobias Mayer's original observations stored in Goettingen and suspects that an observation of his new planet might well be found among them, as was the case with Uranus. It is therefore very reasonable to believe that he therefore inquired at the first source. Some readers will have to suppress a smile when they see in this newspaper the childish and vain way it is assured as a universally known matter that of all German astronomers, Professor Bode was the 'only one' who had an 'exclusive' correspondence with Piazzi. But we really must deny this report, since it is known for certain that Piazzi corresponded with 'three more German' astronomers. We can therefore not permit such absurd and unfounded reports of our highly respected friend, Professor Bode, to be published. Professor Bode himself cannot be indifferent when such incompetent, assiduous minds heap this type of wretchedness on his name in public papers, which could shed an ambiguous and false light on worthy and modest scholars.

In the same newspaper that claims to amend the misunderstanding with Seyffer's letter, some very odd errors occur. For example, it is assured that the German astronomers had conferred the name Hera to this new planet. But – our readers may be unaware of it – this name was a designation 'suggested' by the Duke of Gotha many years before the discovery of this body. We need therefore only refer to our first report of this suspected planet in our Monat. Corr. *Volume III B, p. 621. If we have used this name, it occurred very seldom and merely for the sake of brevity, in order to avoid always having to repeat the long reference of 'Piazzi's newly discovered planet.'*

Since Piazzi has baptized his own child and named it Ceres Ferdinandia, which is entirely within his right as the discoverer, and since all of his correspondents have been asked to use this designation, we on our part also subscribe to this fitting designation with genuine and therefore greater pleasure, because the King of Naples, being an eager protector and patron of astronomy, as well as the magnanimous founder of a new, splendid observatory, indisputably deserves our gratitude, since he not only started to build an observatory but completed it; not only bought the most valuable and splendid English instruments and kept them in boxes and crates in junk rooms but rather put them where they belong, entrusted these splendid instruments not to unskilled and lazy hands, but rather to a scholar of recognized merit and skillfulness, and placed him in a position to promote his work and observations to print at the expense of the King.

Since then, in such a short time, the most helpful and brilliant fruits have come from the Palermo Observatory, the learned world has been given several volumes of the most valuable observations, and this temple of Sicilian Urania has been immortalized, with its founder and priest, for millennia through the remarkable discovery with the coming new century. Piazzi therefore says in his discourse, and rightly so, that Ferdinand IV has more of a right a place in the heavens than some other protectors of astronomy.

We have given Dr. Olbers' reasons in the previous volume (p.367) as to why the calculated positions of this new body within an orbital hypothesis will pretty well maintain the mean between those which we have been able to calculate in an elliptical path. Dr. Olbers' suggestion, that in locating this new planet, one should start out from the points calculated from orbital elements and search through the same latitudinal parallels a few degrees backwards and forwards is indisputably the only and best method which can be recommended and methodically followed. Accordingly, we have decided to calculate the following slight ephemeris of the course of this planet until year's end according to the above orbital elements by Piazzi, which correspond fairly closely with the entire series of his observations, and this will prove a small service to all astronomers and enthusiasts of

Geocentrischer Stand der Ceres Ferdinandea vom 1 Novbr. bis Ende Decbr. 1801.

1801	Geocentrische Länge	Geocentr. Breite nördl.	Gerade Aufsteigung	Abweichung nördl.	Im Meridian	Log. der Entfernung von der Erde	Verhältnis der gesehenen Helligkeit
	Z. ° '	° '	° '	° '	U '		
1 Nov.	5 5 4	8 2	159 58	17 8	19 55	0,473064	0,429
7 —	5 7 2	8 22	161 59	16 42	19 39	0,462163	0,451
13 —	5 8 55	8 44	163 56	16 18	19 23	0,450653	0,476
19 —	5 10 43	9 6	165 47	15 58	19 7	0,438562	0,504
25 —	5 12 24	9 30	167 32	15 41	18 51	0,424905	0,537
1 Dec.	5 13 57	9 56	169 10	15 28	18 34	0,421706	0,569
7 —	5 15 22	10 23	170 41	15 9	18 16	0,399005	0,606
13 —	5 16 38	10 51	172 4	15 15	17 58	0,384882	0,647
19 —	5 17 44	11 21	173 17	15 16	17 40	0,370411	0,692
25 —	5 18 39	11 52	174 21	15 23	17 20	0,355698	0,741
31 —	5 19 21	12 24	175 14	15 36	17 0	0,340855	0,794

Fig. 10.7 Geocentric position of Ceres Ferdinandia from November 1 until the end of December 1801

astronomy. The inscription of the columns makes it easy to recognize the contents. But one, which carries the inscription – "Relation of the Observed Luminosity" – needs to be dealt with in more detail (Fig. 10.7).

At the beginning, as our readers have seen above, Piazzi and [Nicola] *Carioti judged the luminosity of this new planet to be similar to that of a magnitude 7 or 8 star. Later, towards February 11, it appeared to Piazzi as being smaller and visibly diminished, which he partly attributed to the murky and hazy condition of the atmosphere. But when we calculate the distance of this planet from Earth for both epochs in the orbit, it follows that the distance of this body was 1.924 from us on January 1 and 2.432 on February 11 – that is, in recent times approximately ¼ of the first distance further removed from us. The light, or rather, the 'observed luminosity' of this planet, must accordingly also have visibly lessened. The observed luminosity with which we see an innately not radiant planetary body depends on the amount of light which the planet 'receives' from the Sun and from the amount of light it 'sends' to us. The former is in direct proportion to the illuminated surface and in the inverse squared distance from the Sun. The latter is in the inverted proportion of the square of the distance from Earth. We abstract here from the modifications to which the falling and reflected light may be subjected, depending on whether the particular physical characteristics of the surface of the planet catches, absorbs and reflects light. Our intention is to merely compare the present visible luminosity of the new planet with the past, which Piazzi had judged as a magnitude 7 - 8 star upon its first appearance. Let therefore the diameter on Ceres on January 1, 1801 = D; its distance from the Sun = S; from Earth = R; its visible luminosity = H. Also, for any other epoch the diameter = d; distance from the Sun = s; from Earth = r; visible luminosity = h: thus: H: h:: $D^2/S^2R^2 : d^2/s^2r^2$*

If we allow the greatest luminosity yet seen H = 1, then, since we must assume the diameter of the planet within the orbital hypothesis equally great, the relation/proportion of luminosity is $h = S^2 R^2/s^2 r^2$

According to this formula, we have calculated the proportion of the visible luminosity. On January 1, it = 1.000 and on February 11 = 0.625. Our short Ephemeride indicates that we can expect the same luminosity as on February 11 only toward December 30, but that by the end of this and by the beginning of the next year, this luminosity will not approach the one with which this body had gleamed at the time of its discovery. This luminosity will only fully return towards the end of January 1802.

From November 19 to 25, this planet will stand in very close proximity to the star 8 Leo; between the 25th and 31st of December, it can be located in the proximity of the very distinguishable star β Leo, and then from November 1 to December 31, it will stop parallel to the above star, and y^1 and y^2 Leo.

Dr. Burckhardt's elliptical elements give exactly the same positions for the 'geocentric latitudes,' but they give the 'geocentric longitudes' from 2 to 3 degrees greater. One would therefore do better to hold exactly on the latitudinal parallel rather than the equatorial, since our instruments are mainly fixed on the latter. If we assume an error or an uncertainty of 2° in geocentric longitude, this will strongly affect the calculated 'declination'. [End of the MC paper for November 1801.]

In London, *The Monthly Magazine* November (1801b) included a comprehensive report on Ceres. It is included here to show what the English reading public was being told about Ceres, as they were not reading the German publications of Zach and Bode:

The celebrated Astronomer M. von Zach, had communicated to Dr. Olbers, of Bremen, M. Piazzi's observations of the 1st and 23d of January; and on the 30th of May received from him a calculation of new elements of the planet's orbit. These elements, however, could not be determined with any great exactness, as the observations are only twenty-two days distant from one-another, and are only given in minutes. Dr. Olbers found, however, from all the data then known, the Diameter of the orbit 2,947465 – Longitude of the ascending node, 2s 21° 55′ 10″ – Inclination of the orbit, 7° 54′ 38″ – Heliocentric longitude on the 1st of January, 1801, 2s 7° 40′ 36″ – Sidereal Revolution, 1841,24 days = 5.04096 years – Diurnal heliocentric motion, 11′ 43,″87 – Annual motion, 71° 24′ 57,″6 – With these elements it would have been difficult to calculate before-hand the course of the planet, so as to be able to find it again on its re-appearing in the morning in August, if it be not at first light distinguishable from a star of the 8th magnitude; "for, probably, (says Dr. Olbers) it has a considerable eccentricity. In opposition it may, perhaps, increase in luminousness, so as to equal a star of the 6th magnitude. I have little doubt that it will be found in La Lande's [star] Catalogue."

On the 16th of May Professor Bode writes to M. von Zach, "That it gave him great pleasure to find, that M. von Zach agreed with him in opinion respecting the Piazzian comet, and that Oriani and Piazzi himself incline towards the same opinion. –How often (continues he) have I wished that I might live to witness this discovery – I have been several times laughed at by others about my ideas of the harmonic progression in the distances of the planets. Adopting 2,75 for the distance, I find the heliocentric difference of longitude, betwixt the 1st and 23d of Jan. very well corresponding with the observations; the planet goes to its node, which I placed in Taurus: its inclination must exceed 6°; and this I think was one of the causes why it was not sooner discovered."

Till towards the end of May M. von Zach received no farther accounts relative to this star. He had communicated to his friends the Parisian astronomers the observations and elements calculated: and, not doubting that La Lande, to whom Piazzi had sent the first account of the discovery of the comet, had likewise been made acquainted with the subsequent observations and conjectures, he requested him to send to him an account of all the particulars that had come to his knowledge relative to the new planet.

But to his no small surprise he received, in the beginning of June, several letters from Paris; one from the Senator La Place, dated the 29th of May; from La Lande and Burckhard [sic], of the 26th of May; from De Lambre, of the 24th of May from Méchain, of the 26th of

May; from Henry, of the 28th of May; in which none of the six astronomers, who had communicated several important observations and new discoveries, writes even a single syllable about the new planet! Méchain only makes mention of Piazzi's comet;—from which it appears, that so late as the end of May they knew nothing of the conjecture of its being a planet; although the astronomers in Germany had been made acquainted therewith by Professor Bode already in the month of March. – Méchain in his letter to M. von Zach, of the 26th of May, merely says," Have you seen the comet, which the journals announce to have been discovered at Palermo last January? No one here has yet found it. Our astronomers have not discovered any since that of the month of December, 1799. I sometimes look out for them, but without success."

On the 10th of June, M. von Zach received another letter from Professor Bode, in which he says, "Piazzi's first letter I received on the 20th of March, and on the next post-day, the 23d, I answered it. But he did not wait for my reply; and – conceive my joy and at the same time my vexation! – I received a second letter from Piazzi, in which I found only the following few words relative to the newly-discovered planet: "I wrote to you in January, informing you that I had discovered a comet in Taurus, which comet I continued to observe till the 11th of February, when I was attacked by a dangerous disease, from which I have not entirely recovered. As soon as the state of my health will permit, I shall calculate elements for it, and send them to you. In the mean time I have communicated my observations to M. La Lande." It is remarkable that he still calls the star a comet, as in his first letter."

On the 18th of June, M. von Zach received a letter from Dr. Burckhardt, in Paris, from which we learn the following particulars: La Lande had received Piazzi's observation on the 31st of May, when Dr. Burckhardt immediately began to calculate its orbit. Two days later they received Von Zach's and Oriani's investigations, which gave them cause to hope that the supposed comet would prove to be a planet. Dr. Burckhardt had already found that the arc described by it was not considerable. The small geocentric and heliocentric motion of the comet gave him a great deal of trouble in calculating its orbit. He had first chosen for this purpose the observations of the 14th, 21st, and 28th of January: but from this circumstance found himself under the necessity of selecting the observations most distant in time from one another, viz. those of the 1st and 21st of January, and of the 11th of February. During these 42 days the geocentric longitude of the comet varied only 3°, and the heliocentric longitude only 10½°. On attempting to correct, by Laplace's methods, the parabola found by his method, he discovered that nothing in this respect could be effected by the conditional equations. He then tried La Place's method of approximation, but with as little success: the unavoidable errors of observation having too great an influence on the differences of the geocentric longitudes and latitudes. He now proved eight hypotheses by means of La Place's method of correction, but without approximating nearer to the truth. He then calculated the following orbit which agrees with the three observations to within 2½ minutes:

Diameter of the orbit, 2,74. – Epoch, 1801, 2^s 8° 16′ 20″. – Ascending Node, 2^s 20° 15′. – Inclination of the orbit, 11° 21′.– Period of revolution, 4½ years.

However various the trials that had been made: yet, as it did not thence follow, that it was impossible to find a parabola for these observations, he determined to apply a method, which had often proved successful, when all other methods of interpolation failed. Putting the logarithm of the distance from the sun equal 0,378, the smallest error was ± 8′; then putting the logarithm of the distance 0,378, the smallest error was ± 4. It was therefore necessary still more to diminish the distance; and after 20 hypotheses he found the following parabola:

Place of the ascending node, 2^s 20° 50′. –Inclination of the orbit, 9° 41′.– Place of the perihelium, 4^s 8° 38′ 25″. – Smallest distance from the sun, 2,21883, its log 0,3461250. – Logarithm of the diurnal motion, 9,4409408. – Time of the passage through the perihelium, 1801, 30th June, 19h. 1′.

Dr. Burckhardt is of opinion, that there is no other parabola that more nearly agrees with these three observations. The errors in the longitude are on the 14th and 28th of January – 1' 47" and +38. But Piazzi had not mentioned any thing respecting the accuracy with which he was able to observe the comet.

On the 21st of June M. von Zach received the promised continuation of Dr. Burckhardt's researches. He had calculated an ellipsis for the comet, although the arc it had run through was too small for us to expect great accuracy, but he thought he should thereby facilitate the finding of the star.

Place of the ascending node, 2, 20° 58' 30".—Inclination of the path, 10° 47' 0".– Place of the aphelium, 2s 8° 59' 37".– Time of the passage through the aphelium, January, 1801, 1,3328. – Excentricity, 0,0364.– Logarithm of half the great axis, 0,4106586.– Period of sidereal circumvolution, 4,13 years. This ellipsis represents, within a few seconds, the longitudes and latitudes of five observations. It would have been easy to obtain a greater degree of accuracy, but he thought it quite superfluous, as the arc run through is so small." The above ellipsis gave Dr. Burckhardt the following

1801	*Medium Time.*	*Geocentr. Long.*	*Geocentr. Lat.*
20th June	*13h 4'*	*101° 45'*	*30° 26 N.*
17th July	*1 43*	*113 3*	*4 6*
12th August	*10 54*	*124 21*	*4 51*
7th Sept.	*16 19*	*135 28*	*5 41*
12th -----	*22 ---*	*137 40*	*5 52*
18th -----	*3 ---*	*139 50*	*6 3*
23d -----	*8 ---*	*141 58*	*6 15*
28th -----	*13 ---*	*144 5*	*6 27*
3d Oct.	*17 41*	*146 9*	*6 40*
8th ------	*22 ---*	*148 12*	*6 53*
14th ----	*3 ---*	*150 12*	*7 8*
19th ----	*7 ---*	*152 11*	*7 22*
24th ----	*11 ---*	*154 8*	*7 37*
29th ----	*14 45*	*156 3*	*7 53*
3d Nov.	*18 ---*	*157 56*	*8 9*
8th ------	*22 ---*	*159 48*	*8 26*

The Search for the New Planet Ceres by Olbers

This paper, published in the *BAJ* in 1805, contains five letters by Olbers from 1801 and 1802. They are printed here in chronological order, beginning with the first, dated November 14, 1801. The remaining three are in the next book in this series. He is addressing these letters to the editor of the *BAJ*, Johann Bode.

November 14, 1801:

In October as well as November the conditions at your position will make it impossible to observe the new planet. Here (in Bremen) we had only one night in November (from 9th to the 10th) that I could use to mark the tiny stars that I found where I expected the new planet to be. But till now a constant fog prevented me from comparing my observations with the sky and soon the light of the Moon will make it very difficult to observe.

Here are some of my calculations of probable orbit positions in December for Ceres as well as the distance from the Earth.

1801 December	*1*	*h*	*16*	*Longitude*	*5Z*	*12°*	*37'*	*Latitude*	*10°*	*11'*	*Dist.*	*2.5994*
December	*6*	*h*	*16*		*5*	*13*	*45*		*10*	*33*		*2.5321*
December	*11*	*h*	*16*		*5*	*14*	*46*		*10*	*57*		*2.4650*
December	*16*	*h*	*16*		*5*	*15*	*41*		*11*	*22*		*3.3989*

In December Ceres is closer to the Earth than it was on February 11.

December 10, 1801:

The weather seems to be on purpose against us, so we can't discover Ceres. At least it is persistently cloudy here. Only on the night of December 4 and 5 was I able to search without luck.

The two orbits of Gauss show the nature of the Piazzien planet with certainty: but it is the truth that finding it is more and more difficult. To search for and see it as a planet with telescopes and achromats is very boring.

Continued Reports about the Long-Suspected New Primary Planet of our Solar System

MC (December 1801), p. 638.

Since the generally bad weather which has prevailed until now, leading to the coming frost and the imminent beautiful winter nights everywhere, especially in our northern regions of Germany, will very much impede finding Piazzi's new planet, every easing in the weather and every new hope will be doubly welcome. What makes finding this body in this present season so difficult is that a continuous sequence of observations cannot be much relied on, and this is required to recognise this wanderer according to its own movement, since it has no quality which distinguishes it from other telescopic stars. In our regions, observations can be interrupted often by 8, 14, and several days more because of overcast skies during the winter months, a state which is not at all unusual. Every attempted location must be started anew after a long interruption. A great hope of assistance and relief has been granted us by the recently communicated investigations and calculations of Dr. Gauss in Brunswick. They give us both a new and high degree of probability that Piazzi's newly discovered star is truly a planetary body which moves between Mars and Jupiter, according to Kepler's Laws.

We hasten to communicate his calculations all the more since his new elliptical path differs considerably from the elliptical path of Dr. Burckhardt and the two orbital paths of Dr. Olbers and Dr. Piazzi, which we communicated in earlier issues along with the position of the planet calculated in advance. Also, its declination in the current month can extend from 6 to 7° in geocentric longitude from the positions calculated by Gauss. It is therefore of great importance to communicate these remarks to the practical astronomers as soon as possible because they will discover through this that they 'invariably' have to extend the area in space, in which this new and elusive body is to be found, by 6 to 7° to the east. [The reasons are as follows:] 1) According to Dr. Gauss's calculations and as Dr. Burckhardt and Olbers have assumed, Piazzi's observations lie neither near the perihelion or the aphelion, but rather nearly in the centre of the two. 2) If the eccentricity of the path is not as slight as Piazzi believes according to his calculations, then the longitudes calculated in advance according to his elliptical elements can easily differ by approximately 7° from those calculated this month according to the orbital hypothesis. 3) If the elements calculated by Gauss completely justify Piazzi's observations, they will be freed from the suspicion of a lack of exactness as suspected by Dr. Burckhardt and Olbers upon investigating them by their discrepancies.

Gauss's ellipse proves all of this. How much confidence it must arouse will be recognised by astronomers by the exactness with which it represents all of Piazzi's observations. Dr. Gauss was led to these calculations through several investigations of physical astronomy which led him to some not unimportant additions to the theory of the determination of planets in conic sections of all varieties, and of which he was so good as to give us some information. We hope to talk with our readers about this some other time, since this debate would turn the discussion too far from our present subject; we will therefore limit ourselves to what is immediately related to the investigation of the path of Ceres Ferdinandia.

Initially, Dr. Gauss chose the three observations of January 2 and 22 and February 11 to determine the path, in which he took this data exactly as it appears in the September issue of Monat. Corr. p, 280. According to a characteristic [peculiar] *method, he immediately found, on the fourth attempt the following elements:*

Aphelion 330° 14′ 33″
Ω 81 8 50
Inclination of path 10 32 19
Log. of semi-major axis 0.4381058
Eccentricity 0.0832836
Epoch: 31 December 1800 in Palermo
Mean heliocentric longitude 77° 54′ 29″ From this it follows that:
largest focal point equation 9 32 57
Semi-major axis 2.74226
Sidereal period 1658 21 days
Daily mean sidereal movement 781″.355

These elements exactly represent the two outside observations, while the middle one has an error of 2″ in longitude and latitude. Necessary consideration has been given to aberration and precession.

After this first successful test, Dr. Gauss undertook a second calculation of this path. He left out the observations of January 2 and 22 and selected instead those of January 1 and 21 in combination with that of February 11, which he did not want to omit, so that the interim time would remain as great as possible. With the fourth hypothesis, this calculation gave him the following elements:

Aphelion 330° 33′ 20″
Ω 81 2 35
Inclination of path 10 36 30
Log. of semi-major axis 0.4370335
Eccentricity 0.0705553
Epoch: 31 December 1800 in Palermo
Mean heliocentric longitude 76° 28′ 14″.27 From this it follows that:
largest focal point equation 8 5 19
Semi-major axis 2.73548
Sidereal period 1652 ½ days
Daily mean sidereal movement 784″.25

According to these elements, all Piazzi's observations cited in the September issue (p. 280) coincide as follows (Fig. 10.8):

As excellent as this agreement indeed is, Dr. Gauss regards it as highly possible that his elements could considerably deviate from the true ones as they deviate amongst themselves, since the described path is so slight and amounts to only 9° 15′ 35″ from the first to the last observation. In the meantime, since this path is 'possible' though not 'highly probable', practical astronomers have twice the reason to take this into consideration in searching for this body and to choose it as their most favourable connecting thread, since of all attempted and calculated paths, none represents such a close and exact unification of all observations as Gauss's ellipse.

Incidentally, we also remark that Dr. Gauss considered in all these calculations both the printing error of one minute in the position of the Sun indicated in the October issue (p. 365) and a second error occurring on January 13 with the longitude of the Sun, which he assumed to be 9Z 23° 13′ 13″.8.

1801	Berechnete Länge	Berechnete Breite	Fehler der Länge	Fehler der Breite
Jan. 1	53° 22' 58,"42	3° 6' 42,"09	+ 0,"12	— 0,"01
2	53 19 37, 02	3 2 23, 78	— 7, 28	— 1, 12
:: 3	53 16 43, 67	2 58 6, 70	— 14, 93	— 3, 20
4	53 14 14, 03	2 53 51, 16	— 1, 47	— 4, 44
:: 10	53 7 54, 51	2 28 53, 13	— 4, 59	— 7, 47
13	53 10 18, 77	2 16 48, 78	— 18, 83	— 10, 92
14	53 11 55, 25	2 12 51, 23	— 5, 95	— 5, 47
19	53 26 0, 37	1 53 34, 19	+ 1, 17	— 4, 01
21	53 34 22, 68	1 46 5, 98	+ 1, 38	— 0, 02
22	53 39 7, 88	1 42 25, 06	+ 6, 08	— 3, 04
23	53 44 15, 74	1 38 46, 25	+ 0, 04	— 5, 85
28	54 15 20, 88	1 21 4, 07	+ 5, 18	— 2, 83
30	54 30 14, 25	1 14 13, 73	+ 5, 25	— 2, 27
31	54 38 11, 25	1 10 51, 99	+ 3, 95	— 2, 61
::Feb. 1	54 46 28, 37	1 7 32, 09	+ 9, 07	+ 1, 19
2	54 55 5, 09	1 4 14, 25	+ 7, 19	+ 3, 75
5	55 22 50, 25	0 54 32, 88	+ 6, 85	+ 3, 98
8	55 53 23, 15	0 45 9, 20	— 6, 38	+ 4, 20
11	56 26 39, 97	0 36 2, 90	— 0, 03	+ 0, 00

Fig. 10.8 Comparison of Piazzi's observations with the second elements of Gauss

This is how far Dr. Gauss had got in his calculation of the path of Ceres Ferdinandia when he was so good as to inform us of it. In the meantime, we had received Piazzi's discourse along with his improved observations, a faithful excerpt of which we communicated in the November issue. As minor as the these changes were, of which the 15" decrease of the right ascension of February 11 is the most considerable, and through which the previous elements were changed only a little, some more significant errors in printing and calculation in the reduction of these observations snuck in here and there. Since these errors are easily corrected, and the elements could be easily improved after this, we communicated all of Piazzi's observations to Dr. Gauss in the original Italian text. But before he had received our letter, he had already begun calculating a new path. As little as these second elements deviated from the observations, they still had a very remarkable regularity, from which it could be foreseen that these errors could still be appreciably reduced. He made the attempt, and since he had already largely completed this work by the time he received the improved elements, as Piazzi himself had reduced and calculated them, he regarded it as proper to complete them, especially since he knew in advance that the longitude for February 11, according to the new elements, would turn out to be 6" less, and so the corrected observation, naturally unknown to him at this time, would be almost halved by itself. In this way, he found the following "third" elements:

Aphelion 326° 53' 50"
Ω 81 1 44
Inclination of path 10 36 21
Log of semi-major axis 0.4414902
Eccentricity 0.0819603
Epoch: 31 December 1800 in Palermo
Mean heliocentric longitude 77° 34' 28" From this it follows that:
largest focal point equation 9 23 57
Daily mean helio. Movement 772".275
Daily tropical helio. movement 772".413
Sidereal period 1677.8 days

1801		Berechnete Länge	Berechnete Breite	Fehler der Länge	Fehler der Breite
		° ′ ″	° ′ ″	″	″
Jan.	1	53 23 2,34	3 6 43,63	+ 4,04	+1,53
	2	53 19 41,24	3 2 25,68	− 3,06	+0,78
	3	53 16 48,05	2 58 8,97	−10,35	−0,93
	4	53 14 18,47	2 53 53,79	+ 2,97	−1,81
	10	53 7 58,37	2 28 57,12	− 0,73	−3,48
	13	53 10 21,60	2 16 52,89	−16,00	−6,81
	14	53 11 57,70	2 12 55,36	− 3,50	−1,34
	19	53 26 0,59	1 53 38,01	+ 1,32	0,19
	21	53 34 21,99	1 46 9,53	+ 0,69	+3,53
	22	53 39 6,69	1 42 28,45	+ 4,89	+0,35
	23	53 44 14,08	1 38 49,44	− 1,62	,266
	28	54 15 17,11	1 21 5,91	+ 1,41	−0,99
	30	54 30 9,76	1 14 15, 2	+ 0,76	−0,88
	31	54 38 6,44	1 10 52,8	− 0,86	−1,79
Febr.	1	54 46 23,22	1 7 32, 4	+ 3,92	+1,64
	2	54 54 59,71	1 4 14,30	+ 1,8	+3,80
	5	55 22 44,30	0 54 3,72	+ 0,90	+2,82
	8	55 53 17,01	0 45 6,61	−12,49	+1,63
	11	56 26 34,10	0 35 58,96	− 5,90	−3,94

Fig. 10.9 Comparison of Piazzi's observations with the third elements of Gauss

These elements represent those reduced by us and the observations by Piazzi cited in the September issue (p.280) in the following way: The deviations and improvements cited by Piazzi now give the following revisions (Fig. 10.9): Because the right ascension of February 11 is decreased by 15, with the inclination of the ecliptic as 23° 28′ 12″, the longitude becomes 56° 26′ 26″.1, the latitude 35′ 59″.7 – therefore, the deviation of the longitude = + 8″.0 and the deviation of the latitude = –0″.74. After verifying the reduction of the right ascension and the deviation with these observations, where Piazzi's details deviate considerably from ours, Dr. Gauss found the longitude to be 55° 53′ 17″.7; therefore, the error is –0″.7. Dr. Gauss regards what remains, as well as the greatly deviating longitude of January 13, as we do. Since this observation deviates the most from the later as well as the earlier ones, and does not allow itself to be exactly represented without subjecting the remaining observations to constraint, Dr. Gauss suspects that some small error might have occurred. Meanwhile, one sees that the remaining minor improvements (November, p. 573) given by Piazzi changes little in the exactness with which the observations are represented through these final elements.

As modest as these errors are, Dr. Olbers made still a third attempt to somewhat improve the exactness. The errors of longitude of the last elements in February are completely positive if the minor improvements by Piazzi are examined; the errors of latitude also incline to one side towards the end. Dr. Gauss therefore looked for new elements which would make the calculated longitudes and latitudes of February somewhat smaller while the remaining would more or less maintain their size. Whether he did not immediately compare these elements with the complete observations, which is also rather superfluous, he still believes to be able to assure that they maintain the mean between the observations as much as can be, and that the errors of no observation exceeds 5" in longitude and latitude (except those of January 3 and 13), and have absolutely no regularity, but rather have a very irregular change of sign. Dr. Gauss doubts, not without reason, whether a 'noticeably greater' agreement of other elements can be found with 'this data', with which he does not want to have it said or understood that other relevant different elements could not give a similar agreement, especially if, with these delicate calculations, in which a few seconds are a decisive factor, somewhat different determinations [settings] from the Sun would be used. Thus, Piazzi's longitude of the Sun in February differs from ours by half a minute as well as the influence of the solar longitude being less in proximity of the quadrature than in other longitudes. Dr. Gauss therefore believes that it would not be useless to determine the errors in the solar tables through very accurate observations (for these times) and to improve them accordingly. The fourth elements are as follows:

Aphelion 326° 27' 38"
☊ 81 0 44
Inclination of path 10 36 57
Log. of semi-major axis 0.4420527
Eccentricity 0.0825017
Epoch: 31 December 1800 in Palermo
Mean heliocentric longitude 77° 36' 34" From this it follows that:
largest focal point equation 9 27 41
Daily mean heliocentric
tropical movement 770".914

Dr. Gauss estimated the following positions of Ceres from these elements (Fig. 10.10). The mean time is midnight in Palermo. To calculate the location of the planet according to these elements more accurately or for a longer period of time, the following formulas are supplied to this end (Fig. 10.11):

We conclude this report with a comment by Dr. Gauss regarding the inclination of the path of Ceres, which was so striking to astronomers because of its greatness. Whether he agrees with the judgment of a few astronomers that no physical reason justified us in expecting a slight inclination to the ecliptic of all planets of our Solar System having a circular path, it seems to him that the paradox would be significantly lessened and the analogy somewhat saved if this planetary path were related to the solar equator, as Laplace had done in calculating the paths of Uranus's satellites on their own plane (A. G. E. II B., p. 259) to

1801.		Geocentrische Länge			Geocentrische Breite nördl.		Logarith. des Abstandes von der ♁	Logarith. des Abstandes von der ☉	Verhältniss der gesehenen Helligk.
		Z	°	'	°	'			
Nov.	25	5	20	16	9	25	0, 42181	0, 40468	0, 6102
Dec.	1	5	22	15	9	48	0, 40940	0, 40472	0, 6459
	7	5	24	7	10	12	0, 39643	0, 40479	0, 6855
	13	5	25	51	10	37	0, 38296	0, 40488	0, 7290
	19	5	27	27	11	4	0, 36902	0, 40499	0, 7770
	25	5	28	53	11	32	0, 35468	0, 40512	0, 8295
	31	6	0	10	12	1	0, 34000	0, 40528	0, 8869

Fig. 10.10 Comparison of Piazzi's observations with the fourth elements of Gauss

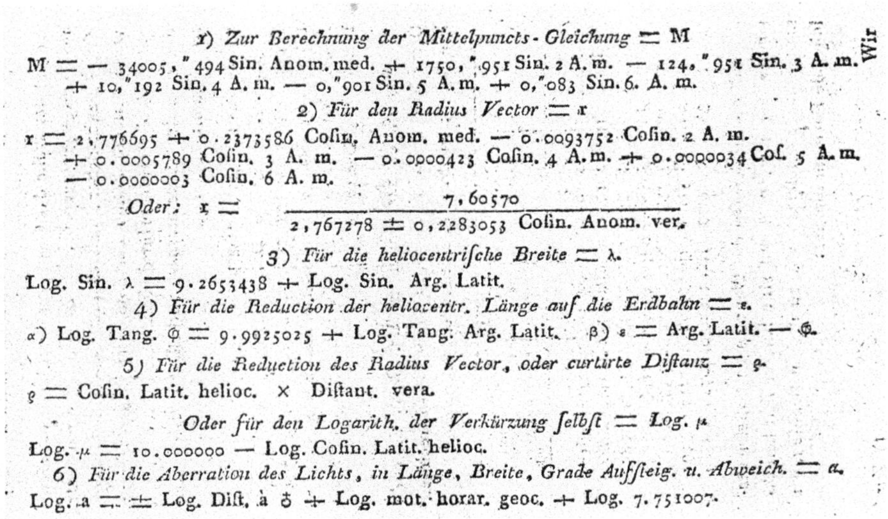

1) *Zur Berechnung der Mittelpuncts-Gleichung* = M

M = − 34005,″494 Sin. Anom. med. + 1750,″951 Sin. 2 A. m. − 124,″951 Sin. 3 A. m.
+ 10,″192 Sin. 4 A. m. − 0,″901 Sin. 5 A. m. + 0,″083 Sin. 6. A. m.

2) *Für den Radius Vector* = r

r = 2,776695 + 0.2373586 Cosin. Anom. med. − 0.0093752 Cosin. 2 A. m.
+ 0.0005789 Cosin. 3 A. m. − 0.0000423 Cosin. 4 A. m. + 0.0000034 Cos. 5 A. m.
− 0.0000003 Cosin. 6 A. m.

Oder: r = 7,60570 / (2,767278 ± 0,2283053 Cosin. Anom. ver.)

3) *Für die heliocentrische Breite* = λ.

Log. Sin. λ = 9.2653438 + Log. Sin. Arg. Latit.

4) *Für die Reduction der heliocentr. Länge auf die Erdbahn* = ε.

α) Log. Tang. φ = 9.9925025 + Log. Tang. Arg. Latit. β) ε = Arg. Latit. − φ.

5) *Für die Reduction des Radius Vector, oder curtirte Distanz* = ρ.

ρ = Cosin. Latit. helioc. × Distant. vera.

Oder für den Logarith. der Verkürzung selbst = Log. μ

Log. μ = 10.000000 − Log. Cosin. Latit. helioc.

6) *Für die Aberration des Lichts, in Länge, Breite, Grade Aufsteig. u. Abweich.* = α.

Log. a = ± Log. Dist. à ♁ + Log. mot. horar. geoc. + Log. 7.751007.

Fig. 10.11 Formulae used to calculate the position of Ceres

which one must relate them. If we compare the planes of each planet with each other, it is revealed then that the path of Ceres is no greater against the paths of other planets than that of Earth. If we compare the path of Ceres with that of Earth, we are comparing both extremes of the solar system. If one thinks of a plane which lies approximately in the middle of the planes of all eight planets, then the inclinations compared with this one will all be small enough. It seems very remarkable that the solar equator is approximately so situated, only with the restriction [qualification], that the path of Ceres, together with the ones of Venus and Mercury, are the least inclined towards it – approximately 3 to 4° – while Earth's, on the other hand, is the most inclined.

Resumed News about the assumed new main Planet of our Solar System

MC (Jan. 1802), p. 89

Every bright morning we have been very eagerly searching Ceres Ferdinandea with our 8-foot meridian telescopes by Ramsden and our powerful night telescopes by Dollond at every station possible according to the different hypotheses – but in vain. We have often observed suspicious small stars on the parallel of Ceres, but the observations of the following nights or our star catalogues proved them as common, rarely observed fixed stars of the 8th or 9th order of magnitude. In this mild winter we have had only a few clear nights in November and December. Oftentimes the evening and the whole night through the sky was bright but during the necessary morning hours when dew was falling the sky became overcast with at least a hazy fog, which made it impossible for us to differentiate between the small stars.

How much the weather teases the practical astronomer at all times of the year in our northern hemisphere is known only to those who observe the sky eagerly and at certain moments. In order to show our readers what difficulties a practical astronomer has to overcome in the present season let me describe our efforts of December. The sky was clear and completely calm. This beautiful night allowed me to bring in a rich harvest of small stars of 7th, 8th and 9th order of magnitude. When the upcoming daylight prevented further observations and I finished my work with the culmination of the planet Uranus, I studied my obtained stock and found four strangers in the region where Gauss' ellipse places Ceres and which were not included in any of the known star catalogues. After having classified them as fixed stars and reduced them accordingly I found their positions for the beginning of 1800 as follows:

No.	*right ascension*	*n. declination*
Nr. 1	*178° 32' 3".1*	*11° 41.'5*
Nr. 2	*178 57 53.9*	*11 33 52".8*
Nr. 3	*preceded Nr. 4*	*11 39*
Nr. 4	*178 15 39.5*	*11 39 5.8*

I only noticed no. 3 but did not observe it since no. 4 followed immediately and I did not want to fail to observe that star that seemed to be of the 9th order of magnitude. Number 3 appeared extremely blurred and slightly misty. On further study I found no. 2 and no. 4 in La Lande's recently published Histoire célèste française *on page 225; his nephew Le Français observed both on April 6, 1796, the first at 37° 8' 44" and the latter 37° 8' 10". But no trace of no. 1 and no. 3. The next morning would show what kind they were. And it was eagerly awaited, but appeared no earlier than on December 18. The skies were not clear, though studded with stars, and veined with stripes of white fog; Jupiter glittered in his veil, and I was unable to discern stars of the 7th order of magnitude with my Dollond searcher. But I hoped to see the strangers at the powerful 8foot meridian telescope, because I determined them on December 7 quite accurately and the night was pitch black and my watch very exact. But when they were to culminate, I could neither find my suspicious stars nor the two of La Lande, despite my precaution of dimming all lights. Previously I had clearly seen stars of the 4th and 5th magnitude, e.g. σ, ι and ε in Leo and even later observed Uranus; but was unable to see the small disc, which I was used to see in clear skies and that I had observed on December 7 for the last time. And in this uncertainty I have remained until this very moment (December 27). Since December 27, for 20 days, there has not been a single clear morning except for some scattered views that would be reasonably suitable for searching for that star or investigating the two other celestial bodies. And now together with the new year thaw has set in and our hopes are vanishing to find an opportunity for a classification or verification of those two stars. That is the reason why we give here the observed apparent right ascension of December 7 of number 1. Other astronomers might be so lucky as to precede us in the examination. On December 7, 1801, at 18h 48' 10".3 mean time the apparent right ascension of no. 1 = 178° 33' 30".6; I estimated the declination 11° 41½. This position matches Gauss' ellipse quite accurately.*

My unfortunate fate was shared by all my foreign friends and correspondents. La Lande wrote that Dr. Herschel sent word to Paris that he, too, was looking for Piazzi's star, but until today in vain.

Méchain, Messier, Le Français, Bouvard and Burckhardt also report that their attempts and efforts are frustrated by bad weather. My German friends Schroeter, Olbers, Harding, von Ende, Bode and many more are complaining about the same.

Dr. Olbers, whose judgments and opinions are valuable, wrote on December 5: "They begin to waver, now that Ceres has been hiding for such a long time. But my belief is firm and steadfast. If Piazzi's observations are correct and true, and there can be no doubt, the consequence is with mathematical certainty the heavenly body observed by Piazzi is a planet roaming between Mars and Jupiter. And why is there already such a doubt? Because we have not yet found it? I am not at all surprised. How small Ceres must be and how awful the weather has been, at least here during October and November. If the skies were only clear this month it appears impossible Ceres could elude us."

Also Dr. Olbers found several suspicious stars on Ceres' parallel but they turned out to be regular fixed stars. At this opportunity he found that no. 328 according to Bode's star catalogue, a star that Professor Bode himself had determined, was missing in the sky [Bode published his first star atlas, *Vorstellung der Gestirne*, in 1782]. *Dr. Olbers said that as far as he knows (not yet verified) Ceres is not among La Lande's stars, and has never been observed. Ceres roamed in other regions than studied.*

Our readers have the possibility to read on p. 66 of this issue a note taken from La Lande's letters that Piazzi had improved his observations of Ceres once again and this now makes him very suspicious of this new planet. But we believe this news is based on a mere misunderstanding; Piazzi hardly corrected or changed *his observations. Our readers might recall that in one of the recent issues some slips of the pen had found their way into Piazzi's calculations of the longitudes and latitudes, the positions of the sun and mean times and thus misprints occurred in the printed version of those calculated tables, which were publicised properly even then. Prof. Piazzi became aware of these errors. And most likely any kind of correction must relate to these alone and not to the original observations. We made those corrections in question long ago. And we have not heard anything of new observations of Piazzi yet. But we can give our improved observations, which will match Piazzi's doubtlessly.*

For a calculation of these observations we used an apparent inclination of the ecliptic 23° 28′ 5″.3 as it was found at this year's summer solstice (June 1801) by Méchain and De Lambre with circles by Borda. For calculating the times and positions of the sun we used our solar tables by subtracting 7″.25 of the longitudes' epoch and added 2′ 27″ to the longitude of the apogee and neglected the perturbation equation for Mars. Thus, the calculated sun's longitudes correspond better with the sky, at least for the present time. All of Piazzi's observations and their elements are as follows: [This table was printed in the September 1801 issue of the *MC*.]

A series of three articles about Ceres were published in the French press in January 1802. The first two given here are by Lalande, the third is by Burckhardt. The French Republican dates are given, followed by the Julian dates.

Journal of Commerce

Pluviose 6, year 10. N 126. Paris Pluviose 5 / January 26, 1802

The small planet discovered one year ago by Mr. Piazzi and which was searched for by astronomers was found by Mr. Olbers in Bremen. It was seen on Pluviose 2 by Mr. Lalande at 187° 11′ RA 12° 17′ above the star Ro.

Moniteur

Pluviose 9, year 10. N129. Paris Pluviose 8/ January 29, 1802

The new planet is today the most curious thing in astronomy, so it is only natural to speak of it. First I must announce that Baron von Zach of Gotha was the first who found it again in the morning on the 8th of December. He only assured himself on the 31st of December since he observed 4 small stars and he was unable to discern which was the planet as soon as he was certain he wrote me about it, and sent me the positions: I communicated those on Pluviose 5 to all astronomers of Paris. But citizen Mechain had already observed 3 or 4 hundred small stars in that region where this small star was expected. But it is so difficult to see that he would not find it if it was among the number. Finally, he recognised it on Pluviose 4 at RA 188° 16′ and 11° 52′ decl. Citizen Delambre found her the next day, Le Francais and Burckhardt observed her on the 6th. Some compared her to an 8th magnitude star and others to a 9th magnitude star. This proves that she does not have two seconds in apparent diameter. Thus not 600 miles of true diameter. She is five times smaller than the Earth, and for that reason it took so long to recognise her. And Piazzi made this curious discovery only by accident.

Burckhardt's account of the new planet

The Moniteur, January 24, 1802

The planet that was discovered by Mr. Piazzi in Palermo on January 1, 1801, was found again by Mr. Olbers in Bremen on January 1, 1802, almost at the same place where it was expected according to several ephemerides calculated by Mr. von Zach. On January 2, 1802 at 11h 58′ 36″ mean time in Bremen it was at 185° 9′ RA and 11° 9′ declination north in the wing of Virgo, close to a star whose position was determined by Mr. Lalande in the Connaissance de temps, year 9, p. 254.

On January 5 at 5.30 p.m. its RA was 185° 43′ and declination north was about 11° 8′.

It appeared as a 9th magnitude star but was still increasing. With a telescope of a magnification of 106 times, one could not distinguish it from a small star.

On January 1 it fortunately formed a triangle with two small stars which are in the Histoire céleste *by Mr. Lalande; and the next day the triangle had already changed shape so the planet was recognised.*

It will be on the parallel of 20 Virgo. The elements of this planet have busied many an astronomer. Oriani, Zach and Bode suspected already very early this star could be a planet because it was observed stationary and without nebulosity. But with only two complete observations they could not confirm their hypothesis. Soon after Mr. Lalande was the first to receive a copy of Mr. Piazzi's complete observations who could not refuse to give them to someone who has been working in astronomy for such a long time. By means of those I will be the first to show that, in a memorandum that I presented to the Institut national, *that there was no parabolic orbit that could represent the observations although they only show an arc of ten degrees. I will show at the same time a circular and an elliptical orbit and I will show how much uncertainty necessarily remains on elements deduced from such a small arc.*

When Mr. Olbers received a more accurate copy of the observations, he for his part wanted to search elliptical elements but he found too great an uncertainty to not prefer circular elements; seeing that it seemed impossible to decide whether the planet was around its aphelion or its perihelion. I tried the first of both cases and Mr. Gauss tried the latter. At the same time he tried to represent Piazzi's observations. He came close except several seconds. Here are his elements:

Epoch 1801 2Z 7° 36′ 34″
Aphelion 10 26 27 38
Node 2 21 0 44
Inclination 10 36 57
Eccentr. 0.0825
Mean anomaly 9° 27′ 41″
Mean dist. 2.7673
Mot. diurn. hel. and trop. 12 50 914
Revolution 1681 days or 4 years 7 months
I found the revolution smaller by 5.5 months.

According to Mr. Lalande's calculations, Mr. Gauss' elements give a longitude greater by one degree than Olbers' observation; according to Mr. von Zach's calculations my elements gave 4° less and those of Piazzi 10° less than the observation.

The idea to search for this planet in the immense collection of observations of the Histoire célèste francaise *was so natural that it came to everyone working on the subject; but it was impossible to perform this search with any hope for success until the elements of this planet would be confirmed by new observations. And I did not hesitate to do so.*

Mr. Piazzi has called his planet Ceres Ferdinandea; Mr. Lalande proposes to call it Piazzi.

Some Particulars respecting the new Planet Ceres Ferdinandea.
Journal of Natural Philosophy (1802), vol. 12, p. 80.

In the two preceding papers [the monograph by Piazzi and a paper by Zach that he wrote in the MC] our readers are put in possession of everything relative to the discovery of this planet. The following are some of the observed places of the Ceres Ferdinandea, upon its being rediscovered lately, and the times noted are mean time (Fig. 10.12).

1801.

Dec. 7. Obſerved by Dr. Zach of Saxe Gotha, under ſome uncertainty of its being the planet.

		M. T.	R. A.	Dec. N.
		at 18h 48′ 10″	178° 33′ 31″	11° 41′ 30″

1802.

Jan.	5.	By Dr. Olbers, at Bremen,		
		17 30	185 43	11 8 0
	11.	Again, by Zach,		
		at 17 3 17	186 45 50	11 10 0
	26.	By Mechain, at Paris,		
		at 16 10 48	188 24 49	12 0 43
Feb.	3.	By the aſtronomer royal, Greenwich,		
		at 16 11 15	188 43 0	12 39 0
	4.	at 17 25 46	188 43 0	12 44 0
	12.	at 15 4 12	188 30 0	13 33 0

Fig. 10.12 Observations of Ceres in late 1801 and early 1802

On the 7th of February and subsequent days, it was observed by Dr. Herschel at Slough, and also by Alexander Aubert, esq at his observatory at Highbury House. The state of the weather at Slough did not admit of Doctor Herschel's seeing the new planet till Monday night the 8th current, he having previously received distinct notice as to its place from his friend Dr. Maskelyne, the astronomer royal, who for the first time observed it, like a star of the eighth magnitude, on the 4th current, in the morning. On the 9th and 10th of February at night, Dr. Herschel again traced the planet, and perceived its motion. It was not however till the morning of the succeeding day, that through a much clearer air, and at a more favourable altitude, with less obstruction from the moon-light, he obtained a distinct view of it for a short time. Through his ten feet reflector, with a magnifying power of six hundred, and higher powers, he perceived its disc, though very small, yet distinctly round and well defined; but saw nothing further which denoted a ring or a coma, or a satellite. The favourable state of the air appearing then very precarious, Dr. Herschel did not attempt to ascertain the apparent diameter of the planet by means of his lamp-micrometer, as some preparations and adjustments would have been necessary before he could have availed himself of that curious apparatus. But in order to form some estimate as to a point of so much importance, he adopted a happy expedient, and very suitable to the urgency of the moment. The Georgian planet being situated near at hand, in the same region of the heavens, he directed his telescope first to it, and then to the other, with his attention fixed upon making a comparative estimate of the apparent diameter of each disc. In order to this, and to avoid as much as possible certain fallacies to which this method is more or less liable, he was careful to form estimates over and over again, according as his telescope was last turned from the greater disc to the smaller; and vice versa. From such observations, frequently repeated, he concludes that the apparent diameter of the Ceres Ferdinandea is about a fourth part only of the apparent diameter of the Georgium Sidus. By applying therefore the proper calculation, Dr. Herschel has inferred that the real diameter of this newly found primary planet, called Ceres Ferdinandea, is only a little more than half of the diameter of our moon and less than 5/3ths *of it. The smallness of the planet, together with the great inclination of its orbit, are peculiarities which may probably lead to other discoveries in the solar system.*

The following are some places of the planet calculated forward in foreign journals, but corrected by the latest observations, showing nearly where the planet may be expected to appear: computed for 12h or 15h Greenwich time.

1802	*R.A.*	*Dec. N.*
Feb. 17	*188° 15'*	*14° 8'*
23	*187 39*	*14 49*
March 1	*186 50*	*15 31*

The planet will be in opposition to the Sun about the 13th of March.

The following are the elements of the planet as calculated by M. Gauss, of Brunswick, for an elliptic orbit.

Epoch 1801, Jan. 0, or Jan. 1, which of the two is uncertain

	2Z 17° 36' 34"
Aphelion	*10 26 27 38*
Ascending node	*2 21 0 44*
Inclination of orbit	*0 10 36 57*
Eccentricity 0.0825017 to its mean distance unity.	
Mean distance from the Sun 2.7673	
Mean diurnal heliocentric and tropical motion, say, 12' 50.914"	
Periodic time 1681 days, or 4 years 7 months.	

Baron von Zach, director of the observatory of Gotha, writes as follows to C. Méchain, administrator of the observatory of Paris:

"M. Schroeter of Lilienthal has seen, with his large telescopes, the new planet Ceres, under a disc of nearly 2". He suspects that it has two satellites. The planet is enveloped in a very thick atmosphere, for it appears to be surrounded with much nebulosity. I am very anxious to learn what Dr. Herschel will tell us respecting it: in the mean time I thought it my duty to write you this in haste."

Elliptic elements of the new planet, corrected by M. Gauss from his last observations:
Diurnal tropical heliocentric motion 770.7376"
Tropical revolution, 1681d 12h 9m.

It is a curious circumstance that the discovery of this planet has been long expected, and even in some measure predicted. Professor Bode, of Berlin, in his Terse Explanation of Astronomy, *Berlin 1794, section 387, has the following passage:*

"Is it probable that Uranus, or the Georgian planet, is really situated at the utmost limit of our solar system? This appears to be very doubtful, considering the immense space interposed between it and the nearest fixed stars. Other planets perhaps may be still more remotely situated, and may perform their revolutions unseen by human eyes. We can scarcely suppose that any planet exists nearer to the Sun than Mercury: but considering the proportions of the distances of the planets from the Sun, we observe between Mars and Jupiter, a distance far greater than a comparison of the other distances would lead us to expect, and this space may perhaps be occupied by a planet yet unknown."

Similar ideas seem to have been entertained in this country, even long anterior to the conjectures of Lambert and the German astronomers, as appears by the following, which is given as a note in a work lately published on astronomy, by Mr. Olinthus Gregory, teacher of mathematics, Cambridge.

"Mr. Maclaurin and other philosophers expected nearly one hundred years ago, that such a discovery as this of Piazzi would be made by some diligent astronomer; and the opin-

ion has been lately revived by Mr. Capel Lofft, a gentleman well known for his attachment both to the sciences and the muses. In the New London Review *for March 1800, this gentleman, in a critique on the Athenian Letters* [published by the Earl of Hardwicke in 1798], *ventured to offer some conjectures respecting an intermediate planet between Mars and Jupiter, the coincidences of which with the new discovery is very remarkable. He supposed that the distance of the intermediate planet from the Sun would be to that of Mars, either as 33 to 15, or as 22 to 15, the midway between which corresponds nearly with the fact. With respect to its diameter, be conceived it might be to that of Mars, as that of Mars to the diameter of the Earth; and then being not much more than half the diameter of Mars, and at five times the perigean distance, it would be seen from the Earth under an angle of 2½" or 3", while Georgium Sidus would appear under an angle of 4". These lucky conjectures were drawn from a certain kind of Pythagorean harmony, and are ingeniously defended in the Review just mentioned."* [The text by Lofft will be given in the next book in this series.]

The Gentleman's Diary; or, The Mathematical Repository, an Almanack for 1802, edited by O. G. Gregory.

On the 1st of January, 1801, M. Piazzi, *astronomer at Palermo, discovered another Primary Planet between the Orbits of* Mars and Jupiter. *In order to preserve the Honour of the Discovery, as well as the Observations, to himself, he kept it secret till the 11th of February, when he fell ill, discontinued his Observations, and so lost its Path; but early this Year, 1802, it was discovered again, by the Rev. Dr.* Maskelyne, *his Majesty's Astronomer at Greenwich, where a competent number of Observations are now making, in order to settle the Elements of its Theory, etc. It is about Half as large as the* Moon, *and One-third of the Distance between* Mars *and* Jupiter, *and performs its Period round the* Sun *in about four Years and a Half. It is named* Ceres Ferdinandea, *in Honour of the Goddess Patroness of* Sicily, *and the reigning Monarch of that Island and Naples. It appears as a Star of the eighth Magnitude, therefore much smaller than the Georgian Planet, which was first discovered by Dr.* Herschell, *in the Year 1781.*

[The original capitalizations and italics from the original are retained, as well as the unusual spelling of 'Herschell.' It is interesting that this British publication attributes the recovery of Ceres to Maskelyne, not Zach or Olbers. The editor, Olinthus Gregory, conducted a feud with Zach stretching over many years. See Gregory (1815)]

Continued Reports regarding the New Primary planet of our Solar System "correctly" suspected between Mars and Jupiter and now "finally" discovered: Ceres Ferdinandia.

MC (February 1802), p. 170.

Finally, the new primary planet of our Solar System– searched for with so much effort, with tireless diligence and enthusiasm, first discovered by Professor Piazzi in Palermo a year ago, doubted by some, expected by others – has again been discovered and found, like a starfish on the beach.

Without knowing it, the readers of Monat. Corr. *have been in possession of the first observation of this primary planet existing between Mars and Jupiter for already four weeks, for our observation of the star which we designated by Nr. 1 and which we had announced as a very suspicious stranger in the previous January issue of* Monat. Corr. *(p. 90) was really none other than the new, so eagerly awaited body – Piazzi's Ceres Ferdinandia itself.*

As early as December 17, I had the first suspicion that the magnitude ten star observed on December 7, Nr. 1, could very well be none other than the sought after planet. But since the sky was not very clear and somewhat streaky, and since other small fixed stars of the same size failed to appear to my meridian telescope, I was not very sure of myself. I therefore did not want to make a fuss as others have done, and therefore kept my suspicion in my bosom until better weather. I revealed my secret to no one other than the "most serene" founder of the Seeberger [Uranien] *Temple (as my duty required) and Professor Buerg who was present and living at the observatory. I had given only one non-local friend, senior appellate official, Baron von Ende, a veiled hint about it. On the advice of the former, in order to take possession of the epoch of the initial discovery, I reduced this stranger as though it really were Ceres by calculating its 'apparent' right ascension with the necessary mean time of the observation and therefore treated it just like a planet, as can be seen on p. 92 of the previous issue of* Monat. Corr. *The observation cited in said place is so accurately calculated that I could not reduce it better or more accurately upon the determination that this star was indeed the sought after planet. From December 17 to 27, the weather was not fair. I could therefore not confirm my suspicion before the end of the January issue of* Monat. Corr. *(December 27). I therefore had to content myself, in order to not mislead others, with drawing this suspected guest to the attention of other astronomers who enjoyed more favourable skies.*

The night of December 31 to January 1, it became first clear again with the coming of the frost, and I acquired on this night the "complete certainty" that the star Nr. 1, observed on December 7, was no longer there, but had moved from its position. There therefore remained no doubt that this small body observed on December 7 could have been none other than Ceres Ferdinandia, since its course and position fit almost exactly in Gauss's ellipse, according to which I gave chase to this small planet. But on this night, I not only verified the non-existence of star Nr. 1, but also, not finding it in its old position, I set my instrument on the parallel of Ceres and observed three strangers in this area, all of magnitude 8 to 9. Their positions were as follows:

		RA			DEC	
No.	*a*	*178°*	*59'*	*30".0*	*11°*	*15'*
No.	*b*	*182*	*27*	*22. 5*	*11*	*5*
No.	*c*	*184*	*44*	*::*	*11*	*4*

Star Nr. c again corresponded to Ceres and its movement was once again in accordance with Gauss' ellipse. But unfortunately, this observation is somewhat doubtful. I had namely only taken it on two lines of my transit instrument and since this dim, small planet hardly illuminated the lines without immediately disappearing, I did not notice in the darkness which of the five lines of the transit instrument were the observed ones; therefore, the observation was uncertain to an interval of a line. The above quotation of the RA of Nr. c is accordingly founded on an uncertain premise in the reduction of this line. The sky was overcast until January 10. On the 11th, it cleared and I could find neither Nr. 1 nor Nr. c at their previous positions, rather, this time the planet was at 17hr 3' 17".4 mean time in 186° 45' 49".95 apparent right ascension and 11° 15' northerly declination: all of which showed at this time a change of position corresponding to the planetary theory, and consequently, there remained no doubt that the so ardently awaited and strenuously sought after Ceres was really in existence and fortunately found.

On the following day, I received a letter from Bremen from Dr. Olbers dated January 6, in which he gave me the pleasant report that he too (for several other astronomers seem to have made the same discovery) had again found the so long sought after Ceres on January 1, 1802, precisely on the anniversary of its first discovery. Following several murky days in Bremen, the weather finally cleared up on January 1. With an excellent comet finder, Dr. Olbers explored the area of the heavens between β Leonis and p Virginis and registered all the stars visible to this instrument in a small chart which was designed for this purpose. On January 2, he saw with glad surprise that one of these stars had changed its position to not far from 20 Virginis, according to Flamsteed. On January 1, he created a right-angled triangle with Nr. 191 of Bode's star catalogue, a star of Lalande's, from his Histoire cèléste *and this newly discovered star. But on January 2, it had moved noticeably closer to these two stars, and its angle had become very obtuse. He compared it on the circle micrometer with Nr, 191 and found, with the mean time being 11hr 58' 36", that the right ascension of Ceres = 185° 7' 40" and the declination 11° 6' 30" northerly.*

On January 3, 4 and 5, it was dull. On the morning of January 6, it cleared up again, and he saw at 5:30 a.m. with pleasure that Ceres had moved from its position as the theory required. From there, he observed the planet on January 5 at 17hr 30' 0" mean time somewhat doubtfully in 185° 43' 7" apparent right ascension and 11° 7' 56" northern declination.

Until January 15, it was dull in Gotha. On the 16th, the sky allowed a repeated observation due to the extreme cold. I immediately found Ceres at 16hr 46' 25".6 mean time in 187° 27' 53".25 right ascension and 11° 26' northern declination.

On neither on these nights, nor on previous ones, during which Professor Buerg had the goodness of assisting, did he succeed in seeing or observing this small planet in the meridian quadrant. The reason for this lay undoubtedly in the not entirely transparent atmosphere which was also filled with ice needles and in the illumination of the lines of the telescope. Only with great effort could I see this body in my transit instrument opened to 4 inches with great light transmitting capacity, and recognise the cross-hairs. The weaker quadrant telescope, which has only a 1½ inch aperture, held the illumination of the lines still less and it was consequently impossible to observe an exact zenith-distance. This is why Dr. Olbers observes the planet with the circle micrometer, which requires no lighting. But we do not doubt that, with a more favourable air quality and with a noticeable increase in the planet's light intensity, it will be visible on the meridian-quadrant, on which we have installed a new device into the bargain, to serve this purpose with more certainty. All declinations of the planet indicated by us are therefore only mere ocular estimates from the location of neighbouring stars and according to the semicircle of the meridian telescope, which is divided only in minutes and serves to approximately direct the telescope. Compared with this, our right ascensions of Ceres and the times of observation are as clearly and precisely determined as possible with our best aids. On January 21, I received a second letter from Dr. Olbers in which he had the goodness of informing me of his continued observations of Ceres on January 10, 13, 14 and 15. He also regards his right ascensions to be very exact. He cannot supply his declinations with as much certainty. We will have the complete observations follow later in a table. On December 7, star Nr. 1, which was later recognised as Ceres, appeared in my eight-foot meridian telescope as a very dull and weak point. I would hardly have classified it as a star of the 10th order. On January 11, it seemed somewhat brighter to me, like a magnitude 9 star. In spite of observing it with a magnification of 120, I could perceive no trace of a disk or any other characteristic, neither of form nor of light. But the latter characteristic seemed to be duller and less gleaming than that of a star of the same magnitude; but I openly admit that I would not have recognised it without a preconceived opinion.

On January 22, on which day I was blessed with the opportunity to once again observe this planet, it seemed to me to have gained remarkably in luminosity, and I estimated it as a magnitude eight star. But on this night, I was able to work with the first clear sky since six weeks, the coldness had eased off and the thermometer stood at only 5° Reaumur under the freezing point. [The Reaumur is an antique unit of temperature measurement. Five degrees R is about 6 degrees C, or 43 degrees F.] *This condition has no minor influence on the assessment of such a delicate object. But this assessment corresponds completely with that of Dr. Olbers. He writes: "Ceres appears to me (from January 1 to 8) as a magnitude 9 star and can be seen well with my comet finder. With a magnification of 106, nothing planetary is noticeable. I have not yet used greater magnifications. Toward the time of opposition, Schroeter and Herschel may find their satellites." And in his second letter he writes: "On January 10, I tried magnifications of 180 and 240 on Ceres but could get no clear picture. The same thing happened in Lilienthal with a 13-foot telescope. Greatly magnified, Ceres always appeared poorly defined and faded, with a pale reddish light. Perhaps the weather was to blame, because in spite of the apparent clarity of the sky, the stars shimmered a great deal. Perhaps Ceres has an atmosphere which poorly reflects the light of the Sun."*

In a letter dated January 13, the observer Harding from Lilienthal informed us that he had observed Ceres on January 11 with a 7-foot Herschel telescope and with a magnification of 84 and could not differentiate it from a magnitude 9 star, but the light seemed to him somewhat pale and reddish. With a magnification of 150, the planet appeared paler still but somewhat greater and had the colour of Mars. With a 13-foot telescope, and a magnification of 136, it remained just as pale; however, Harding recognised a disk of the size of the first or second satellite of Jupiter clearly enough. On its western side, two extremely delicate [splendid] *stars of indeterminate size gleamed, of which the one beside it could have had a distance of approximately 20", the outside one a distance of perhaps 30 to 35". With a magnification of 288, it appeared pale and less defined; of the two neighbouring stars only the one directly beside it was still visible, but its light was so pale that it was recognizable only with great effort. When Harding later directed his telescope on Jupiter, he soon found that the disk of the second satellite positioned close to Jupiter was visibly smaller than Ceres, through which he believed to be able to conclude that its diameter could not be under two seconds.*

Ceres is now easily located and can now and never again be lost, since its path will soon be precisely determined. Since Dr. Gauss's ellipse, communicated in the December issue of Monat. Corr., *has admirably and exactly agreed with the present position of the planet, it will be easy to completely correct these elliptical elements. Dr. Gauss had the goodness, after the reprinting of his last elements in the December issue, to send us others which have been improved for the fifth time. These represent the present observations of Ceres to within half a degree in right ascension and a ¼ degree in the declination. Only those who recognise from the theory how difficult it is to conclude an entire path of 360° from such scanty data, as Piazzi's 40-day-long observations indeed were, and from such a small arc of 9°, and receive only somewhat tolerable elements, which still hold true to within half a degree after such a long period as 10 months – only these will be able to properly treasure and admire the talent, skill and astute powers of reasoning of Dr. Gauss. Even on January 1, Lalande expresses himself in a letter to the publisher regarding Gauss's ellipse in the following way: "The exact agreement of Piazzi's observations with Gauss's elements seems to prove nothing else to me than that a small 'circulus osculator' can adapt itself to every crooked line only when the time interval is very small. The smallest error is enough to represent it this way or that." This only goes to show how very difficult the patriarch of astronomy, to whom the theory of planets owes so much, regarded the determination of paths.*

But Dr. Gauss himself had made this and also other remarks and objections long ago, and only shortly before the rediscovery of Ceres did he write to the publisher again: "Indeed, the agreement of declination [deviation] *of Piazzi's observations amongst themselves is as remarkable as with the calculation, and it is this which reinforces my hope that my elements will suffice for its rediscovery. For it is these observations of latitude which depend the most on deviations, which have the greatest influence on the agreement of elements. A change of 20" of latitude, if it were made positive with the outer one and negative with the middle* [mean] *one, or reversed, the ellipse would be completely changed. This remark is connected with another, the value of which is only fully realised when the rediscovery is successful and the elements derived from the observations are proven, that we will owe it in this case mainly to the great inclination of the path. If the path were to coincide with the ecliptic, one would then have to do without it, at least the uncertainty in the longitude would become extraordinarily great after a year and perhaps amount to a couple of signs, although this zone would then have almost no latitude. The reason for this is easily recognizable, when it is considered that four longitudes would be required to determine the ellipse, that consequently two outer and one middle observation could be completely satisfied through innumerable ellipses, and perhaps also parabolas and hyperbolas, and since, with the narrowness of the series of the observations, it is easy to realise that elements which represent two outer and one middle observation could deviate little from the entire series; therefore, the conic section would be entirely uncertain* [vague]...."

How correctly Dr. Gauss judged in this lies in the rediscovery of Ceres during the day, and the close agreement of its elements with the current observations proves it. Indeed, without the astute efforts and calculations of Dr. Gauss, we would perhaps not have found Ceres again, the greatest credit therefore belongs entirely to him. The worthy Dr. Olbers is of the same opinion as he writes to the publisher: " You will have remarked with pleasure how exactly Dr. Gauss's elements agree with the observations of Ceres. Please inform this worthy scholar of this under attestation of my very exceptional respect. Without his labourious investigations of the elliptical elements of this planet, we would perhaps not have found it at all. I, for my part, would not have looked so far to the east. We therefore want to leave the complete correction of the ellipse to the skillful Dr. Gauss, for it would be very rude to clear up with little effort that which this commendable scholar has developed with so much effort. We therefore owe it to him to lead us to our final goal which, we hope, will perhaps take place in the next issue of Monat. Corr."

Letters from Italy dated December 10, from England dated December 27, from France dated January 1, 2, 9 and 12 do not only mention nothing of the rediscovery of Ceres, but rather make more mention of the many intense and futile attempts, which had so tried the patience of so many, that further searches had been given up. If the French astronomers had had more faith in Gauss's ellipse, then the rediscovery of Ceres could not have escaped them. Until now, as far as we know and up to the conclusion of this issue, Ceres had been observed nowhere other than at the Seeberg Observatory by the publisher and in Bremen by Dr. Olbers. These observations follow (Fig. 10.13):

Beobachtungen der *Ceres Ferdinandea* auf der Seeberger Sternwarte angestellt.

Tag der Beobachtung	Mittlere Zeit	Scheinb. ger. Aufst. der Ceres	Nördl. geschätzte Abweichung der Ceres
1801 7 Dec.	18U 48' 10,"3	178° 33' 30,"60	11° 41½'
— 31 —	17 38⅗ ::	184 44 ::	11 5
1802 11 Jan.	17 3 17, 4	186 45 49, 95	11 15
16 —	16 46 25, 6	187 27 53, 25	11 26
22 —	16 25 23, 9	188 6 25, 80	11 44
25 —	16 14 32, 9	188 20 39, 15	11 56 23,"0 (*)
26 —	16 10 53, 7	188 24 49, 50	11 58
28 —	16 3 29, 0	188 31 37, 85	12 9 41,"3 (*)
29 —	15 59 43, 7	188 34 18, 15	12 14

Beobachtungen der *Ceres Ferdinandea* in Bremen, von Dr. *Olbers* mit dem Kreis-Mikrometer angestellt.

1802	Mittl. Zeit	Scheinbare gerade Aufsteigung der Ceres	Scheinbare nördl. Abweichung der Ceres	Sterne, mit welchen die Ceres verglichen worden ist
	U ' "	° ' "	° ' "	
2 Jan.	11 58 36	185 7 40	11 6 30	Nr. 191 *Bode.* C. d. tems X. Seite 254.
5 —	17 30 0	185 43 7 ::	11 7 56 ::	Nr. 20 *v. Zach*
10 —	12 25 41	186 31 52	11 13 9	ρ Virgo } *v. Zach*
13 —	11 43 38	186 58 56	11 18 56	ρ Virgo
14 —	11 9 3	187 7 11	11 20 57	ρ Virgo
15 —	12 8 9	187 15 27	11 23 25	ρ Virgo
20 —	13 8 0	187 55 ::	11 37 18	ρ Virgo
22 —	12 26 40	188 5 45	11 43 55	ρ Virgo

Wahr-

Fig. 10.13 Observations of Ceres by Zach in Seeberg and in Bremen by Dr. Olbers with an circular micrometer

Stellungen der *Ceres* im Februar 1802 für Mitternacht, mittlere Seeberger Zeit.

1802	Gerade Aufsteig. der Ceres			Abweichung der Ceres		Gerade Aufsteig. in Zeit		
30 Jan.	6 Z	8°	38'	12°	16'	12 U	34'	30"
2 Febr.	6	8	43	12	31	12	34	53
5 —	6	8	45	12	47	12	35	1
8 —	6	8	44	13	4	12	34	55
11 —	6	8	38	13	22	12	34	33
14 —	6	8	29	13	41	12	33	57
17 —	6	8	16	14	1	12	33	5
20 —	6	8	0	14	21	12	31	59
23 —	6	7	40	14	42	12	30	40
26 —	6	7	17	15	3	12	29	8
1 März	6	6	51	15	23	12	27	22

Fig. 10.14 Positions of Ceres in February 1802 for midnight, Seeberg mean time

In the next issue, we will likely be able to communicate to our readers the more detailed elements of the path of Ceres. In the meantime, to facilitate the location of such an innocuously small planet, we will supply a provisional ephemeris of its course for the next month, 'exact to a few minutes', which has been improved according to our observations. We have omitted seconds so that no one will rely on it more than the nature of the matter allows at present. The planet will come to a stand still approximately between February 3 and 4 (Fig. 10.14).

All planets have hieroglyphic indications which, taken from the infancy of the art of drawing, are easy and crude outlines. Ceres is the goddess of the harvest; image of sickle could therefore denote it.

Resumed News about the new main Planet of our Solar System: Ceres Ferdinandea

MC (March 1802), p. 263

I complained already in the last issue about the difficulties that such a small and faint heavenly body like Ceres Ferdinandea poses on astronomical observation and the determination of its positions. It reminded me of the rare case of the comet of 1799 about which I complained in my A.G.E. *vol. 4B, p. 265. This celestial body was also hard to observe because it, too, did not tolerate any illumination of the hairs without disappearing from the telescope. The very tiny and delicate spider's threads in the telescopes of the passage instruments and the meridian quadrant, which are otherwise advantageous, appear to be at such rare occasions disadvantageous. But not the delicateness of these threads and the difficulty of their illumination alone were reasons for the difficulties in observing Ceres, but this heavenly body's own character turned the great magnifications of those instruments into disadvantages. I saw the planet much clearer with a Dollondian comet searcher and much better with a 4-foot Dollondian parallactic telescope at a low magnification than at the 8-foot meridian telescope at a high magnification and at the quadrant's 4-foot telescope. But things had to be that way. Schroeter, Dr. Olbers and Harding saw Ceres very pale. The higher the magnifications the fainter, more blurred and washier the image of this planet appeared.*

But since we know (and the reader is soon to understand) that this planet is not only veiled in a comet-like nebulosity but itself is subject to an odd atmospheric change of light it is no longer a mystery that this celestial body – at least at the beginning of its rediscovery – was so difficult to observe: fast and weird changes of light take place on its surface as are described extensively in Mr. Schroeter's own words in his treatise at the end of this issue. I also became aware of these quick changes in light and shape of this planet but at first I considered them mere coincidences of our atmosphere and accredited it partly to the severe frost.

But this feature became so paramount that I hardly recognised the planet and I was oftentimes deceived by the appearance and sometimes became uncertain whether it was really Ceres I was observing. Often only repeated calculations could reassure me. I was quite weary and anxious about this that I asked Dr. Olbers about this feature. He answered as follows: "I came across similar difficulties. Ceres' size is so variable that I hardly recognised her on several nights. I asked our friend Schroeter to observe Uranus, then in the same region, simultaneously. Thus, we would be able to measure how much our atmosphere influences the observations and a comparison of colour, brightness, size and boundary of both planets can only turn out to be very revealing." Hereunto I received Schroeter's very peculiar treatise, which our reader will read immediately. Now the reader might understand the difficulties in observing this faint planet at the usual magnifications. All right ascensions were taken at the passage instrument and are accurate. The major difficulty lay with the quadrant and the small aperture of this instrument. But these difficulties were overcome by a stronger silver thread and a new eyepiece with a lower magnification. Because the glasses had to be grounded these alterations were only finished by February 9. Only since then accurate observations of zenith distances were possible, although I consider those of January 28, 30 and February 3 not bad either. Below you will find all observations of Ceres made at Seeberg Observatory followed by those of Dr. Olbers.

Our friends from Paris informed us that the new planet Ceres had not been found any earlier than Dr. Olbers' and my notice reached Paris. Méchain and Messier, these two great comet-scouts have searched Ceres in vain. Méchain observed more than 300 small stars without finding Ceres among them. And Messier had been scanning the skies since October but his diligence was not crowned with success. Thus La Lande wrote: "How lucky you are that you did not share my incredulity regarding this new planet!" And he even treated Dr. Gauss justly and repeats what we already mentioned on p. 182 of last month's issue, that this new planet would not have escaped the French astronomers if they had had more trust in Gauss' orbit. La Lande said what we admitted before him in the last issue: "It's because of Mr. Gauss and his excellent elements that you discovered the planet, without them Mr. Olbers and you would have had a lot of trouble finding it because it is so difficult to observe." These remarks on Dr. Gauss are correct; Dr. Olbers and I have already signed them with pleasure. But Mr. Gauss, too, made comments which are not to be neglected here. This commendable scientist said: "I cannot help but mention what merit the existence of such a magazine like the MC is for astronomy. If you had not collected in your magazine all relevant information about Piazzi's discovery it would have been taken in with the utmost indifference, but you spread the news and aroused public interest, weighed the pros and cons and made the planetism of this celestial body most likely. Probably there were only a few astronomers who made an effort to rediscover it since all teachers of today's astronomers doubted the new planet so much?"

After having received the position of the new planet Méchain was the first to find it on the evening of January 22. But he succeeded in observing it properly only on January 25 at 1pm 22', apparent right ascension 188° 20' 15", apparent declination 11° 55' 59". On January 27 Le Français and Burckhardt observed it at the Observatory du Champs de Mars at the wall quadrant. They compared it to Vindemiatrix (ε Virginis). Difference in right ascension to ε Vir = –4° 40' 7", in declination 20", the planet more to the south. From this we calculated on Jan 27 at 4 pm 10' 48.2" mean Parisian time RA Ceres = 188° 24' 49".97 apparent n. declination = 12° 1' 11".44.

Letters from Italy and England do not mention a rediscovery yet, although we sent word to Milan, Palermo and London. Meanwhile Dr. Gauss is busying himself with a temporary improvement of his elements according to my observations.

We informed our readers in last year's December issue on page 647 about all of Gauss' calculated orbital elements, improved for the fourth time according to Piazzi's observations. It belongs to the history of this planet that we keep you posted about the continued efforts. We already mentioned on p. 178 of last month's issue that Dr. Gauss had sent us before Ceres' rediscovery an orbit improved for the fifth time and we would like to advise you of it. According to Piazzi's observations Dr. Gauss created an elliptical orbit by accurate calculations and taking into account nutation, aberration and parallax. From this he developed the following fifth elements:

Aphelion	*324° 37′ 11″*
Ascending node	*80 59 12*
Inclination	*10 37 9.55*
Log semimajor axis	*0.4446804*
Eccentricity	*0.0879111*
Daily mean trop. mov.	*763″.950*
Epoch 1800 Dec. 31	*78 5 16″.6*

It would be unnecessary to show the wonderful correspondence between these elements and Piazzi's observations, which Dr. Gauss had calculated with the utmost accuracy and represented exactly except for several seconds.

After having sent Dr. Gauss the information of the fortunate and eagerly expected rediscovery of this planet together with my first three observations, the first thing he did was calculate them according to his above 5th elements. He found the error in the first observation of December 7 in RA +24′ 8″, in the second of January 11 +30′ 53″, in the third of January 16 +31′ 53″. According to his 5th (improved) elements they vary approximately in the following order: +14½ min, +19¾ minutes, +20½ min. That these elements IV come closer to the truth than the fifth is according to Dr. Gauss pure coincidence. But it might just as well be the influence of the perturbations in Piazzi's observations, especially on the latitudes. As far as Dr. Gauss was able to tell from Dr. Olbers' two first observations, which he had received in the meantime, the latitudes corresponded to both elements except for a few minutes and the difference in declination was mainly caused by the differences in longitudes. Meanwhile I gave Dr. Gauss further observations which he could not resist improving preliminarily. He stuck to my two right ascensions of December 7 and January 16 and thus he found the following elements VI:

Epoch 1801	*77 24 55″.9 Palermo meridian*
Epoch 1802	*155 33 35.1 Palermo meridian*
Aphelion	*326° 14′ 45″*
Ascending node	*80 58 55*
Inclination	*10 37 51*
Log semimajor axis	*0.4421189*
Eccentricity	*0.08080253*
Daily mean trop. mov.	*770″.7376*
Trop. time of orbit	*1681 days 12h 9m*
Mean anomaly	*9° 16′ 23″*

It was in the nature of the method on which this calculation is based that by these elements Piazzi's observations are represented almost as accurately as by the elements V. But it could not be expected that by this first improvement the errors of the two new observations would be decreased to zero from ½ degree. But when Dr. Gauss had calculated the right ascensions, to his great surprise the correspondence was better than expected. The error of the first observation was only +3″.0 and in the last +9″.9. Dr. Gauss had calculated my three new observations, which I had sent him in the meantime, according to these elements and found the following correspondence of the right ascensions:

Seeberg	*Calculated RA*	*Difference*	*Calculated Decl.*
1801 Dec. 7	*178° 33′ 33″.6*	*+3″.0*	*11° 47′ 33″ N*
1802 Jan. 11	*186 46 9.3*	*+19.3*	*11 15 41*
16	*187 28 3.1*	*+ 9.9*	*11 26 40*
22	*188 6 45.9*	*+20.1*	*11 45 18*
25	*188 21 6.5*	*+27.3*	*11 56 49*

Since for determining these elements Piazzi's latitudes were used, which are very close to the node, the inclination cannot be expected to be exact. Dr. Gauss believes that the error can hardly be more than a minute according to a first estimate of Olbers' declinations. This is a new proof of the quality of Piazzi's declinations.

In the future these elements VI will need further improvements and they will be more and more accurate the more distant the observations will be. So it is useless for Dr. Gauss to adapt his elements as good as possible to the present observations. And Dr. Gauss is such a tireless and willing calculator that a determination of new elements does not seem to bother him: before I knew what was happening and before I would have been able to send him my observation of January 26 he surprised me with his 7th orbit of Ceres by using instead of my observation of January 16 that of January 25 which produced the following elements (VII):

Epoch 1801	*77 27 36″.5 Seeberg meridian*
Epoch 1801	*77 27 30.9 Palermo meridian*
Aphelion	*325° 57′ 15″*
Ascending node	*80 58 40*
Inclination	*10 37 56.6*
Log semimajor axis	*0.4424742*
Eccentricity	*0.0814064*
Daily mean trop. mov.	*769″.7924*
Mean anomaly	*9° 20′ 8.0*

And again no other latitudes than Piazzi's have been used. These elements represent all observations of Palermo very accurately and match my observations better as the following table might show (Fig. 10.15):

Seeberg	berechnete Afcenf. rect.	Unterfchied	Berechnete Abweich. N.	Unterfchied
1801 Decbr. 7	178° 33′ 29,″2	— 1,″4		
1802 Januar 11	186 45 47, 6	— 2, 3		
16	187 27 38, 8	— 14, 4		
22	188 6 18, 2	— 7, 6		
25	188 20 37, 2	— 2, 0	11° 56′ 58,″4	+ 35,″4
26	188 24 37, 0	— 12, 5		
28	188 31 25, 7	— 12, 1	12 9 55, 6	+ 14, 3
29	188 34 14, 1	— 4, 0		
30	188 36 38, 4	— 5, 5	12 19 19, 8	+ 19, 1
31	188 38 38, 3	— 7, 1	12 24 15, 3	
Februar 3	188 42 9, 5	— 3, 5	12 39 56, 6	+ 17, 6
4	188 42 30, 7	— 5, 6	12 35 25, 7	
5	188 42 26, 7	— 3, 4	12 51 2, 6	+ 34, 7
6	188 38 1, 6	— 2, 3	13 14 43, 0	+ 22, 0

Fig. 10.15 Observations of Ceres from Seeberg from December 7, 1801 to February 6, 1802

It would be an easy matter for Dr. Gauss to make the above small variation of the calculation of all my observed right ascensions vanish. Dr Gauss believes that it would be wise to continue to compare the observations with one set of elements and to wait until we have some more accurate declinations since these elements are accurate enough to find Ceres again for the time being. Even the above slight variation and the sparse number of reliable declinations show that it is too early for improving the elements VII any further. By the way, Ceres' orbit is not a perfect ellipse because of the perturbations it shows due to the impact of the other planets. These perturbations have to be calculated first before you can obtain a true elliptical orbit. Dr. Olbers agrees: "I think you have to calculate the perturbations first, which this small planet has suffered by Jupiter from January 1, 1801, until today in its arc of its ellipse. The perturbations of the other planets could be neglected for the time being, only Mars would be interesting. (FN: This would be difficult since the mass of that planet is still unknown.) Calculating the perturbations is easily done because only approximately 90 degrees of Ceres' orbit are in question and not an entire theory. So you have to deduct from the observations the amount of the perturbations and you will find an accurate ellipse that would have taken place without these perturbations."

La Place believes that these perturbation equations will be considerable. This great geometer wrote on February 4: [French] *"I am hoping that we will have the elements of this new planet in a few weeks, which are precise enough to determine the perturbations and to know even better its elliptical elements. Its perturbations have to be considerable. Having calculated the perturbations and the new elements you will have excellent tables for the motion of this new planet, which will be known just as well as the others." Dr. Gauss says that the difference between his elements VII and the true elements will be negligible since Ceres has almost completed one quarter of its orbit. Based on all observations of this year until Ceres becomes invisible, he hopes to produce such accurate elements that it would be possible to go back 44 years to Mayer's observations without any error in longitude and only several minutes in latitude. If it is impossible to find Ceres in any of the older star catalogues, as Uranus was found, we will learn of its mean motion only gradually and there will be much to be added and improved in the years to come.*

The hope to find this planet in one of the star catalogues is of course not as great as in the case of Uranus because of smallness and faintness and because large parts of its orbit are outside the old zodiac. But this shall not impede us. It might even be possible to find Ceres among Flamsteed's observations. For Ceres sometimes appears of the 7th order of magnitude and Flamsteed observed such stars also beyond our zodiac, which are missing today. More can be expected of Tobias Mayer because he observed stars up to the 9th order of magnitude among which some are of 10°, 12° and 15° latitude north and south. And it might also be possible that at the time Tobias Mayer compiled his catalogue [he sighted Uranus in 1756] *this planet was not at its largest geocentric latitude. It is hardly imaginable that Ceres is not among the 50,000 stars of La Lande, who wrote: "I think we will find Ceres more than once among the 50,000 stars." He has just finished his new catalogue of 1,500 new stars, which will be published in the next* Conn. des temps Année XIII. *I, for myself am hoping to find Ceres among the observations in my catalogue of the zodiacal stars; because there is a vast number of small stars, which I observed and could not find in any of the known catalogues. Among those are some that are missing. And time will tell whether Ceres was among them or whether they were created by writing or observational errors. And the supplements to Flamsteed's catalogue might prove valuable, which were compiled by Le Monnier, Le Gentil, Darquier, Messier and Méchain on the occasion of the appearance of a comet and which are scattered among the* Parisian Memoires. *Dr. Olbers, whose opinions are always of practical nature and whose studies are proof of his own astuteness, wrote: "According to the orbital period calculated by Dr. Gauss I found that 18 oppositions of Ceres take 23 Julian years. Thus Ceres' geocentric appearances will almost be the same every 23 years and accordingly this small planet wandered through the northern wing of Virgo in 1779 just like today. What a pity that the comet of 1779 did not reach Virgo's wing two months earlier; it would have met Ceres and Messier certainly would have observed Ceres, too, which is of course not among the small stars determined by Messier on that occasion."*

Coincidence was such a great helper in the case of Uranus; we want to hope for the same luck now. Let us observe this new guest as accurately and eagerly as possible, thus here are Dr. Olbers observations:

Observations of Ceres Ferdinandea, made at Seeberg Observatory

Day of Obs.	*Mean Time*	*Apparent RA*	*Apparent Dec.*
Jan. 30	*15h 55′ 57″.5*	*188° 36′ 43″.95*	*12° 19′ 0″.7*
31	*15 52 9.7*	*188 38 45.45*	*12 25*
Feb. 3	*15 40 35.8*	*188 41 13.05*	*12 39 36.0*
4	*15 36 41.4*	*188 42 36.30*	*12 44*
5	*15 32 45.1*	*188 42 30.15*	*12 50 25.0*
9	*15 16 43.7*	*188 38 3.90*	*13 14 18.0*
19	*14 34 46.7*	*187 58 27.90*	*14 20 2.9*

Observations of Ceres Ferdinandea, made in Bremen by Dr. Olbers at a circular micrometer

Jan. 25	*11h 36′ 0″*	*188° 19′ 50″*	*11° 54′ 43″*
26	*11 2 0*	*188 23 50*	*11 59 56*
28	*11 21 0*	*188 31 15*	*12 8 43*
31	*10 44 30*	*188 38 29*	*12 25 8*
Feb. 3	*11 8 0*	*188 42 0*	*12 37 22*
5	*10 40 50*	*188 42 28*	*12 49 6*

Dr. Olbers gives the following notes: "The declination of the observations of Jan. 25 and 31 are not reliable. The reason why the declination of both observations of February might be so unreliable is because I do not know the declination of 34 Vir very exactly and furthermore on Feb. 3 the weather was unfortunate. I derived the declination of 34 Vir 13° 2′ 42″ from the difference of the zenith distances of ε Vir in the Hist. célèste française." In order to make it also in the future easier for our readers to find Ceres here is another ephemerides of its path for the next two months with which Dr. Gauss calculated his elements VII. Hopefully in April these positions will not deviate by more than one minute (Fig. 10.16).

Dr. Gauss accompanies these ephemerides with the following important notes: "The opposition of the planet will be in the afternoon of March 17. At that time it will be best to observe. Shortly before the opposition it will have reached the smallest distance from earth

Stellungen der *Ceres* im März und April 1802 für Mitternacht, mittlere Seeberger Zeit.

1802	AR. ⚳ in Graden	Abweich.	AR. in Zeit	1802	AR. ⚳ in Graden	Abweich.	AR. in Zeit
März 1	186° 41′	15° 30′ N	12ᵘ 26ᵐ 45ˢ	März 28	181° 17′	17° 54′ N	12ᵘ 5ᵐ 7ˢ
4	186 11	15 50	12 24 45	31	180 39	18 1	12 2 37
7	185 39	16 10	12 22 36	April 3	180 3	18 6	12 0 12
10	185 5	16 29	12 20 18	6	179 29	18 9	11 57 54
13	184 28	16 47	12 17 53	9	178 56	18 10	11 55 45
16	183 51	17 4	12 15 24	12	178 26	18 9	11 53 46
19	183 13	17 19	12 12 50	15	177 59	18 5	11 51 57
22	182 34	17 33	12 10 15	18	177 35	17 59½	11 50 20
25	181 55	17 44	12 7 40				

Fig. 10.16 Positions for Ceres in March and April 1802 at midnight, mean time Seeberg

= 1.6025. At the same time (a little earlier) its greatest northern geocentric latitude will be = 17° 9′ and a little later its fastest retrograde motion; every day approximately 13 minutes in RA. The declination north is increasing till the beginning of April; around April 9 the motion will be direct again." Piazzi's name for the new planet, Ceres Ferdinandea, to which he as discoverer is entitled to is not unanimously supported by the astronomers. La Lande, true to his principle wants to name it Piazzi – just as he insists to call Uranus George's planet or Herschel. Some time ago he wrote regarding this matter: [French] *"I will never consent to rip off of this small planet the name of my student Piazzi and replace it by Ceres, who is nothing to me. The rural deities were something in former times but are nothing today. The names had a meaning once but none today." Senator La Place wrote in his latest letter:* [French] *"Bonaparte, to whom I talked about the new planet some days ago, and who has despite all his other obligations a vivid interest in science and especially astronomy and its progress, prefers the name Juno to Ceres, and I agree with him. It is only natural to place Juno close to Jupiter. The German astronomers were the first to give it the name of this Greek goddess, but it certainly is better to give it a Latin name." Well, again a schism in the church of astronomy, just as with Uranus. And our worthy compatriot Georg Szerdahely* [see footnote] *will open his poetic vein for the second time and has to sing*

The astronomers battle, and still the case is before the judge, with which name a new planet should be designated.

He has to exclaim once again:

O Gods above! What would this confusion of voices be? If each voice should offer a name!

We shall wait and see and cry:

It is not ours to construct a troubled quarrel of voices.

[F.N.: Georg Szerdahely (1740–1808), Jesuit and Professor of Rhetoric, taught aesthetics at the University of Buda. He was knight of St. Stephen and mitered abbot at St.Maurice. By a "second time" Zach is referring to a book published by Szerdahely in 1788 (*Historia Urania Musae*) in which he tells of the Muse of Astronomy, Urania. The verses above were translated from Latin]

*It is true that the new Piazzian planet is constantly shrouded in clouds like Juno and Hera: a supposition I already uttered in June, 1801. I then wrote (*MC *vol. III, p. 619): "But is it really proven that planets do not have or are not supposed to have a tail? We have planets with and without satellites; and planets with two or more rings. Why should there not be planets with tails?"*

Here follows the very peculiar treatise by Schroeter.

Ceres Ferdinandea by Dr. Schroeter

Lilienthal, January 26, 1802. (In the original, the date is incorrectly printed as 1801.) *MC* (March 1802), p. 282.

Unusually bad and changeable weather, which is typical for the Lilienthal climate and a resistant catarrh, which is still teasing me and did not permit any observations out of doors, impeded me extremely. But on January 25 when the weather was fine I wanted to combine my observations, despite my catarrh, with those of Harding. Because of my catarrh I chose the 10-foot equatorial Dollond. According to Harding's map of the orbit and surrounding small stars and Mr. Olbers' and von Zach's important observations Ceres must be located east of but near ρ Virgo. And we actually found at that location two small stars of which one had to be Ceres. But mist had again developed in that area and the Dollond was not successful. Thus we hurried to our 13-foot reflector. When I had ρ in the achromatic searcher there was another faint star and when I used a magnification of 136 times after 11 o'clock ρ was westwards out of the field, and Ceres Ferdinandea stood immediately before me. It was so conspicuous, round, steady and had a soft planetary appearance that no doubt was possible. With an aperture of 9½ inch Ceres' image was similar to that of Uranus with its completely white light. It had a considerable diameter, which I estimate at least that of the Georgian planet and its light remained steady while all other small stars were scintillating. This happened while the moon was rising and then in full moonlight.

What I consider very interesting is the fact that at 136 and 288x magnification this planet's disc appeared with clear edges but had a narrow nebulosity through which the planetary sphere shone through.

This exceptional kind of boundary reminded me of the comet of 1799 as described in volume 3 of my Memoirs with the exception that its disc shone through even brighter and more clearly. Because of this I experienced the same when at 12 o'clock I measured the diameter. Measured with nebulosity it appeared almost as big as an illuminated disc of two lines and compared to a disc of 1.5 lines smaller by just one third of this difference than one of two lines, thus 1.834 lines. When I measured the bright disc only the diameter was only 1.334 lines. The distance of both projections was 522.5 lines for both measurements. Everything at 288 x magnification. Accordingly the calculation for

1) the diameter of the mere planetary disc

$$\log 1.334 = 3.1251558$$
$$\log 522.500 = 5.7180863$$
$$7.4070695 = \text{tang}\,\frac{8'\,47''}{288} = 527''.0 = 1''.815$$

2) the entire diameter including the atmospheric nebulosity

$$\log 1.834 = 3.2633993$$
$$\log 522.500 = 5.7180863$$
$$7.5453130 = \text{tang}\,\frac{12'\,4''}{288} = 724''.0 = 2''.514$$

Thus considerably smaller than I had estimated from the planet's appearance.

The last measurement of the whole planetary disc was repeated by Harding at the same distance from the eye and he found the planet 1.750 lines, which is for its entire diameter 2''.330, only 0''.184 or 1/14 less of the entire diameter.

At 1 o'clock I corrected its position according to the searcher alone and found after a comparison with the charts that its right ascension was at most between 188° 20' and 30' and the declination north around 12°.

On January 26 we had fine weather since a quarter to eleven, much better than the previous evening. This could be learned among other things from the achromatic searcher of the 13-foot reflector: the planet had moved and appeared larger than an 8th magnitude star and in fainter but reddish light.

In the telescope at 136 and 288x magnification on the other hand it was a white, slightly bluish but nevertheless pleasant, soft, pale and quite bright planetary light. It was close to an 8th magnitude star and a smaller one, not visible in the searcher.

And again it showed a kind of cometary nebulosity but very astounding was the fact that despite the much more favourable weather today its disc did not shine through the nebulosity as the previous evening but appeared as a comet-like planet or a planet with nebulous boundary which resembled at 288x magnification the planetary nebula ν Aquarius and nothing in its appearance reminded me of the previous evening.

And again I was only able to measure the entire diameter with nebulosity. It matched a disc of two lines but in comparison with a disc of 1.5 lines smaller by 0.125 lines. Harding found it just as large. The distance of the projection disks was 500 lines from the eye. Thus:

$$\log 1.875 = 3.2730013$$
$$\log 500.000 = 5.0989700$$
$$7.5740313 = \text{tang}\,\frac{12'\,54''}{288} = 774''.0 = 2''.687$$

By the way it was agreed upon to continue our observations with the 7 and the 20-foot reflector. But already at 12 o'clock our plans were thwarted by an immense patch of fog.

If you look at the still short history of this planet it seems to me today already that it is a prognosticator of a very interesting series of observations for all of physical astronomy. When Piazzi saw it he estimated the diameter 8". Ceres had a reddish, pale cometary light so it was impossible with an achromatic telescope of less than three inch aperture to discern it and in an equal light von Zach and Olbers found it and even Harding at 288x magnification with the powerful 13-foot reflector. Yesterday it appeared in an achromatic searcher of the 13-foot reflector of at most two inches aperture without much searching immediately and in the 13-foot reflector in a white light as a true planetary disc with clear edges but with a surrounding comet-like mist. Whereas today, in favourable weather it appeared in that searcher at a magnification of 13 times as a round pale planetary disc. I was nevertheless unable to distinguish in the large reflector its clear disc from the nebulous boundary. An odd atmospheric change cannot be denied and I if am not mistaken this will prove to be the case just with the comet of 1799. If this utterly weird planet had come into existence like the Sicilian Ceres by procreation, I would consider it a bastard of a misalliance whose father is a planet and the mother a comet. But one thing is certain: this important discovery falls into a time where we fortunately already know something about the atmospheres of heavenly bodies and this discovery will give us insights into some matters that are shrouded in mystery today.

[Another German publication, *Goettingische gelehrte Anzeigen,* Volume 1, published a report about Ceres in the March 6 issue, pp. 369-372. In a table, it included positional data by Karl Harding for the following dates: January 10, 25, 26, 28 and 31.]

Resumed News about the New Main Planet of our Solar System, Ceres Ferdinandea.

MC (April 1802), p. 379.

Since Ceres Ferdinandea has been found again and its locations have been made public by either our magazine or letters to all astronomers, everyone has been searching for it and tried to observe it astronomically. According to weather and instruments some succeeded earlier than others.

French astronomers were the first foreign astronomers to find the new planet, for instance Méchain, administrator of the National Observatory, and Le Français, Burckhardt and Bouvard. Méchain was so kind as to send us the following observations; among those are only four meridian observations made at the 8-foot Bird wall quadrant (Fig. 10.17); the declinations are improved by refraction but the impact of the parallax and the error in collimation of the quadrant have been neglected; the latter is not more than 3" [F. N.: not said whether + or –. They did not manage to do this observation properly.]

Beobachtungen des neuen Planeten auf der National-Sternwarte in Paris.

1802	Mittlere Zeit	Scheinbare gerade Aufsteigung der ⚳	Scheinbare Abweich. der ⚳ N.	Vergleichung	Beobachter
	U ′ ″	° ′ ″	° ′ ″		
Jan. 24	12 58 0.	188 15 38	11 52 20	mit 27 ♍ *Le Français*	*Méchain*
25	12 23 20	188 19 45	11 55 56 ½	— dito —	— —
	13 21 55	188 20 16	11 55 59	mit ι ♍ *von Zach*	— —
	16 14 28½	188 20 53	11 56 28	mit 3 * *Maskelyne* und ι *von Zach*	— — im Merid.
26	16 10 48	188 24 49	12 0 43	mit 4 * *Maskelyne* und ι *von Zach*	— — — —
29	12 5 10	188 34 11,4	12 13 24,6	mit ι ♍ *von Zach*	— —
31	12 4 0	188 38 40,2	12 23 25,3	— dito —	— —
Feb. 3	11 54 0	188 42 1,1	12 38 13,0	mit 34 ♍ *Le Français*	— —
	15 40 39	188 42 20,9	12 39 28,0	mit 4 * *Maskelyne* und ι *von Zach*	*Bouvard* im Merid.
7	10 57 20	188 41 23,4	13 1 0	mit 34 ♍ *Le Français*	*Méchain*
10	11 21 30	188 36 7,0	13 19 30	mit 34 und 41 ♍ *Le Français*	— —
	15 12 34	188 36 0,7	13 20 29 ½	Mit 4 * *Maskelyne* und ι *von Zach*	— — im Merid.

Fig. 10.17 Observations of the new planet at the National Observatory Paris

We have only received two observations of the Observatory du Champs de Mars by Le Français and Burckhardt at the wall quadrant.

1802	*Mean Time*	*Right ascension*	*North Declination*
Jan. 26	*16h 10′ 48″.2*	*188° 24′ 50″.0*	*12° 1′ 11″.4*
Feb. 27	*13 59 15.0*	*186 58 44.1*	*15 15 54.8*

The president of the Royal Society of Sciences in London, Sir Joseph Banks, was so kind as to tell us that the Astronomer Royal, Dr. Maskelyne, only started to observe this planet after having received the news of it on February 3 in Greenwich and after this all other English astronomers followed. We received two observations from Greenwich (Feb. 3 and 4) at the equatorial sector but since they have not been reduced and this is the observer's business we do not intend to use these observations and furthermore Dr. Maskelyne does not wish them to be published in their present state.

Italy received the news of Ceres' rediscovery rather late because by some coincidence our letters remained unopened in Milan for three months. Oriani was the addressee of all those letters; enclosed were also letters to Piazzi. Oriani had to attend the Cisalpine National Assembly at Lyon on November 29 (last year) and he did not inform us about this any earlier than December 22 from Lyon. But this letter was lost so we continued to send our letters to Milan. Those letters were not forwarded to Lyon because everyone believed that his disputational journey would only last three weeks. But he returned only on Feb. 9 and thus found the entire stock of astronomical news about Ceres, which had temporarily slipped out of his mind due to political quarrels. But we hope to get very accurate observations from that region soon.

Prof. Bode in Berlin observed the new planet on March 1 at 13h 50′ 12″ mean time at the wall quadrant: apparent right ascension 186° 40′ 46″ and apparent decl. n. 15° 29′ 40″. Since then he has sent us the following three successful observations before and after the opposition. He compared the planet at the wall quadrant with θ, β and o Leonis [Z represents 30 degrees].

1802	*Mean Time*	*RA N.*	*Decl.*	*Geocen long.*	*Geocen N. lat.*
March 15	*12h 44′ 44″*	*184° 3′ 7″*	*16° 58′ 24″*	*5Z 26° 44′ 39″*	*17° 8′ 20″*
16	*12 39 56*	*183 50 27*	*17 3 52*	*5 25 30 44*	*17 8 17*
19	*12 25 37*	*183 12 8*	*17 18 52*	*5 25 49 29*	*17 6 41*

From this he calculated that the opposition must have happened incidentally on March 17 at 4h 36' mean time at 5Z 26° 21' 38" of the longitude and 17° 8' latitude north.

The new planet was found in Vienna at the Imperial Observatory only on March 3 because the locations of the stars #87, 111 and 187 Virginis are given erroneously in Prof. Bode's star catalogue (supplement to his Uranographie*), which Vienna used. (the error in #111 is 15 min in RA). Thus the locations determined accordingly had to be erroneous. Consequently, we do not give the observations here.*

On March 15 the Astronomer Royal and Canon David found and observed Ceres at the Prague Observatory. He found the planet close to #147 of Bode's catalogue, but unfortunately this star's position is wrong again. Since we determined that star very accurately we found the error in right ascension to be not less than 1' 38" and the declination an entire minute too small given by Bode. Accordingly, for the same reason we will not give this observation either.

We already mentioned on p. 275 of the last issue how careful the astronomers have to be in their observations and that no direct meridian observations but mere differential observations can be made and that the small stars in the parallel used for this purpose are often badly determined.

A good example happened during the observations in Vienna. The star #87 Virginis in the above mentioned catalogue of Bode is in right ascension too great by 42" and too small by 34" in declination and thus the planet's position determined by this star had to be so erroneous. You can see from this how important and necessary a good catalogue of those stars is, which come into the astronomers' parallels, even for astronomers who do not need to employ this way of observing. But this is not the only reason why observing Ceres is difficult. Oftentimes this planet is so close to unknown stars of similar appearance that it sometimes is hard to tell which is the planet and which is the star. This happened on March 15, 19 and 23 and it would not be a surprise to hear that they had been mixed up. Even greater difficulties encounter those who work with handheld instruments. Dr. Olbers wrote regarding this matter: "It would be very interesting for your readers if you could publish in your MC *a map of the entire region through which Ceres will travel during the following months. The map should start approximately at 174° to 190° RA and from 8° to 20° declination. But it should contain all of La Lande's stars mentioned in his* Histoire céleste française. *Now every astronomer has to make his own map, for it is almost impossible to find Ceres without it, especially after several days of unfavourable weather. Bode's otherwise indispensable maps are not detailed enough for Ceres."*

Such a map, which was indeed very much needed, has been drawn and it would have been published in this issue if the copper engraver had delivered the plate on time. Thus it will be in the next issue. Instead we would like to give our readers and observers a list and exact positions of those stars, which came or will come with Ceres into parallel. We will continue this list in the future because it will be even more necessary then. [Here follows two pages of stellar tables.]

*It has to be said that with my observations, the annual change is given each time according to my determination of the equinox (*MC *vol. II, p. 500). I followed the information, which every observer gave me. The declinations which the astronomers from Mannheim, Henry and Barry* [see footnote], *made at the 8-foot Bird wall quadrant were taken from their handwritten list. La Lande's determinations were taken from the* Histoire célèste française *and the* Connaisssance des temps. *On this occasion Bode's great star catalogue showed the following mistakes, which are too grave to be ignored. The right ascension of the star #III is too great by 15 minutes. The declination of #310 must be north not south and #171 is entirely missing in the sky. #67 13f in Coma Berenices has a declination too great by 10 degrees. We ignore the minor errors of the determinations. For example the right ascension and declination of #87 is wrong by more than half a minute as mentioned above. On March 15 Ceres came very close to a star of the 7th order of magnitude so that a confusion was easily made; it was #147 and we determined its apparent right ascension 184° 0' 58".8, apparent declination 16° 57' 33".9. On March 19 the planet came again close to a 9th magnitude star, which was nowhere mentioned.*

Seeberg 1802		Berechnete Ascensio recta ⚳	Unterschied	Berechnete Abweichung nördlich	Unterschied
Februar	19	187° 58′ 23″8	−4″1	14° 20′ 24″0	+21″1
—	26	187 7 34.4	+1.0	15 9 20.1	+25.8
—	27	186 58 53.2	+5.8	15 16 20.2	+26.3
—	28	186 49 51.9	+3.7	15 23 19.1	+18.1
März	1	186 40 31.4	+3.5	15 30 16.0	+34.9
—	2	186 30 52.2	+3.3	15 37 10.6	. . .
—	3	186 20 55.1	+2.6	15 44 2.0	+19.7

Fig. 10.18 Comparison of Ceres observations at Seeberg with the seventh elements of Gauss, from February 19 to March 3, 1802

Thus, we determined its location: apparent RA 183° 14′ 43″.6 apparent decl. 17° 21′ 5″.0. On March 23 Ceres was again with an unknown and undetermined small star of similar appearance and size so that it was very difficult to distinguish them. The star preceded the planet and stood 3 minutes farther north; we observed its apparent location that day: RA = 182° 16′ 53″.1, apparent declination 17° 39′ 21″.5. These locations will be very useful for those astronomers who observed Ceres on the mentioned days and compared it to stars in the immediate neighbourhood. By the way, the mean position of those stars can be found in the above catalogue.

Gauss' elliptical orbit VII of the new planet (last issue, p. 272) is still correct and matches our observations of this planetary celestial body as you can see from the following calculation and comparison of Dr. Gauss (Fig. 10.18):

The RA are still excellent and the error in declination seems to be the same. If you omit that of March 1 the average of five observations would be +22″.

But soon our observations will be different from this ellipse since the perturbations by Jupiter are considerable and they might amount to ½ degree, as Senator La Place advises. This great geometer does not expect a thorough knowledge of this planet's orbit for two or three years; thus he wrote that he was printing the third edition of his Mécanique céleste *and had reserved the theory of this planet for the fourth edition.*

Dr. Burckhardt has been working on the perturbation equations. This could only be done by a successive approximation because it implies the knowledge of an unknown orbit and this has to be brought closer to the truth by roughly knowing the perturbation equation. These mutual dependent changes have to be made until the final rectified orbital elements correspond to each and every observation with all of their anomalistic equations.

Accordingly Dr. Burckhardt found that the sum of all perturbation equations, which are derived from Jupiter alone, can amount to 27 minutes. But the equations, which come from the squares of the eccentricities, have been neglected because they, at least for now, cannot be significant. For the argument of the most substantial, which depends on the threefold longitude of Jupiter and hardly on Ceres' longitude, increases per year only by 13 degrees. Its influence can therefore be only of minor importance for the observed period. But to verify this Dr. Burckhardt calculated even this equation and found that it has changed only 12″ in one year, that is during the entire period this celestial body has been observed.

Dr. Burckhardt also calculated Ceres' perturbations in latitude. He found that the sum might amount to 1½ minute. And he even studied Saturn's impact on the strange guest but turned out to be negligible. The perturbations caused by Mars can be roughly estimated by the following thought. The ratio of Ceres' semi major axes and Mars' orbit equals almost that of the orbits of Saturn and Jupiter. Thus it is only permitted to multiply the perturbations of these planets by the mass ratio in order to obtain information about the mutual perturbations of Mars and Ceres. Thus you can understand that the impacts are extremely negligible and hardly amount to one second. Only the equations of a very long period could amount to several seconds. But that determination does require a thorough knowledge of the planetary orbit, which is not yet the case today. And the other way round it follows that Ceres' influence on Mars is just as minor and that most likely the Mars tables will remain unchanged. The impact of all other planets can certainly be neglected.

After all these perturbation equations the following rectified elements were found by Dr. Burckhardt:

Epoch 1801	*77° 19' 17" Paris meridian*
Aphelion	*326 42 32*
Ascending node	*81 5 35*
Inclination	*10° 36' 52"*
Eccentricity	*0.0788725*
Trop. time of orbit	*1679.84 days*
Semi major axis	*2.76587*

These elements are only approximate values and will need some improvement in the future; they represent four observations. Dr. Burckhardt is hoping to be able to send tables for the heliocentric path of this planet, which will represent our observations for at least several years and helpful for a future rediscovery so that you have to point the telescope only to that place where the tables indicate.

A consideration that springs to mind regarding the immense progress and quick spread of astronomy these days can hardly be suppressed at the present occasion. What was one of the most difficult problems in astronomy 13 years ago, which solution so many great geometer tried to find and whose success was hard to believe in has been resolved entirely today. Thirteen years ago we did not have a comprehensive, unempirical planetary theory based on the imperishable system of attraction; what we had were incomplete fragments. De Lambre was the first astronomer who calculated according to this new theory the mutual perturbations of Jupiter and Saturn, determined their secular and periodic perturbation equations and accordingly compiled the planetary tables based on the sole law of masses, which are different from the celestial path by a few seconds only, where previously half and quarter degrees were incompatible. Uranus was discovered in 1781 but its perturbations were not calculated until 1789. A decade ago we needed eight years but today this takes us only months and days. Ceres Ferdinandea was discovered only 13 months ago and its perturbations have been calculated, compiled in tables and its orbit calculated accordingly. These perturbation equations, as La Lande wrote, were calculated by Dr. Burckhardt in the course of just one day. Eight years ago there were only four or five capable astronomers in the whole of Europe who could have made such elaborate calculations in several months – today we have more than a dozen young and talented men, who accomplish such a work within days. Unmistakably, pure theory has come to the aid of refined practice to walk hand in hand. Only the extreme accurateness of today's observations enables the calculator to obtain in such a short time such good results. Sometimes theory precedes practice and gives the data for which centuries of observations would be necessary to find them empirically. Thus this noble and thorough science goes ahead of its era and approaches its perfection by the hour to which formerly centuries were necessary. But after a thousand years there is still the opportunity for adding something. [tr. from Latin] *But who can and will be surprised by this? Who dares to think that the human mind could grasp and fathom the inestimable eternal wisdom in this heavenly and divine order? Much work remains* [Latin quote from Seneca, Epistle 27.4.3] *The Creation and the majestic world structure are so immeasurable and infinite and thus the boundaries of our inquiring mind will be equally immeasurable and infinite. And just as their laws are everlasting, the causes will remain hidden*

Tag der Beobachtung	Mittlere Seeberger Zeit	Scheinbare gerade Aufsteig. der ⚳	Scheinbare Abweichung der ⚳
	U ′ ″	° ′ ″	° ′ ″
1802 26 Febr.	14 3 52,2	187 7 33,40	15 8 54,3 nördlich.
27 —	13 59 21,3	186 58 47,40	15 15 52,8
28 —	13 54 49,5	186 49 48,15	15 22 57,9
1 März	13 50 10,4	186 40 27,90	15 29 41,1
2 —	13 45 42,0	186 30 48,40	. . . .
3 —	13 41 6,4	186 20 52,55	15 43 42,3
6 —	13 27 12,8	185 49 18,70	16 3 49,2
7 —	13 22 32,8	185 38 15,90	16 10 15,9
10 —	13 8 27,7	185 3 48,96	16 29 18,9
11 —	13 3 41,3	184 51 53,85	. . . .
15 —	12 44 45,0	184 2 50,00	16 58 30,9
16 —	12 39 58,8	183 50 19,00	17 3 58,9
17 —	12 35 14,3	183 37 34,80	17 9 6,4
18 —	12 30 25,5	183 24 50,72	17 14 5,3
19 —	12 25 38,3	183 11 58,65	17 18 50,8
23 —	12 6 28,9	182 20 23,83	17 30 27,9
27 —	11 47 20,3	181 29 1,05	17 50 29,6
28 —	11 42 33,7	181 16 17,20	17 53 23,1
29 —	11 37 48,0	181 3 49,20	17 56 4,0
30 —	11 33 2,5	180 51 23,85	17 58 25,1
31 —	11 28 17,0	180 38 57,00	18 0 48,4

Fig. 10.19 Observations of Ceres Ferdinandea made at Seeberg Observatory

for us. Thus approximation will always be possible but never perfect knowledge. But how could these mortal Earth's sons hope or expect such an honour – their life and efforts are bound to be asymptotic. Future centuries will see great discoveries but our present generation has at least the honour to get a glimpse of them. And we leave duties for our great-grandsons, which we could only find and describe but not fulfil.

In the most recent issue we gave our astronomical readers all observations made at Seeberg Observatory until February 19. Here are our continued observations of this planet, which we observed most carefully as often as weather was permitting (Fig. 10.19). The right ascensions were made at the passage instrument; the declinations were made by Dr. Buerg at the 4-foot wall quadrant. The planet's brightness has increased to such a degree that Dr. Buerg was able to take the zenith distances at the tiny spider's thread of his telescope since the planet tolerated higher illumination already. The appearance of this heavenly body changed from day to day and it was rather difficult to decide what was due to our atmosphere and what to the weird planet. We leave this question to astronomers equipped with more powerful instruments like Herschel or Schroeter. Several astronomers see this planet slightly darkish and nebulous. Dr. Maskelyne saw a clearly bounded disc, but remarks at the same time that he observed the same with stars, e.g. 34 Virginis and the satellites of Jupiter in clear nights.

Fortunately the above geocentric observations include the entire period in which the planet was at opposition with the sun or close to it. In order to obtain the purely heliocentric location of this star: we also observed and determined each time and equally carefully the position of the sun or better that of the earth in order to find the error of these tables and Ceres' heliocentric locations purely and without any doubt. [Here follows a table of observations of the Sun.]

Thus our above observations of Ceres describe the time when the planet and the sun came into opposition. In order to calculate from this the important heliocentric locations of the planet, we chose the fortunate observations of March 15–19, which include this very special moment. At first we calculated from the observed apparent right ascensions and

1802	Mittlere Seeberger Sonnenzeit	Beob. scheinb. gerade Aufsteig. der ⚳	Beob. scheinb. nördliche Abw. der ⚳	Beobachtete geoc. Länge der ⚳	Beob. geoc. nördl. Br. der ⚳	Unterschied in der Länge	Unterschied in der Breite	Aberration in Länge	Aberration in Breite
März 15	12 U 44′ 45,″0	184° 2′ 52,″0	16° 58′ 34,″0	5 Z 26° 44′ 22,″7	17° 8′ 24,″5	— 31,″8	+ 27,″4	+ 7,″5	+ 0,″0
16	12 39 58, 8	183 50 19, 0	17 4 2, 0	5 26 30 35, 6	17 8 23, 0	— 31, 6	+ 25, 2	+ 7, 5	+ 0, 1
17	12 35 12, 3	183 37 34, 8	17 9 9, 5	5 26 16 45, 8	17 7 58, 9	— 29, 6	+ 27, 7	+ 7, 5	+ 0, 2
18	12 30 25, 5	183 24 50, 7	17 14 8, 4	5 26 3 1, 1	17 7 27, 1	— 31, 3	+ 28, 5	+ 7, 5	+ 0, 3
19	12 25 38, 3	183 11 58, 6	17 18 53, 9	5 25 49 15, 3	17 6 40, 0	— 31, 1	+ 35, 0	+ 7, 5	+ 0, 4

Fig. 10.20 Comparison of Ceres observations at Seeberg with the seventh elements of Gauss, from March 15 to 19, 1802

Fig. 10.21 Solar elements

1802	Länge der Sonne — 4,″4 + 20,″0	Log. Distant. ☉ u. ♁
März 15	11 Z 24° 43′ 11,″5	9. 998[illegible]124
16	11 25 42 57, 1	9. 9981220
17	11 26 42 0, 6	9. 9982437
18	11 27 41 22, 4	9. 9983058
19	11 28 40 42, 0	9. 9984[illegible]

1802	Heliocentrische Länge	Heliocentr. Breite	Elongation	Log. curt. Dist. ☉ ⚳
März 15	5 Z 25° 56′ 14,″5	10° 35′ 36,″6	5 Z 27° 59′ 28,″4	0, 4025038
16	5 26 11 17, 6	10 35 22, 1	5 29 12 40, 6	0, 4025806
17	5 26 26 20, 3	10 35 6, 8	5 29 34 8, 2	0, 4026584
18	5 26 41 23, 1	10 34 50, 9	5 28 20 59, 9	0, 4027360
19	5 26 56 25, 1	10 34 34, 3	5 27 7 54, 7	0, 4028160

Fig. 10.22 Heliocentric data for Ceres, from March 15 to 19, 1802

declinations of Ceres the apparent geocentric longitudes and latitudes and compared those with the last elliptical orbit VII by Gauss and determined the differences. Thus, we found the following data (Fig. 10.20):

The average error of the above mentioned Gaussian ellipse is accordingly –31″.1 in geocentric longitude and +27″.2 in geocentr. latitude; the last observation of the declination was excluded since unreliable.

In order to determine independently of any errors of our solar tables the time and place of the opposition, we improved the latter according to our own above mentioned observations. We took the mean value of five days (March 17–21) and found that we had to decrease every calculated solar longitude by 4″.4. If you take into account the aberration of +20″.0 you get the following solar elements for the above given observational moments of Ceres (Fig. 10.21).

Although our heliocentric data used at present will change when in the future the orbital elements are changed, we give you those derived from Gauss' elements VII. The calculation of the opposition will not be changed significantly (Fig. 10.22).

From this follows that the opposition of Ceres occurred on March 17 at 4h 18′ 0″ mean time Seeberg. For this moment the sun's longitude of the apparent equinox, according to our solar tables 11Z 26° 21′ 11″.0 and after the improvement –4″.4 and the aberration + 20″.0 = 11Z 26° 21′ 26″.6. The elements of Dr. Gauss give the geocentric position of Ceres at that time 5Z 26° 20′ 55″.4. The found improvement is +31″.1; accordingly the improved geocentric longitude of Ceres is 5Z 26° 21′ 26″.5. But the heliocentric longitude of the planet is 5Z 26° 21′ 7″.7: consequently the error of the heliocentric longitude = –18″.8.

1802 Januar 10	2″500	5″16
25	2.514	4.77
26	2.687	5.07
28	2.774	5.18
31	2.930	5.38
Februar 5	3.120	5.59
10	3.342	5.90

Fig. 10.23 The diameter of Ceres, calculated by Gauss

The geocentric latitude of Ceres calculated from these elements is 17° 8′ 36″.2 north; our improvement –17″.2; thus improved geocentric latitude 17° 8′ 9″.0: the heliocentric latitude 10° 34′ 54″.8. The Gaussian elements state the latitude 10° 35′ 12″.2: thus the error in heliocentric latitude +17″.4.

Dr. Schroeter continues to observe the diameter of Ceres. Here are some of his observations, calculated by Dr. Gauss. [Column at left is Apparent Diameter; column at right is Calculated Diameter at distance=1] (Fig. 10.23):

The mean would be 5″.29=0.308 of the earth or 529 geographic miles, including the atmospheric nebulosity. According to Schroeter's measurement of the clearly bounded disc the diameter would be only 3″.44 at the distance=1, thus almost exactly 1/5 of the earth's diameter or significantly smaller than our moon. No wonder that such a little planet has been hidden for such a long time!

Since Gauss' ellipse VII represents Ceres' orbit still very accurately we will give our astronomical readers a continued ephemerides to its path for the following three months, whose calculation was kindly made by Dr. Gauss (Fig. 10.24). It is hardly imaginable that Ceres will be visible since then because we notice already now (March 27/28) such a quick change of light that we cannot blame the condition of our atmosphere or its distance from the earth. But in order that astronomers equipped with equatorial or parallactic instruments might not leave anything undone to find Ceres again, the following may be facilitate this difficult task.

Ceres comes to a standstill and becomes direct again on May 2: longitude 5Z 19′ 45″ and on May 9: RA 176° 15′. This time the extent is in longitude 13° 10′ and lasts 92 days, in RA 12° 27′ and lasts 93 days. The ratio of brightness of the planet, disregarding the phase, is according to Dr. Gauss' calculation as follows (Fig. 10.25):

After this issue's going to press we received two letters from Prof. Piazzi from Palermo of February 2 and 17, that he had not had any idea of Ceres' rediscovery. As postscript he wrote a few lines to tell that he had just read in the papers that in Germany Ceres had been found again and he adds: [French] *"Imagine my satisfaction! I am longing to be assured in further detail by your letters." Of those letters, which do not belong here, another time.*

[F.N.: Henry (1765–1825) was a refugee of the French Revolution, fleeing to Mannheim in July 1789. Early in 1793 he worked for Barry as second astronomer at Mannheim Observatory; however, he left on July 7 that year for St. Petersburg because the French army was approaching. In St. Petersburg Henry was an astronomer until 1800. By the time of his death he was a Colonel of Engineering in Strassbourg. Roger Barry (1752–1813) was director of the observatory in Mannheim from 1788 on. His observations were published in the yearbooks of Bode and the *MC* of Zach. He is honored by asteroid Barry (1703).]

	AR. ⚳ in Graden		Abweich. nördl.		AR. ⚳ in Zeit		
April 21	177°	14′	17°	52′	11 U	48′	55″
24	176	56	17	42	11	47	43
27	176	41	17	30	11	46	44
30	176	29	17	17	11	45	58
May 3	176	21	17	2	11	45	25
6	176	17	16	45	11	45	6
9	176	15	16	27	11	45	2
12	176	17	16	8	11	45	10
15	176	23	15	47	11	45	27
18	176	31	15	25	11	46	5
21	176	43	15	2	11	46	50
24	176	57	14	38	11	47	48
27	177	14	14	13	11	48	57
30	177	34	13	47	11	50	16
Jun. 2	177	57	13	21	11	51	47
5	178	22	12	53	11	53	27
8	178	49	12	25	11	55	17
11	179	19	11	56	11	57	16
14	179	51	11	27	11	59	24
17	180	25	10	57	12	1	40
20	181	1	10	27	12	4	4
23	181	39	9	56	12	6	35
26	182	18	9	25	12	9	13
29	183	0	8	54	12	11	59

Fig. 10.24 Positions of Ceres in April, May and June 1802, midnight, mean time Seeberg

Fig. 10.25 Ratio of the brightness of Ceres

	Abstand von der ⊙	Abstand von der ♁	Lichtstärke
1801 December 7	2.5453	2.4892	0.4227
1802 März 16	2.5706	1.6024	1.0000
Mai 6	2.5961	1.9039	0.6945
Juni 29	2.6305	2.5735	0.3703

Collection of Astronomical Discourses, Observations and Reports

BAJ (1802), p. 102 by Olbers.

[From April 6] *I shall now continue to give you some information of the odd and puzzling body I discovered on the 28th of March. It continues to move with great uniformity, with a movement away which decreases somewhat in RA and declination, and its appearance remains unchanged: namely, that of a small magnitude 7 fixed star, and somewhat fainter than Ceres. Here are my observations of April:* [These observations are in the May issue of the *MC*.]

Meridian Observations of the New Planet Ceres Ferdinandea in the Year 1802, by Dr. Piazzi in Palermo

BAJ (1805), p. 202 (Fig. 10.26).

1802.		Am Paſſage-Inſtrument. M. Z. St. M. S.	Am Paſſage-Inſtrument. Scheinb. ger. Aufſt. in Zeit. St. M. S.	Am ganzen Meridiankreis. Scheinb. ger. Aufſt. in Zeit. St. M. S.	Am ganzen Meridiankreis. Scheinb. Abw. N. corr. durch die Refract. G. M. S.
Febr.	22	14 21 42,96	12 30 35,50		14 48 } circa
	25	8 23,54	29 3,60		15 2 } circa
	26	3 53,96	28 29,77	12 28 30,37	15 8 37,7
	27	13 59 23,87	27 55,55	27 55,57	15 39,7
	28	54 51,98	27 19,51	27 19,83	22 35,7
März	2	45 44,49	26 3,62	26 3,60	36 29,0
	4	36 31,87	24 42,60	24 42,60	50 8,8
	9	13 12,68	21 2,38	21 2,41	16 22 53,4
	11	3 47,13	19 28,37	19 28,77	35 4,9
	12	12 59 3,12	18 40,09	18 40,03	41 14,0
	14	49 33,25	17 1,76	17 1,82	52 53,4
	19	25 41,05	12 48,40	12 48,28	17 18 49,0
	24	1 44,37	8 30,63	8 30,41	40 11,2
	26	11 52 9,52	6 47,34	6 47,45	47 11,3
	27	47 22,70	5 56,33	5 56,35	50 17,4
	31	28 29,59	2 36,30	2 35,90	18 0 44,8
April	1	23 35,05	1 47,45	1 47,54	2 45,4
	2	18 51,09	0 59,35		6 circa
	4	9 25,60	11 59 25,40	11 59 25,35	7 16,9
	5	4 44,36	58 40,00		
	6	0 3,56	57 54,91	57 55,26	9 5,9
	7	10 55 23,70	57 10,90	57 10,98	9 35,9
	8	50 45,07	56 28,00	56 27,82	9 52,0
	9	46 7,07		55 45,85	9 57,0
	10	41 30,53	55 5,05	55 4,79	9 40,5

Fig. 10.26 Observations of Ceres by Piazzi from February 22 to April 10, 1802

Resumed News about the New Main Planet of our Solar System, Ceres Ferdinandea

MC (May 1802), p. 462.

We continue to relay all information about Ceres Ferdinandea to our astronomical readers. We try to keep this collection as complete as possible so astronomers might find here compiled what he had to search for on scattered sheets of paper, or even not at all. At present astronomers are preoccupied with observing the opposition of Ceres and the sun. The latter gives a heliocentric position of the planet. That means: this observation is as if someone had taken it from the centre of the planetary system or made by an observer directly from the centre of the sun. This very important date, which has been observed for the first time for Ceres, can be used to correct and improve its preliminary path. We mentioned our observations of this occurrence already in the last issue on page 398; here follow other astronomers in chronological order as they sent in their observations.

Dr. Burckhardt observed the opposition of this new planet in Paris at Observatory du Champs de Mars. For those he used the observations of March 10, 15, 18 and 19. He also improved De Lambre's solar tables by observations and found the mean value of the error = –11″.0 to be subtracted from the calculated position of the Sun. He calculated Ceres' location according to his elements and perturbation equations published in the MC (April, p. 392). But he had made two improvements prior to this because he took into account two new perturbation equations, which he formerly omitted and which were separate parts dependent on the arguments nt + ε and nt + ε′. [These letters refer to La Place's description in his theory.] *After having considered those he found the inclination 10° 37′ 17″ and the ascending node 81° 2′ 20″. He is inclined to decrease the first by 12″ and increase the latter by 10″. According to this he calculated his four observations, which are as follows (Fig. 10.27):*

From this follows the mean heliocentric error of Burckhardt's tables of Ceres in longitude – 5″.4 and in latitude – 21″.8. The opposition occurred in Paris on March 17 at 3h 46′ 8″ mean time, reduced to the National Observatory, when the planet was at 5Z 26° 21′ 26″.5 of the true heliocentric longitude, that is, freed of any influence of aberration, nutation and parallax and at 17° 7′ 57″.5 of geocentric latitude north. Dr. Burckhardt believes that the error in latitude is 21″.8 and that the radii vectores need to be increased. But he is right in saying that we have to await further observations to be sure of that. For now his tables are more than sufficient to calculate in advance the planet's positions for a very long time. After having informed the Viennese observer of his erroneous locations of the stars used taken from Bode's catalogue, he wrote us on April 3: "I am pleased to hear that you could confirm my suspicion about the wrong location of no. 111 Virgo in Bode's catalogue. Now your determination matches no. 187 Virgo. According to your corrected location of 87 Virgo I corrected my observation of March 7. Also the corrected star no. 147 Virgo now matches your o Leo and 114 Virgo on March 15."

After all those corrections had been applied, the Viennese observations of Ceres are (Fig. 10.28):

From this and Gauss' elements VII the opposition was calculated as follows (Fig. 10.29):

At that time Ceres' observed heliocentric longitude = 5Z 26° 21′ 25″.4. That calculated according to Gauss' elements larger by 3″.5. The observed geocentric latitude north 7° 7′

Paris 1802	Mittlere Zeit	Beobachtete gerade Aufst. der ⚳	Beobachtete Abweich. der ⚳ N.	Beobachtete geocentrische Länge der ⚳	Beob. geoc. Breite der ⚳	Beobachtete heliocentrische Länge ⚳	Fehler der Tafeln	Beob. helio-centr. Breite ⚳ N.	Fehler der Tafeln
März 10	13U 8′ 26,″5	185° 3′ 29,″1	16° 29′ 25,″5	5Z 27° 52′ 18,″3	17° 5′ 59,″9	5Z 24° 41′ 28,″3	— 4,″8	10° 35′ 39,″3	— 18,″9
— 15	12 47 45, 8	184 2 35, 2	16 58 33, 0	5 26 44 8, 0	17 8 16, 5	5 25 56 52, 5	— 4, 5	10 34 56, 8	— 20, 4
— 18	12 30 26, 1	183 24 30, 6	17 14 10, 0	5 26 2 42, 3	17 7 20, 6	5 26 42 2, 0	— 6, 8	10 34 9, 7	— 22, 1
— 19	12 25 39, 3	183 11 43, 5	17 19 3, 5	5 25 48 57, 7	17 6 42, 7	5 26 57 6, 4	— 5, 4	10 33 51, 2	— 24, 1

Fig. 10.27 Observations of Ceres by Burckhardt in March 1802

1802	Mittl. Zeit	Beobachtete gerade Aufsteig. der ⚳	Beobacht. Abweich. der ⚳	Beobachtete geocentrische Länge der ⚳	Beobachtete geocentrische Breite der ⚳	Verglichen mit
März 3	13U 41′ 11,″1	186° 20′ 58,″7	15° 43′ 45,″0	5Z 29° 22′ 58,″1	16° 55′ 17,″4 N.	Nro. 111. 187 ♍
— 4	13 36 35, 2	186 11 0, 2	15 50 35, 0	5 29 10 52, 6	16 57 32, 2 —	— — —
— 7	13 22 39, 2	185 38 30, 3	16 10 0, 4	5 28 32 43, 9	17 2 16, 6 —	Nro. 87 ♍
— 12	12 59 5, 0	184 39 55, 8	16 41 32, 1	5 27 15 34, 1	17 7 39, 1 —	Nro. 114 147 ♍
— 15	12 44 49, 2	184 2 44, 0	16 58 21, 2	5 26 44 20, 3	17 8 11, 7 —	o ♌. Nro. 114. 147 ♍
— 16	12 40 3, 3	183 50 18, 6	17 3 32, 2	5 26 30 48, 4	17 7 55, 9 —	Nro. 147 ♍
— 17	12 35 14, 6	183 37 35, 4	17 8 38, 6	5 26 16 59, 6	17 7 33, 9 —	Nro. 476 ♌
— 19	12 25 42, 1	183 12 32, 5	17 18 33, 5	5 25 49 54, 6	17 6 37, 7 —	— — —
— 20	12 20 54, 2	182 59 25, 6	17 23 19, 9	5 25 35 59, 6	17 5 37, 3 —	— — —

Fig. 10.28 The Viennese observations of Ceres

Fig. 10.29 The opposition of Ceres

Aus der Beobachtung des

15 März	der 17 März	4U	37′	49,″7	M. Z. in Wien
16 —	——	4	40	30, 1	
17 —	——	4	38	11, 2	
19 —	——	4	41	50, 0	
20 —	——	4	37	30, 9	

im Mittel ☍ den 17 März
4U 39′ 10,″4

Cracau 1802	Mittlere Zeit	Beob. ger. Aufsteig. der ⚳	Beob. Abw. der ⚳ N.	Vergleichung mit den Sternen nach den Bode'schen Numern
März 1	13U 50′ 24″	186° 40′ 43″	15° 30′ : . .	. . . .
2	13 45 50	186 31 5	15 36 44″	Nro. 167 304 407 ♌
15	12 44 53	184 3 9	16 58 34	Nro. 107 ♌ 147 ♍
16	12 40 8	183 50 35	. . . .	Nro. 2 141 ♋ 107 ♌
17	12 35 21	183 37 51	17 9 8	Nr. 2 10 ♋ 190 ♌ 147 ♍
19	12 25 47	183 12 15	17 18 53	Nr. 125 190 476 ♌
20	12 51 0	182 59 20	17 23 47	Nr. 12 84 425 190 ♌

Fig. 10.30 Improved observations of Ceres from Cracow in March 1802

42″.0; the heliocentric 10° 34′ 38″.6; that calculated from the elements VII larger by 33″.0. The planet's opposition was observed in Cracow by Prof. Sniadecki. He found Ceres already on March 1 and started to observe the planet. But he, too, was misguided by Bode's new star catalogue. To illustrate this fact here are some examples, which might also serve as warning.

For instance the star #407 of Bode's catalogue is too great by 1 min. 5½ sec. in RA and in declination too small by 13″. Unfortunately Prof. Sniadecki employed two very poorly determined stars of Cancer. No. 10 is too small by 6½ sec. in RA and too large by 2′ 20″ in declination; no. 141 Cancer on the other hand is too small in RA by 1′ 9″.1 and by 6″.8 too great in declination. The other erroneous positions were already mentioned in the last issue. You can see from this how little reliability there is in our current star catalogues; they were not made with the necessary elaborateness and contain other determinations than those made by contemporary astronomers. If Prof. Piazzi had relied on such positions of stars at his first sighting of Ceres and did not make very accurate meridian observations it would have been impossible for Dr. Gauss to produce such an accurate ellipse into which Piazzi's observations fit in perfectly and maybe we would not know anything about an eighth main planet of our solar system. And it almost was just what the mockers needed (who became audible), but it is easier for them to crack jokes than to find a sound argument. The first requires only presumptuousness, the latter thorough knowledge.

If the observations from Cracow are improved according to those corrected positions of the stars the observations are as follows (Fig. 10.30):

Dr. Maskelyne had made the following observations in Greenwich, which he kindly sent Dr. Gauss and us (Fig. 10.31):

All of these are made at the meridian except for the first, which was made at the equatorial sector (MC, vol. V, p. 381) The observation of February 19 obviously contains a writing error: Ceres' RA of that day should be 187° 58′ 17″.6. Prof. Bode in Berlin observed Ceres on March 27. His observations are as follows:

	Mean time	*RA*	*Decl.*	*Longitude*	*Latitude*
Mar. 27	*11h 47m 21s*	*181° 29′ 15″*	*17° 50′ 48″N*	*5Z 24° 2′ 14″*	*16° 54′ 54″N*

Prof. Piazzi sent us two letters from Palermo, telling that he had found the planet using Gauss' ephemerides on February 23 for the first time without any effort but could only observe

Greenwich 1802	Mittl. Zeit			Beob. gerade Aufft. der ⚳			Beob. Abweichung der ⚳		
Febr. 4	17U	25′	40″	188°	42′	56, 2	12°	44′	45,″0 N.
— 12	15	4	18	188	30	28. 7	13	33	8, 1 —
— 19	14	34	38	187	44	17, 6	14	20	0, 9 —
März 6	13	27	4	185	48	58, 8	16	3	49, 5 —

Fig. 10.31 Observations of Ceres by Maskelyne in February and March 1802

Beobachtungen der *Ceres Ferdinandea* auf der Seeberger Sternwarte angeſtellt.

Tag d. Beobacht.	Mittlere Seeberger Zeit			Scheinbare gerade Aufſteig. der ⚳			Scheinbare Abweichung der ⚳			
	U	′	″	°	′	″	°	′	″	
April 1	11	23	32,4	180	26	45,45	18	2	49,8 nördlich	Den 4 und 5 beobachtete Prof. *Pasquich* die AR. der ⚳, den 12 und 13 April Prof. *Bürg*.
— 2	11	18	48,7	180	14	46,50	18	4	34,8	
— 3	11	14	6,0	180	3	2,40	18	6	10,4	
— 4	11	9	24,1	179	51	30,90	18	7	29,4	
— 5	11	4	41,9	179	39	54,10	18	8	27,3	
— 7	10	55	21,3	179	17	39 70	18	9	47,1	
— 12	10	32	18,0	178	26	34,60	18	8	54,0	
— 13	10	27	44,7	178	17	12,30	18	7	42,6	
— 15	10	18	41,8	177	59	23,70	18	5	7,5	
— 18	10	5	17,0	177	35	4,00	17	59	29,1	
— 19	10	0	50,9	177	27	37,20	17	57	8,7	

Fig. 10.32 Observations of Ceres at Seeberg Observatory in April 1802

it properly on the 26th due to bad weather. He wrote: [French] *"I congratulate you with all my heart to be the first to have rediscovered Ceres. Since you saw the planet before me it is useless to tell you about its appearance, which resembles that of the first days when I discovered it. Please convey my regards to Mr. Gauss who saved us a lot of trouble and without whom it might have been impossible to verify my discovery." He plans to observe the planet as long as it will be visible in the meridian with the utmost diligence. Oriani in Milan wrote that he had found Ceres on February 24 but due to bad weather was only able to observe it properly on March 10, 11 and 13 together with his colleagues Reggio and Cesaris* [Angelo de Cesaris, 1749–1832]. *He now works on calculating the perturbations and is waiting for the observation of the opposition in order to correct the orbital elements accordingly.*

Prof. Buerg and the editor have observed Ceres Ferdinandea every clear night. Our observations until March 31 were published in the last issue, here are those of April (Fig. 10.32).

The correspondence between our observations and Gauss' ellipse VII can only be called continuous, although recently a divergence has occurred, which will increase in the future. This had to be expected. Dr. Gauss called his elements temporary since he very well knows that the perturbations of the other planets, especially Jupiter, have a great influence on Ceres, which will show more and more. His elements VII were only made to find the planet any time without problems or doubt.

And you, Dr. Gauss, really did this job and what astronomer will not be grateful for it? We publicly and with pleasure bestow praise on him and admit that his ephemerides has always led us to this small heavenly body and has saved us much trouble. We are convinced that every honest astronomer joins in this praise. For, once again: Without Dr. Gauss and his work and our magazine there still might be no Ceres. Most astronomers think the same. Thus, Prof. Sniadecki begins his letter about his observation of Ceres with the following words: [French] *"With your tireless zeal you worked on the astronomers so that you forced them to search for Piazzi's star." On the other hand there were different people too, who mocked all these efforts. No one knows this better than the editor, who has received satirical letters in which he was accused of wasting his time and energy with the search for such a chimera and to encourage others to do the same. Some even gave me to understand how eager they were to see how he would honourably get out of it and close the featured article series "Continued News..." which has been published for eight months. On April 1 we received a letter from far away mocking our efforts and advising us to give up our castle in the air, while whispering confidentially into our ear that the new planet matched Cicero's words written to a friend: That about which you are writing is nothing* [*De eo quod scribis nihil est;* Even though this purports to be from a letter of Cicero to Atticus, it does not appear to be in any works of Classical Latin.] *We can only reply to this well-meaning correspondent with Virgil: whether he be Trojan or Rutulian.* [From *Aeneid*, Book 10 line 108; the Rutulians opposed the settlement of the Trojans in Latium. Here, it is intended to introduce the next remark with the preface "whether friend or foe."] *It would be much easier not to believe in any planet and sit back and do nothing and to follow the humble Capuchin rule of wisdom* [The Capuchins are one of the three independent branches of male Franciscans. The Capuchin order was founded in 1528 as a reform movement by Franciscan friars who wanted to stress the practice of contemplation and to live a stricter interpretation of the rule of St. Francis.] *There were enough people and even scientists who did not believe in such a planet. And some even declared it not biblical and atheistic. Those people and their attitudes revealed themselves. He is black hearted, watch out for him, you Roman!* [from Horace's *Satires*, Book 1, Satire 4, line 85. Horace is describing a person who takes malicious pleasure in making fun of and wounding others. This fits the critic of Zach who ridicules Zach's apparent failure – and even madness – in defending what the critic viewed as an absurd position.]

[There is a section here about Gauss that was included in Chap. 1 of this book.]

We mentioned in the last issue on page 389 our correspondence between our observations and Gauss' elements.

Since then Dr. Gauss continued this comparison as follows (Fig. 10.33):

Dr. Gauss also compared two observations from Greenwich with his elements with the result that the right ascensions matched ours perfectly if you assume that on that of February 19 happened a writing error (Fig. 10.34).

This comparison shows that around the end of March the observations of the planet will be different than Gauss' ellipse. But should they, what is not to be expected, be wrong by five minutes in June, this has nothing to say, for the error can only slowly increase and it will be known sufficiently in advance if you compare previous observations to the elements or the ephemerides. Dr. Gauss also wrote that he had made such preparations that he was able to change the elements according to new observations as soon as the error would amount to one minute and then a good correspondence could be expected for quite some time. As soon as the observations will be done for this year he intends to calculate the true ellipse taking into account all perturbation equations.

Prof. Wurm in Blaubeuren tried to calculate Ceres' perturbations according to Kluegel's method, mentioned in volume X of the Goettinger Commentarien. *He decided to do so right after having read in the February issue of the MC about Ceres' rediscovery. He then did not know Gauss' improved ellipse, whose elements Gauss made by taking into account the latest observations. Prof. Wurm based his calculations on the mean value of the elements*

Seeberg 1802	Berechnete AR der ⚳	Unter-schied	Berechnete Abweich. der ⚳	Unter-schied
März 6	185° 49′ 23″3	+ 4″6	16° 4′ 12″2	+ 23″0
— 7	185 38 22.0	+ 6.1	16 10 45.6	+ 29.7
— 10	185 3 57.2	+ 8.3	16 29 48.4	+ 29.5
— 11	184 52 5.2	+ 9.4	16 35 55.2	. . . . .
— 15	184 3 2.9	+ 10.9	16 58 58.2	+ 27.5
— 16	183 50 28.4	+ 9.4	17 4 20.6	+ 21.7
— 17	183 37 48.5	+ 13.7	17 9 32.6	+ 26.2
— 18	183 25 3.2	+ 12.5	17 14 34.4	+ 29.1
— 19	183 12 14.4	+ 15.8	17 19 25.1	+ 34.3

Fig. 10.33 Comparison of Ceres observations at Seeberg with the calculations of Gauss, from March 6 to 19, 1802

Greenwich 1802	Berechete AR der ⚳	Unterſchied bey Dr. Maskelyne	Unterſchied bey v. Zach	Berechnete Abweich. der ⚳	Unterſchied bey Dr. Maskelyne	Unterſchied bey v. Zach
Febr. 19	187° 58′ 13,″3	— 4,″3	— 4,″1	14° 20′ 36,″2	+ 35,″3	+ 21,″1
März 6	185 49 3, 8	+ 5, 0	+ 4, 6	16 4 24, 0	+ 34, 5	+ 23, 0

Fig. 10.34 Comparison of Ceres observations at Greenwich with the calculations of Gauss

mentioned in the December 1801 issue, so that he assumed the mean distance of Ceres = 2.75 and its eccentricity = 0.0829. He justifies his procedure by saying: "It would have been useless if I had taken improved elements by Gauss but disregarding Jupiter's impact. For because of the considerable magnitude of the perturbations the orbital elements have to be determined right from the beginning by taking those into account. If this is once done the corrected elements can be used to determine more accurately Jupiter's perturbations. Certain erroneous positions are inevitable. By the way I have not disregarded any perturbation, which might be of great importance and which does not include the powers of the eccentricity and there can be no doubt that these powers will produce quite some equations. [F.N.: Kluegel's method does not transcend the power of one of the eccentricities.]

"But it is equally certain that those calculated by me are and will remain the most important. As soon as I have received Dr. Gauss' improved elements of Ceres, I will improve my perturbation calculations accordingly. Maybe Ceres does also have some kind of grande inegalité *like Jupiter and Saturn, since Ceres suffers so much from Jupiter's vicinity. Only this periodic inequality correcting the mean motion cannot be as 'reciproque' as in the case of those two planets."*

Perturbations of Ceres, calculated by Prof. Wurm (Fig. 10.35)

Set ψ = the mean heliocentric longitude of Ceres minus the mean heliocentric longitude of Jupiter. The mean anomaly of Ceres = ω of Jupiter = ω' thus Ceres' perturbations are:

In order to transform this value of the perturbation of the 'radius vector' into parts of the true mean distance of Ceres, the above expression in seconds is multiplied by the factor

= Mean distance of Ceres/value of the radius in seconds

This issue contains the chart promised in the last issue of those regions where Ceres has been wandering since its last sighting and visibility; we mapped the entire apparent orbit from Dec. 7, 1801 until June 29, 1802, which represents a kind of epicycle. We have entered approximately 600 stars taken from La Lande's Histoire célèste française, *which cannot be found in Bode's new star charts. We do not have enough space in our magazine to print this catalogue. We content ourselves to remark that we are willing to give everyone upon request*

In der Lange:

$$-223,''40 \sin\psi + 459,''95 \sin 2\psi + 41,''54 \sin 3\psi + 9,''44 \sin 4\psi + 2,''85 \sin 5\psi + 0,''99 \sin 6\psi$$
$$+ 24,''63 \sin(\psi+\omega) + 41,''65 \sin(\psi-\omega)$$
$$- 51,''03 \sin(2\psi+\omega) - 494,''12 \sin(2\psi-\omega)$$
$$- 6,''42 \sin(3\psi+\omega) - 234,''57 \sin(3\psi-\omega)$$
$$- 1,''36 \sin(4\psi+\omega) + 11,''09 \sin(4\psi-\omega)$$
$$+ 100,''18 \sin(\psi-\omega') + 1,''42 \sin(2\psi+\omega') + 243,''29 \sin(2\psi-\omega')$$
$$+ 0,''80 \sin(3\psi+\omega') - 28,''07 \sin(3\psi-\omega') + 0,''36 \sin(4\psi+\omega') - 5,''36 \sin(4\psi-\omega')$$

Im Radius vector:

$$+ 74,''35 \cos\psi - 263,''29 \cos 2\psi - 29,''56 \cos 3\psi - 7,''54 \cos 4\psi - 2,''43 \cos 5\psi - 0,''88 \cos 6\psi$$
$$- 10,''81 \cos(\psi+\omega) + 14,''94 \cos(\psi-\omega)$$
$$+ 21,''30 \cos(2\psi+\omega) + 49,''74 \cos(2\psi-\omega)$$
$$+ 3,''02 \cos(3\psi+\omega) + 101,''23 \cos(3\psi-\omega)$$
$$+ 0,''84 \cos(4\psi+\omega) - 5,''68 \cos(4\psi-\omega)$$
$$- 13,''89 \cos(\psi-\omega') - 0,''90 \cos(2\psi+\omega') - 108,''04 \cos(2\psi-\omega') - 0,''60 \cos(3\psi+\omega')$$
$$+ 17,''76 \cos(3\psi-\omega') - 0,''30 \cos(4\psi+\omega') + 4,''02 \cos(4\psi-\omega').$$

Fig. 10.35 Perturbations of Ceres

*the true location of each star, which he might need or might have used for a comparison with Ceres. From Prof. Sniadecki's observations of Ceres we learn that some astronomers use sometimes very distant stars in the parallel of the planet. It would require a much greater catalogue which would go beyond the scope of this magazine if we wanted to list all of them. Since Prof. Sniadecki used some stars of Cancer and Dr. Triesnecker and Prof. David some of Leo for a comparison, we want to give at least some very accurately determined stars in the parallel of Ceres. All right ascensions were entirely determined by us at Seeberg Observatory and the declinations by Barry and Henry at the Mannheim Observatory (except for those marked with an asterisk *, which we also made).* [Here follows two tables of star positions.]

Until now all efforts to find the new planet in any of the older star catalogues have been fruitless. La Lande wrote that he believed for a short moment to have found Ceres in the zone of March 13, 1797. The planet was most likely in the telescope but since it did not belong to the zone, which was intended to be observed, it went unobserved. La Lande reports further that the secretary of the Royal Society of Sciences in London, Sir Charles Blagden, who has just arrived in Paris, brought news with him that Dr. Herschel had found Ceres' diameter to be only one second at most. In reply to the injustice to deny the discoverer the right to name his planet La Lande wrote: [French] *"It seems to me that the name Ceres is being taken on, but I, who would like to call it Piazzi, will use the name Ceres as late as possible." Every astronomer should appreciate the chance to show his gratitude by using the discoverer's name like the German and English astronomers did. (F.N. by Zach: Sir Joseph Banks and Dr. Maskelyne always call the new planet Ceres Ferdinandea.) Most astronomers often use in their letters our proposed sign of a sickle ?. Prof. Piazzi wrote in this matter:* [French] *"Bode wrote me that he would like to assign to Ceres such an ear, it appears to me that we should make or adopt your sign of the sickle ?. Come to an arrangement with Bode and I will follow you."*

About perturbations of the novel planet (Ceres) through the effect of Jupiter

Councillor and Knight Schubert in Petersburg. Submitted on May 1, 1802. [Incorrectly dated as 1801 in the original.]

BAJ (1805), p. 166.

The elliptical elements, calculated by Dr. Gauss, of the planet discovered by Mr. Piazzi have passed the most vigorous examination: only through them, the German astronomers have been able to find again this planet which otherwise probably would have been lost for a long time until some lucky chance might have brought it to light again. Hence, these elements certainly have the degree of precision that is required to determine without considerable error the perturbations affecting this planet. As Mr. von Zach rightly noted, it is time now to calculate these perturbations, in order to arrive at a more precise approximation of the elements. Only then shall we be able to correct the elements by means of more accurately calculated perturbations and go back step by step until this for astronomy so immensely useful, indirect method will lead to the highest degree of precision for both elements and perturbations. With Ceres, it is necessary to do this calculation right away, because it is easy to predict that this planet will be affected by considerable perturbations.

I therefore have undertaken this calculation, wishing that the same has been done by others, in order to be able to test the results by comparison; and I have found that the perturbations of Ceres are the most significant ones occurring among the main planets of our system, making the discovery of Mr. Piazzi one of the most important discoveries ever made in astronomy. My time permitted only to calculate the perturbations due to Jupiter, and only to the single eccentricity; although I suppose that other perturbations will be considerable, too: the one from the square of the eccentricity and another one depending on the cube of the eccentricity and on the argument (2nd longitude Ceres – 5th longitude Jupiter) as well as the one originating from the effect of Saturn. (Fig. 10.36; Fig. 10.37) When correcting the elliptical elements, it is most important to reduce the geocentric observations to heliocentric longitudes, whereby the true distance represents a fundamental part of the calculation; hence, I have also calculated the perturbations of the radius vector as well as those of the eccentricity and the position of the apsides. By the way, I have used once again the excellent method of Mr. Laplace, which has been completely developed by this great surveyor in his Mecanique celeste, Tom. I. Liv. II. Ch. VI. and which I assume is well-known among the readers of this yearbook. The elements that I have used are for Ceres the 7th elements of Gauss which Mr. von Zach published in his Monthly Correspondence, March 1802, *and for Jupiter those which can be found in Mr. Laplace's* Exposit. du Syst. du Monde. *They are as follows:*

semimajor axis of the orbit of Ceres = 2.7699644;
of Jupiter = 5.202792;
daily mean motion of Ceres = 769."7924;
of Jupiter = 299."2673;
eccentricity of Ceres = 0.0814064;
of Jupiter = 0.048144;
distance from Sun on Jan. 1, 1802,
of Ceres = Ω = 10Z 25° 58′ 0″;
of Jupiter = Ω′ = 6Z 11° 10′ 20″;
mass of Jupiter in parts of Sun mass = 1/1067.09.

From these data result the following equations for the perturbations due to the effect of Jupiter, if the mean longitudes of Ceres and Jupiter are called l and ♃

[Ed: Schubert also published a much lengthier mathematical treatise: *Perturbations of the New Planet caused by the action of Jupiter.* It was presented at a conference May 2, 1802, and published in *Nova Acta Academiae Scientiarum Petropolitanae* (1805) vol. 14, 714. Its concluding sentence is important]:

The last expression of $\delta\nu$ (section 8) makes one see that the longitude of Ceres, produced by Jupiter's action can rise to 20′ and that in this regard the discovery of Mr. Piazzi is one of the most important ever made in astronomy, because the motion of this planet provides one of the strongest proofs in favour of the theory of universal gravitation.

I. *Jährliche Abnahme der Eccentricité,* in Secunden $= 1{,}''21787$; in Theilen des Halbmessers $= 0{,}0000059045$.

II. *Jährliches Vorrücken der Apsiden gegen die Fixsterne* $= 70{,}''756$.

III. *Periodische Gleichung des Radius Vector* $= \delta r = -94{,}6$
$+ 1033{,}93 \cdot \cos(♃ - l) - 3814 \cdot \cos 2(♃ - l)$
$- 421{,}87 \cdot \cos 3(♃ - l) - 107{,}75 \cdot \cos 4(♃ - l) - 33{,}97 \cdot \cos 5(♃ - l) + 52{,}53 \cdot \cos(l - \omega) - 14{,}1 \cdot \cos(l - \omega')$
$- 199{,}14 \cdot \cos(♃ - \omega) + 62{,}47 \cdot \cos(♃ - \omega') - 709{,}44 \cos(2♃ - l - \omega) + 198{,}66 \cdot \cos(2♃ - l - \omega')$
$- 1387{,}39 \cdot \cos(3♃ - 2l - \omega) + 1408{,}92 \cdot \cos(3♃ - 2l - \omega') + 263{,}66 \cdot \cos(4♃ - 3l - \omega) - 259{,}98 \cdot \cos(4♃ - 3l - \omega')$
$+ 58{,}24 \cos(5♃ - 4l - \omega) - 46{,}48 \cdot \cos(5♃ - 4l - \omega') + 145{,}31 \cdot \cos(2l - ♃ - \omega) + 12{,}38 \cdot \cos(2l - ♃ - \omega')$
$- 299{,}97 \cdot \cos(3l - 2♃ - \omega) + 8{,}77 \cdot \cos(3l - 2♃ - \omega') - 38{,}38 \cdot \cos(4l - 3♃ - \omega) + 4{,}25 \cdot \cos(4l - 3♃ - \omega')$

V. *Periodische Gleichung der Länge* $= \delta v = + 232{,}''59 \cdot \sin(♃ - l) - 498{,}''40 \cdot \sin 2(♃ - l)$
$- 44{,}''45 \cdot \sin 3(♃ - l) - 10{,}''30 \cdot \sin 4(♃ - l) - 3{,}''30 \cdot \sin 5(♃ - l) + 40{,}''91 \cdot \sin(♃ - \omega) - 23{,}''93 \sin(♃ - \omega')$
$- 540{,}''62 \cdot \sin(2♃ - l - \omega) + 110{,}''49 \cdot \sin(2♃ - l - \omega') - 240{,}''35 \cdot \sin(3♃ - 2l - \omega) + 238{,}''94 \cdot \sin(3♃ - 2l - \omega')$
$+ 30{,}48 \sin(4♃ - 3l - \omega) - 30{,}''84 \cdot \sin(4♃ - 3l - \omega') + 5{,}''73 \cdot \sin(5♃ - 4l - \omega) - 5{,}''87 \cdot \sin(5♃ - 4l - \omega')$
$- 23{,}''76 \cdot \sin(2l - ♃ - \omega) - 1{,}''47 \cdot \sin(2l - ♃ - \omega') + 53{,}''53 \cdot \sin(3l - 2♃ - \omega) - 0{,}''88 \cdot \sin(3l - 2♃ - \omega')$
$+ 5{,}''74 \cdot \sin(4l - 3♃ - \omega) - 0{,}''43 \cdot \sin(4l - 3♃ - \omega')$.

Sieht man die Lage der Apsiden als unveränderlich an, so kann man die 2 letztern Gleichungen auch so ausdrücken:

V. $\delta r = - 94{,}6 + 1033{,}93 \cdot \cos(♃ - l) - 3814 \cdot \cos 2(♃ - l) - 421{,}87 \cdot \cos 3(♃ - l) - 107{,}75 \cdot \cos 4(♃ - l)$

— — —

Fig. 10.36 Equations for the perturbations of Ceres due to Jupiter

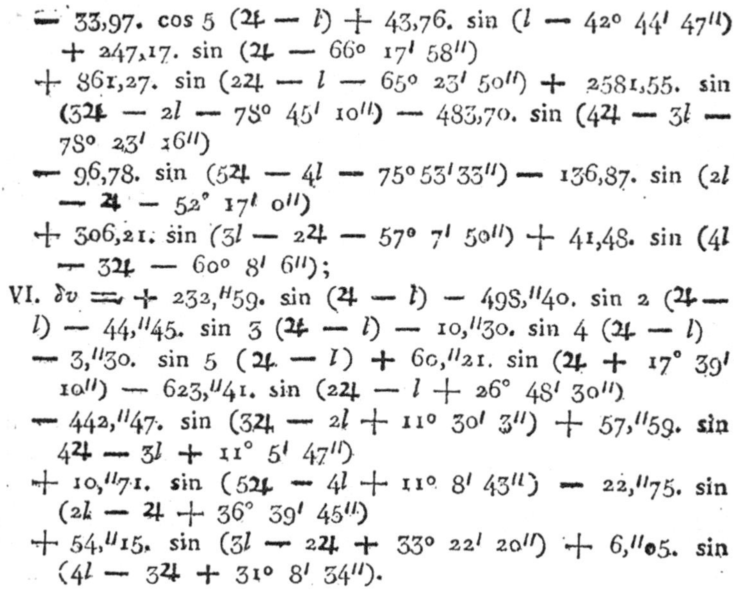

$$- 33{,}97.\ \cos 5\,(\text{♃} - l) + 43{,}76.\ \sin\,(l - 42^\circ\ 44'\ 47'')$$
$$+ 247{,}17.\ \sin\,(\text{♃} - 66^\circ\ 17'\ 58'')$$
$$+ 861{,}27.\ \sin\,(2\text{♃} - l - 65^\circ\ 23'\ 50'') + 2581{,}55.\ \sin\,(3\text{♃} - 2l - 78^\circ\ 45'\ 10'') - 483{,}70.\ \sin\,(4\text{♃} - 3l - 78^\circ\ 23'\ 16'')$$
$$- 96{,}78.\ \sin\,(5\text{♃} - 4l - 75^\circ 53' 33'') - 136{,}87.\ \sin\,(2l - \text{♃} - 52^\circ\ 17'\ 0'')$$
$$+ 306{,}21.\ \sin\,(3l - 2\text{♃} - 57^\circ\ 7'\ 50'') + 41{,}48.\ \sin\,(4l - 3\text{♃} - 60^\circ\ 8'\ 6'');$$

VI. $$\delta v = + 232{,}''59.\ \sin\,(\text{♃} - l) - 498{,}''40.\ \sin 2\,(\text{♃} - l) - 44{,}''45.\ \sin 3\,(\text{♃} - l) - 10{,}''30.\ \sin 4\,(\text{♃} - l) - 3{,}''30.\ \sin 5\,(\text{♃} - l) + 60{,}''21.\ \sin\,(\text{♃} + 17^\circ\ 39'\ 10'') - 623{,}''41.\ \sin\,(2\text{♃} - l + 26^\circ\ 48'\ 30'')$$
$$- 442{,}''47.\ \sin\,(3\text{♃} - 2l + 11^\circ\ 30'\ 3'') + 57{,}''59.\ \sin\ 4\text{♃} - 3l + 11^\circ\ 5'\ 47'')$$
$$+ 10{,}''71.\ \sin\,(5\text{♃} - 4l + 11^\circ\ 8'\ 43'') - 22{,}''75.\ \sin\,(2l - \text{♃} + 36^\circ\ 39'\ 45'')$$
$$+ 54{,}''15.\ \sin\,(3l - 2\text{♃} + 33^\circ\ 22'\ 20'') + 6{,}''05.\ \sin\,(4l - 3\text{♃} + 31^\circ\ 8'\ 34'').$$

Fig. 10.37 Continuation of the perturbation equations

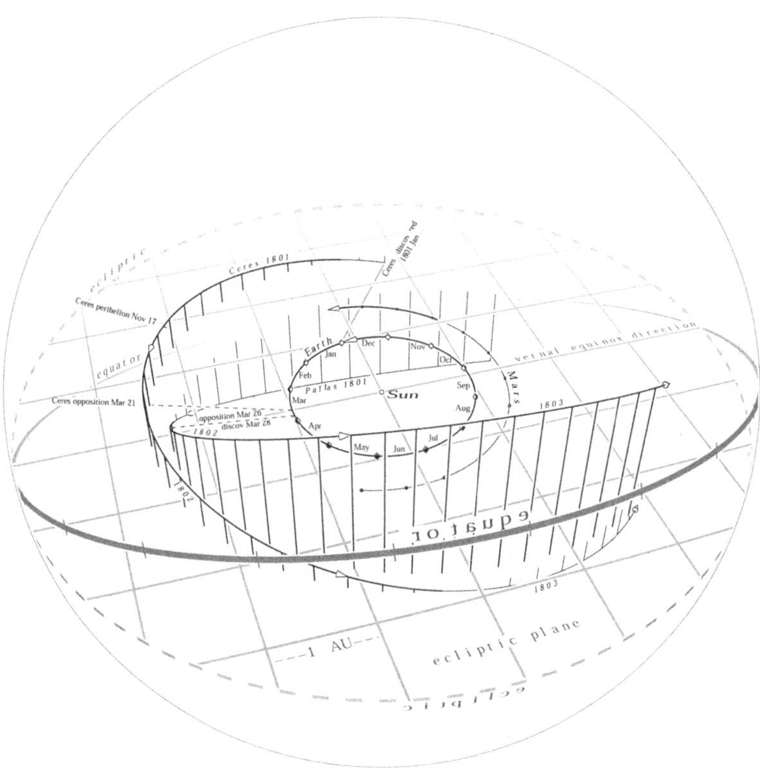

Fig. 10.38 3D plot of the orbits of Ceres and Pallas

[Fig. 10.38: The 3-D plot was specially prepared for this book by Guy Ottewell, editor of *The Astronomical Calendar*. It plots the course of Ceres and Pallas for the years 1801, 1802 and 1803. Their orbits come very close near solar longitude 180, and it was in this part of their respective orbits that both asteroids were located in early 1802. In other words, Earth, Ceres and Pallas were roughly along a line, a major factor leading to the discovery of Pallas. As noted, Ceres came to opposition on November 17, 1801, which also helped to bring this part of its orbit closer to that of Pallas. The orbital elements for Ceres and Pallas in this epoch were kindly supplied by the late Brian Marsden, Director of the Minor Planet Center.]

Resumed News about the new main Planet of our Solar System Ceres Ferdinandea.

MC (June 1802), p. 576.

We finally have the good fortune of being able to now communicate to our astronomical readers the first observations of this new planet conducted by its first discoverer since its rediscovery. We received them in a duplicate copy, once by mail and the second time included for Oriani in Milan. Both arrived at almost the same time; the copies are exactly identical. With it, Professor Piazzi writes us on April 18: "I have observed Ceres so often since February 18 to the present when it was visible on its course through the meridian. You will find a copy of these observations enclosed, which, I hope, will be still more exact than my first ones. In converting sidereal time to mean solar time, I made use of observations on the transit instrument, in which there are five lines; in the entire circle I have but one line. Your solar tables give 0".4 more in this time conversion than those I have used; I just didn't want to change anything, but will do so in future." For the time being, the observations Professor Piazzi was so good to give us will be presented exactly the same way (Fig. 10.39).

Beobachtungen der Ceres Ferdinandea, auf der k. Palermoer Sternwarte vom Prof. Piazzi angestellt.

1802	Mittlere Zeit	Scheinbare Abweich. N.	Scheinbare AR. der Ceres am Kreis	Scheinbare AR. der Ceres am Passagen-Instrument		1800	... horal.	Namen und Grösse
Februar 22	14U 21' 42,"96	. . . . .	. . . . .	12U 30' 35,"50	. 2 3 4 .	12U 28' 30,"46	14° 54' 34,"4	* . . . 7 8
25	8 23, 54	. . . . .	. . . . .	29 3, 60	. . 3 . 5			
26	3 53, 96	15° 8' 37,"7	12U 28' 30,"37	28 29, 77	. 2 3 4 5	12 9 55, 22	14 58 2, 1	* . . . 7 8
27	13 59 23, 87	15 39, 7	27 55, 57	27 55, 55	alle			
28	54 51, 98	22 35, 7	27 19, 83	27 19, 51	alle			
März 2	45 44, 49	36 29, 0	26 3, 60	26 3, 62	alle	11 38 50, 43	15 41 28, 2	β Leonis
4	36 31, 87	50 8, 8	24 42, 60	24 42, 60	alle			
9	13 12, 68	16 22 53, 4	21 2, 41	21 2 38	alle	12 28 30, 46	14 54 34, 4	* . . . 7 8
11	3 47, 13	35 4, 9	19 28, 77	19 28 37	alle			
12	12 59 3, 12	44 14, 0	18 40, 03	18 40, 09	. . 2 3 4 5	. . . . .	. . . . .	β Leonis
14	49 33, 25	52 53, 4	17 1, 82	17 1, 76	. 2 3 4 5	11 55 0, 57	9 50 47, 0	o Virginis
19	25 41, 05	17 18 49, 7	12 48, 28	12 48, 40	alle			
24	1 44, 37	40 11, 2	8 30, 41	8 30, 63	alle			
26	11 52 9, 52	47 11, 3	6 47, 45	6 47, 34	alle	. . . . .	. . . . .	β Leonis
27	47 22, 70	50 17, 4	5 56, 35	5 56, 33	alle			
31	28 29, 59	18 0 44, 8	2 35, 90	2 36, 30	alle			
April 1	23 35, 05	2 45, 4	1 47, 54	1 47, 45	alle	. . . . .	. . . . .	o Virginis
2	18 51, 09	. . . . .	. . . . .	0 59, 35	alle			
4	9 25, 60	7 26, 9	11 59 25, 35	11 59 25, 40	1 2 3 4 .			
5	4 44, 36	. . . . .	. . . . .	58 40, 0	 5	. . . . .	. . . . .	β Leonis
6	0 3, 56	9 5, 9	57 55, 28	57 54, 91	. 2 3 4 5	12 9 55, 22	14 58 2, 1	* . . . 7 8
7	10 55 23, 70	9 35, 9	57 10, 98	57 10, 90	alle			
8	50 45, 07	9 52, [illegible]	56 27, 82	56 28, 00	alle			
9	46 7, 07	9 57, 0	55 45, 85	. . . . .	. . . . .	. . . . .	. . . . .	β Leonis
10	41 30, 53	9 40, 5	55 4, 79	55 5, 05	1 2 3 . 5			
15	18 44, 30	[illegible] 3, 2	51 58, 00	51 57, 90	alle			
16	14 14, 87	[illegible] 25, 2	51 24, 15	51 24, 32	alle	12 28 30, 46	14 54 34, 4	* . . . 7 8

Fig. 10.39 Observations by Piazzi

Since Dr. Gauss made use of our and Piazzi's observations in calculating the elements of this planet's path, according to our reduction and conversion of time, we have reduced the present observations in the same way for the sake of conformity. Some differences, from half to a full second, have turned up in this conversion. Regarding the observation of March 31, there seems to have occured a writing error of 10" mean time. We have, as is fitting, given preference to the right ascensions observed on the transit instrument. Only on April 9 did we borrow it from the entire orbit, since it was not observed on the meridian instrument. According to this, Piazzi's observations now follow (Fig. 10.40):

In Paris, from the National Observatory, Ceres was continuously observed. We published a part of these observations in the April issue (p. 380). The continuation of these supplied to us by Méchain follows (Fig. 10.41)

1802	Mittlere Zeit in Palermo			Beobachtete scheinbare gerade Aufst. der Ceres			Beobachtete scheinbare Abweich. der Ceres N.		
Febr. 22	14 U	21′	42,″49	187°	38′	52,″50	.	.	.
25	14	8	22, 46	187	15	51, 00	.	.	.
26	14	3	52, 82	187	7	26, 55	15°	8′	37,″7
27	13	59	22, 76	186	58	53, 25	15	15	39, 7
28	13	54	50, 92	186	49	52, 65	15	22	35, 7
März 2	13	45	44, 06	186	30	54, 30	15	36	29, 0
4	13	36	31, 45	186	10	39, 00	15	50	8, 8
9	13	13	12, 27	185	15	35, 70	16	22	53, 4
11	13	3	46, 70	184	52	5, 55	16	35	4, 9
12	12	59	2, 64	184	40	1, 35	16	41	14, 0
14	12	49	32, 76	184	15	26, 40	16	52	53, 4
19	12	25	40, 54	183	12	6, 00	17	18	49, 7
24	12	1	43, 92	182	7	39, 45	17	40	11, 2
26	11	52	9, 09	181	41	50, 10	17	47	11, 3
27	11	47	22, 31	181	29	4, 95	17	50	17, 4
31	11	28	19, 18	180	39	4, 50	18	0	44, 8
April 1	11	23	34, 56	180	26	51, 73	18	2	45, 4
2	11	18	50, 67	180	14	50, 25	.	.	.
4	11	9	25, 16	179	51	21, 00	18	7	16, 9
5	11	4	43, 98	179	40	0, 00	.	.	.
6	11	0	3, 10	179	28	43, 65	18	9	5, 9
7	10	55	23, 30	179	17	43, 50	18	9	35, 9
8	10	50	44, 94	179	7	0, 00	18	9	52, 0
9	10	46	6, 66	178	56	27, 75	18	9	57, 0
10	10	41	30, 06	178	46	15, 75	18	9	40, 5
15	10	18	43, 87	177	59	28, 50	18	5	2, 2
16	10	14	14, 47	177	51	4, 80	18	3	26, 2

Fig. 10.40 Observations of Ceres by Piazzi from February 22 to April 16, 1802

1802	Mittlere Zeit in Paris	Beobachtete scheinbare gerade Aufst. der Ceres	Beobachtete Abweich. der Ceres nördl.	Beobachter
Febr. 26	14h 3' 36,"3	187° 7' 25,"35	15° 8' 57,"1	Méchain
27	13 59 15, 6	186 58 41, 70	15 15 58, 3	
März 5	13 31 45, 6	185 59 54, 00	15 57 11, 4	
6	13 27 6, 4	185 49 2, 70	16 3 48, 0	
9	13 13 3, 9	185 15 15, 45	16 23 9, 0	
10	13 8 21, 4	185 3 31, 80	16 29 22, 5	
14	12 49 24, 1	184 15 1, 80	16 52 53, 8	Méchain u. Bouvard
15	12 44 38, 7	184 2 37, 2[illegible]	16 58 35, 4	— — —
17	12 35 6, 0	183 37 20, 00	17 8 55, 5	— — —
18	12 30 19, 0	183 24 31, 35	17 14 10, 3	— — —
19	12 25 32, 1	183 11 44, 25	17 19 2, 1	— — —
April 7	10 55 14, 9	179 17 23, 40	18 9 38, 6	— — —
8	10 50 36, 0	179 6 39, 00	18 9 50, 0	Méchain
12	10 32 11, 8	178 26 22, 05	18 8 31, 0	Bouvard
13	10 27 38, 5	178 16 58, 50	18 7 33, 4	Méchain u. Bouvard
15	10 18 35, 6	177 59 9, 90	18 5 7, 0	— — —
16	10 14 6, 3	177 50 48, 00	18 3 25, 6	— — —
17	10 9 38, 0	177 42 40, 54	18 1 33, 4	— — —

Fig. 10.41 Observations of Ceres by Méchain from February 26 to April 17, 1802

In the May issue, we have communicated to our readers (p. 467) four observations of Ceres by Greenwich. Sincethen Dr. Maskelyne sent us two more as follows:

	Mean time in Greenwich	*RA of Ceres*	*N. Dec of Ceres*
March 18	*12h 30' 17"*	*188° 24' 27"*	*17 14' 10".6*
April 6	*10 59 52*	*179 28 19*	*18 9 10.2*

In Cracow, Professor Sniadecki continued tirelessly to observe this new planet with his meridian telescope and in a quadrant placed on the meridian area. He had the goodness of sending us all his observations from March 2 to May 3, of which we immediately published a few in the May issue (p. 467). Because he used very small and poorly determined stars for comparison, he repeated all his reductions according to the improved positions of these stars, especially of Nr. 165 and 415 Leo (May, p. 479). After these revisions, his observations stand as follows (Fig. 10.42):

Professor Sniade ckiindeed sent us his complete observations to May 3, but since he reserved their revision, we will only be able to print the result in the next issue. Meanwhile, he calculated the opposition of this planet from March 15 to 25 from his observations and discovered that this occurred in Cracow on March 17 at 4hr 56' 45".8 mean time, in 5Z 26° 21' 22".78 geocentric longitude + and 17° 7' 54".0 geocentric latitude + northerly. [Sniadecki (1802a) also reported on Ceres in a Polish-language publication]

The next hope of receiving exact observations of Ceres comes from Vilnius [capital of Lithuania]. *Professor Sniadecki forwarded information of it to Director Poczobut; according to his report, Ceres has been observed there since April 3 on the superb Ramsden 8-foot Wall quadrant and an excellent meridian telescope.* [A w a l l quadrant was an instrument for measuring altitudes consisting commonly of a graduated arc of 90 degrees with an index.]

According to Cremsmuenster, our report of the rediscovery of Ceres arrived only on March 15. During this time, Professor Derfflinger [Thaddeus Derfflinger] *was observing this planet on his wall quadrant and sent us his observations, which will need improvement because of the erroneous positions of the stars he used for comparison, which is made all*

1802		Mittl. Zeit in Cracau	Scheinbare AR. der Ceres	Scheinbare Abweichung der Ceres nördl.	Sterne womit verglichen worden
März	2	13u 45' 4"	186° 30' 27,"00	15° 36' 45,"0	167 304 407 ♌
	15	12 44 53	184 2 51, 50	16 58 37, 0	107 309 ♌
	16	12 40 7	183 50 10, 00	. . . . .	2 ♋ 167 ♌
	17	12 35 21	183 37 40, 00	17 9 10, 0	2 10 ♋ 190 ♌
	19	12 25 46	183 11 54, 39	17 18 53, 0	α ♌ AR 125 190 476 ♌
	20	12 21 1	182 59 15, 17	17 23 58, 8	α ♌ AR 12 125 476 190 ♌
	25	11 57 4	181 54 41, 30	17 43 36, 8	η 372 415 ♌ 66 ♍
	27	11 47 30	181 29 15, 00	17 50 7, 3:	η 165 212 415 ♌
	28	11 42 40	181 16 15, 00	17 53 29, 0	η 165 415 ♌ 66 ♍
	31	11 28 25	180 39 4, 74	18 0 51, 0	. . . .
April	2	11 18 58	180 14 56, 00	18 3 48, 2	165 415 ♌ 66 ♍
	3	11 14 13	180 3 1, 25	18 6 9, 8	η 165 415 ♌ 66 ♍
	4	11 9 32	179 51 34, 00	18 7 38, 0	η β 165 415 ♌ 66 ♍
	5	11 4 49	179 39 58, 20	18 8 54, 0	β ♌ AR 66 ♍

Fig. 10.42 Observations of Ceres by Sniadecki from March 2 to April 5, 1802

Beobachtungen der Ceres auf der Cremsmünster Sternwarte.

1802		Mittl. Zeit in Cremsmünster	Scheinb AR der Ceres	Scheinb Abweich. der Ceres nördl.	vergli-chen mit	Scheinb. AR. des ✳	Scheinb. Abweich. des ✳ N.
		u ' "	° ' "	° ' "		° ' "	° ' "
März	17	12 35 15	183 37 47	17 8 52	λ ♊	10 40 54,4	10 53 15, 5
	18	12 30 30	183 25 31	17 13 29 1/2	474 ♌	174 43 14,8	17 13 35, 1
	19	12 25 43	183 12 43	17 18 32	476 ♌	174 45 54,0	17 20 24, 1
	20	12 20 50	182 59 26	17 23 6 1/2	η ♌	149 8 17,4	17 43 21, 0
			182 59 37	17 22 48	476 ♌	174 45 54,0	17 20 24, 1
	31	11 28 19	180 38 51	18 0 27 1/2	η ♌	149 8 16,7	17 43 21, 7
April	1	11 23 34	180 26 41	18 2 51 1/2	η ♌	149 8 16,6	17 43 21, 5
	2	11 18 50	180 14 40	18 3 44	222 ♌	159 43 1,6	. . . .
	3	11 14 9	180 2 15	18 0 11	222 ♌	159 43 1,0	18 11 20, 4
	11	10 36 54	178 35 39	18 9 44	222 ♌	159 43 1,2	18 11 21, 0

Fig. 10.43 Observations of Ceres by Derfflinger from March 17 to April 11, 1802

the easier since Professor Derfflinger took the praiseworthy caution of indicating the apparent position of the stars whenever he used them. Meanwhile, Professor Derfflinger will not guarantee his observed declinations within a quarter minute because he also found this planet very difficult to observe, partly because of the lack of light making it difficult to observe this body, and partly because of the unclear sky and the low light intensity of this planet (Fig. 10.43).

1802		Mittl. Zeit in Seeberg			Scheinbare gerade Aufst. der Ceres			Scheinbare Abweich. der Ceres nördl.			
April	24	9U	39′	4,″8	176°	55′	46,″00	17°	42′	14,″5	
	25	9	34	47, 3	176	50	21, 00	17	38	36, 4	
	26	9	30	31, 5	176	45	22, 60	17	34	46, 3	
	27	9	26	17, 5	176	40	49, 90	17	30	51, 8:	
	29	9	17	53, 2	176	32	41, 40	17	22	7, 7	
May	1	9	9	35, 2	176	26	8, 10	17	12	47, 7	
	2	9	5	28, 3	176	23	22, 30	17	7	51, 1	AR. Decl. Bürg
	3	9	31	23, 4	176	21	6, 90	17	2	42, 5	
	5	8	53	17, 2	176	17	30, 40:	16	52	28, 9:	AR. Decl. Bürg
	6	8	49	16, 3	176	16	16, 50	16	46	15, 1	AR B. Decl. Harding
	7	8	45	17, 0	176	15	24, 00	16	40	19, 4	AR. Pasquich Decl. B.
	8	8	41	19, 1	176	14	54, 70	16	34	26, 0	AR. Pasquich Decl. B.

Fig. 10.44 Observations of Ceres by Zach

Seeberg 1802		Berechnete AR. der Ceres			Berechnete Abweich. der Ceres			Unterschied	
								in AR.	in Decl.
März	23	182°	20′	39,″9	17°	36′	51,″4	+ 16,″1	+ 23,″5
	27	181	29	21, 2	17	50	55, 5	+ 20, 1	+ 25, 9
	28	181	16	41, 4	17	53	52, 9	+ 23, 7	+ 29, 8
	29	181	4	7, 5	17	56	36, 3	+ 18, 3	+ 32, 3
	30	180	51	40, 3	17	59	5, 5	+ 16, 4	+ 36, 5
	31	180	39	21, 1	18	1	20, 3	+ 24, 1	+ 31, 9
April	1	180	27	10, 7	18	3	20, 0	+ 25, 2	+ 30, 8
	2	180	15	10, 1	18	5	6, 2	+ 23, 6	+ 31, 4
	3	180	3	20, 1	18	6	37, 4	+ 17, 7	+ 27, 0
	4	179	51	41, 7	18	7	53, 5	+ 10, 8	+ 24, 1
	5	179	40	15, 6	18	8	54, 9	+ 21, 5	+ 27, 6
	7	179	18	3, 9	18	10	12, 5	+ 24, 2	+ 25, 4

Fig. 10.45 Comparison of the positions of Ceres with Elements VII of Gauss

At the Seeberg Observatory, Ceres was continuously observed in clear skies. In the previous issue, our observations of this planet went to April 19. Since then, the following ones have been added (Fig. 10.44):

Dr. Gauss also continued, as before, to compare all our observations of this planet with the VII elements of its path. This comparison does not only prove that the deviations o f t h e s e elements change very slowly a n d b y small degrees, a s a consequence representing the position of the planet during its entire period of visibility with exceeding exactness, rather, this comparison will come in very useful in the future in more accurate corrections and if perturbation equations are considered. Until March 19, Dr. Gauss continued these comparisons in the May issue, p. 470. The results of the twelve subsequent observations follow (Fig. 10.45):

Dr. Gauss calculates the opposition of the planet with the Sun from our observations of March 15, 16, 17, 18, 19 and found it had occurred on March 17 at 4hr 21′ 10″ mean Seeberg time at 5S 26° 21′ 27″.3 longitude and 17° 7′ 59″.6 geocentric latitude.

In the previous issue, we communicated the perturbation equations that Professor Wurm calculated for Ceres according to Gauss's older elements of the path. When he received the March issue of our Monat. Corr. *and found the VII elements of this planet's*

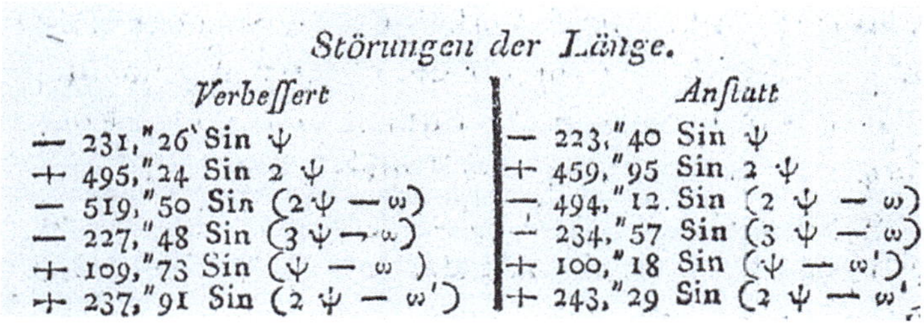

Störungen der Länge.

Verbessert	Anstatt
− 231,"26 Sin ψ	− 223,"40 Sin ψ
+ 495,"24 Sin 2 ψ	+ 459,"95 Sin 2 ψ
− 519,"50 Sin (2 ψ − ω)	− 494,"12 Sin (2 ψ − ω)
− 227,"48 Sin (3 ψ − ω)	− 234,"57 Sin (3 ψ − ω)
+ 109,"73 Sin (ψ − ω)	+ 100,"18 Sin (ψ − ω')
+ 237,"91 Sin (2 ψ − ω')	+ 243,"29 Sin (2 ψ − ω')

Fig. 10.46 Perturbation equations of Ceres by Wurm

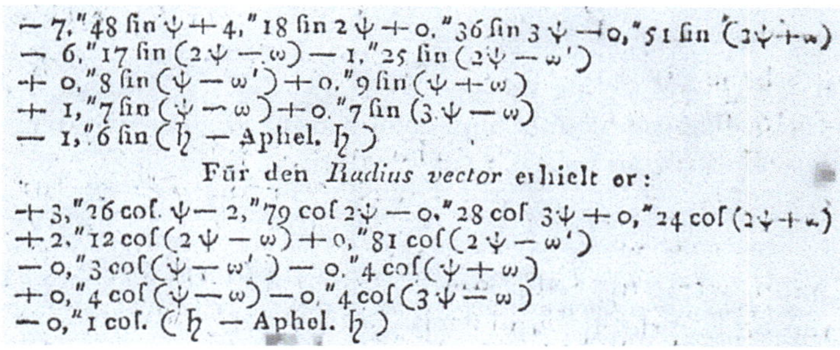

− 7,"48 sin ψ + 4,"18 sin 2 ψ + 0,"36 sin 3 ψ − 0,"51 sin (2ψ + ω)
− 6,"17 sin (2ψ − ω) − 1,"25 sin (2ψ − ω')
+ 0,"8 sin (ψ − ω') + 0,"9 sin (ψ + ω)
+ 1,"7 sin (ψ − ω) + 0,"7 sin (3 ψ − ω)
− 1,"6 sin (♄ − Aphel. ♄)

Für den Radius vector erhielt er:

+ 3,"26 cos. ψ − 2,"79 cos 2ψ − 0,"28 cos 3ψ + 0,"24 cos (2ψ + ω)
+ 2,"12 cos (2 ψ − ω) + 0,"81 cos (2 ψ − ω')
− 0,"3 cos (ψ − ω') − 0,"4 cos (ψ + ω)
+ 0,"4 cos (ψ − ω) − 0,"4 cos (3 ψ − ω)
− 0,"1 cos. (♄ − Aphel. ♄)

Fig. 10.47 Perturbations of Ceres by Saturn, calculated by Wurm

path by Gauss, he immediately made it his primary business to repeat not all, but a few of his perturbation calculations with these elements, wherever the coefficients were considerable. The coefficients, which he calculated a second time with the VII elements, would sustain a change of one or two seconds here and there. However, he didn't want to neglect anything so as to determine the perturbations as accurately as possible on the first attempt. These changes are as follows:

Mean distance of Ceres = 2.76996 and eccentricity = 0.081406

Professor Wurm likewise investigated the perturbations of Ceres through Saturn and found the following equations "for the longitude" (Fig. 10.46; Fig. 10.47):

In this connexion ψ = longitude of Ceres – longitude of Saturn's; ω = mean anomaly of Ceres, and ω' that of Saturn. Professor Wurm also examined the perturbation of Ceres through Earth in greater detail. It was foreseeable that they would be very slight and could be neglected, since their sum would not amount to very much over half a second. In order to become convinced of this, he did not regard the elaborate calculations of perturbations as being superfluous, and thus got the following for the perturbation of Ceres through Earth:

$$-0''.442 \sin \psi + 0''.012 \sin 2\psi + 0''.003 \sin 3\psi \ldots$$
$$+0.056 \sin (\psi + \omega) + 0''.014 \sin (\psi - \omega) \ldots$$

in which ψ = mean heliocentric longitude of Ceres – longitude of Earth, ω = mean anomaly of Ceres. In relation, the perturbations coming from Earth are just as slight for the remaining arguments.

Professor Wurm likewise wanted to devote a separate investigation of the perturbations of Ceres through Mars. He reassured himself about the insignificant value of it entirely with the observation expressed by Dr. Burckhardt in the April issue of Monat. Corr., *p. 391. Since Professor Wurm finds in his present book (p. 546) in a discourse over the masses of the planets, that that of Mars is approximately half the size of what Lagrange and Laplace assume it to be; therefore, the influence of Mars on Ceres is decreased even more and is altogether only half as great as Burckhardt previously believed. However, Oriani finds, as our readers will now see, that Mars could bring about a perturbation of 2".4, and although we have expressed our displeasure at this, he insists, according to repeated calculations, that the influence is really that great. This equation will remain uncertain since the mass of Mars itself is so; meanwhile, it suffices to say at any rate that it has no particular effect on the movement of Ceres. All these calculations are mere approximations for now, and an approximation will have to be improved through others in future anyway. Oriani undertook the calculation of perturbation for Ceres according to Laplace's theory and method and conducted it at the same time according to Gauss's VII elements of this planetary path. The perturbation equations communicated to us are the following. The epochs of the year 1800 are assumed for the arguments, so that Saturn = 4Z 2° 17', and Jupiter = 2Z 22° 9'. For future times, these arguments are maintained if one adds its sidereal movement from 1800 to the present to the epochs for Saturn Jupiter Ceres.*

Further notices respecting the two NEW PLANETS, with some remarks tending to shew that they cannot both belong to the Planetary System

By Brewster (1802b, his italics). *Edinburgh Review,* dated June 22, 1802.

Having transmitted to you, on former occasions, all the information respecting the two newly-discovered planets, which my situation enabled me to collect, I now trouble you a third time, with some additional notices on the same subject, and with a few loose observations calculated to disprove the prevailing opinion, that the two stars lately discovered belong to the planetary system.

"The excentricity of Pallas, it appears, is a little greater than that of Mercury; the inclination of its orbit to the ecliptic 33° .39; its mean distance a little less than that of Ceres, and its periodic time four years and five months, about two months less than that of Ceres. But the most remarkable circumstance concerning it is, that it crosses the orbit of Ceres, approaching the sun nearer in its perihelium, and receding further from him in his aphelium than Ceres does. Dr. Herschel has made some curious observations on the apparent diameters both of Ceres and Pallas, from which he infers the real diameter of Pallas to be 95 miles and that of Ceres 162 miles. He considers them a different species from the known planets. In their smallness and motion they resemble comets; but in the clearness of their light, they resemble the other planets." (footnote: see the Monthly Magazine *for May 1802, from which the above paragraph is extracted.)* [This is an error, as the article was in the June issue: The Monthly Magazine, 1802a.]

From these facts we shall now deduce a few conclusions to support the opinion which has already been advanced; and though our arguments will be drawn chiefly from analogy, the only source to which we can at present apply, yet they should nevertheless have their due influence in the formation of our opinion, till certainty and experience can be substituted in their place. For it must be recollected, that, with respect to the present question, we are in a situation where the maxim of Terence most pointedly applies: "Dum in dubio est animus, paulo momento huc illuc impellitur." [While the mind is in doubt, the smallest impulse sways it one way or another; *Andria.* I. 5. 32.]

By attending to the facts which have already been made known respecting Pallas, the star discovered by Dr. Olbers, we will find many circumstances of resemblance between it and comets, and many marks of dissimilarity to the bodies of the planetary system.

1. *Its excentricity is said to be greater than that of Mercury, the ellipticity of whose orbit is almost double that of any other planet. But as very excentric orbits are peculiar to comets, it is reasonable to believe that Pallas has no connection with the planetary system. This argument would be more powerful, if, on account of Mercury's proximity to the Sun, we could assign for his superior excentricity, as final cause that would not apply to bodies situated at a greater distance.*

2. The next circumstance of dissimilarity to the planets, is the inclination of its orbit. The orbits of all these bodies are inclined to the ecliptic at very small angles, and may all be comprehended between two planes situated at the distance of only fourteen degrees, but on the supposition that Pallas is a planet, these planes, in order to comprehend its orbit, must be removed to the enormous distance of 67°..18', its inclination being 33°..39'. This circumstance affords a strong presumption that this star should be numbered among the comets whose orbits are inclined to the plane of the ecliptic at various angles, and are often perpendicular to each other.
3. The next circumstance of dissimilarity is the smallness of its diameter. As all the planets of our system are bodies of immense magnitude, the diameter of the least of them amounting to several thousand miles, is it not a striking breach of the uniform regularity which pervades that system, to suppose that a star 95 miles in diameter, and infinitely less than the smallest secondary, should revolve round the Sun as a planet; while it has a nearer resemblance to comets whose general characteristic is the smallness of their diameters.
4. Before the discovery of Piazzi, or Ceres, astronomers suspected the existence of a star between Mars and Jupiter, as the distances of the other planets would then increase in a regular progression; and after a star was discovered in that situation, the very same argument, drawn from an idea on uniformity in the system, strengthened them in the belief that this star should be ranked among the number of the planets. Now, by the very same argument, with this difference only, that in the present case it comes with tenfold force, might it be proved, that Ceres cannot be a planet. Can there be a greater breach of uniformity, than to suppose two heavenly bodies revolving round the Sun almost at equal distances? Can there be a greater breach of uniformity, than to suppose the orbits of two planets crossing one another, one of them being nearest the Sun in its perihelium, and farthest from him in his aphelium? Would they not run the risk of meeting each other in the heavens? Or if this did not happen, would not their reciprocal action, increased by their frequent proximity, produce in their movements the most enormous irregularities? It may be said, however, that the great inclination of Pallas's orbit is intended to prevent those effects which might otherwise arise from the necessary proximity of these planets. It is true indeed, that this circumstance must prevent, in a great degree, the effects that will result from their mutual action; but if the line of Pallas's nodes should coincide at any time with the nodes of the crossing orbits, i.e., the points where the orbits of Pallas and Ceres cross each other, and if the two planets should happen to be in or near these points, the greatest irregularities would still be produced, and the two planets would actually run counter to each other.
5. The opinion of Dr Herschel also strengthens these arguments. He "considers them of different species from the known planets. In their smallness and motions they resemble comets; but in the clearness of their light they resemble the other planets." But in these words there is an objection to the opinion which we have advanced, as well as a strong confirmation of the reasoning which has been employed to support it. They resemble the other planets, it is said, in the clearness of their light. Now, in order to obviate this objection, we must enquire whether or not any comets have existed which resembled the other planets in the "clearness of their light;" and if this question can be decided in the affirmative, the objection must completely fall to the ground.

Comets are generally believed to be opaque bodies, illuminated by the Sun, and surrounded by large atmospheres, by means of which their tails are produced. There are comets, however, which have no tails, and which, as Mr [Patrick] Brydone [1741–1818] remarks, "seem to be of a very different species from those which have tails, having a less resemblance to these than to the other planets." Here then seems to be a species of comets without tails, having a considerable resemblance to the other planets. But in addition to this we are informed by other observers, that the light and apparent bigness of comets are sometimes like those of a small clouded star, and sometimes like the satellites of Jupiter. And Cassini, in particular, affirms, that he has seen through a glass, comets whose disc was a pure, big, and clear, as that of Jupiter. Such was the second comet of 1665, and that of 1682. From these facts, then, we actually see that comets do resemble the other planets in the *"clearness of their light;" and since the star Pallas resembles a comet in its motion, in its* smallness, *in its* orbit, *and in the* inclination *of that orbit, we are authorized to rank it among the number of these heavenly bodies, till astronomers, who are of a contrary opinion, shall have actually traced it through the different parts of its orbit round the Sun.*

Several of the arguments which we have used for proving Pallas to be a comet, authorize, in some measure, the same conclusion respecting Piazzi, *or Ceres. On this point, however, we shall not pretend to call in question the opinion of philosophers. But it is proper to remark, that we ought to pause a while for farther observations before we pronounce with confidence, and exalt to the dignity of a planet, an obscure star 162 miles in diameter!*

The force of these arguments, Sir, you will easily perceive, depends on the accuracy of the facts which astronomers have given to the public respecting these two stars. If these be erroneous, the reasoning which I have used must lie under the same imputation. Such as they are, however, they are submitted to your consideration, and left to your disposal.

Resumed News about the new main Planet of our Solar System Ceres Ferdinandea

MC (July 1802), p. 60.

While we are continuing to make all observations of this planet known to our astronomical readers, we are, at the same time, trying to indicate all errors of writing, printing and calculation, which are unavoidable here and there in such large numerical calculations. Thus we have, in the May issue (p.467) and the June issue (p. 579), communicated some of the observations of this planet conducted by the royal astronomer Dr. Maskelyne in Greenwich. But even at that time, we had suspected, not without reason, an obvious error in writing or calculation, and indicated an improvement accordingly. This suspicion was not only proved true, but Dr. Maskelyne has, in a new copy of his complete observations of Ceres, increased its observed right ascension by 3".8.

The reason for this increase is made more important because it stems from a common cause and is of such significance for all practical astronomers, that it cannot be made known to them quickly enough. Namely, Dr. Maskelyne had quite precisely determined the right ascension of the star a Aquilae, on the one hand from the comparison with the Sun, and on the other hand from the observed declination in the opposing equinoxes for a number of years, and thus concluded that the right ascension of this body, as well as all of those

1802		Mittl. Zeit in Greenwich			Scheinbare gerade Aufsteigung der Ceres			Scheinbare Abweich. der Ceres nördl.		
Februar	4	17U	25′	46″	188°	43′	0,″0	12°	44′	45,″0
—	12	15	4	18	188	30	32, 5	13	33	8, 1
—	19	14	35	34	187	58	19, 1	14	20	0, 9
März	6	13	27	4	185	49	2, 6	16	3	49, 5
—	14	12	49	22	184	15	3, 8	16	52	0, 0
—	18	12	30	17	183	24	30, 8	17	14	10, 6
—	25	11	56	46	181	54	18, 8	17	43	49, 8
April	6	10	59	53	179	28	22, 8	18	9	10, 2
—	21	9	51	56	177	13	37, 2	17	51	41, 4
May	1	9	9	27	176	26	1, 2	17	12	34, 7
—	4	8	57	12	176	19	5, 0	16	57	7, 7
—	11	8	29	28	176	15	48, 0	16	12	41, 8
—	13	8	21	45	176	18	6, 9	16	1	48, 6

Fig. 10.48 Observations of Ceres by Maskelyne from February 4 to May 13, 1802

in his catalogue of 36 fixed stars (all of which are based on α Aquilae) had to be increased by 3″.8. For the same reason, all our previous statements of the right ascension of Ceres must also be increased by the same amount since Maskelyne's catalogue of 36 fixed stars also forms the basis our zodiacal star catalogues. Whether 3″.8 has to be added to the right ascensions of all other astronomers who have observed Ceres depends on the star catalogue which they have used as a basis for comparison and on the extent to which they have been used in citing sidereal time. For the present, only the observations in Palermo of Professor Piazzi seem to require this improvement. According to this, there follow the complete improved observations from Greenwich (Fig. 10.48):

In Milan, Professor Cesaris observed Ceres with a splendid eight-foot Ramsden wall-quadrant, as follows (Fig. 10.49):

From Vilnius in Lithuania, we received the following observations of Ceres from the director of the royal observatory, Martin Odlanicki Poczobut, Knight of the White Eagle and recipient of the Order of St. Stanislaus (Fig. 10.50). These observations were made on a splendid 5½ foot meridian telescope with a 4 inch aperture and on an eight foot wall-quadrant. We will communicate these observations in the same form in which they were sent to us. Although Professor Sniadecki brought it to our attention that no refraction seems to have been displayed in the observed declinations of this planet because he always tries to give his observations in this apparent form, we have still not yet ventured to make these improvements since the temperature and barometric levels, and consequently also the correction of the mean refraction according to the density and temperature of the air were unknown to us at the time of these observations. Since it also seems to be agreed that the complete declinations have been improved through no refraction and we still hope to improve the state of meteorological instruments, it is up to us to indicate these improvements in the future. Director Poczobut, an honourable old man who is presently entering his 73rd year, is still possessed by such a youthful zeal for his science that he did not rest until he had found Ceres. During bad weather, which he found very disadvantageous and restricting, he nonetheless

1802	Mittl. Zeit in Mailand			Gerade Aufsteig. der Ceres			Abweich. der Ceres nördl.		
May 4	8U	57′	18,″9	176°	19′	10″	16°	57′	13″
5	8	53	16, [illegible]	176	17	37	16	51	43
6	8	49	15, 0	176	16	23	16	46	0
7	8	45	15, 7	176	15	29	16	40	9
8	8	41	17, 9	176	14	59	16	34	9
9	8	37	21, 7	176	14	50	16	28	1
10	8	33	26, 7	176	15	4	16	21	45
11	8	29	33, 7	176	15	44	16	15	17
12	8	25	41, 4	176	16	43	16	8	41
17	8	6	43, 5	176	26	59	15	33	43
18	8	2	59, 5	176	30	4	15	26	19
19	7	59	17, 2	176	33	26	15	18	47
20	7	55	36, 2	176	37	10	15	11	10
21	7	51	56, 1	176	41	16	15	3	25

Fig. 10.49 Observations of Ceres by Cesaris from May 4 to 21, 1802

continued to search for and tirelessly observe this body with such persistence and effort that he lost consciousness several times during these observations. What an example and what a shame for our younger astronomers! The planetary symbol of the sickle for Ceres won his approval, and he then composed the following Latin verses: [Translated as: Thou hast taught her to cut the stalks of standing corn with a sickle. The toothed sickle shall become for you the consecrated garland of Ceres.]

In the previous issue (p. 580), we promised our readers the continuation of the revised observations of this planet from Cracow by Professor Sniadecki himself. Since we have received them, we will communicate them here. With it, professor Sniadecki remarks that he employed the stars a and β Leonis according to Dr. Maskelyne, and μ and 8 Leonis according to myself, since wherever he had to use smaller stars from time to time, he made use of the positions of these stars which I had cited in the previous editions. Consequently, the above mentioned 3".8 has to be added to all the right ascensions of Ceres observed by Professor Sniadecki, which we have not done however, in order to not change anything on the original observations sent in to us and to leave everyone the freedom conducting these improvements (Fig. 10.51).

Professor Sniadecki had already calculated the opposition of this planet from his earlier observations, as we had also indicated in the previous issue (p. 580). After he had revised all of his observations and after he had improved the position of the stars used for this, he calculated the time and the position of opposition once again, as follows (Fig. 10.52):

From this it further follows from the observations of March 16 and 17 that the time of opposition of Ceres and the Sun was 4hr 53' 22" Cracow mean time, in 5Z 26° 21' 16."8 geocentric longitude and 17° 8' 5."2 north geocentric latitude.

In Prague, the royal astronomer and canon David observed this new planet from March 16 to May 8 (Fig. 10.54). Since he names the star and its location with which he compared the planet every time, his positions can therefore be more easily improved in the future since canon David always included the ascensional difference in time. Incidentally, he had also

a

1802	Mittlere Zeiten der Culminationen		Scheinbare gerade Aufsteig. der Ceres	Scheinb. Zenith-Distanz der Ceres	Nördliche scheinbare Abweich. der Ceres
	von η ♌	von der ⚳			
April 9	8U 47' 21"	10U 46' 16"	178° 56 50"	36° 30' 25"	18° 10' 37"
10	8 43 25	10 41 38	178 46 18	36 30 35	18 10 27
12	8 35 33	10 32 31	178 27 31	36 31 46	18 9 16
13	8 31 37	10 27 56	178 17 44	36 32 43	18 8 19
15	8 23 45	10 18 53	177 59 57	36 35 13	18 5 49
18	8 11 58	10 5 28	177 35 23	36 40 50	18 0 12

b

1802	Mittlere Zeiten der Culminationen		Scheinbare gerade Aufsteig. der Ceres	Scheinb. Zenith-Distanz der Ceres	Nördliche scheinbare Abweich. der Ceres
	von η ♌	von der ⚳			
April 22	7U 56' 14"	9U 47' 53"	177° 7' 34"	36° 51' 30"	17° 49' 32"
23	7 52 18	9 43 34	177 1 48	36 54 48	17 46 14
24	7 48 22	9 39 15	176 56 2	36 58 16	17 42 46
25	7 44 27	9 34 57	176 50 16	37 1 51	17 39 11
27	7 36 35	9 26 27	176 40 44	37 9 43	17 31 19
28	7 32 39	9 22 15	176 36 44	37 13 44	17 27 18
29	7 28 43	9 18 4	176 32 58	37 18 10	17 22 52
30	7 24 47	9 13 54	176 29 28	37 22 43	17 18 19
May 1	7 20 51	9 9 45	176 26 13	37 27 27	17 13 35
2	7 16 55	9 5 39	176 23 42	37 32 27	17 8 35

Fig. 10.50 Observations of Ceres conducted by Martin Odlanicki Poczobut from the Royal Russian Observatory of Vilnius in Lithuania

1802	Mittlere Zeit in Cracau	Scheinbare AR. der Ceres	Scheinbare Abweich. der Ceres	Sterne womit verglichen worden.
April	U ′ ″	° ′ ″	° ′ ″	
6	11 0 9	179 28 38,8	18 9 41,5	η, 415, 165 ♌ 66 ♍
7	10 55 30	179 17 43,7	18 9 59,0	η, 415 ♌ 66 ♍
10	10 41 36	178 46 21,4	18 10 5,0	η, 165, 415, 222 ♌
14	10 23 21	178 8 3,8	18 6 32,5	η, 165, 415, 222 ♌
22	9 47 52	177 7 29,3	17 48 39,8	165, 415, 222 ♌
23	9 43 31	177 1 20,5	17 45 46,0	β, η, 165, 415 ♌ 66 ♍
24	9 39 11	176 55 36,0	17 42 22,5	α, η, β, 415, 222 ♌ 66 ♍
25	9 34 54	176 50 13,0	17 38 32,3	β, η, 165, 222, 415 ♌
26	9 30 38	176 45 20,7	17 34 37,3	α, η, 165, 415, 222 ♌
27	9 26 24	176 40 34,5	17 30 49,6	α, η, 372, 415, 476 ♌
28	9 22 11	176 36 23,6	17 26 28,4	η, 372, 415, 476 ♌
29	9 18 0	176 32 31,4	17 22 25,0	α, η, 372, 415, 476 ♌
30	9 13 50	176 29 2,7	17 17 20,6	α, η, 372, 476 ♌
May 1	9 9 41	176 26 0,0	17 12 43 :	α, η, 372, 476 ♌
2	9 5 36	176 23 22,5	17 8 7,0	θ, 372, 476 ♌
3	9 1 29	176 20 48,0	17 2 53,0	α, θ, 372, 476 ♌
7	8 45 21	176 14 51,0	16 42 44,0	β, θ ♌
8	8 41 24	176 14 43,7	16 34 31,1	α, β, θ ♌
9	8 37 30	176 14 43,7	16 27 53,8	α, β, θ ♌
11	8 29 42	176 15 30,5	16 18 :	β, θ, ♌

Fig. 10.51 Observations of Ceres from the Royal Observatory in Cracow, by Professor Sniadecki

1802	Geocentrische Länge der Ceres	Geocentrische Breite der Ceres	Länge der Sonne nach v. Zach's Tafeln
März 15	5Z 26° 44′ 23,″7	17° 8′ 27″	11Z 24° 41′ 32,″25
16	5 26 30 32,7	17 8 12	11 25 40 51, 99
17	5 26 16 53, 8	17 8 2	11 26 40 21, 80
19	5 26 49 15, 0	17 8 38	11 28 38 56, 90

Fig. 10.52 Time and position of the opposition of Ceres, by Sniadecki

Nro. nach Bode	Flamsteed Nro.	Grösse	Mittlere gerade Aufsteig. 1800	Jährl. Veränder. +	Beobachter	Mittl. Abweichung nördl. 1800	Jährliche Veränd. —	Beobachter
			° ′ ″	″		° ′ ″	″	
6	3 ♋	6	117 19 34,8	51,62	v. Zach	17 50 40	9, 17	Henry u. Barry
372	81 ♌	6	168 47 21,2	47,14	—	17 33 19	19, 65	— —
490	95 ○♌	6	176 20 35,4	46,29	—	16 45 34	20, 00	— —

Fig. 10.53 Stars used by David in his Ceres observations

made use of our star catalogue for the most part. The stars that he used, no. 66 Virginis, 147 Virginis and 476 Leonis appear in the November issue (p. 386), no. 165 Leonis in the May edition (p. 479) and no. 103 Com. Ber. (or no. 25 according to Flamsteed) in the June edition (p. 602). Of late, he has made use of somewhat more dubious determinations of three stars; we therefore cite the more exact ones here (Fig. 10.53).

Beobachtungen der Ceres auf der k. Prager Sternwarte vom Can. David angestellt.

1802		Mittlere Prager Zeit	Ascensional-Diff. in m. Z.	Scheinbare ger. Aufst. der Ceres	Scheinbare Abweich. der Ceres N.	Sterne womit verglichen worden
März	16	12U 39′ 36,″5	0′ 42″	183° 50′ 4″	. . . .	476 ♌ und 147 ♍ nach v. Zach
—	17	12 34 56, 5	1 32	183 37 27	. . . .	
—	18	12 30 5, 5	2 23	183 24 45	. . . .	
—	19	12 25 18, 5	3 14	183 11 58	17° 20′ ::	
—	27	11 47 1, 0	5 28	181 29 10	17 50 25″	3 ♋ 66 ♍ nach v. Zach
—	28	11 42 14, 0	4 38	181 16 39	17 53 13	
—	29	11 37 27, 0	3 48	181 4 7	17 55 58	
—	30	11 32 41, 0	2 58	180 51 34	17 58 27	
—	31	11 27 56, 5	2 8	180 39 2	18 0 37	
April	1	11 23 13, 0	3St. 19′ 12″	180 27 6	18 3 44	150 ♋ nach Tob. Mayer
			1 43 27	26 56	3 7	165 ♌ nach La Lande
—	2	11 18 29, 0	3 18 24	180 15 4	18 5 10	150 ♋
			1 42 39	14 53	4 47	165 ♌
—	3	11 13 46, 0	3 17 37	180 3 16	18 6 35	150 ♋
			1 41 52	3 7	5 59	165 ♌
—	4	11 9 4, 0	3 16 51	179 51 44	18 7 47	150 ♋
			1 41 6	51 35	7 10	165 ♌
—	5	11 4 23, 0	3 16 6	179 40 27	18 8 50	150 ♋
			1 40 21	40 18	8 28	155 ♌
—	6	11 59 44, 0	29 6	179 28 28	18 9 30	103 Com. Beren. nach La Lande
—	20	9 56 7, 0	11 3	177 20 45	17 54 52	66 ♍ nach v. Zach
—	25	9 34 26, 0	1St. 50′ 30″	176 50 17	17 39 10	[illegible] ♌ nach v. Zach
—	26	9 30 10, 0	31 39	176 45 32	17 35 9	[illegible] ♌ nach v. Zach
—	27	9 25 57, 0	31 11	176 41 2	17 31 6	
May	2	9 8 12, 8	. . .	176 23 26	17 7 36	474 ♌ nach v. Zach
—	3	9 4 14, 5	. . .	176 21 11	17 2 34	— Diese Beobachtungen vom
—	4	9 0 18, 0	. . .	176 19 18	. . . .	— Monat May sind an ei-
—	5	8 56 21, 6	. . .	176 18 2	. . . .	— nem Kreis-Mikrometer
—	6	8 52 25, 1	. . .	176 16 47	. . . .	— angestellt.
—	7	8 48 29, 7	. . .	176 15 47	16 39 0	—
—	8	8 44 32, 3	. . .	176 15 1	16 33 2	490 ♌. θ ♌

Fig. 10.54 Observations of Ceres by David in Prague, from March 16 to May 8, 1802

Only No. 150 Cancer can we not find among our observations, and Canon David therefore determines its apparent position according to Tobias Mayer, for April 1, 1801 as RA = 130° 30′ 55″.1, northern declination = 18° 7′ 32″.2. The position of this star can be improved in the future.

As tireless as the observers were in the determination of the apparent geocentric course of this new planet, the calculators were equally tireless in their determination of those elements that are supposed to represent the true heliocentric course of this planet. Oriani, this so skillful mathematician in the intricate calculation of perturbations, spared no effort in repeating the entire calculation of perturbation of Ceres in a new hypothesis of mean distance and to verify his first results therewith. Here is the result of the entire calculation that this great astronomer had the goodness of sharing with us:

Let D = mean longitude of Ceres – mean longitude of Jupiter;
A′ = mean anomaly of Jupiter; A mean anomaly of Ceres.
H′ = mean longitude of Jupiter – mean longitude Jupiter's node
H = mean longitude of Ceres – mean longitude of Ceres' node

It is also to be mentioned that the perturbation equations for each other eccentricity e of the path are maintained when the sections containing A are multiplied with

e/0.081406

the sections containing 2 A are multiplied with

$(e/0.081406)^2$

the sections containing 3 A are multiplied with

$(e/0.081406)^3$

The perturbation-equations of Ceres through Jupiter and its changes are therefore as follows: (Fig. 10.55)

Nach den VII Elementen der Bahn nach Dr. *Gauſs* in der Länge	Wenn d. mittlere jährl. Beweg. ♃ um 20′ vermehrt wird	Für den Radius vector: Nach VII Elementen des Dr. *Gauſs*	mit der vermehrt. jährl. Bewegung
− 228,″78 ſin D	−231,″07	− 0,000095	− 93
+ 494, 71 ſin 2 D	+482, 33	+ 0,001030 coſ D	+1025
+ 44, 15 ſin 3 D	+ 43, 07	− 0,003802 coſ 2 D	−3688
+ 10, 07 ſin 4 D	+ 9, 74	− 0,000421 coſ 3 D	− 409
+ 3, 05 ſin 5 D	+ 2, 97	− 0,000108 coſ 4 D	− 104
+ 1, 07 ſin 6 D	+ 1, 04	− 0,000035 coſ 5 D	− 34
+ 0, 41 ſin 7 D	+ 0, 40	− 0,000013 coſ 6 D	− 13
		− 0,000005 coſ 7 D	− 5
+ 23, 45 ſin A′	+ 33, 70	− 0,000062 coſ A′	− 61
− 40, 98 ſin (A − D)	− 40, 81	+ 0,000199 coſ (A − D)	+ 193
+ 110, 21 ſin (D − A′)	+106, 29	− 0,000198 coſ D − A′	− 193
− 538, 92 ſin (2D − A)	−526, 86	+ 0,000708 coſ 2 D − A	+ 703
+ 238, 96 ſin (2D − A′)	+241, 12	− 0,001409 coſ 2 D − A′	−1424
− 241, 57 ſin (3D − A)	−242, 48	+ 0,001394 coſ 3 D − A	+1403
− 30, 83 ſin (3D − A′)	− 29, 76	+ 0,000260 coſ 3 D − A′	+ 251
+ 30, 21 ſin (4D − A)	+ 28, 73	− 0,000262 coſ 4 D − A	− 249
− 5, 81 ſin (4D − A′)	− 5, 62	+ 0,000058 coſ 4 D − A′	+ 56
+ 5, 60 ſin (5D − A)	+ 5, 40	− 0,000058 coſ 5 D − A	− 56
− 1, 85 ſin (5D − A′)	− 1, 77	+ 0,000020 coſ 5 D − A′	+ 20
+ 1, 73 ſin (6D − A)	+ 1, 65	− 0,000020 coſ 6 D − A	− 20
+ 1, 45 ſin (2D + A′)	+ 1, 44	− 0,000012 coſ 2 D + A′	− 12
+ 24, 62 ſin (D + A)	+ 24, 34	− 0,000145 coſ D + A	− 144
+ 0, 84 ſin (3D + A′)	+ 0, 83	− 0,000008 coſ 3 D + A′	− 8
− 53, 34 ſin (2D + A)	− 51, 84	+ 0,000299 coſ 2 D + A	+ 290
+ 0, 40 ſin (4D + A′)	+ 0, 39	− 0,000004 coſ 4 D + A′	− 4
− 5, 67 ſin (3D + A)	− 5, 58	+ 0,000038 coſ 3 D + A	+ 37
+ 0, 19 ſin (5D + A′)	+ 0, 19	− 0,000002 coſ 5 D + A′	− 2
− 1, 57 ſin (4D + A)	− 1, 50	+ 0,000012 coſ 4 D + A	+ 12
− 0, 48 ſin 2 A′	− 0, 45	− 0,000002 coſ 2 A′	− 2
+ 7, 85 ſin (A′ + A − D)	+ 7, 49	− 57 coſ (A′ + A − D)	− 54
− 35, 13 ſin (2A − 2D)	− 34, 01	+ 308 coſ (2 A − 2 D)	+ 300
− 2, 68 ſin (2H − 2D + 4° 45′)	− 2, 48	− 9 coſ (2H − 2 D + 4° 45′)	− 9
+ 8, 29 ſin (2 A′ − D)	+ 8, 66	− 0,000014 coſ (2 A′ − D)	− 15
− 92, 92 ſin (A′ + A − 2D)	− 98, 94	− 26 coſ (A′ + A − 2 D)	− 24
+ 39, 86 ſin (2 A − 3 D)	+ 42, 67	+ 111 coſ (2A − 3 D)	+ 110
+ 21, 09 ſin (2 H − 3D + 4° 45′)	+ 22, 68	− 38 coſ (2H − 3D + 4° 45′)	− 39
− 29, 48 ſin (2D − 2 A′)	− 28, 95	+ 0,000105 coſ (2D − 2A′)	+ 103
+ 60, 87 ſin (3D − A′ − A)	+ 59, 58	− 185 coſ (3D − A′ − A)	− 182
− 31, 00 ſin (4D − 2A)	− 29, 92	+ 80 coſ (4 D − 2 A)	+ 78
− 15, 45 ſin (4D − 2H − 4° 45′)	− 15, 04	+ 53 coſ 4 D − 2H − 4° 45′)	+ 52
+ 66, 91 ſin (3D − 2 A′)	+ 56, 85	− 0,000466 coſ (3D − 2A′)	− 397
− 133, 61 ſin (4D − A′ − A)	−111, 84	+ 938 coſ (4 D − A′ − A)	+ 788
+ 66, 36 ſin (5D − 2A)	+ 55, 93	− 468 coſ (5 D − 2 A)	− 397
+ 25, 67 ſin (5D − 2H − 4° 45′)	+ 21, 79	− 180 coſ (5D − 2H − 4° 45′)	− 153
− 6, 81 ſin (3D − 2A′ + A)	− 5, 79	+ 0,000018 coſ 3D − 2A′ + A)	+ 15
+ 13, 70 ſin (4D − A′)	+ 11, 48	− 35 coſ − 4 D − A′	− 29
− 6, 84 ſin (5D − A)	− 5, 78	+ 18 coſ 5 D − A	+ 15
− 2, 63 ſin (5D + A − 2H − 4° 45′)	− 2, 23	− 7 coſ 5D + A − 2H − 4° 45′)	+ 6
+ 110, 99 ſin (2D − 3A′)	+ 84, 46	− 0,000056 coſ (2 D − 3 A′)	− 48
− 340, 28 ſin (3D − 2A′ − A)	−258, 45	+ 171 coſ (3D − 2A′ − A)	+ 148
+ 344, 03 ſin (4D − A′ − 2A)	+258, 24	− 173 coſ (4D − A′ − 2A)	− 148
− 115, 10 ſin (5 D − 3 A)	− 87, 23	+ 58 coſ (5 D − 3 A)	+ 49
+ 156, 57 ſin (4D − A′ − 2H − 4° 45′)	+118, 45	− 78 coſ (4D − A′ − 2H − 4° 45′)	− 67
− 135, 85 ſin (5D − A − 2H − 4° 45′)	−103, 53	+ 67 coſ (5D − A − 2H − 4° 43′)	− 59

D

Fig. 10.55 The perturbations of Ceres, based on the seventh elements of Gauss

Die Störungen in der heliocentriſchen Breite ſind:

Nach den VII Gauſs'iſchen Elementen der Bahn in der Länge	Wenn die mittlere jährliche Bewegung der Ceres um 20 Min. vermehrt wird.
— 13,"19 ſin (H—D)	— 12,"97
+ 1, 62 ſin H'	+ 1, 59
+ 16, 21 ſin (2D — H)	+ 16, 03
— 1, 99 ſin (D—H')	— 1, 96
+ 32, 38 ſin (3D—H)	+ 32, 78
— 3, 97 ſin (2D—H')	— 4, 01
— 5, 50 ſin (4D—H)	— 5, 31
+ 0, 67 ſin (3D—H')	+ 0, 65
+ 6, 17 ſin (D+H)	+ 6, 06
— 0, 76 ſin (2D+H')	— 0, 74
+ 1, 52 ſin (2D+H)	+ 1, 49
— 0, 19 ſin (5D+H')	— 0, 18
+ 16, 15 ſin (4D—A'—H—2° 33')	+ 13, 83
— 13, 86 ſin (5D—A—H—2° 33')	— 11, 92

Fig. 10.56 The perturbations in heliocentric latitude

Nach den VII Elementen in der Länge	Die mittlere jährliche Bewegung der Ceres um 20 Min. vermehrt.
+ 0,"54 ſin (3A—2d)	+ 0,"38
— 1, 17 ſin (2A+a—3d)	— 0, 82
+ 0, 82 ſin (A+2a—4d)	+ 0, 59
— 0, 19 ſin (3a—5d)	— 0, 14
+ 0, 28 ſin (A+2H—12° 29'—2d)	+ 0, 20
— 0, 17 ſin (a+2H—12° 29'—3d)	— 0, 12

Fig. 10.57 The perturbations acting on Ceres through Mars

The perturbations in heliocentric latitude are (Fig. 10.56):

The perturbations acting on Ceres through Mars are, when d = Mars - Ceres and when a = the mean anomaly Mars (Fig. 10.57)

One sees that the equations are so slight that they can be justifiably entirely disregarded. In this way, the inequalities affected by Saturn remain almost the same when the mean yearly movement of Ceres is increased by about 20".

Continued Reports regarding the New Primary Planet Ceres Ferdinandea
MC (August 1802), p. 180.

We pointed our readers' attention in the last issue (p. 62) to the fact that probably no refraction has been applied to the observations of Ceres' declination from Vilnius and we also said we would improve those observed declinations according to the readings of the meteorological tools. This supposition proved correct and we got the reading of thermometer and barometer of each day an observation was made (Fig. 10.59). According to this and La Lande's refraction tables (3rd edition of his Astronomie*) we calculated the true refraction and thus improved all declinations:*

[Columns from left to right:
1802/Barometer French inch/Meas. Line
Thermometer Reaumur
True refraction
Improved decl. n. of Ceres]

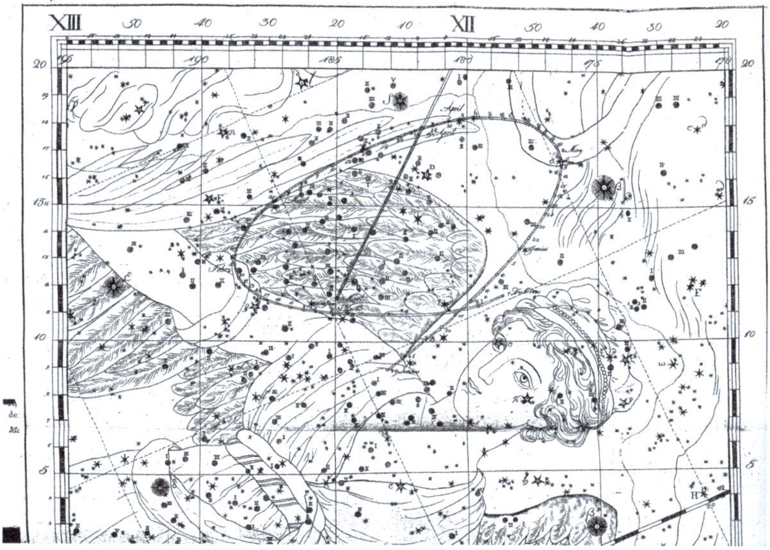

Fig. 10.58 Detail from a chart published in the *Monthly Correspondence*, showing the path of Ceres through the wing of Virgo from 7 December 7, 1801 to the end of June 1802

1802		Barometer franz. Mafs Zoll	Lin.	Thermomet. Reaumur	Wahre Strahlenbrechung	Verbefferte nördliche Abweich. der ⚳		
April	9	27	8,2	+ 2,°5	43,"1	18°	9′	53,"9
	10	27	7,6	+ 3,7	42,8	18	9	44,2
	12	27	4,3	+ 2,7	42,6	18	8	33,4
	13	27	8,2	+ 1,0	43,5	18	7	35,5
	15	27	7,7	+ 3,2	43,0	18	5	6,0
	18	28	0,4	+ 4,5	43,2	17	59	28,8
	22	28	0,2	+ 3,2	44,0	17	48	48,0
	23	27	11,8	+ 8,0	42,7	17	45	31,3
	24	28	0,3	+ 5,7	43,5	17	42	2,5
	25	28	0,2	+ 8,6	42,8	17	38	28,2
	27	28	0,4	+ 11,7	42,4	17	30	36,6
	28	28	0,4	+ 12,7	42,2	17	26	35,8
	29	28	0,4	+ 13,2	42,0	17	22	10,0
	30	27	11,1	+ 13,5	41,8	17	17	37,2
May	1	27	10,8	+ 8,6	43,0	17	12	52,0
	2	27	10,4	+ 8,0	43,2	17	7	31,8

Fig. 10.59 Improved declinations of Ceres, observed at the Russian Imperial Observatory Vilnius, Lithuania (add. to p. 63 and 64 of the last issue)

At this opportunity we also received several observations of Ceres made by Sir Poczubut in Vilnius, which follow together with all improvements (Fig. 10.60):

The observation of the RA of May 4 is not reliable due to evolving clouds.

The declinations of this day and of May 12 were made only at the passage instrument and not at the wall quadrant. Since with the beginning of May all meridian observations of this planet had to cease we can only expect those made at great equatorial sectors. Oriani in Milan made observations with such an instrument of excellent quality. Here is the entire series of his observations of this planet since its rediscovery. [See his Italian paper for these observations; see Fig. 10.58 for the path of Ceres up to June 1802.]

Since the last observations around the end of June/beginning of July are particularly important for improving the orbital elements of this planet, Oriani kindly relayed his last original observations in extenso so we could reduce those according to the latest and best data. This carefulness regarding observations, that have a certain impact, is praiseworthy and preferably all astronomers would follow his example. For not everyone has every resource handy to reduce accurately according to reliable data. If the original observations are communicated in their pure and pristine form they remain comprehensible and even in the future these observations can be improved and corrected. Oriani, for instance, used for his last observations of Ceres the stars o and s of Virgo. The right ascensions of these stars according to La Caille and Tob. Mayer match except for 3" but those determinations are nonetheless erroneous by 18" to 21" as I could convince myself during more than 30 observations during 1794 to 1802. Is the error caused by older observations or is the difference due to a falsely assumed annual motion or in a still unknown inherent motion of these stars? Enough, for me it is an incomprehensible fact that during my work on my star catalogue I nowhere encountered greater differences to older catalogues than in the sign of Virgo or in the region of RA 180°. The course of these differences is too noticeable and regular to accredit it to mere coincidence or a constant error in the observations. We content ourselves with this general remark because the opportunity presented itself; maybe the future will bring more light into this matter. [There was a section here about Oriani's observations.]

A thorough calculation results in the following apparent positions and we leave it to everyone to use the mean of the observations of any day or to exclude any unreliable observation. Necessarily, our reductions had to be slightly different from Oriani's since he employed the older star catalogues by La Caille or Tob. Mayer. Apparently, Oriani made a mistake when writing the mean time; but we do not want to alter the original sent to us and want to publish it with diplomatic accuracy. Now everyone can understand and become convinced of the rightfulness of these improvements.

[Another German publication offered a survey of the discovery of Ceres and Pallas in August. See: Strombecf (1802).]

1802	Mittl. Zeiten der Culminationen v. η ♌	Mittl. Zeiten der Culminationen v. d. ⚳	Scheinb. ger. Aufst. der ⚳	Scheinbare nördl. Abw. der ⚳	Barometer Zoll L.	Thermometer Réaumur	Wahre Strahlenbrechung	Wahre nördliche Abweich. der ⚳
May	U ′ ″	U ′ ″	° ′ ″	° ′ ″		°	″	° ′ ″
4	7 9 3	8 57 32	176 20 11,5	16 57 46	27 8,5	+7,5	43,8	16 57 2,2
	v. θ ♌							
7	8 4 24	8 45 26	176 15 41,0	16 41 32	28 0,4	+5,7	45,9	16 40 46,1
8	8 0 28	8 41 28	176 15 10,5	16 35 12	28 1,0	+6,2	45,2	16 34 26,8
12	7 44 43	8 25 49	176 16 55,0	16 10 0	27 9,7	+5,2	45,9	16 9 24 1

Fig. 10.60 Observations of Ceres by Poczubut, in May 1802

Resumed News about the new main Planet of our Solar System, Ceres Ferdinandea

MC (Sept. 1802), p. 290.

On August 20 we received from Prof. Piazzi in Palermo the brochure mentioned in the last issue on page 192 with the following title: Della Scoperta del…

A vignette on the cover emblematically depicts the city and port of Palermo with its molo, *with moored ships and Monte Pellegrino in the background. In the sky are three heavenly bodies: Mars, Jupiter and its four satellites and in the centre Ceres. Below the sphere, signifying the newly discovered planet, is the goddess Ceres in her chariot drawn by snakes. In the foreground is a cherub with a telescope pointed at Ceres, with the following words engraved: Ceres addita coelis (Ceres added to the sky). Next to it a shield with Palermo's coat of arms leaning against a tree, one head with three legs.* [The cover just described is pictured in Discovery of The First Asteroid, Ceres.] *The brochure begins with a short and practical letter of sympathy addressed to the King of Naples. Then follows the outer architectural view of Palermo Observatory, where Prof. Piazzi discovered on January 1, 1802, the new planet and which was built between 1791 and 1792, sponsored by then Viceroy of Sicily Prince Caramanico, on an old tower of the royal palace. A detailed description of this observatory and its instruments can be found in our review of the great Piazzian work:* Della specola astronomica de' regi studi di Palermo. *Palermo 1792, in Professor* [Carl Friedrich] *Hindenburg's* Archive of abstract and applied mathematics, *vol. I, book III 1795, p. 364f. Since the treatise is, according to Prof. Piazzi, almost entirely borrowed from our* MC *and our correspondence and it is merely a translation for Italian readers, it would be a repetition of our own words, published in various issues of the MC, if we published a summary. Thus, we restrain ourselves to a short announcement of what happened in Palermo regarding this matter that our readers do not know yet. Prof. Piazzi is rightly puzzled by the fact that this planet was seen last in Italy with its more favourable climate, where it would have been only natural to detect it first. But the observatories in Bologna, Padua, Pisa and Florence are not the most active. Prof. P. excuses this fact by the lack of instruments and some astronomers are more involved in theoretic than practical astronomy. But most of all this winter's bad weather is to blame. Winter in Lombardy is usually rainy and foggy so that observations were impossible for several months in Milan. But Oriani was able to observe this planet on February 24 upon his return from Lyon. In April it was observed at the Collegio Romano in Rome. But the observations were not relayed. Piazzi wrote: "And what is even more astonishing – I did not find the planet any earlier, here in Palermo, the place of its discovery with its mild climate, than in the night of February 22." But you must know that Piazzi neither possesses an equatorial sector nor a parallactic telescope and thus is reduced to the meridian circle.*

He made some attempts in November with his circle in order to find the planet by azimuth and zenith distances. But very soon he learned how difficult and insufficient this method of observing is. He was forced to wait until this body would be visible in the meridian circle again. But this could not happen any earlier than December 22 according to Burckhardt's elements. But he searched for his planet on December 23, 24 and 26 all the same, but in vain. Although Prof. P. had received Gauss' elements on January 10 and the ephemeris of this planet's orbit – they were of no use to him until February 22. For all of January and also a good part of February (except for two or three days) the weather was so bad and accompanied by gale force winds and heavy showers that Prof. P. had given up hope to see the planet before March. At last the sky was clear on February 22. After Piazzi had calculated the planet's position according to Gauss' elements he pointed his telescope to the calculated zenith distance but approximately 10 minutes further south. Carioti, who was observing at the meridian telescope, pointed the same 10 minutes further north. Thus, both telescopes covered an area of almost one degree polar distance, where they shared an area of ten minutes. So if the heavenly body was within these boundaries, what was most likely, it must appear in one of those telescopes.

With these telescopes Piazzi and Carioti observed all stars that passed 13 time minutes before and after the calculated time of the planet's culmination. They hoped to get hold of the planet this way. On February 24 it was overcast again but the following day was bright; the stars' observations were being repeated and one, which was observed between two stars of La Lande, was missing and thus had to be the sought-after planet. And this was confirmed on the night of the 26th. In this way Ceres was rediscovered, where it first was found, and observed until May 23 continuously. [Here follows an extract from Piazzi's monograph, printed in *Discovery of the First Asteroid, Ceres.*]

On March 9, after Prof. Piazzi had applied a 130x magnification to the telescope of his circle, and observed Ceres and eliminated all light for illuminating the hairs, the planet appeared larger and of dark ruddy colour but without clear edges. He observed the same the following night and could not discern any trace of an atmosphere. But he was puzzled by the change of colour as soon as the threads were illuminated – the planet was a pale ashen. On that evening he also observed two very tiny stars, which were close to the planet but had moved when he observed them again and then vanished. And he further wrote that he did not understand how the quick changes in light and size might be explained by a supposed atmosphere or nebulosity. He asked whether such an atmosphere really existed and what was one to think of it? What extreme and violent movements, what fast and intense changes had to be assumed in order to explain those changes in brightness and colour. Was this atmosphere composed of several layers of different density? In this case it would rather be a dense ring than an atmosphere. We had heavenly bodies of different and strange appearances: with spots and stripes, rings and with or without satellites. It would not surprise him, if someone regarded it as a comet, which has entered our system and is being kept there by the impact of the other planets. But Prof. Piazzi abandons all those hypotheses and assumptions, which must be far away from astronomical sobriety; according to him only observations will bring true knowledge.

The apparent diameter of this planet is still an important question without answer. And he admits that his first measurement on January 2, 3 and 4, 1801, was much too great and that it was rather a rough estimate than a real measurement since he did not have a

1802	Mittl. Zeit	Scheinbare Abweich. nördl.	Scheinbare AR. des Ceres am Kreis	Scheinbare AR. des Ceres am Passagen-Instrument	Beobachtete Fäden
April 15	10U 18' 44,"21	18° 5' 5,"0	11U 51' 57,"80	11U 51' 57,"80	alle
10	14 14, 63	3 27, 4	51 24, 07	51 24, 28	alle
21	9 52 7, 09	17 51 55, 1	48 55, 65	48 55, 52	alle
26	30 33, 70	34 48, 9	47 1, 65	47 1, 53	alle
27	26 19, 28	30 42, 3	46 42, 95	46 43, 26	alle
28	22 7, 04	26 33, 7	. . . . .	46 26, 53	1 3 4 5
30	13 45, 76	17 32, 4	45 57, 00	45 56, 95	alle
May 1	9 37, 15	12 45, 2	45 43, 80	45 44, 30	alle
2	5 30, 52	7 46, 1	45 33, 51	45 33, 60	alle
5	8 53 19, 19	16 51 49, 4	45 9, 86	45 9, 87	alle
6	49 18, 11	46 5, 7	45 4, 74	45 4, 72	alle
7	45 19, 14	40 16, 7	45 1, 90	45 1, 00	alle
10	33 29, 84	21 51, 4	44 59, 60	44 0, 00	alle
12	25 44, 29	8 47, 1	45 0, 90	45 6, 34	alle
15	14 16, 85	15 48 ::	45 25 [illegible]	45 26, 72	2
16	10 30, 04	41 4, 1	45 36, 55	45 36, 68	alle
20	7 55 38, 08	11 10, 0	46 28, 20	. . . .	. . . . .
21	51 59, 15	3 30, 9	46 44, 70	. . . .	. . . . .
22	48 20, 65	14 55 38, 7	47 2, 30	47 2, 18	alle
23	44 43, 55	47 42, 5	47 21, 0 :	. . . .	. . . . .

Fig. 10.61 Observations of Ceres by Piazzi in April and May 1802

micrometer for such a delicate measuring. He estimated the apparent diameter between March 11 and 24 to be 4" and hopes that Herschel obtains a much more accurate result with his lamp-micrometer.

Prof. Piazzi, too, does our journal in his treatise justice insofar as without it the rediscovery of this new main planet would not have been guaranteed. He wrote: "Probably, it would have been treated half-heartedly and with negligence and only a few would have searched for the new celestial body since even the fathers of astronomy doubted its existence." He defends his proposed name with dignity.

This brochure closes with a complete listing of all observations made by Prof. Piazzi in 1802 of Ceres Ferdinandea at Palermo Observatory together with the positions of 13 stars used for comparison. We have already published in the June issue on page 577 a considerable part of these observations, which Prof. Piazzi kindly had sent us (until April 16). A comparison showed there was no mistake in the MC *except for the last two observations of April 15 and 16, where the professor had made tiny changes. Here are the resumed observations of this planet made in Palermo of this epoch (Fig. 10.61).*

Again, we have reduced the current Palermo observations. The differences are negligible. Of course the RA observed at the meridian telescope are preferred (Fig. 10.62); only on May 20, 21 and 23 they were substituted by circle-observations because they were missing. And in Piazzi's text the stars are not mentioned with which he compared the planet, but which were given in the first handwritten observations as you can see from the June issue on page 577.

Reducirte Palermer Beobachtungen.

1802	Mittlere Zeit in Palermo			Beobachtete scheinbare gerade Aufsteigung der ⚳			Beobachtete scheinbare Abweich. der ⚳ nördlich		
April 15	10U	18'	43,"83	177°	59'	27,"90	18°	5'	5,"0
16	10	14	11, 43	177	51	4, 30	18	3	27, 4
21	9	52	6, 53	177	13	52, 80	17	51	55, 1
26	9	30	33, 31	176	45	22, 95	17	34	48, 9
27	9	26	19, 19	176	40	48, 90	17	30	42, 3
28	9	22	6, 59	176	36	37, 95	17	26	33, 7
30	9	13	45, 26	176	29	14, 25	17	17	32, 4
May 1	9	9	36, 77	176	26	4, 50	17	12	45, 2
2	9	5	30, 16	176	23	24, 00	17	7	46, 1
5	8	53	18, 76	176	17	28, 05	16	51	49, 4
6	8	49	17, 71	176	16	10, 80	16	46	5, 7
7	8	45	18, 69	176	15	24, 00	16	40	16, 7
10	8	33	29, 36	176	15	0, 00	16	21	51, 4
12	8	25	43, 87	176	16	35, 10	16	8	47, 1
15	8	14	16, 45	175	21	40, 80	15	48	::
16	8	10	30, 48	176	24	10, 20	15	41	4, 1
20	7	55	38, 21	176	37	3, 00	15	11	10, 0
21	7	51	58, 76	176	41	10, 50	15	3	30, 9
22	7	48	20, 27	176	45	32, 70	14	55	38, 7
23	7	44	43, 14	176	50	15, 00	14	47	42, 5

Fig. 10.62 Observations of Ceres by Piazzi, reduced by Zach

Resumed News about the new main Planet of our Solar System Ceres Ferdinandea

MC (October 1802), p. 382.

Never before had anything so pressing happened to practical astronomy, never before had the need for equatorial sectors in observatories been so pronounced than since the discovery of the two planets, Ceres and Pallas. Those astronomers who were equipped with only meridian instruments had to give up their observations of these planets at the beginning of May. With parallactic telescopes, with filar and circle micrometers, it was possible to track them for a longer period, but these instruments were hitherto just makeshift substitutes with ' comets', where sharpness is neither necessary nor attainable. But with such small 'planets' which perennate in our Solar System and which are so difficult to observe and even more difficult to locate, with which the greatest exactness is requisite because their theory proves that their entire path must be inferred from a very slight described arc, very exact observations, balanced with meridian observations, are doubly necessary.

It is seldom the case that a comet, after its conjunction with the Sun (as, for example, the comet of 1759), appears again from the rays of the Sun and must be located. If this does occur, approximate elements of its path will more than suffice to locate it, for all comets distinguish themselves from other bodies through a characteristic form, through their nebulous appearance and through their rapid movement. They can therefore be located again with little effort.

Such is not the case with our two new planets. Nothing much distinguishes them from magnitude 8 – 9 fixed stars. It is therefore extremely difficult to locate these small bodies from a myriad of such stars if their whereabouts cannot be indicated fairly exactly. Only those astronomers equipped with good equatorial sectors could accurately follow these planets until late August, and they will also be the first to find them when they emerge from the rays of the Sun once again.

Oriani, who was able to use a five-foot Sisson [Jeremiah Sisson, 1720–1783, instrument maker] *equatorial sector at the Milan Observatory, observed these new planets until August 8. He tells us that he observed Pallas in the field of his telescope until August 17 and 18, but they were so small and weak that it was not within his abilities to conduct an actual observation. These long observations ensure us as to the true path of this planet and will contribute not a little to the future, certain location of this body. Incidentally, these precious observations of Oriani reveal what advantage the beautiful Italian sky and the brevity of the dusk there can afford astronomers. Rightly so, Oriani concludes that we will 'certainly' see Pallas again in the coming year.*

Since the observatory in Greenwich is equipped with two splendid equatorial sectors, we hope to receive later and very exact observations of this planet from there. The Sicilian climate allowed Professor Piazzi to follow Ceres until May 23 in the meridian, whereas the twilight in Germany forced us to give up our observations on May 11. Meanwhile, he felt the need for a good equatorial instrument so much that he used this opportunity to request of his king the acquisition of such a valuable instrument, and was immediately authorized to have a six-foot sector sent from England. Piazzi then wrote: "The surname of my planet, Ferdinandia, which so many astronomers deemed as unnecessary, brought me a splendid equatorial sector and a yearly salary increase of 100 [ounces] *(50 Louis d'or). I received permission from the king to use the money which was first intended for the minting of a medal upon the discovery of Ceres for the acquisition of a six-foot English equatorial sector." We didn't let the opportunity of remarkable and historical cause slip away unused and received, upon our suggestion, the most gracious approval for the purchase of a ten-foot equatorial sector and a new Arnold Regulator from our magnanimous founder and preserver of the Seeberg Uranien Temple* [Duke Ernst II].

In this year's August issue (p. 184), we included Oriani's equatorial observations of Ceres until July 8; the continuation of these, until the planet's complete disappearance as follows (Fig. 10.63):

Fig. 10.63 Observations of Ceres from the Milan Observatory, conducted by Oriani with an Equatorial Sector

1802		Mittlere Zeit in Mailand			Scheinbare AR der Ceres			Scheinb. nördl. Abweichung der Ceres		
Jul.	10	9U	17′	13″	185°	41″	41″	6°	56′	34″
—	16	9	6	57	187	19	22	5	50	39
—	18	9	12	32	187	52	57	5	28	28
—	24	9	8	31	189	36	42	4	21	33
—	25	9	9	10	189	54	20	4	10	50
—	29	9	37	57	191	6	49	3	25	27
Aug.	5	8	50	51	193	17	14:	2	6	58:

Also this time, Oriani did us the favour of including the journal of his observations, according to which a more exact reduction of these may be carried out, since he only made use of the details from Bode's star catalogue in his calculations. Since there are several inexactly determined stars among them, we reserve the reduction of these for the next issue, we will offer, in the meantime, a fragment of this astronomical journal: [Refer to table in *MC*, p. 386, top.] *The following continuation of the position and action of the clock running according to mean solar time serves for the reduction of the cited observations:* [Refer to table in *MC*, p. 386, bottom.]

The perturbations of longitude and distance of Ceres by Jupiter have indeed been subjected to calculation by several astronomers, but such complicated difficult calculations cannot be repeated enough. Just recently, Professor Wurm wrote us about this subject and on the occasion of a number of improvements and elucidated, additional remarks sent in regarding the formulae of the perturbations of Mars, which we will include in our issues next time: "I believe that whoever has never encountered thorns on the road of calculating perturbations has certainly never walked the road himself, for coming across thorns now and again in doing these calculations, and to assume a slight quid pro quo until one has become better oriented, is not to be avoided."

For this reason, Dr. Gauss repeated these perturbation-calculations according to his 7th elements of the path of Ceres, and discovered a few errors here and there. The analytic formulae that he used in these calculations were all first developed by him. They have turned out slightly differently from Laplace's in form.

Perturbations of Ceres through Jupiter, calculated by Dr. Gauss
Eccentricity 0.000005909
annual movement of the Sun +70."15
against the fixed stars

[Tables of perturbation calculations and an ephemerides of Ceres for 1803 follow.]

Resumed News about the new main Planet of our Solar System, Ceres Ferdinandea

MC (Nov. 1802), pg. 492

In the last issue on page 385 we announced the intention to calculate the Milanese observations of Ceres according to Oriani's diary; we kept our promise and here follow the reduced observations in the same way as on p. 186 of the August issue. Now every single observation can be judged and everyone can choose a helpful one or exclude one from the 'arithmetic mean' (Fig. 10.64).

1802	Mittlere Zeit	Scheinbare gerade Aufsteigung der Ceres	Scheinbare nördl. Abweichung der Ceres	Sterne womit verglichen
Jul. 10	9h 17′ 13.″4	18[illegible]° 41′ 38.″6	6° 56′ 31.″3	5 Virginis
— 16	9 6 56, 6	187 19 1, 1	5 50 48, 4	5 Virginis
		19 40, 5	50 27, 5	Nr. 113 —
		19 33, 0	50 9, 5	Nr. 168 —
— 18	9 12 32, 5	187 52 53, 2	5 28 20, 4	Nr. 113 —
		52 54, 5	28 20, 5	Nr. 168 —
— 24	9 8 31, 3	189 36 42, 4	4 21 33, 8	Nr. 113 —
— 25	9 9 10, [illegible]	189 54 19, 8	4 10 51, 8	Nr. 113 —
— 29	9 37 56, 9	191 6 53, 1	3 25 23, 9	Nr. 226 —
Aug. 5	8 50 51, 6	193 17 10, 9	2 6 52, 7	Nr. 394 —

Fig. 10.64 Observations of Ceres by Oriani, reduced by Zach

The mean positions of the stars used for Ceres and Pallas were taken from our and Henry's and Barry's star catalogues, or calculated from La Lande's Histoire céléste française *and the result is the following small star catalogue. Dr. Gauss was so kind as to send us two of Dr. Maskelyne's observations made in Greenwich at the equatorial sector. By the way, he informed us that these observations were not properly reduced and the mean position of the compared stars was according to Wollaston's* General Catalogue *without nutation or aberration. Although the original observations are not attached, the above note and the mentioning of each compared star make up for this so that we were able to deal with the true reduction. Here is Dr. Maskelyne's data:*

1802	*Mean Time*	*RA of Ceres*	*N. Decl. of Ceres*	*star*
June 20	*11h 46′ 41″*	*181° 2′ 7″.8*	*10° 26′ 15″*	*12 Virg*
July 3	*11 0 44*	*183 55 53.4*	*8 10 29*	*17 Virg*

In Wollaston's star catalogue is the position of 12 Virginis given only according to Flamsteed but 17 Virginis according to Flamsteed and Tob. Mayer. Thus, it is very likely that Dr. Maskelyne used the indication of the latter astronomer. Under this assumption we reduced the above observations. The mean positions of the stars are: [he prints a table of star data]. *From this the newly reduced apparent positions of the planets were obtained, while taking the stars' precession, aberration and nutation into account.*

1802	*Mean Time*	*RA of Ceres*	*N. Decl. of Ceres*
June 20	*11h 46′ 41″*	*181° 1′ 28″.7*	*10° 27′ 2″.4*
July 3	*11 0 44*	*183 55 38.3*	*6 10 28.6*

At the end Dr. Maskelyne added the following remarks and estimates of Ceres' brightness.

Feb. 3	*8th magnitude*
March 4	*9*
April 22	*9*
May 17	*9*
June 20	*10*
July 3	*11*

Meanwhile Dr. Gauss has worked on the planet's orbit. Regarding the perturbations as he calculated them according to his elements VII (October issue, p. 387) and the latitude equations, whose method of calculation is almost equal to Oriani's (cf. June issue, p. 586; July issue, p. 68), he determined the following elements:

Epoch 1801 for Seeberg	*77° 19′ 38″.4*
Daily tropical movement	*770″.764*
Period	*1681 days 11 hours*
Eccentricity	*0.0788132*
Log semimajor axis	*0.4421085*
Aphelion 1801	*326° 33′ 10″*
Node 1801	*80 54 52*
Inclination	*10 37 48*

[There was a section here of Ceres perturbation formulae.]

Finally, Dr. Gauss calculated according to these formulae the numerical value of the perturbations for the observations anew and found the following orbital elements, which we would like to mark (VIII) (Fig. 10.65).

1804	Mittlere Zeit in Seeberg			Scheinbare gerade Aufst. der ⚳			Scheinbare südl. Decl. der ⚳		
Sept. 13	13u	20′	21,″662	12°	52′	30,″14	11°	24′	27,″3
14	13	15	43, 741	12	41	58, 26	11	30	23, 5
15	13	11	4, 617	12	31	8, 27	.	. .	.
17	13	1	44, 035	12	8	53, 22	11	47	22, 7
18	12	57	2, 488	11	57	26, 79	11	53	0, 3
23	12	33	26, 349	10	58	8, 19	12	19	29, 8
27	12	14	24, 535	10	8	27, 45	12	38	37, 4
28	12	9	38, 646	9	55	55, 71	12	43	5, 1
30	12	0	5, 514	9	30	31, 95	12	47	43, 3
Octob. 2	11	59	32, 154	9	5	4, 58	12	59	37, 9
4	11	40	58, 996	8	39	40, 37	13	7	12, 9
5	11	36	12, 377	8	26	57, 65	13	10	41, 5
6	11	31	26, 204	8	14	21, 65	13	13	44, 9
10	11	12	23, 647	7	24	29, 67	13	24	55, 0
12	11	2	55, 895	7	0	26, 78	13	29	33, 8
20	10	25	27, 844	5	30	0, 48	13	39	41, 7
21	10	20	50, 617	5	19	39, 02	13	39	49, 4

Fig. 10.65 Elements VIII of Ceres, by Gauss

These new elements are so similar to the previous that one could regard a new calculation of the perturbations according to these as redundant. Out of curiosity, Dr. Gauss wanted to determine how much these new elements VIII differ from the elements VII regarding the conjunction next year. He found, omitting the aberration, for June 28, 1803 at 12 o'clock mean time Seeberg:

Elements	*Longitude of Ceres*	*Latitude of Ceres*
VII	*280° 7' 43"*	*5° 1' 21"*
VIII	*280 17 58*	*5 4 53*
Difference	*10 15*	*3 32*

Dr. Gauss closed his calculation with the following remark: "Since it took so much time and effort to calculate one position of the planet according to the above perturbation equation, it appears not advisable to increase their numbers by Saturn's equations and by those depending on the squares and products of the eccentricities of Jupiter, whose consideration would have no other difficulty than the tedium of the mechanic calculus especially since after almost six months we will be able to obtain more accurate elements with the aid of new observations. Since the calculation of the equations' amount is so tedious, it appears important to me, to think about a shortening and simplification of the tables for those. Soon I will send you my ideas regarding this matter together with a specimen."

Resumed News about the new main Planet of our Solar System Ceres Ferdinandea

MC (December 1802), p. 575.

On page 186 of the August issue on the occasion of Oriani's observations we expressed our doubt about the given mean time of the observation of June 28, but we only noticed this and did not change it because it is our law not to change the submitted original observations on our own authority but to treat them with diplomatic accuracy. But since then Oriani himself has confirmed our suggested change and consequently the indicated erroneous mean time of June 28 on page 182, should be 9h 45' 48 instead of 10h 3' 10", as we already mentioned on p. 186.

We also pointed Oriani's attention to some variants between his and Dr. Gauss' perturbation formulae regarding Ceres, and he provided the following information. If you summon all inequities of the mean distance of Ceres (see July issue, p. 68) found by the first hypothesis, which depend on the eccentricities and their products and which have the same variable angle in their argument, that is, all elements according to the seven first of the first and third column of page 69 of the July issue, Oriani obtains the following results: [Oriani's complex perturbation formulae are given.]

These results correspond fairly well with those of Dr. Gauss. Only the fifth inequality is different – because Oriani brought the following six elements together in one (Fig. 10.66):

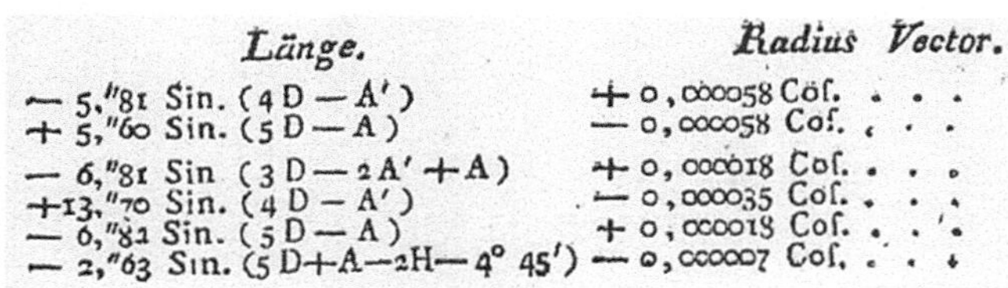

Länge.	Radius Vector.
— 5,"81 Sin. (4 D — A')	+ 0,000058 Cos. . . .
+ 5,"60 Sin. (5 D — A)	— 0,000058 Cos. . . .
— 6,"81 Sin (3 D — 2 A' + A)	+ 0,000018 Cos. . . .
+13,"70 Sin. (4 D — A')	— 0,000035 Cos. . . .
— 6,"82 Sin. (5 D — A)	+ 0,000013 Cos. . . .
— 2,"63 Sin. (5 D+A—2H—4° 45')	— 0,000007 Cos. . . .

Fig. 10.66 The six elements of Oriani's perturbation calculations

Dr. Gauss on the other hand has summarized only the first two elements. We can see that Oriani's formulae, published in the July issue, are now free of any wrong signs, what was not the case in the June issue besides other minor mistakes. But Oriani indicates the following two changes for p. 69 of the July issue:

+23."95 Sin A' +23."70 –0.000062 Cos A' –52
–40."98 Sin (A-D) –40. 53 +0.000199 Cos (A-D) +195
We have to add the following two elements for the radius vector:
–0.000053 Cos A –52
+0.000014 Cos (D + A') +14

Brougham's Article in the *Edinburgh Review*, Jan. 1803

Our astronomical readers are acquainted with the interesting discoveries which have, within the space of a few months, introduced to our acquaintance two new celestial bodies; the one names Ceres, by its discoverer Piazzi; the other called Pallas, by its discoverer Olbers. Our own indefatigable astronomer Dr. Herschell [sic], who has himself, by his numerous and accurate observations, so far extended the bounds of human knowledge, appears to have directed his attention, without loss of time, to the new and interesting field of observation opened to him by his brethren on the Continent. The results of his first inquiries were, as might be expected, extremely interesting. He found that the magnitude of these supposed planets, or, as he calls them, moving stars, was much inferior to that of the other primary planets, or even of their satellites. Thus he found that Ceres has a diameter only three eighths the diameter of the moon. In the present paper, besides extending the same observation, and the same conclusions to Pallas also, this excellent astronomer has given us a set of new and accurate observations, tending to establish some very singular and interesting facts. We hold it to be a duty indispensably incumbent on us to present our readers with a sketch of this valuable paper.

The first remarkable circumstance that strikes us in all the observations, is the great difference between the real magnitudes and the lucid disks. By one measurement with the most delicate micrometer, expressly adapted for the purpose of such experiments, the real diameter of Ceres was found to be only three fourths of the lucid disc; and that of Pallas only two thirds. The angle which the former subtends, was found to be only 0".38; that of the latter no more than 0".13. He calculates by a rough estimate, that the diameter of Ceres is only 161.6 miles, and that the diameter of Pallas is no more than 110 ⅓ *miles.*

From the very small quantity of matter which these bodies contain, we cannot expect that they can have any satellites; accordingly various observations concurred to convince Dr. Herschell that this is consistent with truth. He also determined that Ceres has a visible disc, but that Pallas cannot be discovered to have any. The last set of observations are extremely important for ascertaining the precise nature of the two new bodies. By them it is ascertained, that both the stars have at all times a small coma or haziness, which grows denser near the nucleus.

Our author next proceeds to make his observations upon the results of these inquiries. He begins by defining planets to be celestial bodies of a considerable size and small eccentricity of orbit, moving in planes not very different from that of the earth, in direct curves at considerable distances from each other, with no atmospheres that bear any proportion to their diameters, and of bulk sufficient to retain satellites in their orbits. It is evident that, with this definition, the new stars but ill agree. Our author then defines comets to be very small celestial bodies, moving in directions wholly underdetermined and in most eccentric orbits, situated in every variety of position, and having very extensive atmospheres. Although the definition agrees in most particulars with the circumstances of the new stars, it differs in that of the atmosphere, which, in the comets, is at the very least a hundred times greater than the diameter of the nucleus, and in the new stars is only a few times greater. Dr. Herschell therefore maintains, that these bodies are neither referable to the class of comets nor planets, but he gives them the name of Asteroids, which he thus defines: —

Asteroids are celestial bodies, which move in orbits either of little of or considerable eccentricity, round the Sun, the plane of which may be inclined to the ecliptic in any angle whatsoever. Their motion may be direct or retrograde, and they may or may not have considerable atmospheres, very small comas, disks, or nuclei.

Having thus followed the Doctor through his very interesting speculations, we must now proceed to the more invidious, but equally necessary part of our office, and offer a few remarks upon the Doctor's theory; premising, that we rely with the most implicit confidence on the accuracy of his observations, from long experience of his great skill, patience, and fidelity, and from our knowledge of the unrivalled excellence of his instruments. It is to his conclusions alone that we object; and, with all possible deference, we hold ourselves as well qualified to judge of the truth of these, as if we had ourselves made or verified the observations upon which they are founded.

And, first, we must positively object to the unnecessary introduction of new terms into Philosophy. The science of Astronomy is, beyond any other branch of the mixed mathematics, loaded with an obscure and difficult technology. As all nations have been observers of the heavenly bodies, so all languages have contributed to form the nomenclature of the astronomer. Not only are the same bodies indifferently known by a variety of names, but, so defective is the phraseology, that no one list can be given in two or three languages, or according to two or three systems of mythology. To a person who had resided in ancient Italy and Greece, on the banks of the Nile, of the Ganges and Euphrates, in modern Europe, and amongst the Gothic nations, the astronomical technology might be natural and simple, as it is composed of all the languages spoken, all the mythologies received, and many of the court calendars published in these various countries and distant ages. Knowing, as we do, the great power of words in misleading and perplexing our ideas, we cannot allow the unnecessary introduction of a new term to escape unnoticed. Where a new object has been discovered, we cheerfully admit the right of the discoverer to give it a new name; but we will not allow a needless multiplication of terms, or an unnecessary alteration in the old classification of things, to be either justifiable or harmless, a substitute for real discovery, or a means of facilitating the progress of invention. It remains, therefore, to inquire, whether the circumstances of Ceres, or of Pallas, distinguish them from the bodies formerly known?

We cannot admit the difference of magnitude to be of any importance, while the largest and the smallest planets, Jupiter and Mercury for instance,– the largest and smallest satellites,– the largest and smallest comets, between which the difference of magnitude is still more remarkable,– while all these bodies are severally arranged under the same class, from considerations wholly independent of their size, it is but a clumsy and cumbersome invention, to arrange a new body under a separate class, from the mere difference of its bulk. The same remark applies, though certainly with diminished force, to the other criterion assumed by the Doctor, the difference in the position of the planes of motion; and, most unquestionably, the mere circumstance of wanting satellites, is no distinguishing mark, while so many of the acknowledged planets have none; nor, indeed, is it by any means certain that, as the Doctor seems to think, the mass of matter in the new planets is insufficient to retain secondary bodies in their orbits. The proportion of their distances from the centre of the system, or their proximity to each other, is evidently no better criterion.

In short, if it shall be admitted that comets move in ellipses; that the chief difference between those bodies and planets, consists in the greater eccentricity of the cometic orbits, in the perceptible atmosphere which accompanies them, and in the state of ignition which we have every reason to believe is the cause of that atmosphere; the more philosophical view of the subject would certainly be, to consider both planets and comets as bodies of the same nature, forming different parts of one great system. Indeed, Dr. Herschell himself admits the probability of the comets cooling in the process of time; and their atmosphere diminishing, so as to reduce them to a state of planets in every thing but their magnitude and eccentricity; and he applies the same remark to the case of the new bodies. Such an observation is obviously destructive of the principle of arrangement for which he contends. But whatever may be our opinion upon this subject, or however much we may be disposed to admit the propriety of distinguishing comets from planets; in the present state of our knowledge, the grand circumstance of concentricity is evidently sufficient to authorise a classification of the new bodies under the head of planets; and the discovery of them is chiefly valuable, on account of their coincidence, in certain particulars, with the nature of comets, and their differing from those bodies in the extent of their atmospheres, probably in decreased ignition. If it shall be found demonstrated, that the cometary orbits are elliptical, and not parabolic, these new planets will form a sort of link in the system, in consequence of an intermediate step between the greater and the smaller, the concentric and eccentric heavenly bodies. In the mean time we must enter our protest to the formation of a separate class, distinguished by a new and uncouth name.

Such being our opinion, it is of much less consequence to inquire, whether the new name of Asteroid is the most appropriate that could be imagined. To us, that name presents the idea of some body resembling fixed stars; whereas the two new planets have no one circumstance in common with those distant bodies. If a new name must be found, why not call them by some appellation which shall, in some degree, be descriptive of, or at least consistent with, their properties? Why not, for instance, call them Concentric Comets, or Planetary Comets, or Cometary Planets? Or, if a single term must be found, why may we not coin such a phrase as Planetoid or Cometoid?

Dr. Herschell's passion for coining words and idioms, has often struck us as a weakness wholly unworthy of him. The invention of a name, is but a poor achievement in him who has discovered whole worlds. Why, for instance, do we hear him talking, on page 220 of this volume, of the space-penetrating power of his instrument- a compound epithet and metaphor which he ought to have left to the poets, who, in some future age, shall acquire glory by celebrating his name? The greatest discoverers have scarcely ever immortalized their deeds by efforts of nomenclature. Columbus, Cabral, Gama, and Cook, left the honour of being attached to the regions which they had penetrated, to the imposters who succeeded them, or the princes and saints whom they served. [The paper concludes with two paragraphs that attack Herschel's other recent papers in the *Philosophical Transactions.*]

[Brougham mentions here Christopher Columbus, Pedro Cabral (discoverer of Brazil), Vasco de Gama (first European to reach India), and Capt. James Cook (who circumnavigated the globe).]

Tables for the Perturbations of Ceres by Dr. C. F. Gauss Brunswick, December 26, 1802

MC (Mar. 1803), p. 259.

Some time ago, I wrote to you that I have thought about a shortening of the tables for the perturbations of Ceres and that I would want to submit a sample of it to you in future. I now have the honour of being able to explain this further and to send you not just a sample, but rather a complete [in extenso] *table (Fig. 10.67).*

Although I have limited myself to the perturbations through Jupiter and to the first powers of the eccentricities, the number of equations is already so considerable that it would be extremely labourious to calculate the numerical values for a great number of positions without tables, from the formulas themselves. But even with tables, arranged according to custom, this work is truly arduous. For 'all' equations printed in the November 1802 issue of M. C., *'forty' tables would be required, and almost 'thirty' if one were to leave out the smaller equations, some of which amount to under 2". To form so many arguments, to go into so many tables and to make so many additions still requires much patience if it has to be done often. I have therefore tried to shorten these tables into a more workable form.*

In this way, we have only five tables in which, admittedly, some trigonometric calculations are required, but which seems much easier to me than having to take all the single equations summarized here from the special tables. One would also be able to make some revisions for the better, for instance setting log. A instead of A, 'or also in other ways'. But I will save this for the future since these tables will only be used until future observations

Fig. 10.67 Tables for the perturbations of Ceres, by Gauss

I TAFEL. *Störung der Länge.* Erster Theil.
Argumentum ♀ — ♃.

°	0^{s} +	I^{s} +	II^{s} ±	III^{s} —	IV^{s} —	V^{s} —
0	0″	364″,6	215″,6	272″,2	620″,2	505″,2
1	16,6	368,6	202,3	288,7	624,6	493,1
2	33,1	372,0	188,7	305,0	628,5	480,6
3	49,6	374,7	174,7	321,1	631,8	467,6
4	66,0	376,8	160,4	336,9	634,5	454,2
5	82,3	378,2	145,7	352,6	636,6	440,3
6	98,3	378,9	130,8	367,9	638,2	426,1
7	114,2	379,0	115,6	383,0	639,1	411,4
8	129,9	378,4	100,0	397,8	639,5	396,3
9	145,3	377,2	84,3	412,2	639,4	380,8
10	160,4	375,4	68,3	426,4	638,6	365,1
11	175,2	372,8	52,0	440,2	637,3	348,9
12	189,6	369,7	35,6	453,6	635,3	332,5
13	203,6	365,9	18,9	466,7	632,8	315,8
14	217,3	361,5	2,1	479,4	629,7	298,8
15	230,5	356,6	14,8	491,7	626,1	281,5
16	243,3	350,9	31,9	503,5	621,8	263,9
17	255,6	344,8	49,0	515,0	617,0	246,1
18	267,4	338,0	66,3	526,0	611,7	228,2
19	278,7	330,6	83,6	536,6	605,7	209,8
20	289,5	322,7	100,9	546,7	599,2	191,4
21	299,7	314,3	118,3	556,3	592,2	172,8
22	309,4	305,3	135,7	565,4	584,6	154,0
23	318,5	295,7	153,0	574,1	576,5	135,0
24	326,9	285,7	170,3	582,2	567,8	116,0
25	334,8	275,2	187,6	589,8	558,7	96,8
26	342,1	264,2	204,7	597,0	549,0	77,6
27	348,6	252,7	221,8	603,6	538,8	58,2
28	354,6	240,7	238,7	609,7	528,1	38,8
29	359,9	228,4	255,6	615,2	516,9	19,4
30	364,6	215,6	272,2	620,2	505,2	0
°	XI^{s} —	X^{s} —	IX^{s} ∓	VIII^{s} +	VII^{s} +	VI^{s} +

correct the elements still better, thereby allowing the perturbations to be calculated more precisely and completely. For the time being, I think that these tables will serve me well, especially in the calculation of the elements of the coming year. [Gauss here prints secular and periodical equations.]

Continued News on the new main Planet Ceres

MC (August 1803), p. 192.

We published in our June issue the first observations of this year of this planet made by Mr. Piazzi in Palermo. In Germany, Dr. Olbers was the first to find her again. On June 29 Dr. Gauss wrote us from Bremen regarding this matter: "From here, where I am spending a couple of wonderful days at Dr. Olbers', it is my honour to share some information on our new planet. It is my pleasure to be able to inform you about some practical exercises

I undertook. Dr. Olbers had not been observing Pallas until the beginning of this month because without doubt there had already been made meridian observations. He started observing Ceres of which no foreign observations have become known. I wanted to do this for a long time because my studies of last year convinced me that there must remain a certain degree of uncertainty regarding the orbit of this year. And that is why I did not publicise the result of further examinations of all observations made in 1801 and 1802, the derived elements (which did not differ much from the elements VIII) and an ephemeris based on this because the difference between those and the elements VIII and VII was not distinctive enough. I do not have those elements on me and thus cannot give any more details but I brought the ephemeris accurately calculated to the second; according to that I can give the difference between our observations and the calculation. Upon my return to Brunswick I will use the observations that have become known to improve the elements. Here follow the observations from Bremen with the difference of the calculation:

Date	*Mean time*	*RA*	*N. Dec.*	*Difference*	
	Bremen			*RA*	*Dec*
June 22	*12h 40' 10"*	*283° 0'43"*	*27° 40' 11"*	*–5' 28"*	*+34"*
24	*12 28 10*	*282 33 10*	*27 48 57*	*–5 54*	*+49*
25	*12 43 8*	*282 18 47*	*27 53 37*	*–5 47*	*+37*
26	*12 27 26*	*282 4 51*	*27 58 7*	*–5 57*	*+31*
27	*12 16 51*	*280 50 25*	*28 2 23*	*–5 51*	*+36*

Ceres has always been compared with τ Sagittarius. The observation of the 25th is the mean value of three of which two were made by me; that of the 26th is by me and all others by Dr. Olbers. On the 27th I observed Ceres together with Inspector Harding at Lilienthal; she was compared four times with τ Sag and the observations correspond well with each other and the one made by Dr. Olbers. But they have not been reduced properly yet since the time of the clock had to be determined by Inspector Harding the following day and has not become known to me yet."

*At Seeberg Observatory Prof. Buerg obtained the following observations of Ceres of which four have been marked by me by an *, which were made by me during my brief presence at home.*

1803	*Mean time*	*RA of Ceres*	*Dec. of Ceres*
	in Seeberg		
July 1	*12h 7' 17."3*	*280° 51' 13. "2*	----------
2	*12 2 23.2*	*280 37 33.2*	----------
5	*11 47 40.9*	*279 53 52.2*	----------
*7**	*11 37 53.5*	*279 24 53.7*	----------
*8**	*11 31 59.8*	*279 10 24.9*	*18° 45' 54"S*
*9**	*11 28 6.3*	*278 55 53.6*	*23 49 2*
11	*11 18 22.5*	*278 27 55.1*	----------
12	*11 14 31.0*	*278 13 53.6*	----------
18	*10 44 36.4*	*276 53 58.0*	----------
19	*10 39 50.6*	*276 41 26.8*	----------
20	*10 35 5.0*	*276 29 0.7*	----------
*22**	*10 25 37.1*	*276 4 55.0*	*29 26 21*
23	*10 20 56.0*	*275 53 35.5*	----------

Continued News on the new main Planet Ceres

MC (October 1803), p. 369.

The observations of Ceres had to be interrupted around the beginning of August due to her low position. We publish here some remaining observations made at Seeberg Observatory.

1803	*mean time*	*RA of Ceres*
July 27	*10h 2′ 20.″9*	*9z 5° 10′ 35.″4*
28	*9 57 44.9*	*5 0 32.5*
29	*9 53 11.1*	*4 51 1.9*
30	*9 48 38.0*	*4 41 43.0*

We published in the last issue on page 290 the elements of this planetary orbit improved by Dr. Gauss for the IX time. He calculated the following ephemeris from these for a future search and to follow Ceres that we would like to publicise as soon as possible (Fig. 10.68 and Fig. 10.69).

Fig. 10.68 Geocentric positions of Ceres based on Elements IX of Gauss, April to August 1804

Geocentrifcher Lauf der Ceres 1804.
Nach den IX Elementen.
Von Dr. *Gaufs.*

Mitternacht in Seeberg		Gerade Auffteig.		Abweich. füdl.		Im Merid. mittl. Zeit		Lichtftärke
April	30	355°	11′	11°	32′	21U	5′	0,00899
May	3	356	9	11	18	20	57	0,00916
	6	357	6	10	55		49	0,00933
	9	358	3	10	38		41	0,00951
	12	358	59	10	21		33	0,00971
	15	359	54	10	4		25	0,00992
	18	0	48	9	48		17	0,01013
	21	1	40	9	33		8	0,01035
	24	2	33	9	19		0	0,01059
	27	3	24	9	5	19	51	0,01084
	30	4	13	8	52		43	0,01111
Jun.	2	5	2	8	39		34	0,01139
	5	5	50	8	27		26	0,01168
	8	6	36	8	17		17	0,01199
	11	7	21	8	7		8	0,01231
	14	8	5	7	57	18	59	0,01265
	17	8	47	7	49		50	0,01300
	20	9	28	7	42		41	0,0 337
	23	10	7	7	36		32	0,01376
	26	10	45	7	30		23	0,01417
	29	11	21	7	26		13	0,01460
Jul.	2	11	55	7	23		4	0,01505
	5	12	27	7	20	17	54	0,01552
	8	12	57	7	19		44	0,01601
	11	13	25	7	19		34	0,01651
	14	13	51	7	21		24	0,01703
	17	14	15	7	23		14	0,01758
	20	14	36	7	26		4	0,01815
	23	14	55	7	31	16	53	0,01873
	26	15	11	7	37		42	0,01933
	29	15	25	7	44		31	0,01996
Aug.	1	15	35	7	52		20	0,02060
	4	15	43	8	2		9	0,02125
	7	15	48	8	13	15	58	0,02191
	10	15	51	8	24		46	0,02258
	13	15	50	8	37		34	0,02325
	16	15	46	8	51		22	0,02393
	19	15	39	9	5		10	0,02460
	22	15	28	9	21	14	57	0,02527
	25	15	15	9	37		45	0,02592
	28	14	59	9	54		32	0,02655
	31	14	39	10	11		18	0,02715

Fig. 10.69 Geocentric positions of Ceres based on Elements IX of Gauss, Sept 1804 to January 1805

Mitternacht in Seeberg	Gerade Aufsteig.		Abweich. südl.		Im Merid. mittl. Zeit		Lichtstärke
Sept. 3	14°	17'	10°	29'	14 U	5'	0,02772
6	13	52	10	46	13	52	0,02825
9	13	25	11	5		38	0,02873
12	12	55	11	22		24	0,02915
15	12	23	11	40		10	0,02952
18	11	49	11	57	12	56	0,02982
21	11	14	12	13		42	0,03006
24	10	37	12	28		28	0,03022
27	10	0	12	43		14	0,03030
30	9	22	12	56		0	0,03031
Oct. 3	8	44	13	8	11	45	0,03024
6	8	6	13	18		31	0,03009
9	7	28	13	26		17	0,02987
12	6	51	13	33		2	0,02959
15	6	16	13	39	10	48	0,02924
18	5	42	13	42		34	0,02883
21	5	11	13	44		20	0,02836
24	4	41	13	44		7	0,02785
27	4	14	13	42	9	53	0,02731
30	3	49	13	38		39	0,02674
Nov. 2	3	27	13	32		26	0,02613
5	3	8	13	25		13	0,02551
8	2	52	13	16		0	0,02487
11	2	39	13	5	8	48	0,02423
14	2	29	12	53		35	0,02358
17	2	23	12	40		23	0,02293
20	2	19	12	25		11	0,02228
23	2	19	12	9	7	59	0,02168
26	2	21	11	52		47	0,02105
29	2	27	11	33		36	0,02042
Dec. 2	2	35	11	13		25	0,01983
5	2	47	10	53		14	0,01926
8	3	1	10	31		3	0,01870
11	3	18	10	9	6	52	0,01816
14	3	38	9	45		42	0,01764
17	4	0	9	21		31	0,01714
20	4	24	8	56		21	0,01667
23	4	51	8	31		11	0,01621
26	5	20	8	5		1	c,01577
29	5	51	7	38	5	51	0,01535
Jan. 1	6	24	7	11		42	0,01494
(1805) 4	6	59	6	43		33	0,01456
7	7	36	6	15		23	0,01421
10	8	15	5	47		14	0,01391
13	8	55	5	18		5	5,01365
16	9	37	4	49	4	56	0,01342
19	11	21	4	19		47	0,01322

Continued News on the new main Planet Ceres

MC (December 1803), p. 533.

Several observations and news have been handed in regarding this planet that we would like to communicate.

Adjunct Bittner [Adam Bittner (1777–1844), 5th director of the observatory in Prague. Asteroid 6596 is named in his honour] *observed at Prague Observatory Ceres' opposition with the Sun; he compared the planet with τ, no. 38 and 234 Sagittarius according to Bode's catalogue; the apparent right ascension of τ Sag was according to my star catalogue 283° 40′ 31.″9, decl. s. 27° 56′ 27.″2. The comparison of the planet with this star gave for July 2 at 12 o'clock 6' 6."8 mean time Ceres' apparent right ascension 280° 37' 52."9, declination s. 28° 23' 9."3. The comparison with 38 Sag whose mean position was taken from Bode's catalogue and its apparent right ascension 268° 55' 14."7 declination 28° 27' 48."0 gives RA Ceres 280° 37' 40."7, decl. Ceres 28° 23' 11."4 south. The star 234 delivered slightly different results: but because it appears to have a faulty determination in Bode's catalogue, Adjunct Bittner compared it with τ Sag and the derived apparent right ascension was 288° 4' 26."5, declination s. 28° 13' 36."1; this gives for RA Ceres 280° 37' 47."5, decl. Ceres 28° 23' 9."1.*

The above stars employed by Bittner are taken from La Caille; nr. 38 according to Bode is no. 1495 and no. 234 is nr. 1593 of the southern star catalogue (Coelum Australe) of this French astronomer; according to Piazzi's latest star catalogue the true determination of these stars is as follows:

After Bode	*After L. Caille (Coel Austr.)*	*RA 1800*	*Annual change*	*Dec south*	*Annual change*
No. 38	*No. 1495*	*208° 51' 13."2*	*+56."75*	*28° 27' 46."2*	*+0."40*
No. 234	*No. 1593*	*288 0 17.3*	*+56.16*	*28 14 15.74*	*-6. 19*

There is an error of 33" in RA with no. 234 and one of 1' 5" in declination.

The mean value of the above three determinations is for July 2, 1803, at 12 o'clock 6' 6."8 mean time, RA Ceres 280° 37' 49."0, declination Ceres 28° 23' 10."0 south. Bittner calculated from these the geocentric longitude 9Z 9° 22' 47."0, geocentric latitude 5° 5' 55."0 south, the longitude of the Sun for this time is according to my solar tables with the improvement mentioned on p. 94 of the January issue of my MC 1802: 3Z 10° 0' 28", its daily motion 57' 10."5; this and the daily motion of this planet in longitude 13' 12", in latitude 3' 4", which result from the ephemeris of the orbit of this planet, for the opposition on July 1, 1803 at 11 pm 24' 57."2 m. t. the geo- and heliocentric longitude of Ceres 9Z 9° 29' 52."0, her geocentr. latitude 5° 14' 17."0 s.

The ephemeris of the geocentric path of Ceres for the following year 1804, which we published in the October issue of the MC on page 370, the following typos need to be corrected: the decl. of Ceres on May 3 is too great by 5', instead of 11° 18' it should be 11° 13'; in January 13 1805 luminosity 5.01365 should be 0.01365; on January 19, 1805, instead of RA 11° 21', 10° 21'.

We will publish in the next issue a map of Ceres' orbit in 1804 by Inspector Harding of Lilienthal, sketched with the greatest carefulness. In this issue we communicate a list of all those stars which will be in Ceres' path or neighbouring her which are determined by Prof. Piazzi, taken from his excellent star catalogue. Since this book is probably hardly available to foreign astronomers, we believe to have done something good for practical astronomy. In the future we will try to add to this small catalogue by our own observations.

Observations of the new planet Ceres Ferdinandea done by Equatorial Sector

By Barnaba Oriani. *Ephemerides Astronomicae Meridianum Mediolanensem* (1803), p. 3.

Already for some years the famous professor Piazzi, astronomer of Palermo, with his excellent instruments was verifying the positions of the fixed stars in the modern catalogues. It was during this useful task that he discovered the new planet Ceres, because, while he was looking for the 87th star of Tobias Mayer, according to Wollaston, he wanted to determine the situation of all the smaller neighbouring stars. Among these, on the 1st of January 1801, he observed one which the day after had moved position. On the 3rd of January he made sure that the star was moving for real; it was moving 4' each day in right ascension with retrograde motion, and increasing its declination about 3.5'. He continued its obs and on January 24 he wrote to the famous Berlin astronomer Bode and to me, announcing his discovery. In the letter I received he explained the situation of the new planet as he observed the 1st of January, and the 23rd. Mentioning that between the 11th and the 13th its motion changed from retrograde to direct. Adding further that he suspected it was not a comet but a planet, because it had no halo or tail around it. I received this letter two months later, so it was too late to deduce its position from the two positions given. I tried then to calculate its orbit by supposing it was a primary planet moving in a circle. I tried to calculate its position under this hypothesis, but in vain even though I extended my search a few degrees ahead and behind the calculated positions. In order to announce immediately such a discovery to Baron von Zach of the Ducal Obs. in Seeberg near Gotha, I sent all the information from Piazzi and the elements of the circular orbit which I had calculated. From them it derived that the new discovery was a primary planet with an orbit between the one of Jupiter and the one of Mars. He received my letter almost simultaneously with one from Bode from Berlin, telling him about the same discovery. Zach immediately made a summary of both letters and published them in his well-known newspaper in the month of June, 1801. Due to its angular motion, the planet approached the Sun too closely to observe until after the month of September or Oct, 1801. Shortly after, in May 1801, I received from Piazzi all his obs. of the new planet, named by him later Ceres Ferdinandea. The obs. began the 1st of January and ended on February 11th. He said also he sent a copy of them to Bode, and another to Lalande saying he did not wish anything published before having computed the results. Under the same condition I sent a copy to Zach, who privately sent them to other astronomers in order to calculate the orbit more precisely than the one calculated based on only 2 obs (January 1 and 23).

In the meantime the news of the new planet had become known to all the most important German astronomers. Two years before, a society of astronomers was founded. Baron von Zach was the secretary and Schroeter of Lilienthal was the president. The object of this society was to search for a planet which was suspected to be between Mars and Jupiter. The foundation of this supposition was an empirical law of analogy among the distances of the known planets from the Sun. This law was pointed out by Lambert 40 yrs before, and confirmed later by Bode and Prof. Wurm. In fact if we suppose the distance of Saturn from the Sun divided into 100 parts, we obtain all the other planets' distances in the following form [see the mathematical progression as published in the MC, June 1801] *If we call a the avg. distance of Mercury from the Sun, and a' the one of the second planet we have the distance of the nth planet* $= a + 2^{n-2}(a' - a)$. *The planet Ceres conveniently fills the empty space between Jupiter and Mars; the before mentioned law of analogy was confirmed and the members of the society were totally satisfied. All the experts in calculation were looking for circular, parabolic and elliptic orbits, which fit most of the Piazzi obs. The arc of the planet's motion around the Sun was only 9°, so all of the determined orbits were very close to the obs., and the ones showing an error (a difference with the obs.) of only 30–40 seconds were considered the best. Someone though, not being able to totally eliminate errors or minimize them within a few seconds, began to doubt Piazzi's obs., claiming they were not precise. Some others believed it was strange and inconvenient to list among the planets one with an orbit inclined more than 10°. Consequently most of the time it was outside the zodiac.*

Fortunately Baron von Zach with admirable determination continued to support the existence of the planet and the value of Piazzi's obs. in his highly scientific publication. Every month he published the results of calculations which theoretically were better representing the observations; he printed Piazzi's original obs. In September 1801 and again, the more corrected ones in November. Even more than that, to make it easy to find the planet, in the July and November newspaper he wrote a little ephemeris of the calculated places of the elliptic orbit by Burckhardt and the circular orbit by Olbers of Bremen. Most astronomers got busy searching for the new planet in the places indicated by the ephemeris but without result in September, October and November. Finally, Dr. Gauss, an experienced astronomer in Brunswick, announced to Baron von Zach the elements of different elliptic orbits which admirably corresponded within a few seconds to all the obs. of Piazzi. Gauss himself deduced a new ephemeris from the determined elements. These results were published by Zach in the journal in December, 1801. Seeing the surprising concordance between Piazzi's obs. and the calculations of Gauss, the German astronomers set out to find the new planet with more zeal. In fact it was observed at first on December 7 by Zach himself, but because of bad weather, he could not see the planet again until December 31 to his satisfaction. Olbers observed it on January 1, 1802, precisely one year after the first sighting by Piazzi, and later it was seen and observed by many astronomers.

Baron von Zach promptly sent me the news of the finding of Ceres and the elements computed by Gauss, but because I had to go to Leon at the end of November, where I stayed until the end of January, I didn't have his letter or his journal till February 9, when I returned. The cloudy sky didn't allow me to search for the planet for the whole month of February. Only on the night of the 24th could I see 10 little stars, which are situated around it, but because of bad weather on the following days, I couldn't verify if anyone of them had changed position. Finally, March 10 the sky cleared and reviewing the same stars I noticed that one of them had disappeared. I went to search for it right away in the places where Gauss assigned to the planet, and to be cautious I observed, in the surrounding, a dozen small stars some of which, such as the 34th of Corona Borealis, the 93rd of Virgo, and so on were already noted in the catalogue of Bode, and the remaining stars could be the planet Ceres. In fact, March 11 I noticed that one of them had changed position in RA of 12' with retrograde motion, and its DEC was increased 6'. On the same night of March 11 I again observed the planet, lighting the threads of the micrometer to obtain, with the change into 5 threads, a better determination of its RA.

I continued to light the threads for 3 months with good success; later, around the end of June, because of a very clear light of the planet, I stopped the lighting and I had to be happy to get the RA from the instant of the exit of the planet from one bar (on a very smooth brass lamina parallel to the meridian thread) and from the instant of entering a second bar parallel to the first. To find the DEC when the threads were not lighted I used two other bars equidistant and parallel to the equatorial thread, which we could bring near to each other in such a way that will tally along the equatorial thread. In almost every observatory where Ceres was observed, it was seen with instruments situated at the meridian- that is, with a mural quadrant and with transit telescopes. These obs. are more exact than the one done with parallactic instruments or with equatorial sectors. I believe it is not useful therefore to publish the obs. I have done in the months of March April and May, because in Zach's journal there are more exact ones of Zach himself, the ones of Piazzi (which start at February 22 and end May 23), and the ones of my colleague Cesaris taken with our mural quadrant of Ramsden. Because, however, during the month of June it was not possible to observe the planet at the meridian, and in many obs. they didn't observe it at all for lack of good equatorial instruments, I think it would be proper to exhibit my original obs. done the 24th of June and subsequently, a time in which the planet was observed by few, and to give the results of the ones done during the previous months. The RA of the planet can be obtained, as we know, from the difference between the passages in time of the planet and of a fixed star, of which we know the position. This difference converted in equatorial arc and added to the apparent RA of the star, if this comes before the planet, gives the RA of the planet itself. Similarly, adding to the apparent known DEC of the star the difference in DEC between the same star and the planet, we get the DEC of it. When, however, the DEC difference is of many minutes, we have to add to the same the difference of refraction; and if the time angle is of many degrees we have to calculate again the effect of the refraction in RA. An example will show how one reduces the original obs. The obs. of the 16th July was made in the westerly time circle of 4h 13' as to say 63° 15'. The passage at this circle of the star s Virgo is at 8h 25' 46".05 as per the semi-sum of the two instants observed at 8h 24' 47" and 8h 26' 45".1. The passage of the c Virgo star is at 8h 36' 1".9, and the passage of Ceres is at 8h 54' 58".25. The difference of the RA in time between s Virgo and Ceres is 29' 12".2; that means 7° 19' 5".8 in equatorial arc and the difference between c Virgo and Ceres is 18h 56".35; that means 4° 44' 52".4. Now, according to the catalogue of zodiacal stars of Zach the apparent RA of the first star is 179° 59' 48".1, and of the second is 182° 34' 50".6. Therefore if we add to the first 7° 19' 15".8 and to the second 4° 44' 52".4, we will have the apparent RA of Ceres related to s Virgo 187° 19' 3".9, and to c Virgo 187° 19' 43". Similarly the difference in DEC between s Virgo and Ceres is –1° 3' 39", and between c Virgo and Ceres +1° 25' 33", and having, according to that catalogue the apparent DEC of the first star 6° 54' 26".5 and of the second 4° 24' 57", we have as a consequence that Ceres' apparent DEC related to s Virgo is 5° 50' 47".5 and from c Virgo 5° 50' 30".

If we want to remove the effects of refraction, we can use known formulas (refer to Lalande's Astronomy *section 2545 and* [Antonio] *Cagnoli's* [1743–1816] Trigonometry *section 826; keeping in mind that in the equatorial instrument the correction for the RA is only ½ of the one used in the micrometer of the common parallactic machines). We can proceed in the following manner: we calculate the distance from the zenith and the parallactic angle formed by the vertical circle with the hourly circle in the position of the star, or using the known tables or solving the spherical triangle, the 3 angles of which are respectively at the pole, the equator and the zenith of Milan and at the position of the observed star. For this triangle are given two sides, that is the distance of the pole at zenith 44° 32' and the distance of the star from the pole and as well the angle between these two sides, being equal to the hourly angle 63° 15'. It turns out in our case that the apparent distance (from the zenith) of s Virgo 66° 28', of c Virgo 68° 18' and of Ceres 67° 15'; to such distances belong the refractions 2' 22".4; 2' 35".3; 2' 27".4. Because we also obtained from the same triangle the parallactic angle, which is for s Virgo 43° 5', for c Virgo 42° 23' and for Ceres 42° 46', we multiply the refraction by the cosine of this angle and we obtain the refraction in DEC of s Virgo 1' 44".0; of c Virgo 1' 54".7; and of Ceres 1' 48".2. Multiplying then the refraction in the size of the same parallactic angle, and dividing the result by the cosine of the DEC of the other, we obtain the refraction in RA, that is for s Virgo 1' 38".0; for c Virgo 1' 45".0; and for Ceres 1' 40".6. We will correct therefore the RA of Ceres 187° 19' 3".9, deducted from s Virgo subtracting from it the difference 2".6, between 1' 38".0 and 1' 40".6; so it will be 187° 19' 1".3. Again we will correct the RA of Ceres deducted from c Virgo 187° 19' 43".0, adding to it the difference 4".4 between 1' 45".0 and 1' 40".6 and it will be 187° 19' 47".4. We will obtain also the correct DEC of Ceres 5° 50' 43".3 deducted from s Virgo subtracting from 5° 50' 47".5 the difference 4".2 between 1' 44".0 and 1' 48".2; and the same correct DEC 5° 50' 36".5 deducted from C Virgo adding to 5° 50' 30" the difference 6".5, between 1' 54".7 and 1' 48".2. The avg. time of this obs. is given adding to the time of Ceres passage 8h 54' 58".25 the delay 11' 58".5 of the clock on the avg. solar time; as we can get from the table at the end of the obs., it will be 9h 6' 56".75.*

From this example we can see that only rarely will we have to calculate the effect of refraction in RA and DEC, because we took care to compare the planet to those stars having a DEC little different from the planet. We must confess though that frequently changing the reference stars has the drawback of attributing to the position of the planet or to the inexactness of the obs. the mistakes that may exist in the positions of the stars. There are still some doubts in the best catalogues. In the journals of August and September we find published the following obs. of Zach. More than that, at my request he took upon himself to repeat the calculations using the best positions of stars determined by himself or other astronomers, who would like to know either the stars' same positions or the positions of the planet which result from the comparison with each star separately. [Paper concludes with 5 pages of tables.]

Inequalities of the motion of the new planet Ceres due to the attraction of the other planets

By Oriani. *Ephemerides Astronomicae Meridianum Mediolanensem* (1803), p. 35. Excerpt.

The fact that Ceres' orbit is very close to that of Jupiter makes us suspect the probability of very substantial inequalities due to Jupiter's attraction. But because in the calculations of the reciprocal perturbations of the planets, we use as known values the elements of the elliptic orbit, it seems that we have to verify and confirm all these elements before we begin to look into the perturbations. In fact in order to precisely determine the periodicity it would require us to observe the planet at two different times after many revolutions in its orbit. Then comparing the space covered by the planet to the time interval between two observa-

tions we could easily calculate the time of a complete revolution; and even if at the two times of the observations the periodic inequalities due to the perturbations are not the same, the difference affects results less as the time interval is bigger. From the periodic time we can calculate the avg. distance from the Sun, using Kepler's theorem, and the avg. motion which are the most important quantities necessary for the calculations of the perturbations. If we want to deduce the periodic time from 2 obs. so close that the planet in the meantime could not complete an entire revolution, we would have to know the position observed from the perturbations due to the other planet, in order to obtain the elliptic position. But how can we determine the perturbations if they cannot be calculated by ignoring the periodic time? It is evident that in this case the problem remains uncertain.

Continued News on the new main Planet Ceres

MC (November 1804), p. 472.

As soon as Ceres had returned from the rays of the Sun and the twilight it was being observed again for the fourth year. Inspector Harding at Lilienthal has since then made the following observations at the circular micrometer (Fig. 10.70):

At Seeberg Observatory the following meridian observations were made (Fig. 10.71):

1804	Mittl. Zeit in Lilienthal			Scheinb. gerade Aufsteig. der ♀			Scheinbare südl. Decl. der ♀		
August 2	13 U	8′	40″	15°	45′	0″	7°	51′	16″
3	12	22	8	15	47	28	7	54	37
5	14	14	30	15	52	16	.	.	.
6	12	10	58	15	53	47	.	.	.
9	11	58	54	15	57	14	8	14	28
10	13	10	27	15	57	22	8	19	50
11	10	53	25	15	57	44	8	24	1
14	11	56	20	15	54	44	8	37	39
16	14	27	22	15	52	40	8	46	55
17	12	31	14	15	50	55	8	52	48
25	11	42	4	15	23	12	9	32	40
Septb. 4	10	59	20	14	16	48	10	30	19
5	11	40	44	14	8	5	10	36	21

Fig. 10.70 Observations of Ceres by Harding at Lilienthal, August and September 1804

	Epoche für Seeberg	Sonnenferne	Knoten
1801	77° 19′ 34″9	326° 33′ 37″	80° 54′ 59″
1802	155 28 35.1	326 35 39	80 55 0
1803	233 37 35.3	326 37 40	80 55 1

tägliche tropische Bewegung	770″7951
Tropische Umlaufszeit	1681 Tage 9 Stunden
Excentricität (1801)	0,0788352
Neigung (1801)	10° 37′ 56″0
Logarithmus der halben Axe	0,4420971

Fig. 10.71 Observations of Ceres by Zach at Seeberg, September and October 1804

1803	Mittl. Zeit in Mailand	Scheinbare gerade Aufsteig. der ⚳	Scheinbare südl. Abweichung der ⚳
April 17	15u. 28' 31"	286° 26' 39"	23° 28' 31"
May 20	14 56 3	285 50 30	25 7 11
Jun. 12	12 29 48	285 9 34	26 52 49
15	11 43 34	284 33 44	27 6 22
16	11 37 52	284 21 22	27 12 0
18	11 24 1	283 55 51	27 20 55
19	11 43 38	283 42 28	27 26 1
23	11 24 56	282 47 15	27 44 14
26	11 53 9	282 5 7	27 59 17

Fig. 10.72 Observations of Ceres by Oriani at Milan, April to June 1804

I. Excentricität = 0,077	II. Excentricit. = 0,079	III. Excentricit. = 0,081
Mittelpuncts - Gleichung.		
− 31741,"28 Sin. Anom. med.	− 32564,"45	− 33387,"54
+ 1525, 47 Sin. 2 Anom. med.	+ 1605,"55	+ 1687, 69
− 101, 64 Sin. 3 Anom. med.	− 109, 74	− 118, 27
+ 7, 74 Sin. 4 Anom. med.	+ 8, 57	+ 9, 48
− 0, 64 Sin. 5 Anom. med.	− 0, 73	− 0, 82
+ 0, 05 Sin. 6 Anom. med.	+ 0, 06	+ 0, 07

Fig. 10.73 General mean equations for the orbit of Ceres, by Oriani

It follows that Gauss' ephemeris of this planet which we communicated in volume VIII of our MC *p. 371 gives the right ascension too small by about 9 min, the declination too great by about 4 min, which could not impede the finding of this planet, though. We would like to add the following observations from Milan by Oriani to last year's observations on this planet (Fig. 10.72):*

Since the elements of Ceres' orbit will need further changes and improvements with continued observations, Oriani calculated according to three hypotheses of eccentricity the following general mean equations according to which the one can easily be calculated corresponding to the true eccentricity (Fig. 10.73).

For example, assume the found eccentricity of Ceres' orbit = 0.0788410. If one demands the corresponding mean equation; take the first and second differences between the three elements which correspond to the three assumed eccentricities 0.077, 0.079, and 0.081, the interpolation can be as follows:

Of the given eccentricity = ...	*0.0788410*
subtract the eccentricity of the first hypothesis	*- 0.077*
	0.0018410

The multiplicator of the first difference will be
X = 0.001841 / 0.002 = 0.9205
and the multiplier of the second difference
x (x-1)/2 = -0.0366

[The paper concludes with detailed perturbation calculations.]

Opposition of the New Planet Ceres to the Sun in 1804

By Oriani, *Ephemerides Astronomicae Meridianum Mediolanensem* (1806), p. 32

All the following obs. have been done with the mural quadrant of Ramsden. The no. of obs. would have been larger if, during October, November and December, the bad weather had allowed me to see the sky. I was able to see the planet only 11 times in those 3 months. Sometimes the stars and the planet used in determining the position of the planet were so obfuscated by the fog that it was difficult to determine their passage in the micrometer hairs. The obs. done in September are less dubious, so later we will use them in determining the opposition time of the planet to the Sun. The clock used in these obs. was calibrated on the sidereal time and we came to see its course by confronting the passages at the meridian for the same time on subsequent days. It has to be said that on November 21 and 22 the pendulum was lengthened in order to better fit sidereal time. On October 1 two hairs of the micrometer were replaced, the third and fourth, because they were loose. There are no notations about the barometer and thermometer for lack of space and because the DEC differences were so small (seldom more than a degree). Thus, the refraction difference is subject to no substantial deviations. [Oriani here includes three pages of tables.]

From the great star catalogue of Piazzi we have the following avg. positions for the year 1800 [table of star positions]. *The DEC of η Cetus is established by us one minute less than in the catalogue – Piazzi confirmed a printing error of 1′. If we reduce the avg. positions to the true and apparent one relative to the day of obs. by means of precession, nutation and light aberration we obtain the following apparent positions for Ceres (Fig. 10.74):*

Giorni 1804	Tempo medio			Aſcenſione retta oſſervata di Ginnone			Declinazione Auſtrale di Ginnone		
	or	′	″	°	′	″	°	′	″
Settembre 19	12	52	18,9	11	45	45,3	11	58	24,3
20	12	47	36,7	11	34	7,3	12	3	53,6
21	12	42	53,2	11	22	12,6	12	9	8,3
22	12	38	9,3	11	10	9,4	12	14	28,2
23	12	33	24,9	10	58	0,2	12	19	24,4
24	12	28	40,2	10	45	47,0	12	24	22,[illegible]
26	12	19	9,6	10	21	0,4	12	34	6,6
27	12	14	22,8	10	8	14,8	12	38	41,5
28	12	9	37,2	9	55	46,2	12	43	25,8
Ottobre 6	11	31	24,6	8	14	16,[illegible]	13	13	46,4
7	11	26	37,9	8	1	33,0	13	16	47,1
17	10	39	22,8	6	2	11,3	13	37	27,6
31	9	55	32,7	3	49	56,1	13	32	38,3
Novembre 9	8	56	31,5	2	55	16,7	13	9	43,1
11	8	48	5,7	2	46	45,9	13	2	32,7
12	8	43	55,6	2	43	14,2	12	58	47,1
18	8	19	20,7	2	28	20,0	12	32	26,4
19	8	15	19,8	2	27	4,1	12	27	35,2
Dicembre 16	6	35	18,6	3	59	24,6	9	25	46,0

Fig. 10.74 Positions of Ceres from September to December 1804

At the end of September the Ceres-Sun opposition took place, so I calculated the longitudes and latitudes from the RA and DEC in that month. I used the ecliptic obliquity of 23° 27′ 59″.5 and by evaluating the parallax = 3″.8. Afterward I reduced the apparent longitudes to the real one calculated from the avg. equinox applying to them the light aberration -8″.5 and the nutation -15″.4. After I calculated from the tables in the 3rd edition of Astronomy *by Lalande, the positions of the Sun for every moment of Ceres' obs. during September. Finally I calculated the corresponding positions of the planet from the elements of Ceres' orbit corrected for the 8th time by Gauss, which were published in our ephemerides of 1804. To these positions I then applied the perturbations due to Jupiter and here are the results.*

The moment of opposition is deducted from the position of Ceres the 26th and 27th of September, reduced to the avg. error. The longitude difference between Ceres and the Sun, September 26, increased of 6″ line, is 46′ 6″.6 = 2.766″.6. By reason of this opposition and the other two years (1802 and 1803) Gauss corrected for the 10th time the elliptic elements of Ceres in the following way [see table in Oriani's paper on p. 40]. *The position of Pallas in conjunction with the Sun brought me to observe it at the same time with the mural quadrant of 8 feet and with the equatorial sector. If with the first instrument we have the advantage of a more solid machine, and precision of the divisions, with the 2nd instrument we have the advantage of being able to obs. the planet without weakening its light with light inside the telescope. Using the two brass bars situated in this object in the focus of the 2 lenses; for the declination I always used the equatorial thread under which it would occult the other object.*

Observations of the new Planet Ceres

MC (January 1805), p. 81.

Already in the November issue of 1804 we publicised the observations of Ceres made at the Ernestinian Observatory, which included the time of its opposition with the Sun. This opposition occurred between September 27 and 28, 1804, and was also observed at the observatories in Milan and Prague, by Oriani and the adjunct Bittner. We received from both the observation and calculation of this opposition, which we are communicating here together with all previous observations of Ceres. From September 19 through November 12, Oriani obtained at the mural quadrant the following observations: [These are the same observations printed in the paper just given by Oriani.]

He compared Ceres with φ^1 φ^3 and η Ceti, whose positions for 1800 are as follows according to Piazzi (Fig. 10.75):

Namen d. Sterne	Gerade Aufsteigung 1800 ° ′ ″	Jährl. Veränderung ″	Südliche Decl 1800 ° ′ ″	Jährl. Veränderung ″
φ¹ Ceti	8 31 17,7	+ 45, 31	11 42 1, 2	− 19, 80
φ³ Ceti	11 29 48,6	+ 45, 05	12 21 2, 9	− 19, 62
η Ceti	14 37 57,4	+ 44, 92	11 14 39, 3	− 19, 37

Fig. 10.75 Three star positions used by Oriani

1804	Beobachtete geoc. Länge der ⚳ vom mittl. Aequin an gezählt	Beobachtete füdliche Breite ⚳	Fehler d. VIII Elemente in der Länge	Fehler d. VIII Elemente in der Breite
Sept. 19	0S 5° 58′ 24,″8	15° 38′ 45,″7	− 3′ 59,″8	+ 4,″6
20	5 45 17, 9	15 39 10, 5	− 4 2, 5	− 4, 6
21	5 32 2, 1	15 39 15, 2	− 4 2, 8	− 4, 7
22	5 18 33, 5	15 39 21, 9	− 3 55, 2	− 9, 3
23	5 5 15, 1	15 39 2, 5	− 4 2, 1	− 2, 0
24	4 51 50, 1	15 38 43, 4	− 4 6, 6	− 1, 5
26	4 24 46, 1	15 37 47, 6	− 4 8, 6	− 15, 0
27	4 11 1, 3	15 36 54, 6	− 3 57, 4	− 15, 9
28	3 57 38, 5	15 36 17, 1	− 4 8, 0	− 10, 9
		Mittl. Fehler	− 4′ 2,″5	− 6,″5

Fig. 10.76 Latitude and longitude of Ceres, compared with Elements VIII, calculated by Oriani

1804	Mittl. Zeit in Mailand	Länge der ⚳	Länge der Sonne aus La Lande's Taf.	Geoc. Breite der ⚳
Sept. 26	12U 19′ 9,″6	0S 4° 24′ 40,″6	6S 3° 38′ 34,″0	15° 37′ 39,″1
27	12 14 22, 8	0 4 11 6, 4	6 4 37 21, 1	15 36 45, 2
Unterfch.	23U 55′ 13,″2	− 13′ 34,″2	+ 58′ 47,″1	− 53,″9

Fig. 10.77 Oriani's results for September 26 and 27, 1804

Regarding η Ceti, Oriani discovered a major typo in Piazzi's star catalogue – the declination was given by one minute too great. We have corrected this error in the previous data. Oriani derived the following longitudes and latitudes from these observations (Fig. 10.76):
Accordingly (Fig. 10.77)
from which is found the time of the opposition of Ceres and the Sun:
27 Sept. 3h 33′ 47.″4 mean time Milan
At this time was:

Longitude of Ceres = 0Z 4° 14′ 1.″7
geocentric latitude 15 37 4.8
heliocentric latitude 10 20 46.2

Accordingly then was:
The VIII Elements of Ceres
in heliocentric longitude = 2′ 38.″4
in heliocentric latitude = 4.2

Oriani used the following elements for his calculation:
Parallax of Ceres = 3.″8
Apparent angle to Ecliptic = 23° 27′ 59.3
Aberration in longitude = -8.5
Nutation = -15.4

At Prague Observatory, Adjunct Bittner has been observing this new planet just as eagerly since the last rediscovery of Ceres, and sent the following observations (Fig. 10.78):

All observations until September 13 were made at the meridian and also at a parallactic machine which was equipped with a brass diamond in order to avoid illumination. From September 28 onward, Ceres was observed in the meridian at a quadrant. Since it was no longer possible after September 28 to compare Ceres with the stars suggested in the December issue of 1803, Adjunct Bittner chose for this purpose the stars no. 365 and 374 in Aquarius of Bode's star catalogue and no. 52, 95 and 101 of Cetus.

Adjunct Bittner calculated from the observations of September 28 and 30 and October 4 and 5 the following longitudes and latitudes of Ceres (Fig. 10.79):

from which the following data for the calculation of the opposition yielded (Fig. 10.80):

1804	Mittl. Zeit in Prag			Scheinb. gerade Aufſteig. ⚳			Scheinb. ſüdl. Declin. ⚳			Verglichene Sterne aus dem Wallfiſche
Aug. 28	11U	26'	4,"5	15°	7'	0"	9°	50'	0"	32 Zeich. n. Flamſt.
29	10	30	51	15	0	19	9	57	24	32
30	10	30	34	14	54	22	10	1	3	32
31	10	14	2	14	47	54, 5	10	6	45	32
Sept. 2	11	4	17	14	35	26, 5	10	18	41	32
3	10	27	10	14	25	56, 2	10	24	26	32
4	11	16	16	14	17	33	10	30	7	30
5	10	48	18	14	9	23	10	36	0	30
6	11	7	9	14	0	36, 5	10	42	4	30
7	10	45	36	13	51	54	10	48	14	30
8	10	20	56	13	42	55	10	54	44	28
11	10	20	24, 5	13	14	2, 6	11	13	0	31
13	9	54	26	12	54	2, 3	11	25	7	31 u. 22 φ 3
28	12	12	37, 6	9	55	56, 7	12	43	14	23 φ 4
30	12	2	53, 5	9	30	51	12	51	31	
Oct. 4	11	43	53, 6	8	39	53	13	7	0	
5	11	39	5, 2	8	27	4	13	10	35	

Fig. 10.78 Observations of Ceres by Bittner in Prague, from August to October 1804

1804	Mittl. Zeit in Prag			Geoc. Länge der ⚳				Südl. geocen. Breite der ⚳		
Sept. 28	12U	12'	37,"6	0 S	3°	58'	2,"3	15°	36'	14,"5
30	12	2	53, 5		3	34	24, 2	15	33	49, 4
Oct. 4	11	43	53, 6		2	37	47, 4	15	27	41, 6
5	11	39	5, 2		2	24	27, 0	15	25	51, 8

Fig. 10.79 Longitudes and latitudes of Ceres, calculated by Bittner

Länge der Sonne nach *v. Zach*'s neuen Sonnen-Taf. für d. 27 Sept. 1804 12U 12' 37,"6

mittlere Prager Zeit	6 S	4°	36'	24,"3
Länge der Ceres für dieſe Zeit = . . .	0	4	11	24, 3
Unterſchied			25'	0"
tägliche Bewegung der Sonne = ...			59'	1"
— — der Ceres = ...			13	21
Motus relativus = ...		1°	12'	22"

Fig. 10.80 Data for the calculation of the opposition of Ceres

1804		Mittlere Zeit auf Seeberg			Scheinb. ger. Aufſt. ♀			Scheinb. ſüdl. Abweich. ♀		
Octbr.	23	10U	11'	38,"769	4°	59'	35,"22	13°	40'	16,"0
	24	10	7	4, 295	4	49	55, 22	13	39	51, 8
	30	9	40	0, 855	3	57	46, 91	13	34	28, 3
Nov.	6	9	9	21, 750	3	10	43, 20	13	19	0; 6
	18	8	19	22, 794	2	28	35, 81	12	32	25, 0
Dec.	3	7	21	28, 326	2	44	41, 06	. .	. .	.
	4	7	17	47, 705	2	48	31, 05	10	57	35, 4

Fig. 10.81 Observations of Ceres by Zach, October to December 1804

Those 25' were thus completed in 8h 17' 41."8, and the opposition of Ceres occurred on September 27 12h 12' 37."6 – 8h 17' 41."8 = Sept. 3h 54' 55" mean time Prague. Both observed oppositions correspond very well.

The observations that have been made at the Ernestinian Observatory since October 20 are the following (Fig. 10.81):

The error of Gauss' latest elements for Ceres is constantly decreasing, so that the latest observations are well described by the ephemerides mentioned in MC 1803, p. 370f.

Observations of Ceres

MC (March 1805), p. 283.

Also for Ceres the epoch of visibility is over now and even at the greatest equatorial instruments its observation should be associated with difficulties since it must be made at a considerable distance from the meridian. For the following six to seven months we have to abstain from observing this planet in the skies and only strive to find it at the next reappearance. This will be not difficult due to the newly improved elements by Dr. Gauss and the accordingly calculated ephemerides since Ceres' orbit can be regarded as well determined and we are delivering here everything else needed for facilitating its future observation.

Since the ephemeris of Ceres' geocentric path, calculated according to the IX elements of Dr. Gauss, was too small by nine minutes in September and October regarding its RA and the declination too great by four minutes, Dr. Gauss used the three oppositions of Ceres of 1802, 1803 and 1804 to base his elements X on those.

Epoch Seeberg Meridian 1804	*312° 1' 33."5*
Daily movement	*771."0524*
Annual Movement	*326° 26' 3."1*
Eccentricity	*0.0784757*
Logarithm of the semimajor axis	*0.4420004*
Ascending Node 1804	*80° 59' 12"*
Inclination	*10° 37' 45"*

According to these, Dr. Gauss calculated the geocentric path of this planet from July 28, 1805 through May 24, 1806. He wrote that Bessel, famously known by his treatise on the orbit of the comet of 1607 (MC 1804, vol X, p. 425f), helped him in calculating this ephemeris because he delivered all the necessary positions of the Sun. Certainly, everyone, who knows the calculations necessary for a determination of planetary elements and the position derived from these, must be amazed at how one man was able to complete in such short intervals so many tedious calculations.

In order to facilitate a comparison of Ceres with stars in the next observations, we give here the positions of those which are almost in one parallel with the planet, taken from Piazzi. [Table of star positions follows.]

Excerpts from a Letter by Dr. Gauss in Brunswick; January 3, 1807
MC (February 1807), p. 152.

I have the honour of now being able to send you the ephemeris for the course of Ceres, the elements of the calculation of which I have, for the lack of anything else, based on Professor Pasquich's observation of the last opposition. I must therefore await more reliable ones to further improve it. They were the following:

Epoch in longitude in the meridian of Seeberg, 1806	*108°*	*19'*	*34".7*
Longitude of the perihelion	*326*	*37*	*59*
Longitude of the ascending node 1806	*80*	*53*	*23*
Inclination of path	*10*	*37*	*33.7*
Mean annual tropical movement	*78*	*9*	*23.3*
Mean daily tropical movement	*770".8584*		
Eccentricity	*0.0783486*		
Log. of the semi major axis	*0.4420723*		

My little work on the determination of the paths of planets etc., to which I have dedicated most of my time, is nearing completion. Until now, I have written a 22 sheet manuscript and am presently occupied in entering the detailed examples. One is taken from Juno, in which the time interval is 22 days; in the second, from Pallas, it is 71 days; the third, from Ceres, in which it was 118 days, had to be replaced by me in order to use another, amounting to 260 days, which would show the public the method in a better light. In all these examples, they can be directly applied in such a way that absolutely nothing of the path can be regarded as known. The entire work will amount to approximately 30 printed sheets. Given the present conditions, I fear that it will take a great deal of trouble to find a solid [respectable] *publisher.*

[Gauss then prints tables of the geocentric course of Ceres from December 19, 1806, to September 1807.]

Observations of Ceres by Santini in Padua
MC (April 1808), p. 375.

We can hardly expect in our latitudes before the month of July any meridian observations on Ceres, due to her southern declination of this year and thus we hurry to communicate several Italian observations which reached us later, made around last year's opposition.

This opposition was observed and calculated completely at Padua at the university observatory by its adjunct, Giovanni Santini. The observations were made with an 8ft mural quadrant by Ramsden and a pendulum clock after mean solar time by Lepaute. The original observations were the following (Fig. 10.82):

1807.	μ Virginis. Zeit der Uhr.	μ Virginis. Zenith-Distanz.	ζ Librae. Zeit der Uhr.	ζ Librae. Zenith-Distanz.	Ceres. Zeit der Uhr.	Ceres. Zenith-Distanz.
	u ′ ″	° ′ ″	u ′ ″	° ′ ″	u ′ ″	° ′ ″
April 7	13 26 3,0	50 11 8,0	13 37 0,1	56 28 19,0	14 5 20,4	51 29 15
8	13 22 8,0	50 11 13,0	13 33 6,1	56 23 19,0	14 0 55,1	51 27 16
9	13 18 13,0	50 11 16,0	13 29 11,6	56 23 16,0	13 56 20,7	50 25 15
10	13 14 17,9	50 11 12,0	13 25 16,1	56 28 22,0	13 51 49,3	51 22 34
11	13 10 22,8	50 11 12,7	13 21 20,9	56 28 21,0	13 47 14,6	51 20 29
12	13 6 28,1	50 11 15,0	13 6 28,1	56 28 18,0	13 42 40,7	51 18 20
23	12 55 32,2	50 11 15,0	12 46 30,7	56 28 21,0	13 3 18,0	50 26 45
24	12 51 37,3	50 11 15,0	12 42 36,1	56 28 19,0	12 58 32,5	50 54 54
27	12 19 52,5	50 11 15,0	i Virginis	. . . .	12 44 11,5	50 50 14
28	12 15 56,0	50 11 7,0	11 49 0,4	50 26 55,0	12 39 21,2	50 48 43
30	12 8 3,6	50 11 10,0	11 41 8,6	50 26 55,0	12 29 41,9	50 46 13
May 2	12 0 11,8	50 11 11,0	. . . .	. . . .	12 26 6,8	50 43 50
5	11 56 15,0	50 11 11,0	. . . .	. . . .	12 15 10,5	50 42 47
7	11 40 23,8	50 11 10,0	11 15 59,0	50 26 56,0	11 55 50,5	50 39 24
8	11 36 38,1	50 11 11,0	11 9 43,3	50 26 57,0	11 51 0,7	50 39 0
11	11 24 62,7	50 11 5,0	10 57 57,7	50 26 50,0	11 36 35,0	50 38 20
13	. . . .	. . . .	10 50 8,0	50 26 50,0	11 26 59,9	50 38 29
15	. . . .	. . . .	10 42 17,5	50 26 56,0	11 17 26,3	50 38 50
17	. . . .	. . . .	10 34 25,3	50 27 1,0	11 7 53,1	50 39 54
18	. . . .	. . . .	10 30 31,3	50 26 52,0	11 3 9,5	50 40 40
19	10 53 58,5	50 11 10,0	10 26 43,2	50 26 54,0	10 8 33,0	50 41 6
20	10 49 45,5	50 10 54,0	10 22 50,8	50 26 58,0	10 53 52,5	50 42 2
23	. . . .	. . . .	10 11 5,4	50 26 40,0	10 39 48,2	50 45 34
24	. . . .	. . . .	10 7 7,5	50 26 39,5	10 35 8,2	50 46 43
28	. . . .	. . . .	9 52 0,5	50 26 55,0	10 17 14,5	50 54 8

Fig. 10.82 Observations of Ceres by Santini, April and May 1807

From this follows, the position of Ceres on May 2 was inclined to the ecliptic 23° 27′ 50″.

Ceres' positions were calculated after the elements X by Dr. Gauss (Fig. 10.83). Ceres' opposition occurred between May 3 and 4 and if one applies the above found mean error of the elements to the geocentric positions of the planet, it follows:

1807	*Mean time In Padua*	*Geoc. Latitude of Ceres*	*Geoc. Long. of Ceres*	*Long. of Sun*
May 2	*12h 13′ 4″*	*222° 23′ 9.″2*	*10° 42′ 27.″2*	*221° 35′ 8.″2*
3	*12 8 14*	*222 9 40.7*	*10 39 23.5*	*222 33 1.8*

From this follows
Hourly motion of Ceres in longitude = –33.″80
Hourly motion of the Sun +1′ 45.24
Relative motion 179.04
Hourly motion of Ceres in latitude – 7.68

1807.	Mittl. Zeit in Padua.	Scheinb. ger. Auffteig. ♀	Sch. füdl. Declin. ♀	Länge der ♀ v. mitt. Aequi.	Nördl. Breite der ♀	Fehl. d. X Gaufsifch. Elem. in d. Länge.	Fehl. d. X Gaufsifch. Elem. in d. Breite.
April 7	14h 10' 48"	228° 5' 10",4	6° 6' 55",5	227° 20' 29",6	11° 21' 19",2	− 3' 42",6	+ 1' 41",0
8	14 6 2	227 57 2,9	6 4 53,9	227 12 19,3	11 21 4,0	− 4 28,4	+ 1 4,0
9	14 1 47	227 47 43,4	6 2 51,4	227 2 40,3	11 20 29,5	− 4 25,0	+ 1 15,5
10	13 57 19	227 38 35,9	6 0 10,9	226 53 1,3	11 20 36,0	− 4 31,9	+ 0 55,8
11	13 52 39	227 28 41,9	6 58 6,1	226 42 44,6	11 19 54,0	− 4 7,7	+ 0 31,3
12	13 48 4	227 18 49,4	5 55 57,2	226 32 33,5	11 19 17,0	− 4 17,1	+ 0 13,6
23	12 56 23	225 11 51,0	5 34 18,9	224 22 41,0	11 4 43,4	− 4 14,2	+ 0 56,0
24	12 51 35	224 59 9,5	5 32 29,9	224 10 53,0	11 2 46,2	− 5 23,4	+ 0 46,9
27	12 37 2	224 20 1,0	5 27 43,4	223 30 29,1	10 56 2,1	− 4 24,5	+ 0 52,0
28	12 32 22	224 6 34,6	5 26 16,5	223 17 2,0	10 53 25,8	− 4 23,1	+ 0 49,9
30	12 24 44	223 39 43,4	5 23 44,8	222 50 16,3	10 48 8,6	− 4 28,5	+ 0 40,3
May 2	12 13 4	223 12 19,7	5 21 25,7	222 23 0,0	10 42 19,6	− 4 12,7	+ 0 46,6
3	12 8 14	222 58 48,2	5 20 19,7	222 9 35,1	10 39 24,0	− 4 22,8	+ 0 38,5
7	11 48 50	222 4 0,3	5 16 55,5	221 15 24,9	10 26 21,4	− 3 44,8	+ 0 21,2
8	11 44 1	221 50 27,4	5 16 29,5	221 2 21,8	10 22 42,6	− 3 56,5	+ 0 21,8
11	11 29 31	221 10 16,9	5 15 56,6	220 23 26,6	10 11 6,6	− 4 6,8	+ 0 33,1
13	11 19 54	220 43 54,0	5 16 2,9	219 58 4,3	10 2 59,8	− 4 16,5	+ 0 40,3
15	11 10 29	220 18 1,6	5 16 18,0	219 33 15,3	9 54 53,7	− 4 3,1	+ 0 19,4
17	11 0 47	219 52 43,6	5 17 17,1	219 9 14,5	9 46 6,1	− 3 51,1	+ 0 26,0
18	10 56 1	219 40 19,7	5 18 11,8	218 57 36,6	9 41 24,0	− 3 51,0	+ 0 34,0
19	10 51 17	219 28 9,2	5 18 47,2	218 46 31,2	9 37 4,2	− 4 28,2	+ 0 33,4
20	10 46 32	219 15 9,2	5 19 49,4	218 35 19,4	9 32 31,3	− 4 32,7	+ 0 46,1
23	10 32 28	218 41 46,7	5 23 17,8	218 3 32,0	9 18 14,7	− 4 26,3	+ 0 30,3
24	10 27 48	218 30 43,8	5 24 27,3	217 53 17,4	9 13 44,7	− 4 21,1	+ 0 3,8
28	10 8 23	217 48 58,9	5 31 57,0	217 16 57,0	8 53 27,5	− 4 51,5	+ 0 40,0
Mittlerer Fehler	.	.	.	.	.	− 4 17,4	+ 0 39,0

Fig. 10.83 Observations of Ceres by Santini based on Elements X of Gauss

Thus the opposition occurred on May 3 at 4h 18' 35" mean time. For this moment is:

Helioc. longitude of Ceres 222° 14' 5."2

Geoc. Latitude -- ---- 10 40 23.6 north

Error of elliptical elements in longitude -2 40.4

in latitude +0 24.5

Helioc. latitude of Ceres 6° 41 51.4

In order to find the true error of the elements, Santini calculated the perturbations of Ceres resulting from Jupiter's attraction for May 2 and 3, using the formulae given in the Milan Ephemerides of 1803 by Oriani, and found through the numerical expansion in two hypotheses, the perturbation on longitude on

2 May .... +4' 21."6

in latitude –3.8

in longitude on May 3 +4' 26.7

in latitude –3.5

If one applies these perturbations to the calculated elliptical positions of May 2 and 3, the mean error in longitude is +1 55."5, in latitude +20."7. The conjunction as well as the above determinations remain the same. At the same time we received from Milan the following observations made by Carlini at Ramsden's mural quadrant (Fig. 10.84):

1807.	Stern im Parall.	Durchg. d. Sterns im Meridian.	Durchg. d. Ceres im Meridian.	Zenith-Distanz d. Sterns.	Zenith-Distanz d. Ceres.
		U ′ ″	U ′ ″	° ′ ″	° ′ ″
April 10	β ♎	15 5 57,52	15 9 50,40	54 6 37,0	51 27 24,0
19	37 ♎	15 22 53,02	15 3 15,98	54 50 23,4	51 8 35,0
21	37 ♎	15 22 51,80	15 1 38,00	54 50 27,2	51 4 48,6
26	β ♎	15 5 46,38	14 57 17,58	54 6 42,5	50 56 23,0
27		15 5 45,50	14 56 23,90	54 6 37,7	50 54 50,8
28		15 5 44,70	14 55 29,60	54 6 42,6	50 53 29,2
29		15 5 43,98	14 54 35,16	54 6 44,3	50 52 6,6
30		15 5 43,12	14 53 40,36	54 6 39,4	50 50 50,8
May 1	μ ♍	14 51 58,74	14 52 45,90	50 15 49,9	50 49 42,0
2		14 31 58,20	14 51 50,48	50 15 45,8	50 48 32,3
7		14 31 56,84	14 47 15,78	50 15 46,3	80 44 24,5
8		14 31 56,80	14 46 21,15	50 13 45,4	50 43 59,3
10		14 31 56,30	14 44 33,20	50 15 43,5	50 43 8,5
11		14 31 56,30	14 43 39,58	50 15 49,2	50 43 6,6
12		14 31 55,86	14 42 45,80	50 15 48,1	50 43 9,0
13		14 31 55,94	14 41 53,34	50 15 49,7	50 43 2,6

Fig. 10.84 Observations of Ceres by Carlini in April and May 1807

Among the observations from Paris for 1804, printed in the Connaissance de tems for 1808, are two observations on Ceres, which have, as we believe, not been publicised and from which the following positions for Ceres can be obtained:

1804	Mean time in Paris	Apparent RA of Ceres	Apparent south Declin. of Ceres
Sept. 13	13h 20′ 14.″8	12° 52′ 11.″1	11° 23′ 55.″1
14	13 15 37.1	12 41 41.6	11 30 18.4

Due to the lovely correspondence between the sky and Dr. Gauss' elements XI this year's rediscovery of this celestial body, if faint, will not be difficult.

Acknowledgments

Thanks must first go to the translation experts who provided me with such a vast amount of material to work with: Telse Wokersien (German and French text), Dr. John Ramsey and Christopher Gordon (Latin text), Marcin Sawicki (Polish text), Piero Sicoli (Italian text). Additional linguistic assistance was provided by Lise Jobin and the famous author and impresario David Leddick, who generously wrote an artistic preface.

Thanks to the archivists at the Paris Observatory, and the Academy of Science in Paris, who were a great help during my visit there in 2002; to Adam Perkins of Cambridge University for supplying me with the Nevil Maskelyne material; to Dr. Owen Gingerich for access to the Herschel archives on microfilm many years ago; Raza Ansari for the Persian map of the solar system; and to Sharon Hanna, librarian at NRC, Victoria for a great deal of archival research. Thanks also to Earl Ogden for his expertise on electronic file issues, and Guy Ottewell for the 3D plot. Finally to Dr. Roger Ceragioli for his linguistic expertise including the proper printed way to spell Johann Schroeter, as he himself rarely, if ever, signed his name using the o with an umlaut.

I began archival research for this four-volume work in 1989, at which time I met with Dr. Peter Brosche at the University of Bonn (the leading expert on Baron von Zach) and Dr. Viktor Shor of the Russian Academy of Sciences in St. Petersburg. Both have been a great assistance to me throughout the years. Regarding Dr. Shor, I can only say that the honour he conferred upon me by writing the foreword to this book is one I will always cherish. Portions of this book are based on my unpublished 2014 PhD thesis at the University of Southern Queensland.

C. Cunningham, *Early Investigations of Ceres and the Discovery of Pallas*,
DOI 10.1007/978-3-319-28815-4

Appendix A: Positional data on Ceres and Pallas for 1801 and 1802

PLACE OF OBS. PALERMO	(1.0)		(PIAZZI 1845, VON ZACH 1801)			
YEAR MO DAY	JULIAN DATE (UT)	APP. R.A. H M S	(O-C) ARCS.	APP. DECL. D M S	(O-C) ARCS.	DIST.
1801 1 11.3	2378872.29783	3 25 30.41	+ 5.3	(+16 14 36.0	+ 0.0)	2.041
01 1 14.3	8875.28966	3 25 32.31	+ 1.5	+16 27 7.3	+ 1.3	2.074
01 1 19.3	8880.27643	3 26 8.83	- 1.6	+16 49 17.6	+ 0.9	2.131
01 1 21.3	8882.27127	3 26 35.01	+ 0.4	+16 58 37.4	+ 2.0	2.154
01 1 23.3	8884.26619	3 27 7.44	+ 1.4	+17 8 6.9	- 1.2	2.178
01 1 28.3	8889.25378	3 28 55.37	+ 4.1	+17 32 55.5	- 0.3	2.238
01 1 30.2	8891.24894	3 29 48.69	+ 1.7	+17 43 12.4	+ 1.1	2.263
01 1 31.2	8892.24654	3 30 17.76	+ 3.9	+17 48 22.8	- 0.3	2.275
1801 2 2.2	2378894.24180	3 31 19.62	+ 1.0	+17 58 58.8	+ 4.7	2.300
1802 2 22.6	2379279.56131	12 30 36.13	+ 1.2	(+14 40 20.0	-11.1)	1.665
02 2 27.5	9284.54581	12 27 56.05	- 1.6	+15 15 41.8	+ 6.5	1.640
02 2 28.5	9285.54266	12 27 19.98	- 1.1	+15 22 37.8	+ 3.5	1.636
02 3 2.5	9287.53633	12 26 3.95	- 1.3	+15 36 31.1	+ 5.2	1.628
02 3 4.5	9289.52994	12 24 43.07	- 2.3	+15 50 10.9	+ 5.5	1.621
02 3 9.5	9294.51374	12 21 2.73	- 2.9	+16 22 55.4	+ 4.4	1.608
02 3 11.5	9296.50720	12 19 28.74	+ 0.8	+16 35 6.9	- 5.5	1.605
02 3 12.5	9297.50391	12 18 40.60	+ 1.6	+16 41 16.0	+ 4.5	1.604
02 3 14.5	9299.49731	12 17 2.17	- 1.3	+16 52 55.4	+11.5	1.602
02 3 19.5	9304.48074	12 12 48.85	- 0.2	+17 19 51.6	+ 7.2	1.603
02 3 24.5	9309.46411	12 8 30.97	+ 3.6	+17 40 13.1	+10.9	1.610
02 3 26.5	9311.45745	12 6 47.83	- 1.6	+17 47 13.2	+ 8.3	1.614
02 3 27.5	9312.45414	12 5 56.75	- 2.3	+17 50 19.3	+ 3.1	1.617
02 3 31.4	9316.44090	12 2 36.77	+ 0.3	+18 0 46.6	+ 4.6	1.630
02 4 1.4	9317.43761	12 1 47.92	- 1.4	+18 2 47.2	+ 4.5	1.634
02 4 2.4	9318.43432	12 0 59.77	- 2.3	(+18 4 32.0	+ 3.3)	1.638
02 4 4.4	9320.42778	11 59 25.64	- 4.5	+18 7 18.7	+ 2.2	1.647
02 4 6.4	9322.42127	11 57 55.21	- 1.0	+18 9 7.7	+ 3.0	1.657
02 4 7.4	9323.41804	11 57 11.24	- 1.0	+18 9 37.7	+ 1.1	1.663
02 4 8.4	9324.41481	11 56 28.41	+ 1.3	+18 9 53.8	+ 0.3	1.668
02 4 10.4	9326.40839	11 55 5.36	- 1.2	+18 9 42.3	- 0.8	1.680
02 4 15.4	9331.39258	11 51 58.30	+ 0.6	+18 5 6.8	+ 4.3	1.713
02 4 16.4	9332.38946	11 51 24.52	+ 0.0	+18 3 29.1	+ 5.3	1.721
02 4 21.4	9337.37409	11 48 55.86	+ 1.5	+17 51 56.8	+11.0	1.760
02 4 26.4	9342.35912	11 47 1.95	- 0.3	+17 34 50.6	+ 6.7	1.803
02 4 27.4	9343.35618	11 46 43.64	+ 0.9	+17 30 44.0	+ 1.3	1.812
02 4 28.4	9344.35326	11 46 26.88	+ 3.1	+17 26 35.4	+ 5.8	1.822
02 4 30.3	9346.34746	11 45 57.28	- 0.6	+17 17 34.1	+ 5.3	1.841
02 5 1.3	9347.34458	11 45 44.86	- 0.6	+17 12 46.9	+ 5.4	1.851
02 5 2.3	9348.34173	11 45 34.00	+ 0.2	+17 7 47.8	+ 4.7	1.861
02 5 5.3	9351.33326	11 45 10.41	+ 1.9	+16 51 51.1	+ 6.8	1.891
02 5 6.3	9352.33047	11 45 5.17	- 3.0	+16 46 7.4	+ 3.2	1.902
02 5 7.3	9353.32770	11 45 2.15	+ 2.6	+16 40 18.4	+ 4.1	1.912
02 5 10.3	9356.31950	11 45 0.63	+ 0.1	+16 21 53.1	+ 5.4	1.945
02 5 12.3	9358.31411	11 45 6.80	- 3.2	+16 8 48.7	+ 3.7	1.967
02 5 16.3	9362.30354	11 45 37.14	+ 3.8	+15 41 5.7	+ 6.0	2.012
1802 5 22.3	2379368.28814	11 47 2.65	+ 1.1	+14 55 40.3	+ 4.2	2.083

PLACE OF OBS. GREENWICH	(1.2)			(MASKELYNE 1811)		
YEAR MO DAY	JULIAN DATE (UT)	APP. R.A. H M S	(O-C) ARCS.	APP. DECL. D M S	(O-C) ARCS.	DIST.
1802 2 12.6	2379269.62800	12 34 2.10	+ 1.0	+13 33 7.2	+ 4.8	1.731
02 2 19.6	9276.60740	12 31 53.38	- 2.4	+14 20 0.0	+ 5.4	1.682
02 3 6.6	9291.56047	12 23 16.38	- 0.3	+16 3 48.6	+ 5.7	1.615
02 3 14.5	9299.53429	12 17 0.45	+ 0.5	+16 52 59.6	+ 3.2	1.602
02 3 18.5	9303.52103	12 13 38.19	- 1.3	+17 14 10.2	+ 5.7	1.602
02 3 25.5	9310.49776	12 7 37.36	+ 0.6	+17 43 49.3	+ 1.3	1.612
02 4 6.5	9322.45825	11 57 53.62	- 0.2	+18 9 9.6	+ 3.4	1.658
02 4 21.4	9337.41108	11 48 54.84	+ 0.9	+17 51 40.6	+ 1.2	1.760
02 5 1.4	9347.38157	11 45 44.28	- 2.7	+17 12 33.9	+ 3.3	1.851
02 5 4.4	9350.37306	11 45 16.42	+ 0.1	+16 57 6.8	+ 4.6	1.881
02 5 11.4	9357.35380	11 45 3.28	+ 0.8	(+16 15 10.0	+ 3.8)	1.956
1802 5 13.3	2379359.34844	11 45 12.61	+ 0.1	+16 1 47.7	+ 2.0	1.978

Fig. A1 Positional observations of Ceres

C. Cunningham, *Early Investigations of Ceres and the Discovery of Pallas*,
DOI 10.1007/978-3-319-28815-4

PLACE OF OBS.		PARIS	(1.0)		(BOUVARD 1821, LE VERRIER 1858)		
		(UT)	H M S	ARCS.	D M S	ARCS.	
1802	2 26.6	2379283.57946	12 28 30.00	+ 1.7	+15 8 50.0	+ 2.1	1.644
02	2 27.6	9284.57632	12 27 54.92	- 2.2	+15 15 52.5	+ 4.3	1.640
02	3 5.6	9290.55723	12 23 59.91	+ 1.0	+15 57 5.6	+ 4.2	1.618
02	3 6.6	9291.55400	12 23 16.49	- 2.8	+16 3 42.3	+ 2.0	1.615
02	3 26.5	9311.48796	12 6 46.23	- 2.0	+17 47 7.7	- 3.3	1.614
02	3 27.5	9312.48464	12 5 55.19	- 2.4	+17 50 32.1	+10.3	1.617
02	3 28.5	9313.48132	12 5 4.52	- 2.1	+17 53 15.4	- 3.6	1.620
02	4 7.4	9323.44854	11 57 9.89	- 1.3	+18 9 31.1	- 6.2	1.663
02	4 8.4	9324.44532	11 56 27.01	- 0.1	+18 9 51.4	- 2.4	1.668
02	4 12.4	9328.43253	11 53 45.87	+ 0.5	+18 8 23.3	- 9.2	1.693
02	4 13.4	9329.42937	11 53 8.34	- 0.9	+18 7 25.7	-10.1	1.700
02	4 15.4	9331.42309	11 51 56.92	- 4.1	+18 4 59.8	+ 0.1	1.713
02	4 17.4	9333.41686	11 50 50.95	- 2.2	+19 1 25.9	- 1.6	1.728
02	4 19.4	9335.41070	11 49 49.97	- 4.6	+17 56 54.1	- 6.5	1.744
02	4 24.4	9340.39557	11 47 42.46	- 0.3	+17 42 7.8	+ 4.2	1.785
02	4 27.4	9343.38670	11 46 42.92	- 1.7	+17 30 25.9	- 9.2	1.813
02	4 30.4	9346.37797	11 45 56.52	- 5.8	+17 17 14.3	- 5.9	1.841
02	5 3.4	9349.36940	11 45 24.29	- 0.4	+17 2 21.3	- 3.1	1.871
02	5 5.4	9351.36378	11 45 10.03	- 1.1	+16 51 31.4	- 2.6	1.891
02	5 7.4	9353.35822	11 45 1.87	- 0.3	+16 39 59.0	- 4.4	1.913
02	5 10.4	9356.35002	11 45 0.61	- 1.0	+16 21 35.8	- 0.2	1.945
02	5 11.3	9357.34732	11 45 3.06	- 2.0	+16 15 10.7	+ 2.0	1.956
1802	5 15.3	2379361.33668	11 45 27.65	+ 0.9	+15 47 54.2	- 0.8	2.001
PLACE OF ORS.		PALERMO	(1.1)		(PIAZZI 1845)		
		(UT)	H M S	ARCS.	D M S	ARCS.	
1803	6 29.5	2379771.47479	18 45 25.96	- 0.7	-28 11 5.0	+ 1.0	1.896
03	7 1.5	9773.46799	18 43 29.44	- 0.3	-28 19 20.4	+ 1.4	1.897
03	7 2.5	9774.46458	18 42 30.96	- 1.1	-28 23 27.3	- 4.0	1.897
03	7 3.5	9775.46118	18 41 32.37	- 3.4	-28 27 18.2	+ 2.0	1.898
03	7 4.5	9776.45777	18 40 34.03	- 2.9	-28 31 17.6	- 5.1	1.900
03	7 5.5	9777.45437	18 39 35.74	- 3.4	-28 35 2.0	- 1.9	1.901
1803	7 6.5	2379778.45097	18 38 37.83	- 1.1	-28 38 43.9	- 1.1	1.903
1809	11 11.4	2382098.43208	2 38 9.09	- 0.2	+ 4 44 49.8	- 0.3	1.841
1809	11 12.4	2382099.42872	2 37 14.65	- 1.0	+ 4 44 8.2	- 1.2	1.843

Positional observations of Ceres and Pallas for 1801–1802. All the original observations have been put in standard modern notation with a Julian date and distance in AU. Tables are from Schubart (1976).

PLACE OF OBS. GREENWICH (1.2) (MASKELYNE 1811)

YEAR	MO	DAY	JULIAN DATE (UT)	APP. R.A. H M S	(O-C) ARCS.	APP. DECL. D M S	(O-C) ARCS.	DIST.
1802	4	22.4	2379338.42039	12 5 53.40	- 0.2	(+18 21 48.0	+ 5.6)	1.562
02	4	23.4	9339.41715	12 5 34.90	+ 0.5	+18 32 27.5	+ 4.6	1.572
02	4	25.4	9341.41132	12 5 2.03	+ 1.3	+18 52 29.2	+ 1.1	1.594
02	4	26.4	9342.40841	12 4 47.50	- 1.2	+19 2 2.0	+ 8.8	1.605
02	5	1.4	9347.39419	12 3 57.46	+ 0.1	+19 42 53.7	- 0.1	1.664
02	5	2.4	9348.39139	12 3 51.89	+ 1.0	+19 49 55.9	+ 0.8	1.676
02	5	4.4	9350.38585	12 3 44.94	- 0.3	+20 2 52.5	+ 2.9	1.700
02	5	7.4	9353.37766	12 3 45.73	+ 0.2	+20 19 29.0	+ 1.8	1.738
02	5	9.4	9355.37229	12 3 53.50	- 0.2	+20 28 41.5	- 5.9	1.764
02	5	11.4	9357.36699	12 4 7.06	- 0.1	+20 36 47.2	- 0.3	1.790
02	5	13.4	9359.36176	12 4 26.35	+ 0.9	+20 43 32.1	+ 1.2	1.817
02	5	14.4	9360.35916	12 4 38.10	+ 1.6	+20 46 25.6	+ 0.7	1.831
02	5	16.4	9362.35402	12 5 5.85	+ 4.8	+20 51 17.8	- 1.5	1.858
1802	5	17.4	2379363.35147	12 5 21.11	- 2.6	(+20 53 17.0	- 3.5)	1.872

PLACE OF OBS. PARIS (1.0) (BOUVARD 1821, LE VERRIER 1858)

			(UT)	H M S	ARCS.	D M S	ARCS.	
1802	4	12.4	2379328.44390	12 10 10.49	- 1.2	+16 10 53.0	- 7.5	1.467
02	4	13.4	9329.44081	12 9 39.37	- 0.4	+16 26 1.0	- 1.6	1.475
02	4	17.4	9333.42858	12 7 46.26	- 1.9	+17 21 43.8	- 3.4	1.511
02	4	19.4	9335.42256	12 6 57.39	+ 1.1	+17 46 59.7	- 1.9	1.531
02	4	22.4	9338.41363	12 5 53.57	+ 0.5	+18 21 31.5	- 6.6	1.562
02	4	24.4	9340.40776	12 5 17.84	+ 0.4	+18 42 45.4	+11.2	1.583
02	4	27.4	9343.39907	12 4 34.78	+ 0.6	+19 10 46.0	- 4.2	1.617
02	4	30.4	9346.39052	12 4 4.26	- 5.0	+19 35 21.6	- 4.8	1.652
02	5	3.4	9349.38214	12 3 47.83	+ 2.2	+19 56 29.2	- 1.9	1.688
02	5	5.4	9351.37664	12 3 43.77	+ 0.3	+20 8 35.8	- 5.6	1.713
02	5	7.4	9353.37120	12 3 45.71	+ 0.2	+20 19 19.6	- 5.6	1.738
02	5	10.4	9356.36317	12 3 59.43	- 1.4	+20 32 54.3	- 1.4	1.777
02	5	11.4	9357.36052	12 4 7.10	+ 1.2	+20 36 42.2	- 3.9	1.790
02	5	12.4	9358.35789	12 4 16.16	+ 3.5	+20 40 15.0	- 2.3	1.804
1802	5	15.4	2379361.35011	12 4 50.95	- 0.4	+20 49 1.4	+ 1.5	1.844

PLACE OF OBS. SEEBERG (2.0) (VON ZACH 1802)

			(UT)	H M S	ARCS.	D M S	ARCS.	
1802	4	4.4	2379320.44579	12 14 56.57	- 3.1	(+13 54 53.5	+ 4.8)	1.407
02	4	5.4	9321.44262	12 14 17.71	- 5.0	(+14 13 20.2	+ 4.9)	1.414
02	4	7.4	9323.43628	12 13 2.70	- 3.6	(+14 48 59.3	+ 4.4)	1.427
02	4	8.4	9324.43314	12 12 26.65	- 0.4	(+15 6 12.8	+ 5.6)	1.435
02	4	15.4	9331.41142	12 8 41.23	+ 1.1	(+16 54 33.7	+ 5.6)	1.493
02	4	18.4	9334.40233	12 7 22.17	+ 5.5	(+17 34 25.3	+ 5.4)	1.521
02	4	19.4	9335.39931	12 6 57.82	- 0.5	(+17 46 50.0	+ 5.4)	1.531
02	4	24.4	9340.38451	12 5 18.65	+ 6.5	(+18 42 25.2	+ 5.0)	1.583
02	4	25.4	9341.38160	12 5 2.28	- 1.5	(+18 52 15.9	+ 5.0)	1.594
02	4	26.4	9342.37670	12 4 48.25	+ 3.7	(+19 1 41.6	+ 4.9)	1.605
02	4	27.4	9343.37582	12 4 35.17	+ 2.1	(+19 10 42.7	+ 4.9)	1.616
02	4	29.4	9345.37010	12 4 13.24	- 2.5	(+19 27 31.5	+ 4.5)	1.639
02	4	30.4	9346.36728	12 4 4.79	- 0.2	(+19 35 19.6	+ 3.9)	1.651
02	5	1.4	9347.36447	12 3 57.63	- 0.2	(+19 42 44.8	+ 3.9)	1.663
02	5	2.4	9348.36167	12 3 51.59	- 5.3	(+19 49 46.5	+ 3.6)	1.675
02	5	3.4	9349.35889	12 3 47.84	+ 1.2	(+19 56 25.9	+ 3.9)	1.688
02	5	5.4	9351.35338	12 3 43.77	+ 0.1	(+20 8 36.2	+ 2.8)	1.712
02	5	6.4	9352.35065	12 3 43.74	- 3.2	(+20 14 8.8	+ 2.4)	1.725
02	5	7.3	9353.34794	12 3 45.72	+ 1.1	(+20 19 20.4	+ 2.2)	1.738
02	5	8.3	9354.34526	12 3 48.80	+ 0.4	(+20 24 10.8	+ 1.7)	1.751
1902	5	11.3	2379357.33727	12 4 6.91	+ 1.3	(+20 36 41.9	+ 1.0)	1.790

Fig. A2 Positional observations of Pallas

Appendix B: The Elements of Ceres by Encke

*By Professor Encke (*1831*)*

Since the completion of the first calculations for newly determining the orbit of Ceres, one of my respected astronomical friends has given me the hope that the investigations on this subject will be more completely and more accurately performed by another hand. It will therefore be sufficient in this place to explain the ground-work of my determination, in order the better to form an estimate of the confidence to which the places derived from it are entitled.

The perturbations were developed in the same manner as for the other small planets, in regard to the elements themselves, and not to the places of the planets in space. A review of the last determination of Professor Gauss (Zach's Monthly Correspond. *1809, May) on which all places of the planet hitherto given were founded, and some trials made at the latest oppositions, seem to prove sufficiently that the equations for the perturbations, if developed as is usual for the old planets, would require to be extended considerably beyond the first power of the eccentricity, if great accordance is intended. In the same proportion, however, the calculation of a single place would have become irksome, even taking into consideration the facility afforded by the excellent construction of the tables of perturbations (Zach's* Corresp. *1803, March); and therefore, even if every part had already been perfectly developed, still this method would hardly have deserved the preference on the score of brevity of computation.*

As an interval of time, the number of one hundred days was selected for this first approximation, and only the attraction of Jupiter was taken into account. The mass of this planet was taken, according to Nicolai, at 1/1053.924.

This value, which is one-eightieth part more than the old determination by Laplace, appears in the cases of Pallas, Juno, and Vesta, to agree better with the observations, and therefore seems likewise for Ceres to deserve the preference. The four oppositions necessary for deducing the elements were found to be, from the observations published, as follows (Fig. A3):

And proceeding from the elements at the moment of the first opposition, the computations of the perturbations for the following ones gave the following corrections of

C. Cunningham, *Early Investigations of Ceres and the Discovery of Pallas*,
DOI 10.1007/978-3-319-28815-4

☍ ⚳	Mean Time at Göttingen.	Heliocentric Longitude.	Geocentric Latitude.
	h ′ ″	° ′ ″	° ′ ″
1820 Jan. 25	3 43 10	124 38 29·6	+11 58 35·2
1821 May 22	5 43 47	241 12 36·4	+ 5 41 46·0
1822 Aug. 22	8 28 23	329 5 15·6	−14 53 14·6
1825 Mar. 14	11 5 56	174 4 50·4	+17 10 32·9

Fig. A3 The four required oppositions

Correction.	1821. May 21·5.	1822. Aug. 22.	1825. March 14.
Mean Longitude ... = L =	− 1′ 30″·3	+ 2′ 36″·0	+ 1′ 36″·8
Long. of Perihelion = ϖ =	− 16 17·2	−32 47·9	−107 55·7
Node................ = ☊ =	− 50·6	− 41·9	− 2 20·3
Inclination......... = i =	+ 2·4	+ 8·3	− 19·1
Angle of Eccentricity = φ =	+ 12·5	+ 1 12·9	− 1 9·5
Daily Sider. Motion = μ =	+ 0·19072	+ 0·70887	+ 0·16263

Fig. A4 Corrections of the elements

the elements, in which however the precession is still to be added to all the longitudes (Fig. A4).

These determinations require, perhaps, a repetition, being calculated with elements which give for the single oppositions places erroneous by fifteen minutes. For this very reason I did not deem it necessary to produce a perfect accordance of the elements with the oppositions, but was satisfied with such as gave errors in longitude less than 3″. The elements thus deduced, and true for the moment of the epoch, the longitude being referred, for the sake of agreement with the other small planets, to the mean equinox of 1810, are as follows:

Elements of Ceres.
Mass of Jupiter 1/1053.294
Epoch 1822. Jan. 22. 0h mean time at Goettingen.
$L = 127° 36' 51''.6$
$\pi = 147^{o}\ 36\ 57'\ 6\ S$ *Mean equinox 1810.*
$\Omega = 80\ 41\ 55.0$
$i = 10\ 38\ 7.7$
$\varphi = 4\ 31\ 18.0$
$\mu = 770.72468$ *(sidereal).*

A rigorous comparison with the geocentric observations at the times of the above four oppositions, has presented the following differences (Fig. A5).

Date.			Right Ascension.	Declin.	Place of Observation.
1820.	Jan.	31	−5″·8	+1″·1	Mannheim.
	Febr.	2	−6·2	−1·7	——
1821.	May	16	+5·1	−0·6	Königsberg.
		23	+0·1	+0·5	——
		28	+1·9	−0·9	——
1822.	Aug.	18	−4·7	+7·0	——
		19	−5·6	+2·0	——
		22	−3·0	+8·4	——
		23	−4·8	+6·0	——
1825.	March	9	−4·3	−0·3	Göttingen.
		10	−6·0	+1·2	——
		18	−1·7	+0·1	——
		19	−1·9	+0·4	——
		20	−3·0	+2·1	——

Fig. A5 Differences in the observations

1827.		Mean Time at Göttingen.	Rt. Ascension ⚳	Declination ⚳
Sept.	27	12h 11′ 28″·7	8° 50′ 36″·1	−13° 17′ 6″·1
Oct.	3	11 42 50·2	7 34 37·7	41 34·8
	4	38 3·7	7 21 56·4	45 6·0
	5	33 17·7	7 9 22·8	48 24·9
	6	28 31·9	6 56 53·0	51 37·1

Fig. A6 Observations of Ceres by Gauss in 1827

The two subsequent oppositions of 1827 and 1829 served as a test of the accuracy of the elements. For the former one Prof. Gauss had the kindness to communicate to me the following observations (Fig. A6). Hence the oppositions of Ceres will be deduced as follows (Fig. A9):

For the opposition of 1829, I received the following excellent observations from Professors Schwerdt at Speyer [in Bavaria] *and* [Friedrich] *Argelander at Abo* [in Finland], *which are the more creditable as at the time of that opposition it was difficult to find this planet (Fig. A7):*

The computations of the perturbations, taken in the same sense as above, gave the following corrections of the elements (Fig. A8).

Opposition Ceres 1827. Sept. 26. 9h 30′ 45″. Mean time at Goettingen.

Heliocentric longitude... 2° 58′ 19″.5

Geocentric latitude −15 41 56 .4

Opposition Ceres 1829. Jan. 1. 4h 8′ 47″. Mean time at Goettingen.

Heliocentric longitude... 101° 3′ 13″.5

Geocentric latitude +5 56 5.5

The mean geocentric errors in all six oppositions are consequently (Fig. A10*):*

1828–1829.	Mean Time.	Right Ascension ⚳	Declination ⚳	Place of Observation
Dec. 23	12h 49′ 20″	104° 47′ 50″·3	+28° 6′ 2″·0	Åbo.
28	24 42	103 32 53·6	34 37·0	—
29	19 44	103 17 25·6	40 7·2	—
Jan. 13	11 5 16	99 25 0·0	29 51 35·0	Speyer.
15	10 55 29	98 56 13·5	59 12·0	—

Fig. A7 Observations from the opposition of 1829

Correction.	1827. Sept. 26.	1829. Jan. 17.
Δ L	+ 5′ 8″·4	+ 2′ 51″·1
Δ π	−99 33·5	−94 28·6
Δ ☊	− 3 16·9	− 3 35·7
Δ i	− 24·8	− 20·4
Δ φ	− 3 30·4	− 4 27·4
Δ μ	−0·11458	+0·03944

From which we obtain the following comparison:

1822.

Fig. A8 Corrections of the elements

Date.	Right Ascen. ⚳	Declination ⚳	Place of Observation.
1827. Sept. 27	− 2″·1	+ 1″·4	Göttingen.
Oct. 3	− 3·5	+ 0·8	——
4	0·0	+ 0·6	——
5	− 1·9	− 1·6	——
6	− 4·9	− 0·4	——
1828. Dec. 23	−25·6	−11·5	Åbo.
28	−28·3	−11·7	——
29	−27·0	−10·8	——
1829. Jan. 13	−26·4	− 9·3	Speyer.
15	−26·4	−10·6	——

Fig. A9 Data from 1827, 1828 and 1829

where the last somewhat more considerable difference answers to heliocentric errors of 13″ in longitude, and 7″ in latitude; so that it is to be hoped, even if these errors are chargeable to the elements only, and not, perhaps, also partly to the perturbation caused by Saturn and Mars, which have been neglected, that these errors will not render the finding of Ceres difficult for the approaching years, until the orbit shall have been more accurately determined.

	Right Ascen. ⚳	Declination ⚳
	″	″
1820	− 6·0	− 0·3
1821	+ 2·4	− 0·3
1822	− 4·5	+ 5·9
1825	− 3·4	+ 0·9
1827	− 2·5	+ 0·2
1829	−26·7	−10·8

Fig. A10 The mean geocentric errors

The early development of the perturbations of Ceres, which was almost contemporary with the determination of the orbit itself, and the certainty, thereby obtained, of always being, for the future, sure of its position within ten or fifteen minutes of a degree, would appear to have been the cause that this planet, the first-discovered of the small ones, has been least observed in recent years; – at any rate I have not succeeded in obtaining accurate observations of Ceres at the times of its opposition, even since the period that the oppositions of the other small planets have been regularly observed by meridian instruments in German observatories. It is possible that some oppositions have really been entirely neglected. It is the more desirable that the future oppositions should not pass by unnoticed, as Ceres might likewise afford additional means of determining the mass of Jupiter, or might assist in answering the question which has lately been agitated, Whether for all planets, the attraction is rigorously proportional to their mass? [This is the end of the paper by Encke.]

Further to Encke's work on the asteroids at this time, we have two contributions from British periodicals, both of which involve *White's Ephemeris*, disparagingly described at the time as "the especial favourite of all *small* amateurs of astronomy, who may possess a *small* equatorial, a *small* 'achromatic Dollond,' a pocket chronometer by the 'elder Arnold,' and a *small* pair of Carey's old globes, in a *small* attic observatory; whither they love to retire, when all their neighbours are going to bed; with serenity in their hearts, and *White* in their hand." (*The Magazine of Popular Science* 1836, vol. 2, 430.) It was begun by Robert White in 1750; he died in 1773, but it continued for many years. In 1831 it became *White's celestial atlas*, edited by Olinthus Gregory. A valuable insight into how people received information about the positions of the asteroids during this era comes from a letter to the editor of *The Mechanics' Magazine* (1830b):

> *An esteemed correspondent reminds us that for general, and even nautical purposes, we need not have recourse to 'Encke's Ephemeris' to supply the deficiencies of the 'Nautical Almanac' with regard to the new planets, Pallas and Ceres. He adds: 'There is a regular ephemeris of them in 'White's Ephemeris' for every 6th day in the year, and, indeed, for every day near their respective oppositions. You will also find that even the renowned astrologer, Francis Moore, gives the southing and declination of each of these four interesting bodies, for five days near their respective oppositions, and specifies the fixed stars which*

they will then be near. For example, he states, that on April 30th, Ceres in opposition will be 'about 3° east of the star μ, on the right foot of Virgo, and very near that marked 16 Librae.' The correspondent to The Times should have specified that the right ascensions and declinations of Ceres and Pallas in 'Encke' are for the midnights, not the noons of the respective days.

The second periodical is the *Monthly Notices of the Astronomical Society of London* (Dec. 11, 1829, no. 23, 143–144), which published this:

> *The next communication was a letter from Dr. Gregory, pointing out some errors in the places of the planet Ceres, at p 36 of* White's Ephemeris *for the year 1830. Dr. Gregory states, that the elements of Ceres having received farther correction from recent observations, the right ascensions and declinations of that planet, at and near the opposition in 1830, will be nearly as follow:–*

		RA			*Dec.*		
April 28,	*noon*	*14h*	*46m*	*17s*	*–4°*	*12′*	*54″*
29,	—	*14*	*45*	*23*	*4*	*11*	*31*
30,	*opp.*	*14*	*44*	*28*	*4*	*10*	*13*
May 1,	*noon*	*14*	*43*	*34*	*4*	*9*	*1*
2,	—	*14*	*42*	*39*	*–4*	*7*	*55*

The geocentric places of Ceres, for every day at midnight (Berlin time), from April 14th to May 16th, are given at p. 248 of Encke's Jahrbuch *for 1831.* Encke's Ephemeris, and his study of the asteroids, will be examined in detail in the next book in this series.

Fig. A.11 Ceres riding her chariot, a commemorative illustration from 1901, celebrating the 100th anniversary of Piazzi's discovery

References

Abbadie, J. (1692). *L'Art de se connoitre soy-meme. Chez Pierre vander Slaert, Rotterdam* [English edition: 1695, *The art of knowing one-self: Or, an enquiry into the sources of morality*]. Oxford: Leonard Lichfield.

A Journal of Natural Philosophy 4, 120–128 (1803a, January).

A Journal of Natural Philosophy 4, 142–147 (1803b, February).

Althaus, H. (2000). *Hegel: An intellectual biography*. Cambridge: Polity.

Andrewes, L. (1657). *Aposmasmatia sacra: Or a collection of posthumous and orphan Lectures: Delivered at St. Paul's and St. Giles his Church*. London: R. Hodgkinsonne.

Anon. (1837). Essay on astronomy. *The Naval Magazine, 2*(2), 119–129.

Ansari, R. (2002). European astronomy in Indo-Persian writings. In S. M. R. Ansari (Ed.), *History of oriental astronomy* (pp. 133–144). Dordrecht: Springer Science.

Arago, M. (1871). Herschel. In *Annual report of the Board of Regents of the Smithsonian Institution.*

Aspray, W., & Kitcher, P. (Eds.). (1988). History of philosophy of modern mathematics. In *Minnesota Studies in the Philosophy of Science* (Vol. XI). Minneapolis: University of Minnesota Press.

Aubert, A. (1786, June 19). Letter to William Herschel. RAS, W1/13, A3.

Balinski, M. (1865). *Pamiętniki o Janie Sniadeckim* (*Memoirs about Jan Sniadecki*), two volumes. Vilnius: J. Zawadzki.

Ball, J. (1985). *Paul and Thomas Sandby, Royal Academicians*. Bath: Charles Skilton.

Banks, J. (1792, February 27). Letter to Jean Florimond Boudon de Saint-Amans. Dawson Turner Collection VIII. London: 13, Natural History Museum.

Banks, J. (1802, June 7). Letter to William Herschel. RAS, B.42.

Banks, J. (1807, June 3). Letter to William Herschel. RAS, B.47.

Bardi, J. (2006). *The calculus wars: Newton, Leibniz, and the greatest mathematical clash of all time*. New York: Thunder's Mouth Press.

Becker, C. (1932). *The heavenly city of the eighteenth century philosophers*. New Haven: Yale University Press.

Bell, E. T. (1937). *Men of mathematics*. New York: Simon & Schuster.

Benichou, P. (1948). *Morales du grand siècle*. Paris: Gallimard.

Bennett, J. A. (1976). "On the power of penetrating into space": The telescopes of William Herschel. *Journal for the History of Astronomy, 7*, 75–108.

Berger, H. (2000). *Fictions of the pose*. Stanford: Stanford University Press.

Bermingham, A. (2005). *Sensation and sensibility*. New Haven: Yale Center for British Art.

Beuermann, K. (2005). Die Herschels—eine hannoveranische Astronomenfamilie in England. In E. Mittler (Ed.), *Eine Welt allein ist nicht genug: Grossbritannien, Hannover und Göttingen* (pp. 245–259). Göttingen: Niedersächische Staats- und Universitätsbibliothek Göttingen.

C. Cunningham, *Early Investigations of Ceres and the Discovery of Pallas*,
DOI 10.1007/978-3-319-28815-4

Biancani, G. (1615). De *Mathematicarum Natura dissertatio*, an appendix in *Aristotelis Loca Mathematica* [English version in: Mancuso, P. (1995), *Philosophy of mathematics and mathematical practice in the seventeenth century*]. Oxford: Oxford University Press.
Bigourdan, G. (1887). Histoire des observatoires de l'Ecole Royale Militaire. *Bulletin Astronomiqe, 4*, 497–504. and 5, 30–40.
Blanning, T. C. W. (1977). "That Horrid Electorate" or "Ma Patrie Germanique"? George III, Hanover, and the Fürstenbund of 1785. *The Historical Journal, 20*(2), 311–344.
Bode, J. (1772). *Anleitung zur Kenntnis des gesirnten Himmels* [Introduction to the knowledge of the starry heavens] (2nd ed.). Hamburg: Dieterich Harmsen.
Bode, J. (1802a, August 5). Letter to Herschel. RAS, B.119.
Bode, J. (1802b). *Dem neuen zwischen Mars und Jupiter entdeckten achten Hauptplaneten des Sonnensystems [The new planet between Mars and Jupiter]*. Berlin: Himburgische Buchhandlung.
Bode, J. (1802c, May 4). Letter to Olbers. *Bremen University Archives*.
Bode, J. (1803). Ueber die in den Jahren 1801 und 1802 gemachten neuen entdeckungen im Planatensystem Der Sonne. *Der Gesselschaft Naturforschender Freunde zu Berlin Neue Schriften*, 147–160.
Bode, J. (1804). Aus einem Schreiben des Hrn. Hofrath Huth aus Frankfurt an der Oder vom 21 Sept. 1804. Berlin Astronomical Yearbook for 1807, pp. 265–267.
Bode, J. (1805). *Berliner Astronomische Jahrbuch* (p. 248).
Bolles, W. (1845). *An explanatory and phonographic pronouncing dictionary of the English language*. New London: Bolles & Williams.
Bolyai, F. (1851). *Kurzer Grundriss eines Versuchs*. No publisher given, M. Vaserhely, Romania.
Borden, A. R. (1982). *A comprehensive old-English dictionary*. Washington, DC: University Press of America.
Bourtembourg, R. (2012). Messier's missed discovery of Pallas in April 1779. *Journal for the History of Astronomy, 43*, 209–214.
Bovaird, T. (2015). Using the inclinations of Kepler systems to prioritize new Titius–Bode-based exoplanet predictions. *MNRAS, 448*(4), 3608–3627.
Brewster, D. (1802a, April 19). Notice concerning another NEW PLANET discovered by Dr. Olbers of Bremen, in March 1802. *The Edinburgh Magazine*, New Series, 287.
Brewster, D. (1802b, June 19). Further notices respecting the two NEW PLANETS, with some remarks tending to show that they cannot both belong to the Planetary System. *The Edinburgh Magazine*, New Series, 445–448.
Brewster, D. (1805, January 7). Letter to William Herschel. RAS, B.126 (2).
Brewster, D. (1806). *An examination of the letter addressed to Principal Hill*. Edinburgh. Published anonymously. See also Larder (1970) p. 303, footnote 35.
Brewster, D. (1823). Biographical account of Sir William Herschel, Knight Guelph. *The Edinburgh Philosophical Journal, 8*(16), 209–226.
Brougham, H. (1796). Experiments and observations on the inflection, reflection and colours of light. *Philosophical Transactions, 86*, 227–277.
Brougham, H. (1803). Observations on the two lately discovered celestial bodies. *Philosophical Transactions* 1802. *The Edinburgh Review* 1, October 1802 ... January 1803, second edition, 426–431. Printed by D. Willison for Archibald Constable, Edinburgh; and T.N. Longman & O. Rees, London.
Bryan, M. (1799). *A compendious system of astronomy*. London: H.L. Galabin.
Buchan, J. (2003). *Capital of the mind: How Edinburgh changed the world*. London: John Murray.
Bucke, C. (1823). *On the beauties, harmonies and sublimities of nature* (Vol. 3). London: G. & W.B. Whittaker.
Buehler, W. K. (1981). *Gauss: A biographical study*. New York: Springer.
Bujarski, G. T. (1972). Polish liberalism, 1815–1823. *The Polish Review, 17*(2), 3–37.
Bullynck, M. (2006). Aspects of 18th Century Mathematical Socialisation. http://www.kuttaka.org/Gauss1.pdf.

Burckhardt, J. (1802a, January 24). Astronomy. *Gazette nationale ou le Moniteur Universel, 26*, 495.

Burckhardt, J. (1802b, April). Astronomy. *Gazette nationale ou le Moniteur Universel, 26*, 944.

Burgess, J. P. (2008). *Mathematics, models and modality: Selected philosophical essays*. Cambridge: Cambridge University Press.

Burja, A. (1786). *Der selbstlernende Algebrist*. Berlin: Lagarde and Friedrich.

Burney, C., Sn. (1802a). Letter to Charles Burney Jr., 10 May (likely correct date is 3 May). Yale University Archives, Beinecke Rare Book and Manuscript Library, OSB MSS 3.

Burney, C., Sn. (1802b). Letter to Frances Crewe, 7 December. Yale University Archives, Beinecke Rare Book and Manuscript Library, OSB MSS 3.

Bynum, C. (2001). *Metamorphosis and identity*. New York: Zone Books.

Cacciatore, G., & Schiaparelli, G. (Eds.). (1874). *Corrispondenza Astronomica fra Giuseppe Piazzi e Barnaba Oriani*. Milan: Univ. of Hoepli.

Cajori, F. (1897). *A history of elementary mathematics with hints on methods of teaching*. New York: Macmillan.

Cann, C. (1828). *A scriptural and allegorical glossary to Milton's Paradise Lost*. London: Printed for the authoress.

Cantor, G. N. (1971). Henry Brougham and the Scottish methodological tradition. *Studies in History and Philosophy of Science, Part A, 2*(1), 69–89.

Chaikin, A. (2003, September 26). A Bizarro "Cometoid". *Sky & Telescope* (online article). http://www.skyandtelescope.com/news/3308121.html?page=1&c=y.

Chambers, R. (1856). *Biographical dictionary of Eminent Scotsmen*. Glasgow: Blackie & Son.

Chênedollé, C.-J. (1807). *Le genie de l'homme, poëme*. Paris: H. Nicolle.

Chesterton, G. K. (1913). *The Victorian age in literature*. London: Williams and Norgate.

Clark, R. N. (1990). *Visual astronomy of the deep sky*. Cambridge: Cambridge University Press.

Clerke, A. (1901). *The Herschels and modern astronomy*. London: Cassell & Co.

Conybeare, W. D. (1836). *An elementary course of theological lectures*. London: Sherwood, Gilbert and Piper.

Cooke, T. (1830). A discourse on the theory of the Planetary System. *The Imperial Magazine, 12*, 916–923.

Corry, J. (1802). *The detector of quackery; or, analyser of medical, political, dramatic, and literary imposture*. London: Hurst.

Cotton, R. W. (1892). Stephen Weston. *Notes and Gleanings, 5*(49–50), 6–9.

Couteau, P. (1981). *Observing visual double stars*. Cambridge: MIT Press.

Craig, H. (1936). *The enchanted glass: The Elizabethan mind in literature*. New York: Oxford University Press.

Crocker, L. (1959). *An age of crisis: Man & world in eighteenth century French thought*. Baltimore: Johns Hopkins Press.

Cross, J. (1983). Euler's contribution to potential theory 1730–1755 (p. 331). In *Leonhard Euler 1707–1783*. Basel: Birkhauser Verlag.

Crum, W. (1855). Sketch of the life and labours of Dr Thomas Thomson. *Proceedings of the Philosophical Society of Glasgow, 3*, 250–264.

Cunningham, C. J. (2003). Piazzi's deception. *Mercury, 32*(4), 8.

Cunningham, C. J. (2004). Discovery of the missing correspondence between Carl Friedrich Gauss and the Rev. Nevil Maskelyne. *Annals of Science, 61*(4), 469–481.

Cunningham, C. J. (2007). *The planetary chess match: Giuseppe Piazzi vs Europe*. Australia: Astronomy colloquia given at James Cook University.

Cunningham, C. J., & Orchiston, W. (2011). Who invented the word asteroid, William Herschel or Stephen Weston? *Journal of Astronomical History and Heritage, 14*(3), 230–234.

Cunningham, C. J., & Oestmann, G. (2013). Classical deities in astronomy, the employment of verse to commemorate the discovery of the planets Uranus, Ceres, Pallas, Juno and Vesta. *Culture and Cosmos, 17*(1), 3–29.

Cunningham, C. J., & Orchiston, W. (2015). The clash between William Herschel and the great "amateur" astronomer Johann Schröter. In W. Orchiston, D. Green, & R. Strom (Eds.), *New insights from recent studies in the history of astronomy: Following in the footsteps of F. Richard Stephenson* (pp. 205–222). New York: Springer.

Cunningham, C. J. (2015). Discovery of the Origin of the Word *asteroid* and the Related Terms *asteroidal*, *planetoid*, *planetkin*, *planetule*, and *cometoid*. *Studia Etymologica Cracoviensia, 20*, 47–62.

Cunningham, C. J. (2016). Milton's *Paradise Lost*: Previously unrecognized allusions to the Aurora Borealis, and a solution to the comet conundrum in book 2. *Renaissance and Reformation, 39*(1), 5–34.

D'Alembert, J. (1751). Preliminary discourse. In *Encyclopedie, ou Dictionnaire raisonne des sciences, des arts et metiers* (Vol. 1). Paris: Briasson.

D'Alembert, J. (1766). *Mélanges de Litterature, d'Histoire et de Philosophie*. Amsterdam: Zacharie Chatelain & Sons.

Danielson, D. (2011). Ramus, Rheticus, and the Copernican connection. In S. J. Reid & E. A. Wilson (Eds.), *Ramus, pedagogy and the liberal arts: Ramism in Britain and the wider world* (pp. 153–170). Farnham: Ashgate.

Daston, L. (1988). *Classical probability in the enlightenment*. Princeton: Princeton University Press.

Davis, G. N. (1969). *German thought and culture in England 1700–1770*. Chapel Hill: The University of North Carolina Press.

Débarbat, S. (1998). Parisian astronomy at the time of the Gotha meeting. In P. Brosche, W. Dick, O. Schwarz, & R. Wielen (Eds.), *The message of the angles—Astrometry from 1798 to 1998* (pp. 31–42). Acta Historica Astronomiae 3. Deutsch: Frankfurt am Main.

Débarbat, S., & Dumont, S. (2014). un Allemand de Gotha a Paris, Johann Karl Burckhardt. *Almagest, 5*(2), 8–25.

Delambre, J. (1802). Extrait de la dissertation italienne publiee par cet astronome, et ju journal allemande de Zach. November 1801. *Bulletin des Sciences de la Societe Philomathique de Paris, 3*(59), 84–86.

DeQuincey, T. (1873). *The last days of Immanuel Kant*. Boston: Shepard and Gill.

Devlin, K. (2002). Kurt Gödel—Separating truth from proof in mathematics. *Science, 298*, 1899–1900.

Dick, S. J. (2013). *Discovery and classification in astronomy*. Cambridge: Cambridge University Press.

Diderot, D. (1754). *Pensées sur l'Interpretation de la nature*. Paris: Londresno.

Diogo, M. P., Carneiro, A., & Simoes, A. (2001). The Portuguese naturalist Correia da Serra (1751–1823) and his impact on early nineteenth-century botany. *Journal of the History of Biology, 34*(2), 353–393.

Dixon, T. (2008). *The invention of altruism, making moral meanings in Victorian Britain*. London: British Academy.

Dreyer, J. L. E. (1912). *The scientific papers of Sir William Herschel* (Vol. 1). London: Royal Society and Royal Astronomical Society.

Drobyshevski, E. (2008). Stardust findings favor not only the planetary origin of comets but the underlying close-binary cosmogony of the Solar System as well. *Icarus, 197*, 203–210.

Dubyago, A. (1961). *The determination of orbits*. New York: Macmillan.

Duncombe, R. L. (1969). Heliocentric coordinates of Ceres, Pallas, Juno, Vesta 1928–2000. In *Astronomical papers prepared for the use of the American Ephemeris and Nautical Almanac, Pt II* (Vol. XX). Washington DC: Nautical Almanac Office.

Dunnington, G. W. (1937). *Carl Friedrich Gauss, inaugural lecture on astronomy*. Baton Rouge: Louisiana State University Press.

Dyck, M. (1980). Math and literature in the German enlightenment. In *Studies on Voltaire and the eighteenth century* (Vol. 190, pp. 508–512). Oxford: Voltaire Foundation.

Encke, J. F. (1831). Elemente der Ceres. *Berliner Astronomische Jahrbuch*, 275–280.

Encke, J. F. (1846). On the newly discovered planet. *Philosophical Magazine and Journal of Science, 30*, 181–185.

Euler, L. (1744). *Theoria motuum planetarum et cometarum*. Berlin: Ambrosius Haude.

Euler, L. (1748a). Letter to Johann Wettstein, 5 March. Letter 257 in the Euler Archives; Translated by John Glaus. http://eulerarchive.maa.org/correspondence/letters/OO2758t.pdf.

Euler, L. (1748b). Letter to Johann Wettstein, 29 June. Letter 259 in the Euler Archives; Translated by John Glaus. http://eulerarchive.maa.org/correspondence/letters/OO2760t.pdf.

Fader, D., & Bornstein, G. (1972). *British periodicals of the 18th and 19th centuries*. Ann Arbor, MI: University Microfilms.

Fauvel, J. (1993). A Saxon mathematician. In J. Fauvel, R. Flood, & R. Wilson (Eds.), *Moebius and his band* (pp. 1–20). Oxford: Oxford University Press.

Fitch, J. (1987). Asteroids and astrophysics. *Cassiopeia*, Canadian Astronomical Society newsletter. Letter to the editor dated 29 June.

Flynn, P. (2002). Francis Jeffrey and the Scottish critical tradition. In D. Wu & D. Massimiliano (Eds.), *British romanticism and the Edinburgh review, bicentenary essays* (pp. 13–32). New York: Palgrave Macmillan.

Forbes, E. G. (1971). Gauss and the discovery of Ceres. *Journal for the History of Astronomy, 2*, 195–199.

Foss, M. (1949). *Symbol and metaphor in human experience*. Princeton: Princeton University Press. Reprinted by University of Nebraska Press, Lincoln, 1966.

Fox, R. (2012). *The savant and the state*. Baltimore: Johns Hopkins University Press.

Freely, J. (2013). *Before Galileo: The birth of modern science in Medieval Europe*. New York and London: Overlook Duckworth.

Gascoigne, J. (1994). *Joseph Banks and the English enlightenment*. Cambridge: Cambridge University Press.

Gauss, C. (1802a, February, 20). Letter to Maskelyne (forwarded to Herschel). *RGO* 119/1.

Gauss, C. (1802b, May 25). Letter to Olbers. Goettingen University Archives.

Gauss, C. (1802c, June 25). Letter to Olbers. Bremen Archives.

Gauss, C. (1802d, October 16). Letter to Zach. *The Monthly Correspondence*. November issue, 503.

Gauss, C. *Werke*. Published originally by the Royal Society of Sciences in Goettingen (Germany), 14 vols. bound in 12. 1861–1933 (Reprinted 1981 by Olms-Weidmann, Hildesheim).

Gell-Mann, M. (2002). *Some adventures among the elementary particles*. Presidential Lecture, Florida International University, Miami. Delivered January 23, 2002.

George, R. (2006). James Jurin Awakens Hume from his dogmatic slumber, with a short tract on visual acuity. *Hume Studies, 31*(1), 141–166.

Gilbert, L. W. (1802). Ist der Ophion, (ein Planet jenseits der Uranusbahn,) ein noch unbekannter Weltkörper? *Annalen der Physik, 11*, 482–485.

Glashow, S. (2002). Immanuel Kant versus the Princes of Serendip: Did science evolve through blind chance or intelligent design? *Contributions to Science, 2*, 251–255.

Godwin, W. (1831). *Thoughts on man, his nature, productions, and discoveries*. London: Effingham Wilson.

Goedel, K. (1961). *Collected works* (Vol. III (1981)). Oxford: Oxford University Press.

Goertz, G., Jones, B., & Diehl, P. (2005). Maintenance processes in international rivalries. *Journal of Conflict Resolution, 49*, 742–769.

Goethe, W. (1994). *From my life: Poetry and truth* (R. R. Heitner, Trans.). Princeton: Princeton University Press.

Goldstein, K. (1948). On naming and pseudonaming. *ETC, V*(3), 191–197.

Golland, L., & Golland, R. (1993). Euler's troublesome series: An early example of the use of trigonometric series. *Historia Mathematica, 20*, 54–67.

Gosse, E. (1897). *A short history of modern English literature*. New York: D. Appleton and Co.

Gould, B. A. (1848). On the orbits of the asteroids. *American Journal of Science, 6*, 28–36.

Grainger, J. D. (2004). *The Amiens Truce: Britain and Bonaparte, 1801–1803*. Woodbridge: Boydell Press.

Grattan-Guinness, I. (1983). Euler's mathematics in French science 1795–1815. In J. J. Burckhardt, E. A. Fellmann, & W. Habicht (Eds.), *Leonhard Euler 1707–1783* (pp. 395–409). Basel: Birkhauser Verlag.

Grattan-Guinness, I. (1990). *Convolutions in French mathematics, 1800–1840*, three volumes. Basel and Berlin: Birkhaeuser.

Grattan-Guinness, I. (2002). *Landmark writings in western mathematics 1640–1940*. Amsterdam: Elsevier.

Gregory, F. (2006). Origins and nature of Fries' philosophy of science. In M. Friedman & A. Nordmann (Eds.), *The Kantian legacy in nineteenth-century science* (pp. 81–100). Cambridge: Massachusetts Institute of Technology.

Gregory, O. (1815). *Dissertations and letters tending to either impugn or to defend the trigonometrical survey of England and Wales*. London: Law & Gilbert.

Grootenboer, H. (2012). *Treasuring the Gaze: Intimate vision in late eighteenth-century eye miniatures*. Chicago: University of Chicago Press.

Grosser, M. (1979). *The discovery of Neptune*. New York: Dover.

Grosseteste, R. (1514). *Opuscula*. Venice: Mondato. The first printed work that includes *De artibus liberalibus*.

Grove, W. (1843, October). Physical science in England. *Blackwood's Edinburgh Magazine 54*(336), 514–525.

Hahn, R. (1998). A scientist responds to his skeptical crisis: Laplace's philosophy of science. In J. van der Zande & R. Popkin (Eds.), *The skeptical tradition around 1800* (pp. 187–201). Boston: Kluwer.

Hanham, A., & Hoskin, M. (2013). The Herschel Knighthoods: Facts and fiction. *Journal for the History of Astronomy, 44*, 149–164.

Harding, K. (1804a, September 13). Letter to Gauss. Goettingen Archives, Cod. Ms. Gauss Briefe A: Harding, 23.

Harding, K. (1804b, September 27). Letter to Gauss. Goettingen Archives, Cod. Ms. Gauss Briefe A: Harding, 24.

Harvey, G. (1577). *Ciceronianus*. London: Henrici Binneman.

Hassenfratz, J. H. (1803). *Cours de Physique Céleste*. Paris: A la Librairie Economique.

Hastie, W. (1900). *Kant's cosmogony*. Glasgow: James Maclehose & Sons.

Hawes, F. (1957). *Henry Brougham*. London: J. Cape.

Hayes, W., & Tremaine, S. (1998). Fitting selected random planetary systems to Titius-Bode laws. *Icarus, 135*, 549–557.

Hegel, G. (1805–1806). *Lectures on the history of philosophy, III* (E. S. Haldane, & F. H. Simson, Trans.). New York: Humanities Press (1974).

Hegel, G. (1817). *Enzyklopädie der philosophischen Wissenschaften*. Heidelberg: Oswald.

Herschel, W. (1782a, January 7). Letter to William Watson. RAS, W 1/1.17.

Herschel, W. (1782b, June 3). Letter to Caroline Herschel. RAS, W.1/8.18.

Herschel, W. (1782c). Letter to Alexander Aubert. RAS W1/1, 21–24.

Herschel, W. (1787). An account of three volcanos in the Moon. *Philosophical Transactions, 77*, 229–232.

Herschel, W. (1802a, April 25). Letter to William Watson. RAS, W.1/1, 247–248.

Herschel, W. (1802b). *Observations of the new Planet*. Published in Dreyer (1912), 1, cix–cxi.

Herschel, W. (1802c). Observations on the two lately discovered celestial Bodies. *Philosophical Transactions, 92*, 213–232. Read 6 May 1802.

Herschel, W. (1802d). Letter of 22 May (to multiple recipients). RAS, W.1/1, 249.

Herschel, W. (1802e, June 10). Letter to Joseph Banks. *Dawson Turner Collection 13*, 163–164. London: Natural History Museum.

Herschel, W. (1803). Account of the changes that have happened, during the last twenty-five years, in the relative situation of double stars; With an investigation of the cause to which they are owing. *Philosophical Transactions, 93*, 339–382.

Herschel, W. (1805). Experiments for ascertaining how far telescopes will enable us to determine very small angles, and to distinguish the real from the spurious diameters of celestial and terrestrial objects, with an application of the result of the experiments to a series of observations on the nature and magnitude of Mr. Harding's lately discovered star. *Philosophical Transactions, 95*, 31–64.

Herschel, W. (1807a, June 1). Letter to Joseph Banks. RAS, W.1/1, 269–270.
Herschel, W. (1807b). Observations on the nature of the new celestial body discovered by Dr. Olbers, and of the comet which was expected to appear last January in its return from the Sun. *Philosophical Transactions, 97*, 260–266. Read 4 June 1807.
Higgitt, R. (2014). *Maskelyne: Astronomer royal*. London: Robert Hale.
Hind, J. (1852). *The solar system*. New York: George P. Putnam.
Holmes, R. (2008). *The age of wonder*. NY: Pantheon Books. footnote 134.
Hooykaas, R. (1999). *Fact, faith and fiction in the development of science*. Dordrecht: Springer.
Hoskin, M. (2003a). Personal communication.
Hoskin, M. (2003b). *The Herschel partnership*. Cambridge: Science History Publications.
Hoskin, M. (2012). *The construction of the heavens: William Herschel's cosmology*. Cambridge: Cambridge University Press.
Howell, W. S. (1971). *Eighteenth-century British logic and rhetoric*. Princeton: Princeton University Press.
Howse, D. (1989). *Nevil Maskelyne: The Seaman's astronomer*. Cambridge: Cambridge University Press.
Huet, P. (1729). *A philosophical treatise concerning human understanding*. London: J. Stone.
Hume, D. (1739). *A treatise of human nature* (Vol. 2). London: John Noon.
Hume, D. (1748). *Enquiries concerning the human understanding and concerning the principles of morals*. London: A. Millar.
Huth, J. (1802, September 10). Letter to Herschel. RAS, H.33.
Huth, J. (1804). Aus einem Schreiben des Hrn. Hofrath Huth vom 21 Sept. 1804. *Astronomisches Jahrbuch für das Jahr, 1807*, 265–268.
Hutton, C. (1815). *A philosophical and mathematical dictionary* (Vol. 2). London: F.C. and J. Ridon.
International Astronomical Union. (2008). 11 June meeting of the IAU Executive Committee, Oslo, Norway.
Jaki, S. L. (1978). *Planets and Planetarians*. Edinburgh: Scottish Academic Press.
Johnson, F. R. (1937). *Astronomical thought in renaissance England*. Johns Hopkins Press. Reprinted 1968 by Octagon Books, New York.
Johnston, J. F. W. (1831, October–April). Meeting of the cultivators of Natural Science and Medicine at Hamburgh [sic], in September 1830. *Edinburgh Journal of Science, 4*, New Series, 189–244.
Johnston, R. (1815). *Travels through part of the Russian Empire*. London: J.J. Stockdale.
Jones, W. P. (1966). *The rhetoric of science*. London: Routledge & Kegan Paul.
Jungnickel, C. (1986). *Intellectual mastery of nature. Theoretical physics from Ohm to Einstein, The torch of mathematics 1800–1870* (Vol. 1). Chicago: University of Chicago Press.
Jungnickel, C., & McCormmach, R. (1996). *Cavendish*. Philadelphia: American Philosophical Society.
Kant, I. (1755). *Allgemeine Naturgeschichte und Theorie des Himmels (Universal natural history and theory of the heavens)*. Koenigsberg and Leipzig: Johann Petersen.
Kant, I. (1787). *Kritik der reinen Vernunft (Critique of pure reason)*. Riga: Johann Friedrich Hartknoch.
Kaoukji, N., & Jardine, N. (2010). "A frontispiece in any sense they please?" On the significance of the engraved title-page of John Wilkins's A Discourse Concerning A New World & Another Planet, 1640. *Word & Image, 26*, 429–447.
Kepler, J. (1609). *Astronomia nova (New astronomy)*. Heidelberg: G. Voegelinus.
Kepler, J. (1627). *Rudolphine tables*. Ulm: Jonae Saurii.
Kirkwood, D. (1888). *The asteroids, or minor planets between Mars and Jupiter*. Philadelphia: J. B. Lippincott.
Klippel, D. (1999). Legal reforms: Changing the law in Germany. In T. Blanning & P. Wende (Eds.), *Reform in Great Britain and Germany 1750–1850* (Vol. 100, pp. 43–59). London: Proceedings of the British Academy.

Kramer, E. (1982). *The nature and growth of modern mathematics*. Princeton: Princeton University Press.
Lalande, J. J. (1787, August 10). Letter to Nevil Maskelyne. London: National Maritime Museum. REG09/000037:7:85.
Lalande, J. (1802a, January 29). *Gazette nationale ou le Moniteur Universel, 129*, 514.
Lalande, J. (1802b). History of astronomy for the Year 1801. *The Philosophical Magazine, 12*, 112–121.
Lalande, J. (1802c). Memoir on the tenth planet discovered by Dr. Olbers. *The Philosophical Magazine, 13*, 279–282.
Lalande, J. (1802d). *Bibliographie Astronomique avec l'Histoire de l'Astronomie depuis 1781 jusqu'a 1802*. Reprinted by J. C. Gieben, Amsterdam, 1970.
Lalande, J. (1802e, July 19). Observations de la nouvelle planète découverte par M. Olbers, de Breme, et de l'opposition Ceres, planète découverte antérieurement par M. Piazzi. *La Décade Philosophique, Littéraire et Politique* 30 (AN X, 30 messidor), 129–132.
Lalande, J. (1803). Astronomy. *The Philosophical Magazine, 16*, 192.
Lalande, J. (1804). History of astronomy for the year 1803. *The Philosophical Magazine, 14*, 209–221.
Laplace, P.-S. (1798–1827). *Traite de Mecanique Celeste*, five volumes. Paris: Chez J.B.M. Duprat.
Lampedusa, G. (1958). *The leopard* (A. Colquhoun, Trans.), 1986. London: Collins Harvill.
Laplace, P.-S. (1802, June 17). Letter to Herschel. RAS, L.34.
Laplace, P.-S. (1808). *Exposition du système du monde* (3rd ed.). Paris: Courcier.
Larder, D. (1970). Thomas Thomson's activities in Edinburgh 1791–1811. *Notes and Records of the Royal Society of London, 24*(2), 295–304.
Leighton, A. (2003, January 17). Blessed bore. *Times Literary Supplement*, no. 5207.
Lemmon, J. G. (1878). Honorary names in scientific nomenclature. *Botanical Gazette, 3*(7), 61–64.
Leipziger Allgemeiner Litterarischer Anzeiger. (1801, May 11). 72. *Unvorgreislicher Vorschlag, nebst einigen Anfragen an Mathematiker und Astronomen*, 694.
Lewis, S. (1842). *The history and topography of the parish of Saint Mary, Islington, in the county of Middlesex*. London: J. H. Jackson.
Linnaeus, C. (1735). *Systema Naturae*. Facsimile of the First Edition. Nieuwkoop: B. DeGraaf (1964).
Linnaeus, C. (1767). *Systema Naturae*, Tomus II. Holmiae. Impensis Direct. Laurentii Salvii.
Lloyd, G. E. R. (1987). *The revolutions of wisdom*. Berkeley: University of California Press.
Locke, J. (1690). *An essay concerning humane understanding*. London: Tho. Basset.
Loemker, L. (1972). *Struggle for synthesis: The seventeenth century background of Leibniz's synthesis of order and freedom*. Cambridge: Harvard University Press.
Lofft, C. (1798). *The Monthly Magazine*, Part II for 1798, p. 406. Letter dated November 23.
Lofft, C. (1800, March). *New London Review*, 226.
Lofft, C. (1802a). Ceres Ferdinandea. *The Monthly Mirror, 13*, 230–231.
Lofft, C. (1802b, October). *The Monthly Magazine, 14*(3), 199–200.
Lofft, C. (1802c, December). *The Monthly Magazine, 14*(5), 375.
Lofft, C. (1803a, January). *The Monthly Magazine, 14*(6), 479.
Lofft, C. (1803b, February). *The Monthly Magazine, 15*(1), 59.
Lofft, C. (1805, July). Whether it would be preferable, according to their Phenomena, to call Ceres, Pallas, and Juno, simply Planets, or Asteroids, or Cometoids. *Monthly Magazine; or British Register, 19*(6), 534–535.
Lofft, C. (1781). *Eudosia*. London: W. Richardson.
Love, J. J. (2013). On the insignificance of Herschel's sunspot correlation. *Geophysical Research Letters, 40*, 4171–4176.
Lynn, W. T. (1900). Alexander Aubert. *The Observatory, 23*, 346–347.
Lynn, W. T. (1904). The third small planet. *The Observatory, 27*, 135.

Lysaght, A. M. (1971). *Joseph Banks in Newfoundland and Labrador, 1766*. Berkeley: University of California Press.
Maas, H. (2003). Where mechanism ends, Thomas Reid on the moral and the animal oeconomy. *History of Political Economy, 35*(Annual Supplement), 338–360.
Magazin füer den neusten Zustand der Naturkunde. (1802). *Nachtrage zu ben neusten Entdecktungen im Planetensystem* (pp. 136–142). Weimar.
Magie, D. (2014). *Historia Augusta*. Cambridge: Harvard University Press.
Maguire, L., & Smith, E. (2012). 'Time's cosmic sparks', The dramaturgy of a mad world, My Masters and Timon of Athens. In G. Taylor & T. Henley (Eds.), *The Oxford handbook of Thomas Middleton* (pp. 181–195). Oxford: Oxford University Press.
Manara, A. (1997). Controversie e curiosità sulla nomenclatura dei piccoli pianeti. *Memorie della Società Astronomica Italiana, 68*, 679–686.
Mandelbrot, B. (2012). *The Fractalist: Memoir of a Scientific Maverick*. NY: Pantheon.
Manley, L. (1980). *Convention 1500–1750*. Cambridge: Harvard University Press.
Markham, F. (1625). *The book of honour*. London: Augustine Matthews & John Norton.
Marsden, B. (1995). Eighteenth- and nineteenth-century developments in the theory and practice of orbit determination. In *General history of astronomy, Part B* (Vol. 2, pp. 181–190). Cambridge: Cambridge University Press.
Maskelyne, N. (1782, August 8). Letter to William Herschel. RAS, M.20.
Maskelyne, N. MS.RGO.4/119 'Correspondence on the minor planets Ceres and Vesta'; held at Cambridge University.
Mathias, T. J. (1801). *The pursuits of literature* (11th ed.). London: T. Becket.
Maurer, A. (1998). A compendium of all known William Herschel telescopes. *Journal of the Antique Telescope Society, 14*, 4–15.
Merz, J. (1896). *A history of European thought in the nineteenth century* (Vol. 1). Edinburgh and London: William Blackwood.
Messier, C. (1779). Memoire contenant les observations de la XVIIe comète. *Histoire de l'Académie Royale des Sciences*, 318–372.
Mill, J. S. (1854). Diary. In J. Robson (Ed.), *The collected works of John Stuart Mill* (Vol. 27). Toronto: University of Toronto 1996.
Monthly Correspondence. (1802, July). Letter from Banks to Zach, 90.
Moorhead, J. K. (1998). *Conversations of Goethe with Johann Peter Eckermann* (J. Oxenford, Trans.). New York: Da Capo Press.
Neal, K. (2002). *From discrete to continuous: The broadening of number concepts in Early Modern England*. Dordrecht: Kluwer.
New, C. (1961). *The life of Henry Brougham to 1830*. Oxford: Oxford University Press.
Newcomb, S. (1887). *Popular astronomy* (6th ed.). New York: Harper Bro.
Newman, J. (1956). *The world of mathematics* (p. 1754). NY: Simon & Schuster.
Newton, I. (1704). *Opticks*. London: Sam. Smith & Benj. Walford, Royal Society.
Newton, I. (1726). *Philosophiæ Naturalis Principia Mathematica* (3rd ed.). Published in English as *Newton's Principia* by N. W. Chittenden, Andrew Motte (Trans.) (1848: 465), Daniel Adee: New York.
Nieto, M. (1972). *The Titius-Bode law of planetary distances*. New York: Pergamon Press.
Norton, C. E. (1887). *Reminiscences by Thomas Carlyle* (Vol. 1). New York: Macmillan.
Olbers, W. (1797). *Abhandlung über die leichteste und bequemste Methode die Bahn eines Cometen aus einigen Beobachtungen zu berechnen* (Computation of the paths of comets). Weimar: Verlage des Industrie-Comptoirs. Published in English in: *The Quarterly Journal of Science*. This appeared in installments between 1820 and 1823.
Olbers, W. (1802a, May 23). Letter to Carl Gauss. Goettingen University Archives.
Olbers, W. (1802b, May 24). Letter to J. J. Lalande. Bremen University Archives.
Olbers, W. (1802c, June 17). Letter to William Herschel. RAS, O.2.
Olbers, W. (1802d, March 30). Letter to Johann Bode. In *Berlin astronomical yearbook* (pp. 102–104).

Ong, W. (1958). *Ramus: Method, and the decay of dialogue*. Cambridge: Harvard University Press.

Ong, W. (1962). *The barbarian within*. New York: Macmillan.

Oriani, B. (1802, August 28). Letter to Franz Zach. Archivo storico Osservatorio Astronomico di Brera, Milan.

Patrick, G. T. W. (1889). *The fragments of the work of Heraclitus of Ephesus on Nature*. Baltimore: N. Murray.

Patterson, C. W. (1987). Resonance capture and the evolution of the planets. *Icarus*, 70(2), 319–333.

Paul, C. (1980). *Science and immortality: The Éloges of the Paris Academy of Sciences (1699–1791)*. Berkeley: University of California Press.

Pfaff, C. (1853). *Sammlung von Briefen gewechselt zwischen Johann Friedrich Pfaff und Herzog Carl von Würtemberg, F. Bouterwek, A. v. Humboldt, A. G. Kästner, und Anderen*. Leipzig: J.C. Hinrich.

Phillips, J. S. (1841). Nomenclature of natural science. *Proceedings of the Academy of Natural Sciences of Philadelphia, 1*, 85–88.

Phillips, R. (1836). *A million facts connected with the studies, pursuits, and interests of mankind* (3rd ed.). New York: Connor & Cooke.

Phillips, W. (1817). *Eight familiar lectures on astronomy*. London: William Phillips.

Philo: circa 40AD, *The Decalogue*. English text by F. H. Colson, & G. H. Whitaker. (1929). New York: Putnam.

Piazzi, G. (1802a, July 2). Letter to Oriani. Archivo storico Osservatorio Astronomico di Brera, Milan.

Piazzi, G. (1802b, July 4). Letter to Herschel. RAS, W.1/13.P.20, London.

Piazzi, G. (1845). *Storia Celeste del Osservatorio di Palermo 1792–1813*. Palermo.

Pindar, P. (1816). *The works of Peter Pindar, Esq., with a copious index*. London: Walker and Edwards.

Poe, E. A. (1848). *Eureka*. New York: Putnam.

Poli, G. S. (1805). *Viaggio celeste poema astronomico del tenente colonnello*. Naples.

Pope, A. (1712, November 7). Letter to Richard Steele. Printed in the *Spectator*, November 10.

Pope, A. (1776). *Essay on man*. London: H. Setchel.

Pulte, H. (2006). Kant, Fries, and the expanding universe of science. In M. Friedman & A. Nordmann (Eds.), *The Kantian legacy in nineteenth-century science* (pp. 101–122). Cambridge: Massachusetts Institute of Technology.

Quillet, C. (1655). *Calvidii Leti Callipsedia*. Leyden. *Callipaedia: A poem. Written in Latin, made English by N. Rowe, Esq., 1712*. London: E. Sanger and E. Curil.

Ramus, P. (1569). *Scholarum mathematicarum, libri unus et triginta*. Basel: Episcopius.

Reichenbach, H. G. L. (1822). Vorwort. *Magazin für aesthetische Botanik*, 1.

Reid, A. (1729). *The present state of the republic of letters* (Vol. 3). London: William Innys.

Reid, T. (1785). *Essays on the intellectual powers of man*. Edinburgh: J. Bell.

Reil, J. (1817). *Kleine Schriften wissenschaftlichen und gemeinnuetzigen Inhalts (Brief scientific papers)*. Halle: Curtschen Buchhandlung.

Rheticus, G. (1551). *Canon doctrinae triangulorum*. Leipzig: Wolphigangi Gunteri.

Rice, A. (2001). Inexplicable? The status of complex numbers in Britain, 1750–1850. In J. Lutzen (Ed.), *Around Caspar Wessel and the geometric representation of complex numbers* (pp. 147–180). Denmark: Kgl. Danske Videnskabernes Selskab.

Robertson, R. (2009). *Mock-epic poetry from Pope to Heine*. Oxford: Oxford University Press.

Rothstein, E. (1999, June 12). Connexions. *The New York Times*.

Rousseau, J.-J. (1750). *Discours sur les sciences et les arts (Discourse on the sciences and arts), Part 2*. Geneva: Barillot.

Rousseau, J.-J. (1782). *Les Réveries du promeneur solitaire (The reveries of a solitary walker)*. Lausanne, Switzerland: Grasset.

Saridakis, V. (2014). For 'the present and future happiness of my dear Pupils': The astronomical and educational legacy of Margaret Bryan. *Culture and Cosmos, 16*, 241–253.

Saur, G. (1705). *Difficultates physicae cartesianae, thesibus inauguralibus philosophicis propositae (Difficulties of Cartesian physics subjected to peripatetic inquiry)*. Würzburg: Kleyer.
Savich, A. N. (1855). *Historical view of the discovery of minor planets or asteroids*. St. Petersburg: V.V. Grigor'ev (in Russian).
Schaffer, S. (1989). The nebular hypothesis and the science of progress. In J. R. Moore (Ed.), *History, humanity and evolution* (pp. 131–164). Cambridge: Cambridge University Press.
Schaffer, S. (1998). Leviathan of Parsonstown. In T. Lenoir (Ed.), *Inscribing science, scientific texts and the materiality of communication* (pp. 182–222). Stanford: Stanford University Press.
Schiller, F. (1788, March 17). Letter to C. G. Koerner. In *Schiller's Briefe*, 2 (1893). Stuttgart: Deutsche Verlag Anstalt.
Schilling, C. (1900). *Wilhelm Olbers sein Leben und seine Werke*, two volumes. Berlin: Springer.
Schroeter, J. (1802). Nachtraege zu ben neuesten Entdecktungen im Planetensystem: Ceres Ferdinandea. *Magazin füer den neusten Zustand der Naturkunde*, 136–140.
Schroeter, J. (1805). *Lilienthalische Beobachtungen der neu entdeckten Planeten Ceres, Pallas und Juno*. Goettingen: Vandenhoeck-Ruprecht.
Schubart, J. (1976). New reduction and collection of Meridian observations of Ceres and Pallas. *Astronomy and Astrophysics Supplement, 26*, 405–413.
Schubring, G. (1993). The German mathematical community. In J. Fauvel, R. Flood, & R. Wilson (Eds.), *Moebius and his band* (pp. 21–33). Oxford: Oxford University Press.
Sepper, D. (1993). *Goethe Contra Newton*. Cambridge: Cambridge University Press.
Sidney, P. (1595). In E. Arber (Ed.), *An Apologie for Poetrie, 1868*. London: A. Murray & Son.
Sime, J. (1900). *William Herschel and his work*. New York: C. Scribner's & Son.
Smith, B. (2010). *Natural reflections, human cognition at the nexus of science and religion*. New Haven: Yale University Press.
Smith, J. (1815). *The panorama of science and art* (Vol. 1). Liverpool: Nuttall, Fisher, and Co.
Smolin, L. (2006). *The trouble with physics*. Boston: Houghton Mifflin Co.
Sniadecki, J. (1802a). O nowym planecie Cererze położonym między Marsem I Jowiszem. (New planet Ceres located between Mars and Jupiter). *Nowy Pamietnik warszawski*, 191–201.
Sniadecki, J. (1802b). O nowym planecie położonym między Marsem I Jowiszem oraz O nowej ruchomej gwieździe, nazwanej Pallas. (New planet located between Mars and Jupiter: The new moving star, called Pallas). *Roczniki Towarzystwa Warszawskiego Przyjaciół Nauk (Annals of the Warsaw Society of Friends of the Sciences), 1*, 520–527.
Societé Philomathique. (1802a). *Ceres Ferdinandea ou aster nouveau découvert le 1 Janu. 1802 par Joseph Piazzi, 10*, 84.
Societé Philomathique. (1802b). *Sur la nouvelle planete decouverte par M. Olbers, 10*, 125.
Somerville, M. (1831). *Mechanism of the heavens*. London: John Murray.
Standage, T. (2000). *The Neptune file*. NY: Walker & Co.
Stenger, V. (2013). Atheism and the physical sciences. In S. Bullivant & M. Ruse (Eds.), *The Oxford handbook of atheism* (pp. 432–448). Oxford: Oxford University Press.
Stilling, J. H. (1805). *Siegsgeschichte der christlichen Religion*. Nuernberg: Nam'schen Buchhandlung.
Strombecf, H. v. (1802, August 14). Etwas ueber die Entdectung eines achten Hauptplaneten des Sonnensystems. *Braunschweigisches Magazin, 15*, 515–524.
Swirski, P. (2000). *Between literature and science*. Liverpool: Liverpool University Press.
Sydney, W. C. (1898). *The early days of the nineteenth century in England, 1800–1820* (Vol. 1). London: George Redway.
Szerdahely, G. A. (1788). Historia Uraniae Musae Quam inter Deos, Deasque, Planetarias recens detexit Herschelius. In *Silva Parnassi Pannonii* (pp. 147–174). Vienna: Mathiae Andreae Schmidt.
Tannoch-Bland, J. (1997). Dugald Stewart on intellectual character. *British Journal for the History of Science, 30*, 307–320.
Taylor, E. G. R. (1966). *The mathematical practitioners of Hanoverian England 1714–1840*. Cambridge: Cambridge University Press.

Terdiman, R. (1985). *Discourse/counter-discourse: The theory and practice of symbolic resistance in nineteenth-century France*. Ithaca: Cornell University Press.
Terdiman, R. (2003). The marginality of Michel de Certeau. In T. Stovall & G. Van Den Abbeele (Eds.), *French civilization and its discontents* (pp. 19–40). Lanham, MD: Lexington Books.
Terdiman, R. (2005). *Body & story: The ethics and practice of theoretical conflict*. Baltimore: John Hopkins Press.
(1804, February). *The British Critic, 23,* 101–103.
(1803a). *The Critical Review, 38*, 15–19.
(1803b). *The Critical Review, 39*, 594.
(1807). *The Eclectic Review, 3*, 182–183.
(1803). *The Gentleman's Magazine, 73*, 531.
(1830a). *The Mechanics' Magazine, 13*(349), 110–111.
(1830b). *The Mechanics' Magazine, 13*(350), 127.
(1796). *The Monthly Magazine, 1*, Description of Herschel's forty-foot reflecting telescope, 45.
(1802a). *The Monthly Magazine, 13*(7), Further account of the last new planet, Pallas, 1 June, 514.
(1802b). *The Monthly Magazine, 14*(1), 67.
(1802c). *The Monthly Magazine, 14*(2), An account of a NEW PLANET (PALLAS), lately discovered by Dr. Olbers, of Bremen, 1 September, 117–122.
The Monthly Magazine, 15 (1), 1 February, 64 (1803a).
(1803a). *The Philosophical Magazine, 15*, XXXIV. Intelligence and miscellaneous articles: Astronomy, 190.
(1804a). *The Monthly Magazine*, 16, 25 January, Supplementary number, 647.
(1803b). March 1). *The Monthly Magazine, 15*(2), 155–156.
(1803b). *The Philosophical Magazine, 15*. Astronomy, 288.
(1803c). *The Philosophical Magazine, 16*. Astronomy, 95.
(1803d). *The Philosophical Magazine, 16*. French National Institute: Notice of the labours of the mathematical and physical class of the sciences, 180–181.
Thomson, J. A. (1915). What science owes to German investigators. In W. P. Paterson (Ed.), *German culture* (pp. 65–160). London: T.C. & E.C. Jack.
Thomson, T. (1812). *History of the royal society*. London: Robert Baldwin.
Uglow, J. (2002). *The lunar men*. New York: Farrar, Straus and Giroux.
Valpy, A. J. (1821). Memoir on the antiquity of the zodiacs of Esneh and Dendera. *The Classical Journal*, 24(48), 251–282.
Ventura, F. (1990). Grand Master De Rohan's astronomical observatory. *Melita Historica, 10*, 245–255.
Véricour, L. R. (1842). *Modern French literature*. Edinburgh: William and Robert Chambers.
Vermeir, K., & Deckard, M. (2012). Philosophical enquiries into the science of sensibility. In *The science of sensibility: Reading Burke's philosophical enquiry* (pp. 3–56). NY: Springer.
Vick, B. E. (2014). *The Congress of Vienna: Power and politics after Napoleon*. Cambridge: Harvard University Press.
Voiron de Chambéry. (1810). *Histoire de l'Astronomie, depuis 1781 jusqu's 1811*. Chez Courcier, Imprimeur-Libraire pour les Mathématiques, quai des Augustins, no. 57.
Wagman, M. (2003). *Lost stars*. Blacksburg: McDonald & Woodward.
Wahsner, R. (1993). The philosophical background to Hegel's criticism of Newton. In M. J. Petry (Ed.), *Hegel and Newtonianism. Archives of the history of ideas* (Vol. 136, pp. 81–90).
Walker, W. (1802a). *Gentleman's Magazine, 72*, 294.
Walker, W. (1802b). *The Scots Magazine, 64*, 421.
Walker, W. (1802c, May). *The Monthly Magazine, 13*, 366.
Ward, A. W., & Waller, A. R. (Eds.). (1907–1921). *Cambridge history of English and American literature* (Vol. 12). Cambridge: Cambridge University Press.
Ward, A. W., & Waller, A. R. (Eds.). (1907–1916). *Cambridge history of English and American literature* (Vol. 14). Cambridge: Cambridge University Press.
Watson, P. (2010). *The German genius*. New York: Harper Collins.
Watson, W. (1802, April 27). Letter to William Herschel. RAS, W.76 (1).

Watson, W. (1803, March 2). Letter to William Herschel. RAS, W.77 (1).
Wells, G. (1978). Goethe and the development of science, 1750–1900. *Science in History* (Vol. 5). Alphen an den Rijn: Sijthoff & Noordhoff.
Welsh, D. (1825). *An account of the life and writings of Thomas Brown*. Edinburgh: Tait.
Whatmore, R. (2008). Treason and despotism, the impact of the French Revolution upon Britain. *History of European Ideas, 34*(4), 583–586.
Whewell, W. (1840). *The philosophy of the inductive sciences*. London: John Parker.
Wiener, N. (1954). *The human use of human beings: Cybernetics and society*. Boston: Houghton Mifflin.
Wigner, E. (1960). The unreasonable effectiveness of mathematics in the natural sciences. *Communications in Pure and Applied Mathematics, 13*, 1–14.
Wilde, H. (1910). *Celestial Ejectamenta, the first Halley Lecture*. Oxford: Clarendon Press.
Wilkins, J. (1638). *The discovery of a world in the Moone*. Printed by E.G. for Michael Sparke and Edward Forrest, London.
Wilkins, J. (1640). *A discourse concerning a new world and another planet in two books*. London: John Norton for John Maynard.
Williams, H. S. (1904). *A history of science* (Vol. 3). New York: Harper.
Wilson, E. (1998). *Consilience: The unity of knowledge*. Random House, NY: Vintage Press.
Wilson, P. (1802a, February 17). Letter to Herschel. RAS, W.132 (1).
Wilson, P. (1802b, February 20). Letter to Herschel. RAS, W.133.
Wilson, P. (1802c, March 9). Letter to Herschel. RAS, W.134 (1).
Winch, D. (1983). The system of the north: Dugald Stewart and his pupils. In S. Collini, D. Winch, & J. Burrows (Eds.), *That noble science of politics: A study of nineteenth-century intellectual history* (pp. 23–62). Cambridge: Cambridge University Press.
Wolf, A. (1952). *A history of science, technology and philosophy in the 18th century* (2nd ed.). NY: Harper Torchbooks.
Wollaston, W. (1805). On the discovery of palladium. *Philosophical Transactions, 95*, 316–330.
Wulf, A. (2012). *Chasing Venus*. New York: Alfred Knopf.
Zach, F. X. (1802a, September 17). Letter to Jan Sniadecki. Jagiellonian University Library, Cracow.
Zach, F. X. (1802b, September 17). Letter to Barnaba Oriani. Archivo storico Osservatorio Astronomico di Brera, Milan.
Zach, F. X. (1802c, May 1). Letter to Joseph Banks. Add.MS. 8099, 215–217, British Library.
Zach, F. X. (1802d, March 15). Letter to Joseph Banks. RGO 4/119viii.
Zach, F. X. (1802e, June). *Monthly Correspondence*, 591–606.
Zach, F. X. (1802f, February 7). Letter to Carl Gauss. Göttingen Archives.
Zach, F. X. (1802g, June 24). Letter to Carl Gauss. Göttingen Archives.
Zach, F. X. (1802h, June 26). Letter to Oriani. Brera Observatory Archives.
Zach, F. X. (1818). Letter 29. *Correspondence Astronomique, 1*, 524–525.
Žižek, S. (1997). *The plague of fantasies*. London: Verso.

Index

C. Cunningham, *Early Investigations of Ceres and the Discovery of Pallas*,
DOI 10.1007/978-3-319-28815-4

The manufacturer's authorised representative in the EU is Springer Nature Customer Service Centre GmbH, Europaplatz 3, 69115 Heidelberg, Germany. If you have any concerns regarding our products, please contact ProductSafety@springernature.com

Printed and bound by CPI Group (UK) Ltd, Croydon, CR0 4YY
05/10/2026
02228132-0008